Springer Proceedings in Mathematics & Statistics

Volume 215

Springer Proceedings in Mathematics & Statistics

This book series features volumes composed of selected contributions from workshops and conferences in all areas of current research in mathematics and statistics, including operation research and optimization. In addition to an overall evaluation of the interest, scientific quality, and timeliness of each proposal at the hands of the publisher, individual contributions are all refereed to the high quality standards of leading journals in the field. Thus, this series provides the research community with well-edited, authoritative reports on developments in the most exciting areas of mathematical and statistical research today.

More information about this series at http://www.springer.com/series/10533

Yasunori Maekawa · Shuichi Jimbo
Editors

Mathematics for Nonlinear Phenomena—Analysis and Computation

In Honor of Yoshikazu Giga's 60th Birthday, Sapporo, Japan, August 2015

Springer

Editors
Yasunori Maekawa
Mathematical Institute
Tohoku University
Sendai, Miyagi
Japan

Shuichi Jimbo
Department of Mathematics
Hokkaido University
Sapporo, Hokkaido
Japan

ISSN 2194-1009 ISSN 2194-1017 (electronic)
Springer Proceedings in Mathematics & Statistics
ISBN 978-3-319-66762-1 ISBN 978-3-319-66764-5 (eBook)
DOI 10.1007/978-3-319-66764-5

Library of Congress Control Number: 2017951151

Printed on acid-free paper

This Springer imprint is published by Springer Nature
The registered company is Springer International Publishing AG
The registered company address is: Gewerbestrasse 11, 6330 Cham, Switzerland

Preface

This volume collects research papers and survey articles of participants in the

Mathematics for Nonlinear Phenomena: Analysis and Computation

—International Conference in honor of

Professor Yoshikazu Giga on his 60th birthday—

(Organizers: Shuichi Jimbo, Shunichi Goto, Yoshihito Kohsaka, Hideo Kubo, Yasunori Maekawa, and Masaki Ohnuma)

Which was held from August 16 to 18, 2015, in Sapporo, Japan. The total number of participants is 205 including around 30 persons from outside Japan. We would like to thank all the participants, in particular the invited lecturers of the conference and the researchers who contributed to this volume. We also would like to extend our sincere gratitude to Department of Mathematics of Hokkaido University, Sapporo International Communication Plaza Foundation, Institute for Mathematics in Advanced Interdisciplinary Study, and Hokkaido Kaihatsu Kokusai Kouryu Kikin for their valuable support and commitment to make the conference successful. Our thanks also go to the referees for their efficient work, and to the secretaries, the staff, and the students who kindly helped us during the conference and editing this volume.

The conference was supported in part by the Grant in Aid for Scientific Research of the Japan Society for the Promotion of Science (Nos. 26220702, 25220702, 25247008, 24340024, and 24540200).

Kyoto, Japan
Sapporo, Japan
July 2017

Yasunori Maekawa
Shuichi Jimbo

Mathematics for Nonlinear Phenomena: Analysis and Computation—International Conference in Honor of Professor Yoshikazu Giga on His 60th Birthday

Date: August 16–18, 2015
Venue: Sapporo Convention Center
http://www.math.sci.hokudai.ac.jp/sympo/150816/index_en.html

Program:

Aug. 16 (Sun.)

13:10–13:20	Opening
13:20–14:00	**Robert V. Kohn (Courant Institute)** Prediction without probability: a PDE approach to a model problem from machine learning
14:10–14:50	**Piotr Rybka (University of Warsaw)** The method of viscosity solutions for analysis of singular diffusion problems appearing in crystal growth problems
15:20–15:40	**Takeshi Ohtsuka (Gunma University)** Evolution of spirals by crystalline curvature and eikonal equation
15:40–16:00	**Yasunori Maekawa (Tohoku University)** Stability of scale-critical circular flows in a two-dimensional exterior domain
16:15–17:30	Poster Session
19:00	Banquet

Aug. 17 (Mon.)

10:30–11:10	**Charles M. Elliott (University of Warwick)** Partial differential equations on evolving domains
11:30–12:10	**Zensho Yoshida (The University of Tokyo)** Topological constraints and structures in macro (fluid and plasma) systems

14:00–14:40 **Matthias Hieber (TU Darmstadt)** A journey through the world of incompressible viscous fluid flows: an evolution equation perspective

14:50–15:30 **Alex Mahalov (Arizona State University)** Stochastic three-dimensional rotating Navier-Stokes equations: averaging, convergence, regularity and 3D nonlinear dynamics

16:00–16:40 **Y.-H. Richard Tsai (University of Texas)** Boundary integral methods for implicitly defined interfaces

16:50–17:30 **Ryo Kobayashi (Hiroshima University)** Locomotion of animals, design of robots and mathematics

Aug. 18 (Tue.)

10:00–10:40 **Chun Liu (Penn State University)** Transport of charged particles in biological environments

11:00–11:40 **Jürgen Saal (Düsseldorf University)** Fluid flow and rotation: a fascinating interplay

11:40 Closing

Poster Session

16:15–17:30, August 16, 2015

P1 **Koichi Anada (Waseda University)** Behavior of type II blow-up solutions to a quasi-linear parabolic partial differential equation

P2 **Tomoro Asai (Hiroshima City University)** The self-similar solution for fourth order curvature flow equation

P3 **I-Kun Chen (Kyoto University)** Singularity of macroscopic variables near boundary for gases with cutoff hard potential

P4 **Kiyoko Furuya (Ochanomizu University)** On formally self-adjoint Schrödinger operators with measurable potential

P5 **Mitsuo Higaki (Tohoku University)** Navier wall law for nonstationary viscous incompressible flows

P6 **Tetsuya Ishiwata (Shibaura Institute of Technology)** Behavior of polygons by area-preserving crystalline curvature flow

P7 **Takashi Kagaya (Tokyo Institute of Technology)** Exponential stability of a traveling wave for an area preserving curvature motion

P8 **Yuki Kaneko (Waseda University)** Spreading and vanishing phenomena for a free boundary problem of reaction-diffusion equations

P9 **Kota Kumazaki (Tomakomai National College of Technology)** A mathematical model for concrete carbonation process

P10 **Michal Lasica (University of Warsaw)** On the anisotropic curvature flow of planar curves in the uniformly convex case

P11 **Tatsuhiko Miura (The University of Tokyo)** Zero width limit of the heat equation on moving thin domains

P12 **Tatsuya Miura (The University of Tokyo)** Singular perturbation by bending for an adhesive obstacle problem

P13 **Masashi Mizuno (Nihon University)** Convergence of the Allen-Cahn equation with Neumann boundary conditions

P14 **Atsushi Nakayasu (The University of Tokyo)** On one-dimensional singular diffusion equations with spatially inhomogeneous driving force

P15 **Tokinaga Namba (The University of Tokyo)** On cell problems for Hamilton-Jacobi equations with non-coercive Hamiltonians and its application to homogenization problems

P16 **Masaki Ohnuma (Tokushima University)** Cauchy problem of some nonlinear parabolic equations with uniformly continuous initial data

P17 **Michiaki Onodera (Kyushu University)** Dynamical approach to an elliptic overdetermined problem

P18 **Eugene B. Postnikov (Kursk State University), Anastasia Lavrova (Immanuel Kant Baltic Federal University)** The continuous wavelet transform as an analysis tool for non-linear oscillations [co-authored by E.B. Postnikov, A.I. Lavrova and E.A. Lebedeva]

P19 **Motohiko Sato (Wakkanai Hokusei Gakuen University)** Dynamic boundary conditions for singular degenerate parabolic equations

P20 **Yukihiro Seki (Kyushu University)** Recent results on blow-up for nonlinear heat equations

P21 **Masahiko Shimojo (Okayama University of Science)** On a free boundary problem of a curvature flow with a driving force

P22 **Ken Shirakawa (Chiba University), Hiroshi Watanabe (Salesian Polytechnic)**Energy-dissipations in non-isothermalphase-field systems associated with grain boundary motions

P23 **Takuya Suzuki (The University of Tokyo)** On the Stokes resolvent estimates for cylindrical domains

P24 **Kohtaro Tadaki (Chubu University)** An operational characterization of the notion of probability by algorithmic randomness

P25 **Kazutoshi Taguchi (The University of Tokyo)** On the discrete 1-harmonicflows

P26 **Go Takahashi (Waseda University)** Extension criterion via Morrey type functional on solutions to the Navier-Stokes equations

P27 **Shuji Takahashi (Tokyo Denki University)** On decay rate estimates in subspaces for the Navier-Stokes equations

P28 **Keisuke Takasao (The University of Tokyo)** Existence of weak solution for volume preserving mean curvature flow via phase field method

P29 **Kota Uriya (Tohoku University)** Final state problem for a system of nonlinear Schrödinger equations with mass resonance

P30 **Hiroshi Watanabe (Salesian Polytechnic)**Well-posedness for strongly degenerate parabolic equations

An Introduction of Yoshikazu Giga Through His Contributions to Mathematics and Mathematical Society

1. Academic Career of Yoshikazu Giga

Yoshikazu Giga is a Professor of Mathematics at the University of Tokyo and one of leading experts in mathematical analysis for nonlinear parabolic partial differential equations. In his 37-year research career, his research interests have included analysis of the Navier–Stokes equations, blow-up phenomena for semi-linear heat equations, defect energy surface evolution equations, and singular diffusion equations based on the theory of viscosity solutions.

Yoshikazu Giga was born in Tokyo, Japan, in 1955. He studied at the University of Tokyo before taking his Ph.D. equivalent Master's degree in 1981 under the guidance of Prof. Daisuke Fujiwara. Then, he moved to Nagoya University as a research associate. He officially received the degree of Doctor of Science from the University of Tokyo in 1985. The defense committee included Prof. Seizô Itô and Prof. Hiroshi Fujita. In his earlier career, he stayed at the Courant Institute of Mathematical Sciences as a postdoc and the University of Maryland as a visiting assistant professor. He also stayed at the Institute for Mathematics and its Applications (IMA) of the University of Minnesota as a visiting member, before he moved to the Department of Mathematics of Hokkaido University in 1986. He was invited to many institutions in Japan as an intensive course lecturer, for example, Kyushu University and the University of Tokyo. He became a full professor at Hokkaido University in 1992, after he was invited to Universität Paderborn in 1987 as a visiting professor and IMA in 1990. He was invited to Japan-U.S. Mathematics Institute of Johns Hopkins University in 1994. He was an Andrejewski Lecturer at Humboldt-Universität zu Berlin in 1997 and a Rudolph-Lipschitz Lecturer at Universität Bonn in 1999. He served as a visiting professor at Kyoto University in 2002 and of Research Institute for Mathematical Sciences (RIMS) of Kyoto University in 2004. He was invited to be a plenary lecturer at the International Congress on Industrial and Applied Mathematics (ICIAM) in 2003. He moved to the Graduate School of Mathematical Sciences of the University of Tokyo in 2004. He was invited to many institutions as a course lecturer including the Université de Lyon (2005), Universität zu Köln (2008), University of Victoria (2011), and University of Pittsburgh (2013). He will be invited as a section lecturer at the International Congress of Mathematicians (ICM) in 2018.

He was awarded Nissan Science Prize for his analysis of singularities in nonlinear parabolic equations in 1997. He was awarded the 1999 Autumn Prize by the Mathematical Society of Japan for his contribution on surface evolution equations. He was awarded the Medal of Honour with Purple Ribbon in 2010 by the Government of Japan for his contribution to mathematics. He was selected as a fellow of American Mathematical Society in 2012. He was selected as one of the Highly Cited Researchers of Thomson Scientific in 2002. He has served as an editor of several scientific journals including "Mathematische Annalen". He served as an editor of "SIAM Journal on Mathematical Analysis" for a long time. He was a member of the executive committee of the Mathematical Society of Japan. He was a member of international committees including program committee of ICM and ICIAM.

His research contribution is very influential. He introduced several new methods, new frameworks as well as solving difficult problems. Professor Fujita once described Yoshikazu Giga's research style as "He has very strong originality and at the same time has enough power to realize it in choosing topics, constructing frameworks and developing methods to solve, all of three are very important in the study of modern analysis related to real phenomena." Let us explain his past achievements in a more explicit way in the next section.

2. The Work of Yoshikazu Giga

The research field of Yoshikazu Giga is huge in the nonlinear and linear parabolic partial differential equations, and he has made a lot of pioneering works and established many fundamental theories in each subject of his research area. Yoshikazu Giga has also brought up a large number of students and sometimes postdocs. In his publication list, his (former) students are marked by * and his former postdocs are marked by #. He also contributed to the mathematical community to organize several important international meetings, for example, [Agemi Giga Ozawa (1997) (III-2)], [Adachi Giga Hishida Kozono Ogawa (2011) (III-9)], and [Ambrosio Giga Rybka Tonegawa (2015) (III-11)].

2.1 Navier–Stokes Equations

Yoshikazu Giga started his research by analysis on the Navier–Stokes equations under the guidance of Prof. Daisuke Fujiwara.

(i) Stokes operator. The Stokes operator represents the linearized part of the Navier–Stokes equations. In his first paper [Giga (1981) (I-2)] (announced in [Giga (1981) (I-1)]), he established a resolvent estimate for the Stokes operator on a curved domain in $L^p (1<p<\infty)$ which leads the analyticity of the Stokes semigroup in L^p type space. At that time, there was only one result on analyticity in L^p by V. A. Solonnikov, but the proof in [I-2] is quite general and introduced a new method like calculus of pseudo-differential operators. It is applicable for other boundary condition not necessarily the Dirichlet condition [Giga (1982) (I-3)]. Moreover, it gives an optimal sectorial angle. This opens a new way to analyze the pure imaginary power of the Stokes operator [Giga (1985) (I-15)], [Giga Sohr (1989) (I-39)] along the line of R. Seeley's theory developed for the Laplace operator. It turns out this imaginary power estimate yields L^p-L^q estimate for the

nonstationary Stokes system in [Giga Sohr (1991) (I-54)], [Giga Giga Sohr (1991) (I-49), (1993) (I-60)], which is widely used as maximal regularity property.

Y. Giga later came back to the Stokes operator, for example, in [Giga Matsui Shimizu (1999) (I-98)] and [Geissert Giga (2008) (I-156)]. In later years, by a blow-up method [Giga (1986) (I-20)], he proved in [Abe Giga (2013) (I-183)] the analyticity of the Stokes semigroup in L^∞ type spaces which had been a long-standing open problem. It was extended in several ways, for example, in [Abe Giga (2014) (I-195)], [Abe Giga Hieber (2015) (I-202)], [Abe Giga Schade Suzuki (2015) (I-203)], and [Abe Giga Schade Suzuki (2017) (I-215)]. More recently, he even proved its analyticity in *BMO* type space in [Bolkart Giga (2016) (I-209)], [Bolkart Giga Suzuki (I-221)] as well as [Bolkart Giga Suzuki Tsutsui (II-2)]. In [Bolkart Giga Miura Suzuki Tsutsui (I-220)], it was shown that the existence of L^p Helmholtz decomposition is not a necessary condition for L^p analyticity of the Stokes semigroup although it was known to be sufficient by M. Geissert, H. Heck, M. Hieber, and O. Sawada (2012).

(ii) Navier–Stokes system. The L^p theory developed by Y. Giga turns to be a powerful tool to analyze the full Navier–Stokes system as he himself showed. For example, in [Giga Miyakawa (1985) (I-17)], it was shown that the initial value problem for the Navier–Stokes system in a bounded domain is solvable if the initial data is in L^d where d is the space dimension. A different proof was given by T. Kato (1984) when the domain is a whole space. Although from the point of publication year Kato's paper was published earlier, but if one checks the date of acceptance, the paper [I-17] was earlier and they are independent works. The paper [Giga (1986) (I-19)] is essentially the Doctor's thesis of Y. Giga, which completed L^p theory for semilinear parabolic equations including the Navier–Stokes system. This paper, in particular, includes the result of [Giga (1986) (I-22)]. He is also interested in higher regularity based on his L^p theory. Smoothness is established in [I-17] and analyticity is in [Giga (1983) (I-7)]. He later came back to this analyticity problem in [Giga Sawada (2003) (I-125)] and also in [Giga Jo Mahalov Yoneda (2008) (I-159)].

Of course, Y. Giga is interested in regularity issue of the three-dimensional Navier–Stokes equations. The problem was reviewed in his earlier career by [Giga (1982) (I-4)], [Giga (1983) (I-8)] with a substantial contribution by his thesis [I-19]. In later years, he became interested in a regularity criterion by vorticity direction which was originally introduced by C. Fefferman and P. Constantin (1993). In [Giga Miura (2011) (I-176)], he gave a quite different proof of the regularity criterion saying that if the vorticity direction is continuous in spatial direction uniformly in time, there is no type I blow-up. His proof based on blow-up method [I-20] is close to the intuition of physicists. This result was extended by [Giga Hsu Maekawa (2014) (I-199)], a key mathematical tool established there was a new Liouville-type theorem with the boundary condition. One has to be very careful with the Liouville-type problem when we only assume the boundedness of a solution as pointed out in [Giga (2013) (I-190)].

The local well-posedness theory was extended to much larger class than L^d but having the same scale. This was done by many authors. Y. Giga himself did some in this direction such as [Farwig Giga Hsu (2016) (I-210), (I-223), (II-3)], [Farwig Giga (I-222)].

Besides such work on the fundamental problem on analysis, he introduced several new directions in the study of the Navier–Stokes equations. A typical example is a stability of the two-dimensional Gaussian or Oseen–Lamb vortex, which is also equivalent to the stability of the Burgers vortex with respect to two-dimensional perturbations. In [Giga Kambe (1988) (I-29)] (announced in [Giga Kambe (1988) (I-30)]), it was first proved that small Oseen–Lamb vortex is stable under small perturbation. This is the first rigorous result on the formation of the localized vortex such as the Burgers vortex. It was later proved that no restriction of size in necessary by Th. Gallay and C.E. Wayne (2005). Y. Giga later wrote a book [Giga Giga (1999) (III-3)], [Giga Giga Saal (2010) (III-8)] on this stability of the vortex.

One important aspect of the vortex is that it has infinite energy so that the classical theory of Leray does not apply. Y. Giga prepared a general theory for well-posedness including such vortex in [Giga Miyakawa Osada (1988) (I-31)] and even for the three-dimensional problem in [Giga Miyakawa (1989) (I-38)]. In [I-38], the three-dimensional vorticity equations are solved in a Morrey space, which enables to study the Navier–Stokes equations when the initial vorticity is a vortex filament, i.e., the Radon measure concentrated on the curve in $\mathbb{R}^3$. In this pioneering paper, a systematic approach to construct self-similar solutions is also proposed, and [I-38] seems to be the first paper constructing forward self-similar solutions by starting with homogeneous data. This result elegantly describes the physical objects in fluid mechanics in terms of rigorous mathematics by using the sophisticated real analysis. The idea in [I-38] is used later for surface diffusion equations in [Asai Giga (2014) (I-196)].

Y. Giga is interested in solving the Navier–Stokes system for L^∞ initial data. As is well known, the nondecaying flows are physically important objects, for many typical flows such as constant fields, basic shear flows, Ekman boundary layers, a stretched vortex like the Burgers vortex, etc., possess a certain nondecaying (or even growing) structure. Even in the study of the finite energy flows, the analysis in the nondecaying framework has a mathematical importance. Indeed, in the study of the regularity of solutions using a blow-up argument, the L^∞ theory of the Stokes and the Navier–Stokes equations often plays a crucial role, as observed in [I-176], [I-190], and [I-199].

Besides local well posedness in [Giga Inui Matsui (1999) (I-96)], he proved a global well posedness starting from L^∞ initial data for a planar flow in [Giga Matsui Sawada (2001) (I-108)]. This was not noticed before so it gives a strong influence in a later study. If one allows nondecaying data, uniqueness issue becomes complicated, for example, studied in [Giga Inui Kato Matsui (2001) (I-107)]. He also studied the growing initial data like for the Burgers equation in [Giga Yamada

(2002) (I-117)]. An almost periodic initial data is considered in [Giga Mahalov Nicolaenko (2007) (I-155)], [Giga Mahalov Yoneda (2011) (I-175)].

In his stay in the Courant Institute, Y. Giga met a young plasma physicist Zensho Yoshida. They studied several well-posedness for equations in magneto-hydrodynamics including the Navier–Stokes equations itself in [Yoshida Giga (1983) (I-12)], [Giga Yoshida (1984) (I-13), (1990) (I-41)]. The last problem was recast recently by [Giga Ibrahim Shen Yoneda (I-224)]. There are a few other papers related to plasma physics [Yoshida Giga (1990) (I-42)], [Giga Yoshida (1993) (I-63)]. The first one is very fundamental for spectral analysis on the curl operator.

(iii) Other related problems. Just before he moved to the University of Tokyo, he became interested in geophysical flow at the effect of Alex Mahalov's visit to Sapporo in 2003. He was especially interested in solving the Navier–Stokes equations with the Coriolis force for nondecaying initial data including almost periodic data in the special case. In [Giga Inui Mahalov Matsui (2006) (I-145)], [Giga Inui Mahalov Matsui Saal (2007) (I-152)] established local well-posedness results for such initial data. The space of Fourier images of Radon measures is introduced in [Giga Inui Mahalov Matsui (2005) (I-137)] to assert, for example, global solution for small data, and smallness is independent of the Coriolis parameter in [Giga Inui Mahalov Saal (2007) (I-151)], [Giga Inui Mahalov Saal (2008) (I-158)]; a related problem is studied in [Giga Saal (2011) (I-178)]. This was extended to a boundary value problem in [Giga Saal (2013) (I-194)], [Giga Saal (2015) (I-207)].

Even at the beginning of his study, Y. Giga was interested in scalar conservation laws whose typical example is the Burgers equation. He proposed a kinetic construction of the entropy solutions in [Giga Miyakawa (1983) (I-10)], [Giga Miyakawa Oharu (1985) (I-18)] which is very influential for the field. The characterization of the entropy solution is also studied in later years related to singular diffusion equation in [Giga Giga (2003) (I-121)].

Y. Giga continued to expand his interest to fluid related flows, for example, viscoelastic flow [Endo Giga Götz Liu (2017) (I-216)], [Giga Sauer Schade (II-1)] or fluid equations in a moving surface [Koba Liu Giga (2017) (I-219)], [Miura Giga Liu (II-6)].

2.2 Semilinear Heat Equations

This project was initiated when Y. Giga was staying as a visiting member of the Courant Institute from 1982. At that time, although the existence of blow-up phenomenon is well known for some nonlinear diffusion equations including semilinear heat equations, the behavior of a solution near blow-up time had not yet studied. Y. Giga together with young faculty Robert V. Kohn proposed to study the behavior of solutions using similarity variables. A typical example they considered is the Fujita equation, a semilinear equation with power nonlinearity developing a finite time blow-up. In [Giga Kohn (1985) (I-16)], similarity variables were introduced and derived a monotonicity formula for the rescaled equation. They characterized a behavior near blow-up when the power is critical in Sobolev sense for a type I blow-up, the blow-up having the same growth order for the corresponding problem of ordinary differential equations. This is a very seminal paper. For example, for the mean curvature flow, G. Huisken (1990) derived the famous monotonicity formula reflecting the idea of this paper. The rescaled equation is sometimes called the Giga–Kohn equation.

Y. Giga pushed forward to this approach mainly together with R. V. Kohn. In [Giga Kohn (1987) (I-27)], it is shown that all blow-up is type I for all positive solutions if the nonlinearity power is Sobolev subcritical. It took a long time to remove this positivity assumption until [Giga Matsui Sasayama (2004) (I-133), (2004) (I-134)]. In [Giga Kohn (1989) (I-36)], a non blow-up point is characterized by asymptotics. These papers are by now standard in the field. There are a few announcement and reviewer papers on this subject [Giga (1986) (I-23), (1986) (I-24), (1989) (I-34)], [Giga Kohn (1989) (I-37)] in the process of developing theory.

A blow-up method for parabolic problems to have an a priori maximum bound was first introduced by [I-20] and [I-27]. Nowadays, it is also very standard.

Y. Giga sometimes went back to the problem of blow-up and self-similar solutions other than the Navier–Stokes equations or curvature flows later he studied. Such topics include [Giga (1986) (I-21)], [Friedman Giga (1987) (I-26)]. In [Giga (1995) (I-72)], derivative blow-up example is given for quasi-linear parabolic equations.

After he moved to the University of Tokyo, he was interested in blow-up phenomena at space infinity. A typical result is the blow-up time is the same for ODE starting with a positive constant if the negative initial data tends to this constant at space infinity as its supremum as shown in [Giga Umeda (2006) (I-148)]. This result was extended in several aspects in [Giga Umeda (2005) (I-142), (2008) (I-164)], [Giga Seki Umeda (2008) (I-163), (2011) (I-179)].

2.3 Defect Energy

This project was initiated when Y. Giga met Patricia Aviles, another IMA young visitor in IMA in 1985. It was suggested by J. Ericksen to study singularities of smectic liquid crystals. A Ginzburg–Landau-type energy was proposed for a gradient field not for a vector field in [Aviles Giga (1987) (I-25)]. This energy is now called the Aviles–Giga functional. In [I-25], it was conjectured that the sharp interface limit is expected to be a kind of energy measuring the defect of a gradient field with a constant length in the plane. This conjecture is called the Aviles–Giga conjecture, and it has been almost solved but the original conjecture seems to be still open.

In [Aviles Giga (1989) (I-32)], it was first proved that the singular part of the Hessian measure has rank one under some technical assumptions which were later removed by Alberti (1993). This is fundamental to study singularities of a gradient field when the second derivative (Hessian) of the potential is a measure. In [Aviles Giga (1991) (I-44)], [Aviles Giga (1992) (I-55)] (announced in [Aviles Giga (1990) (I-40)]), a variational integral on mapping of bounded function is considered when the density function is convex in the gradient of the mapping but also depends on the value of mapping explicitly. Such a problem was studied for a scalar function by G. Dal Maso (1979) but it was a completely new topic at that time. The definition itself was not clear at that time. In [I-44], a definition based on graph currents was given. The idea of graph currents is very related to Cartesian currents developed by M. Giaquinta's group, but they are independent works. In [I-55], a relaxed limit was characterized when the density function is in some sense isotropic. Although a different proof was given by I. Fonseca and P. Rybka (1992), this work is very fundamental in the field. This idea using graph current gives a simple representation for the variational integral when the density function is independent of the value of mapping as shown in [Aviles Giga Komuro (1992) (I-56)].

He came back to the original problem in later years. In [Aviles Giga (1996) (I-76)], it was proved that the distance function may not be a minimizer of the defect energy so the mechanism of choosing solution is different from viscosity solutions. For the Aviles–Giga functional, it is proved that its relaxed limit was estimated from below by the proposed defect energy in [Aviles Giga (1999) (I-94)]. Although they stopped working on this problem, the field they started is a still active field of calculus of variation since it is also related to defect of magnetic fields. In [Giga Kubo Tonegawa (2007) (I-153)], various properties of the defect energy (called fold energy) for a unit, dilation-invariant gradient field are studied.

2.4 Surface Evolution Equations

After Y. Giga moved to Hokkaido University in 1986, he was impressed by a beautiful lecture by Prof. Toshio Kuroda on snow crystal growth problem and started to study a surface evolution problem or a front propagation problem, which is an interface controlled problem, a simple example of more general free boundary problems. He was interested in the notion of a weak solution which tracks all evolution after it develops singularities. In his seminal paper [Chen Giga Goto (1991) (I-46)] (announced in [Chen Giga Goto (1989) (I-33)]), a notion of weak solution for the mean curvature flow equation was introduced through a level-set method. This is a rigorous analysis theory based on adjustment of the theory of viscosity solutions. This solution is called a level-setflow nowadays. It gives a new horizon for the theory of viscosity solutions. Almost at the same time, almost the same idea was independently developed by L. C. Evans and J. Spruck (1991). However, one should note that in [I-46], a general flow like anisotropic mean curvature flow with driving force is also studied. These papers gave an analytic foundation of a level-set method proposed by S. Osher and J. A. Sethian for numerical computation in 1989, which Y. Giga apparently did not know before he completed his work.

He studied some useful properties of his level-setflow. For example, the area decreasing properties was established in [Chen Giga Goto (1991) (I-47)], and the consistency with a classical flow was proved in [Giga Goto (1992) (I-57)]. An extinction time estimate was given in [Giga Yama-uchi (1993) (I-62)], and the stability of a stationary solution for general surface evolution equations was established in [Giga Yama-uchi (1994) (I-69)]. For the mean curvature flow equation, a whole evolution of a level-setflow was in some sense "visualized" when the initial surface is axisymmetric in a seminal paper [Altschuler Angenent Giga (1995) (I-70)] (announced in [Altschuler Angenent Giga (1991) (I-43)]). It also clarified the pinching phenomena by asymptotic analysis given in (2). It also gave an example of type II shrinking surface.

He extended this level-set approach for various settings [Chen Giga Goto (1991) (I-48)], [Giga Goto Ishii Sato (1991) (I-52)], [Giga Goto (1992) (I-58)], [Giga (1994) (I-67)] including Neumann problem for a curvature flow [Giga Sato (1991) (I-53)], [Giga Sato (1993) (I-61)] and its large time behavior [Giga Ohnuma Sato (1999) (I-99)], and later its deterministic game interpretation in [Giga Liu (2008) (I-160), (2009) (I-166)]. Fattening phenomena were studied for the Neumann problem in [Giga (1995) (I-73)]. He also extended to this approach to construct global weak solutions, not only interface controlled model like mean curvature flow but also a free boundary problem. The first one was a free boundary value problem obtained as a singular limit of a reaction–diffusion system. A level-setflow solution was given in [Giga Goto Ishii (1992) (I-59)]. The second one was a two-phaseflow problem in fluid dynamics whose level-set solution is given in [Giga Takahashi (1994) (I-68)]. He wrote few surveys on this subject in [Giga (1991) (I-51), (1995) (I-74), (1996) (I-80), (1997) (I-86)] for various societies related crystal growth and materials science as well as mathematics. He eventually wrote a book in Japanese [Chen Giga (1996) (III-1)] with numerical test [Chen Giga Hitaka Honma (1994) (I-65)] and later in English in 2002, as a Lipschitz Lecture Note [Giga (2002) (III-4)]. Its final form was published in 2006 [Giga (2006) (III-5)]. He has been interested in several approaches to construct a level-set solution. He expanded the applicability of what is called the Chambolle scheme in [Eto Giga Ishii (2012) (I-180), (2012) (I-181)]. He also adjusted a level-set approach to study the growth of spirals on a crystal surface in [Ohtsuka Tsai Giga (2015) (I-208)].

The theory of level-setflow is based on the theory of viscosity solutions. Y. Giga contributed to the theory of viscosity solution itself partly through his motivation to crystal growth phenomena. For example, he gave one new way to handle discontinuous viscosity solutions by a level-set approach in [Giga Sato (2001) (I-110)], [Bardi Giga (2003) (I-119)]. The large time behavior of the Hamilton–Jacobi equation is studied in [Yokoyama Giga Rybka (2008) (I-165)], [Giga Liu Mitake (2012) (I-182), (2014) (I-201)], whose Hamiltonian is not coercive, which is outside of the standard theory. A related problem is studied in [Giga Mitake Tran (2016) (I-212)]. Another example is the Hamilton Jacobi equations with discontinuous source term which is very related to the surface of growing crystal. In [Giga Hamamuki (2013) (I-192)], a rigorous notion of solutions was established. Also, a general theory of the Hamilton–Jacobi equation in a general metric space was proposed by [Giga Hamamuki Nakayasu (2015) (I-205)]. In [Elliott Giga Goto (2003) (I-120)], a dynamic boundary condition is treated.

Through these studies, he was interested to study surface evolution equations not necessarily the level-setflow. A simple proof of existence and uniqueness of self-similar shrinking solution for anisotropic curvature flow was given in [Dohmen Giga (1994) (I-66)], [Dohmen Giga Mizoguchi (1996) (I-77)]. In [Giga Mizoguchi (1995) (I-75), (1996) (I-83)], a time-periodic solution was constructed. He also studied a large time behavior of spirals in [Giga Ishimura Kohsaka (2002) (I-115)].

It is well known that an embedded constant mean curvature must be a sphere. Y. Giga proposed its anisotropic version in 2002 in [III-4] and gave a partial answer in [Giga Zhai (2009) (I-169)]. This problem has been solved based on Heintze–Karcher's inequality, which is also studied in [Giga Pisante (2013) (I-193)]. Y. Giga was also interested in higher order surface evolution equations like surface diffusion equations. It is quite different from the second-order problem like mean curvature flow. For example, the self-intersection can be formed and convexity is not prepared as pointed out first in [Giga Ito (1998) (I-92), (1998) (I-93), (1999) (I-97)], [Giga (2010) (I-171)]. The intermediate case was also studied in [Escher Giga Ito (2001) (I-104), (2002) (I-112)].

2.5 Singular Diffusion Equations

During his visit to IMA in 1990, Y. Giga met Morton Gurtin and began to study a curvature flow with very strong anisotropic flow including crystalline flow introduced by S. Angenent and M. Gurtin and J. Taylor independently. From PDE point of view, this is a very singular diffusion equation so that the speed of evolution is determined by a nonlocal quantity. He showed the approximability of crystalline flow by a conventional flow and vice versa by adopting the theory of monotone operators in [Fukui Giga (1996) (I-78)] which was completed in 1992. This was the first work on approximability and was independent of the work of R. V. Kohn and P. Girao which discussed approximability of conventional flow by a crystalline flow. In [Giga Gurtin (1996) (I-82)], a comparison principle of a planar crystalline flow was established. An extension to the three-dimensional case was studied in [Giga Gurtin Matias (1998) (I-91)].

Since the equation is very singular, the notion of solutions itself was not available at that time except special polygonal flows for which the problem is reduced to a system of ODEs and except the work of [Fukui Giga (1996) (I-78)]. Y. Giga tried to extend the theory of viscosity solutions to this situation based on several preparation through the theory of maximum monotone operators with his wife Mi-Ho Giga as in [Giga Giga (1995) (I-71), (1998) (I-88), (1998) (I-90)]. Finally, he successfully established the theory of viscosity solutions for such singular diffusion equations for one space variable in a series of work [Giga Giga (1998) (I-87), (1999) (I-95)] and finally established a level-set method in a planar flow in [Giga Giga (1998) (I-89), (2000) (I-102), (2001) (I-105)]. The papers [I-87], [I-95], and [I-105] are very fundamental in the sense that all second-order singular diffusion equations with one space variable can be handled even if it has no divergence structure, which is essential in the theory of maximum monotone operator. He later extended to the higher-dimensional case which is substantially different from the one-dimensional case in [Giga Giga Pozar (2013) (I-189), (2014) (I-197)], [Giga Pozar (2016) (I-213)]. The main difficulty in higher dimension is in the fact that the nonlocal quantity called crystalline curvature is not necessarily constant over a flat part called a facet.

To handle crystalline growth problems, one has to study spatially inhomogeneous case as well as better understand the crystalline curvature. After [Giga Giga (1998) (I-88)], Y. Giga mainly worked with Piotr Rybka who was a student of Robert V. Kohn. In [Giga Paolini Rybka (2001) (I-109)], it was shown that crystalline curvature may not be written by a gradient field disproving what is called Cahn's conjecture. In the series of papers [Giga Rybka (2002) (I-116), (2004) (I-135), (2005) (I-141), (2006) (I-147), (2008) (I-161)], a quasi-staticStefan-type problem is studied, which is a free boundary problem including crystalline curvature. This is a model of crystal growth grown by supersaturated vapor. This problem is not yet solved globally in time even for the case of smooth anisotropy. The behavior of spatial distribution of supersaturated vapor was studied in [Giga Rybka (2003) (I-124)]. A crystalline flow with given spatially inhomogeneous external force is studied in [Giga Rybka (2008) (I-162), (2009) (I-167)], [Giga Gorka Rybka (2010) (I-172), (2011) (I-173), (2013) (I-191), (2015) (I-204)], where a semi-explicit solution was given. The theory of viscosity solution was extended in [Giga Giga Nakayasu (2013) (I-186)], [Giga Giga Rybka (2014) (I-198)], at least in the level of the paper [Giga Giga (1998) (I-87)] to inhomogeneous driving force. We note that if the singularity of the singular diffusion equation is too strong, it destroys all structure instantaneously as shown in [Chen Giga Sato (1997) (I-85)].

These are several applications of singular diffusions. For example, it is convenient to calculate solutions with shocks (introduced by [Giga (2002) (I-113)]) as numerically studied in [Tsai Giga Osher (2003) (I-128)], [Giga Giga (2003) (I-121)], [Tsai Giga (2003) (I-127)], and [Giga Giga (2005) (I-136)]. An application to image analysis was given in [Hontani Giga Giga Deguchi (2005) (I-143)] (announced in [(2003) (I-126)]). Its analytic foundation is given in [Giga Giga Hontani (2006) (I-144)].

Y. Giga began to work with Ryo Kobayashi who joined faculties of Hokkaido University around 1995 on applications of singular diffusion to grain boundary motion. Ryo Kobayashi visualized numerically what Y. Giga had in mind. There are two review papers [Kobayashi Giga (1999) (I-101)], [(2001) (I-111)]. This is another inhomogeneous situation which was clarified in [Giga Giga Kobayashi (2001) (I-106)]. Also, 1-harmonics flow or total variation type flow with value constrained problem was studied in [Giga Kobayashi (2003) (I-123)], which was a very pioneering work. It led to a study of [Giga Kashima Yamazaki (2004) (I-131)], [Giga Kuroda (2004) (I-132), (2015) (I-206)], and [Giga Kuroda Yamazaki (2005) (I-139), (2007) (I-154)].

More recently, the fourth-order total variation flow was studied because it can be a model of relaxation process of crystal surface by surface diffusion. In [Giga Kohn (2011) (I-174)], [Giga Kuroda Matsuoka (2014) (I-200)], the extinction time estimate is given as well as a rigorous formulation of the problem. There are several survey papers on singular diffusion equations written by Y. Giga including [Giga Giga (2004) (I-129), (2013) (I-185)], [Giga (2000) (I-103), (2004) (I-130), (2007) (I-149), (2010) (I-171)], and [Giga Giga Pozar (2013) (I-188)].

Yasunori Maekawa
Shuichi Jimbo

List of Publications of Yoshikazu Giga

In the following list, HUPSM is the abbreviation of Hokkaido University Preprint Series in Mathematics.

I. Articles including review articles and research announcements

1. Y. Giga, The Stokes operator in L_r spaces, Proceedings of conference on synthetic study for positive operators and its applications (ed. F. Niiro) 39–47, Oct. 5–7, 1980, Hiroshima University, Proc. Japan Acad. **57** (1981), 85–89.
2. Y. Giga, Analyticity of the semigroup generated by the Stokes operator in L_r spaces, Math. Z. **178** (1981), 297–329. DOI: 10.1007/BF01214869
3. Y. Giga, The nonstationary Navier-Stokes system with some first order boundary condition, Proc. Japan Acad. **58** (1982), 101–104. DOI: 10.3792/pjaa.58.101
4. Y. Giga, The nonstationary problem for incompressible viscous and perfect flow (Hiasshukusei nensei ryuutai to kanzen ryuutai no hiteijou mondai), Nagare **1–2** (1982), 126–133 (in Japanese).
5. Y. Giga, The nonstationary problem for two-dimensional incompressible viscous and perfect flow (Nijigen hiasshukusei nensei ryuutai to kanzen ryuutai no hiteijou mondai), Proc. of conference on research for turbulent flow (ed. T. Tatsumi) **94–100** (1982), January 5–9, 1982, Kyoto, (in Japanese).
6. Y. Giga, The Navier-Stokes initial value problem in L^p and related problems, Nonlinear Partial Differential Equations in Applied Science, Proc. of the U.S.-Japan Seminar, Tokyo, 1982, (eds. H. Fujita, P. D. Lax and G. Strang), North Holland Math. Studies **81**, 1984, Lecture Notes in Num. Appl. Anal. **5** (1982), 37–54.
7. Y. Giga, Time and spatial analyticity of solutions of the Navier-Stokes equations, Comm. in Partial Differential Equations **8** (1983), 929–948. DOI: 10.1080/03605308308820290
8. Y. Giga, Weak and strong solutions of the Navier-Stokes initial value problem, Publications of RIMS **19–3** (1983), 887–910.
9. Y. Giga, Y. Hashimoto and Y. Yamada, Parabolic equations with free boundary conditions, Funkcialaj Ekvacioj **26** (1983), 263–279.
10. Y. Giga and T. Miyakawa, A kinetic construction of global solutions of first order quasilinear equations, Duke Math. J. **50** (1983), 505–515. DOI: 10.1215/S0012-7094-83-05022-6
11. Y. Giga and T. Miyakawa, A construction of weak solutions of equations of conservation type via phase space (Hozonsoku wo arawasu houteishiki no jakkai no soukuukan wo mochiite no kowseihou), Proc. of 4th conference on evolution equation **36–45** (1983), Aug. 2–5, 1982, Agematsu, Nagano (in Japanese).
12. Z. Yoshida and Y. Giga, On the Ohm-Navier-Stokes system in magnetohydrodynamics, J. Math. Phys. **24** (1983), 2860–2864. DOI: 10.1063/1.525667

13. Y. Giga and Z. Yoshida, On the equations of the two-component theory in magnetohydrodynamics, Comm. in Partial Differential Equations **9** (1984), 503–522. DOI: 10.1080/03605308408820339
14. Z. Yoshida and Y. Giga, A nonlinear semigroup approach to the Navier-Stokes system, Comm. in Partial Differential Equations **9** (1984), 215–230. DOI: 10.1080/03605308408820331
15. Y. Giga, Domains of fractional powers of the Stokes operator in L_r spaces, Arch. Rational Mech. Anal. **89** (1985), 251–265. DOI: 10.1007/BF00276874
16. Y. Giga and Robert V. Kohn, Asymptotically self-similar blow-up of semilinear heat equations, Comm. Pure Appl. Math. **38** (1985), 297–319. DOI: 10.1002/cpa.3160380304
17. Y. Giga and T. Miyakawa, Solutions in L_r of the Navier-Stokes initial value problem, Arch. Rational Mech. Anal. **89** (1985), 267–281. DOI: 10.1007/BF00276875
18. Y. Giga, T. Miyakawa and S. Oharu, A kinetic approach to general first order quasilinear equations, Trans. Amer. Math. Soc. **287** (1985), 723–743. DOI: 10.1090/S0002-9947-1985-0768737-9
19. Y. Giga, Solutions for semilinear parabolic equations in L^p and regularity of weak solutions of the Navier-Stokes system, J. Differential Equations **62** (1986), 186–212. DOI: 10.1016/0022-0396(86)90096-3
20. Y. Giga, A bound for global solutions of semilinear heat equations, Comm. Math. Phys. **103** (1986), 415–421. DOI: 10.1007/BF01211756
21. Y. Giga, On elliptic equations related to self-similar solutions for nonlinear heat equations, Hiroshima Math. J. **16** (1986), 539–552.
22. Y. Giga, Regularity criteria for weak solutions of the Navier-Stokes system, Proc. of the 1983 AMS Summer Institute on Nonlinear Functional Analysis and its Applications, Proc. Symposia in Pure Math. **45** (1986), 449–453.
23. Y. Giga, Self-similar solutions for semilinear parabolic equations, Proc. of the 1984 CNLS-AMS-SIAM Summer Seminar on Nonlinear systems of PDE in applied mathematics, July 1984, Santa Fe. Lectures in Applied Math. **23** (1986), 247–251.
24. Y. Giga, On blow-up solutions for semilinear heat equations (Hansenkei Netsu Houteishiki no Kai no Bakuhatsu wo Megutte), Proc. of annual conference of Mathematical Society of Japan, Functional Analysis, 1986, (in Japanese).
25. P. Aviles and Y. Giga, A mathematical problem related to the physical theory of liquid crystal configurations, Proc. Miniconference on Geometry / Partial Differential Equations **2**, June, 1986 Canberra. Proc. Centre for Mathematical Analysis, Australian National University **12** (1987), 1–16 (eds. J. Hutchinson and L. Simon).
26. A. Friedman and Y. Giga, A single point blow-up for solutions of semilinear parabolic systems, J. Fac. Sci. Univ. Tokyo Sect IA Math. **34** (1987), 65–79.
27. Y. Giga and Robert V. Kohn, Characterizing blowup using similarity variables, Indiana Univ. Math. J. **36** (1987), 1–40.

28. Y. Giga, (Book Review) The equations of Navier-Stokes and abstract parabolic equations, by W. von Wahl, Bull. Amer. Math. Soc. **19** (1988), 337–340.
29. Y. Giga and T. Kambe, Large time behavior of the vorticity of two-dimensional viscous flow and its application to vortex formation, Comm. Math. Phys. **117** (1988), 549–568. DOI: 10.1007/BF01218384
30. Y. Giga and T. Kambe, Large time behavior of the vorticity of 2D viscous flow and vortex formation in 3D flow, Fluid dynamics Research **3** (1988) 83–86, North Holland.
31. Y. Giga, T. Miyakawa and H. Osada, Two-dimensional Navier-Stokes flow with measures as initial vorticity, Arch. Rational Mech. Anal. **104** (1988), 223–250. DOI: 10.1007/BF00281355
32. P. Aviles and Y. Giga, Singularities and rank one properties of Hessian measures, Duke Math. J. **58** (1989), 441–467. DOI: 10.1215/S0012-7094-89-05820-1
33. Y.-G. Chen[♮], Y. Giga and S. Goto[*], Uniqueness and existence of viscosity solutions of generalized mean curvature flow equations, Proc. Japan Acad. **65** (1989), 207–210. DOI: 10.3792/pjaa.65.207
34. Y. Giga, A local characterization of blowup points of semilinear heat equations, Recent Topics in Nonlinear PDE IV (eds. M. Mimura and T. Nishida), North Holland Math. Studies **160**, 1989 Lecture Notes in Num. Appl. Anal. **10** (1989), 1–14.
35. Y. Giga, Pure imaginary powers of operators and L^p estimates for parabolic equations (Sayouso no junkyo beki to houbutsugata houteishiki no L^p hyouka), Proc. of conference on functional analysis, Hirosaki (1989), (ed. H. Sakata), 84–96, (in Japanese).
36. Y. Giga and Robert V. Kohn, Nondegeneracy of blowup for semilinear heat equations, Comm. Pure Appl. Math. **42** (1989), 845–884. DOI: 10.1002/cpa.3160420607
37. Y. Giga and R. Kohn, Removability of blowup points for semilinear heat equation, Differential Equations (eds. C. M. Dafermos et al), Marcer Dekker, Lecture notes in pure and applied mathematics **118** (1989), 257–264. (Proc. of EQUADIFF 87, Xanthi, Greece, August 1987).
38. Y. Giga and T. Miyakawa, Navier-Stokes flow in R^3 with measures as initial vorticity and Morrey spaces, Comm. in Partial Differential Equations **14** (1989), 577–618. DOI: 10.1080/03605308908820621
39. Y. Giga and H. Sohr, On the Stokes operator in exterior domains, J. Fac. Sci. Univ. Tokyo Sect IA Math. **36** (1989), 103–130.
40. P. Aviles and Y. Giga, Minimal currents and relaxation of variational integrals on mappings of bounded variation, Proc. Japan Acad. **66** (1990), 68–71. DOI: 10.3792/pjaa.66.68
41. Y. Giga and Z. Yoshida, A dynamic free-boundary problem in plasma physics, SIAM J. Math. Anal. **21** (1990), 1118–1138. DOI: 10.1137/0521062
42. Z. Yoshida and Y. Giga, Remarks on spectra of operator rot, Math. Z. **204** (1990), 235–245. DOI: 10.1007/BF02570870

43. S. Altschuler, S. Angenent and Y. Giga, Generalized motion by mean curvature for surfaces of rotation, HUPSM #119 (1991).
44. P. Aviles and Y. Giga, Variational integrals on mappings of bounded variation and their lower semicontinuity, Arch. Rational Mech. Anal. **115** (1991), 201–255. DOI: 10.1007/BF00380769
45. P. Aviles and Y. Giga, Variational integrals on mappings of bounded variation, Recent topics in Nonlinear PDE V. (ed. K. Masuda), North Holland Math. Studies, Lecture Notes in Num. Appl. Anal. **11** (1991), 23–36.
46. Y.-G. Chen[♮], Y. Giga and S. Goto[*], Uniqueness and existence of viscosity solutions of generalized mean curvature flow equations, J. Differential Geometry **33–3** (1991), 749–786.
47. Y.-G. Chen[♮], Y. Giga and S. Goto[*], Analysis toward snow crystal growth, Functional Analysis and Related Topics, (ed. S. Koshi), (Sapporo, September 1990), World Scientific (1991), 43–57.
48. Y.-G. Chen[♮], Y. Giga and S. Goto[*], Remarks on viscosity solutions for evolution equations, Proc. Japan Acad. **67** (1991), 323–328. DOI: 10.3792/pjaa.67.323
49. M. Giga, Y. Giga and H. Sohr, L^p estimate for abstract linear parabolic equations, Proc. Japan Acad. **67** (1991), 197–202. DOI: 10.3792/pjaa.67.197
50. Y. Giga, Evolution equations of surfaces (Kyokumen no hatten houteishiki), Proc. of annual conference of Mathematical Society of Japan, Differential Equation, (1991) (in Japanese).
51. Y. Giga, Motion of phase-boundaries and singularities (Soukyoukai no Undo to Tokuiten), Nippon Kessho Seichou Gakkaishi **18** (1991), 203–208 (in Japanese).
52. Y. Giga, S. Goto[*], H. Ishii and M.-H. Sato[*], Comparison principle and convexity preserving properties for singular degenerate parabolic equations on unbounded domains, Indiana Univ. Math. Journal **40** (1991), 443–470. DOI: 10.1512/iumj.1991.40.40023
53. Y. Giga and M.-H. Sato[*], Generalized interface evolution with the Neumann boundary condition, Proc. Japan Acad. **67** (1991), 263–266. DOI: 10.3792/pjaa.67.263
54. Y. Giga and H. Sohr, Abstract L^p estimates for the Cauchy problem with applications to the Navier-Stokes equations in exterior domains, J. Functional Anal. **102** (1991), 72–94. DOI: 10.1016/0022-1236(91)90136-S
55. P. Aviles and Y. Giga, Minimal currents, geodesics, and relaxation of variational integrals on mappings of bounded variation, Duke Math. J. **67** (1992), 517–538. DOI: 10.1215/S0012-7094-92-06720-2
56. P. Aviles, Y. Giga and N. Komuro, Duality formulas and variational integrals, Adv. Math. Sci. Appl. **1** (1992), 207–228.
57. Y. Giga and S. Goto[*], Geometric evolution of phase-boundaries, Proc. of IMA workshop (Sept. 17–21, 1990) Minneapolis (eds. M. Gurtin and G. McFadden), On the evolution of phase-boundaries, IMA volumes in mathematics and its applications **43**, Springer (1992), 51–65.

58. Y. Giga and S. Goto*, Motion of hypersurfaces and geometric equations, J. Math. Soc. Japan **44** (1992), 99–111. DOI: 10.2969/jmsj/04410099
59. Y. Giga, S. Goto* and H. Ishii, Global existence of weak solutions for interface equations coupled with diffusion equations, SIAM J. Math. Anal. **23** (1992), 821–835. DOI: 10.1137/0523043
60. M. Giga, Y. Giga and H. Sohr, L^p estimates for the Stokes system, Functional Analysis and Related Topics, 1991 ed. H. Komatsu, Springer Lecture Notes in Math. **1540** (1993), 55–67.
61. Y. Giga and M.-H. Sato*, Neumann problem for singular degenerate parabolic equations, Diff. Integral Eqs. **6** (1993), 1217–1230.
62. Y. Giga and K. Yama-uchi*, On a lower bound for the extinction time of surfaces moved by mean curvature, Calculus of Variations and Partial Differential Equations **1** (1993), 417–428. DOI: 10.1007/BF01206960
63. Y. Giga and Z. Yoshida, A bound for the pressure integral in a plasma equilibrium, J. Stat. Phys. **72** (1993), 1375–1389. DOI: 10.1007/BF01048191
64. Z. Yoshida and Y. Giga, Bound for the pressure integral in a toroidal-plasma equilibrium, Physical Review E **48** (1993), 2133–2135. DOI: 10.1103/PhysRevE.48.2133
65. Y.-G. Chen♮, Y. Giga, T. Hitaka* and M. Honma*, A stable difference scheme for computing motion of level surfaces by the mean curvature, Proc of the third GARC symposium on pure and applied mathematics, (eds. D. Kim et al) (1994), 1–19. [announcement: Numerical analysis for motion of a surface by its mean curvature, In: Numerical analysis of partial differential equations that appear in engineering and related topics (Japanese) (Kyoto, 1992), RIMS Kôkyûroku **836** (1993), 147–151.]
66. C. Dohmen♮ and Y. Giga, Selfsimilar shrinking curves for anisotropic curvature flow equations, Proc. Japan Acad. **70** (1994), 252–255. DOI: 10.3792/pjaa.70.252
67. Y. Giga, Motion of a graph by convexified energy, Hokkaido Math. J. **23** (1994), 185–212. DOI: 10.14492/hokmj/1381412492
68. Y. Giga and S. Takahashi*, On global weak solutions of the nonstationary two-phase Stokes flow, SIAM J. Math. Anal. **25** (1994), 876–893. DOI: 10.1137/S0036141092231914
69. Y. Giga and K. Yama-uchi*, On instability of evolving hypersurfaces, Diff. Integral Eqs. **7** (1994), 863–872.
70. S. Altschuler, S. Angenent and Y. Giga, Mean curvature flow through singularities for surfaces of rotation, J. Geom. Anal. **5** (1995), 293–358. DOI: 10.1007/BF02921800
71. M.-H. Giga and Y. Giga, Consistency in evolutions by crystalline curvature, Proc. Free Boundary Problems '95, Theory and Applications (eds. M. Niezgódka and P. Strzelecki), Zakopane, Poland (1995), Longman, pp.186–202.
72. Y. Giga, Interior derivative blow-up for quasilinear parabolic equations, Discrete and Continuous Dynamical Systems **1** (1995) 449–461. DOI: 10.3934/dcds.1995.1.449

73. Y. Giga, Evolving curves with boundary conditions, Proc. of "Curvature flow and related topics" (eds. A. Damlamian, J. Spruck and A. Visintin) Levico 1994, Gakuto, (1995), 99–109.
74. Y. Giga, The method of level surfaces in connection with the equation of evolution of a surface (Kyokumen no hatten houteishiki niokeru toukomen no houhou), Sūgaku **47** (1995), 321–340. English translation: A level set method for surface evolution equations, Sugaku Expositions **10** (1997), 217–241.
75. Y. Giga and N. Mizoguchi, Existence of periodically evolving convex curves moved by anisotropic curvature, Advances in Geometric Analysis and Continuum Mechanics, International Press (eds. P. Concus and K. Lanchester) (1995), 100–108.
76. P. Aviles and Y. Giga, The distance function and defect energy, Proc. Royal Soc. Edinburgh **126** (1996), 923–938. DOI: 10.1017/S0308210500023167
77. C. Dohmen♮, Y. Giga and N. Mizoguchi, Existence of selfsimilar shrinking curves for anisotropic curvature flow equations, Calculus of Variations and Partial Differential Equations **4** (1996), 103–119. DOI: 10.1007/BF01189949
78. T. Fukui* and Y. Giga, Motion of a graph by nonsmooth weighted curvature, Proc. of the first World congress of nonlinear analysists **92**, Walter de Gruyter Publisher, Berlin, (ed. V. Lakshmikantham), vol. I (1996), 47–56.
79. M.-H. Giga and Y. Giga, Geometric evolution by nonsmooth interfacial energy, Proc. of Banach Center Minisemester, Nonlinear Analysis and Applications" (eds. N. Kenmochi, M. Niezgódka and P. Strzelecki), Gakuto, (1996), 125–140.
80. Y. Giga, Equations of motion of phase-boundaries—an approach from nonlinear partial differential equations (Soukyoukai no undou houteishiki-hisenkei henbibun houteishiki ron karano approach), Materia Japan **35** (1996), 772–776 (in Japanese).
81. Y. Giga, Surprising diffusion processes (Fushigina kakusan houteishiki), Sūgaku **49** (1996), 193–196 (in Japanese).
82. Y. Giga and M. Gurtin, A comparison theorem for crystalline evolution in the plane, Quart. of Appl. Math. **54** (1996), 727–737.
83. Y. Giga and N. Mizoguchi, Existence of periodic solutions for equations of evolving curves, SIAM J. Math. Anal. **27** (1996), 5–39. DOI: 10.1137/S0036141094265833
84. Y. Giga and N. Mizoguchi, On time periodic solutions of the Dirichlet problem for degenerate parabolic equations of nondivergence type, J. Math. Anal. Appl. **201** (1996), 396–416. DOI: 10.1006/jmaa.1996.0263
85. Y.-G. Chen♮, Y. Giga and K. Sato*, On instant extinction for very fast diffusion equations, Discrete and Continuous Dynamical Systems **3** (1997), 243–250. DOI: 10.3934/dcds.1997.3.243
86. Y. Giga, Moving interfaces (Ugoku kaimen), Have fun with Mathematics (Sūgaku no tanoshimi) **4** (1997), 69–85 (in Japanese).
87. M.-H. Giga and Y. Giga, Evolving graphs by singular weighted curvature, Arch. Rational Mech. Anal. **141** (1998), 117–198. DOI: 10.1007/s002050050075

88. M.-H. Giga and Y. Giga, A subdifferential interpretation of crystalline motion under nonuniform driving force, Proc. of the International Conference on Dynamical Systems and Differential Equations, Springfield, Missouri (1996), Dynamical Systems and Differential Equations (eds. W.-X. Chen and S.-C. Hu), Southwest Missouri State Univ. **1** (1998), 276–287.
89. M.-H. Giga and Y. Giga, Motion driven by nonlocal curvature in the plane, Proc. of the International Conference on Asymptotics in Nonlinear Diffusive Systems 1997, (eds. Y. Nishiura et al.), Tohoku Math. Publ. #8 (1998), 75–83.
90. M.-H. Giga and Y. Giga, Remarks on convergence of evolving graphs by nonlocal curvature, Progress in Partial Differential Equations (eds. H. Amann et al) Pont-à-Mousson 1997 **1** Longman (1998), 99–116.
91. Y. Giga, M.-E. Gurtin and J. Matias, On the dynamics of crystalline motions, Japan J. Ind. Appl. Math. **15** (1998), 7–50. DOI: 10.1007/BF03167395
92. Y. Giga and K. Ito♮, On pinching of curves moved by surface diffusion, Comm. Appl. Anal. **2** (1998), 393–405.
93. Y. Giga and K. Ito♮, Motion of phase boundaries by surface diffusion, The Third Pacific Rim International Conference on Advanced Materials and Proceeding (PRICM 3) (eds. M. A. Imam et al) The Minerals, Metals & Materials Society (1998), 1363–1368.
94. P. Aviles and Y. Giga, On lower semicontinuity of a defect energy obtained by a singular limit of the Ginzburg-Landau type energy for gradient fields, Proc. Royal Soc. Edinburgh **129** (1999), 1–17. DOI: 10.1017/S0308210500027438
95. M.-H. Giga and Y. Giga, Stability for evolving graphs by nonlocal weighted curvature, Comm. in Partial Differential Equations **24** (1999), 109–184. DOI: 10.1080/03605309908821419
96. Y. Giga, K. Inui* and S. Matsui, On the Cauchy problem for the Navier-Stokes equations with nondecaying initial data, Quaderni di Matematica **4** (1999), 27–68.
97. Y. Giga and K. Ito♮, Loss of convexity of simple closed curves moved by surface diffusion, Topics in Nonlinear Analysis, The Herbert Amann Aniversary volume, (eds. J. Escher and G. Simonett), Progress in Nonlinear Differential Equations, Birkhäuser (1999), 305–320.
98. Y. Giga, S. Matsui and Y. Shimizu*, On estimates in Hardy spaces for the Stokes flow in a half space, Math. Z. **231** (1999), 383–396. DOI: 10.1007/PL00004735
99. Y. Giga, M. Ohnuma* and M.-H. Sato*, On the strong maximum principle and the large time behavior of genaralized mean curvature flow with the Neumann boundary condition, J. Differential Equations **154** (1999), 107–131. DOI: 10.1006/jdeq.1998.3569
100. Y. Giga and M.-H. Sato*, On semicontinuous solutions for general Hamilton-Jacobi equations, Proc. Japan Acad. **75** (1999), 159–162. DOI: 10.3792/pjaa.75.159 In: Singularity theory and differential equations (Kyoto, 1999), RIMS Kôkyûroku **1111** (1999), 117–124.]

101. R. Kobayashi and Y. Giga, Equations with singular diffusivity, J. Stat. Phys. **95** (1999), 1187–1220. DOI: 10.1023/A:1004570921372
102. M.-H. Giga and Y. Giga, Crystalline and level set flow—Convergence of a crystalline algorithm for a general anisotropic curvature flow in the plane, Proc. Free Boundary Problems '99, Theory and Applications (ed. N. Kenmochi), Chiba, Japan (1999), Gakuto International Series, Math. Sci. Appl. **13** (2000), 64–79.
103. Y. Giga, Moving boundary equations with anisotropic curvature (Hitouhouteki kyokuritsu niyoru kaimen undo houteishiki) Sūgaku **52** (2000), 113–127, (in Japanese) English translation : Anisotropic curvature effects in interface dynamics, Sugaku Exposition **16** (2003), 135–152.
104. J. Escher, Y. Giga and K. Ito[♮], On a limiting motion and self-intersections of curves moved by the intermediate surface diffusion flow, Proc. of the third World Congress of Nonlinear Analysts 2000, (ed. V. Lakshmikantham), Nonlinear Analysis **47** (2001), 3717–3728.
105. M.-H. Giga and Y. Giga, Generalized motion by nonlocal curvature in the plane, Arch. Rational Mech. Anal. **159** (2001), 295–333. DOI: 10.1007/s002050100154
106. M.-H. Giga, Y. Giga and R. Kobayashi, Very singular diffusion equations, Proc. of the last Taniguchi Symposium, (eds. M. Maruyama and T. Sunada) Advanced Studies in Pure Math. **31** (2001), 93–125.
107. Y. Giga, K. Inui[*], J. Kato and S. Matsui, Remarks on the uniqueness of bounded solutions of the Navier-Stokes equations, Proc. of the third World Congress of Nonlinear Analysts 2000, (ed. V. Lakshmikantham) Nonlinear Analysis **47** (2001), 4151–4156.
108. Y. Giga, S. Matsui and O. Sawada[*], Global existence of two-dimensional Navier-Stokes flow with nondecaying initial velocity, J. Math. Fluid Mech. **3** (2001), 302–315. DOI: 10.1007/PL00000973
109. Y. Giga, M. Paolini and P. Rybka, On the motion by singular interfacial energy, Japan J. Ind. Appl. Math. **18** (2001), 231–248. DOI: 10.1007/BF03168572
110. Y. Giga and M.-H. Sato[*], A level set approach to semicontinuous viscosity solutions for Cauchy problems, Comm. in Partial Differential Equations **26** (5 & 6) (2001), 813–839. DOI: 10.1081/PDE-100002379
111. R. Kobayashi and Y. Giga, On anisotropy and curvature effects for growing crystals, Japan J. Ind. Appl. Math. **18** (2001), 207–230. DOI: 10.1007/BF03168571
112. J. Escher, Y. Giga and K. Ito[♮], On a limiting motion and self-intersections for the intermediate surface diffusion flow, J. of Evolution Equations **2** (2002), 349–364. DOI: 10.1007/s00028-002-8092-z
113. Y. Giga, Viscosity solutions with shocks, Comm. Pure Appl. Math. **55** (2002), 431–480. DOI: 10.1002/cpa.3015
114. Y. Giga, On the two-dimensional nonstationary vorticity equations, Proc. of the international conference on Tosio Kato's Method and Principle for

Evolution Equations in Methematical Physics", (eds. H. Fujita, S.T. Kuroda and H. Okamoto) pp. 86–97, Univ. of Tokyo Press, Tokyo (2002).

115. Y. Giga, N. Ishimura and Y. Kohsaka*, Spiral solutions for a weakly anisotropic curvature flow equation, Adv. Math. Sci. Appl. **12** (2002), 393–408.
116. Y. Giga and P. Rybka, Quasi-static evolution of 3-D cryctals grown from supersaturated vapor, Differential and Integral Equations **15** (2002), 1–15.
117. Y. Giga and K. Yamada*, On viscous Burgers-like equations with linearly growing initial data, Boletim da Sociedade Paranaense de Matemática **20** (2002), 29–49. DOI: 10.5269/bspm.v20i1-2.7521
118. M. Arisawa and Y. Giga, Anisotropic curvature flow in a very thin domain, Indiana Univ. Math. J. **52** (2003), 257–281. DOI: 10.1512/iumj.2003.52.2099
119. M. Bardi and Y. Giga, Right accessibility of semicontinuous initial data for Hamilton-Jacobi equations, Comm. Pure Appl. Anal. **2** (2003), 447–459.
120. C. M. Elliott, Y. Giga and S. Goto*, Dynamic boundary conditions for Hamilton-Jacobi equations, SIAM J. Math. Anal. **34** (2003), 861–881. DOI: 10.1137/S003614100139957X
121. M.-H. Giga and Y. Giga, Minimal vertical singular diffusion preventing overturning for the Burgers equation, Recent Advances in Scientific Computing and Partial Differential Equations, Contemp. Math. 330 (2003), Amer. Math. Soc. Providence, pp.73–88. [announcement: On minimal vertical singular diffusion preventing overturning, In:Viscosity solutions of differential equations and related topics (Japanese) (Kyoto, 2001), RIMS Kôkyûroku **1287** (2002), 114–126.]
122. Y. Giga, Shocks and very strong vertical diffusion, Lectures on Partial Differential Equations in honor of Louis Nirenberg's, 75th birthday (2003), International Press, Mass. U.S.A, pp. 95–102.
123. Y. Giga and R. Kobayashi, On constrained equations with singular diffusivity, Methods and Applications of Analysis **10** (2003), 253–278. DOI: 10.4310/MAA.2003.v10.n2.a6
124. Y. Giga and P. Rybka, Berg's effect, Adv. Math. Sci. Appl. **13** (2003), 625–637.
125. Y. Giga and O. Sawada*, On regularizing-decay rate estimates for solutions to the Navier-Stokes initial value problem, Nonlinear Analysis and Applications, To V. Lakshmikantham on his 80th Birthday **vol. I** (2003), 549–562, Kluwer Acad. Publ., Dordrecht.
126. H. Hontani, M.-H. Giga, Y. Giga and K. Deguchi, A computation of a crystalline flow starting from non-admissible polygon using expanding self-similar solutions, 11th International Conference DGCI 2003, LNCS 2886, Springer, Naples (2003), 465–474.
127. Y.-H. R. Tsai and Y. Giga, A numerical study of anisotropic crystal growth with bunching under very singular vertical diffusion, HUPSM #591 (2003).
128. Y.-H. R. Tsai, Y. Giga and S. Osher, A level set approach for computing discontinuous solutions of Hamilton-Jacobi equations, Math. Comp. **72** (2003), 159–181. DOI: 10.1090/S0025-5718-02-01438-2

129. M.-H. Giga and Y. Giga, A PDE approach for motion of phase-boundaries by a singular interfacial energy, Stochastic Analysis on Large Scale Interacting Systems, Advanced Studies in Pure Mathematics, 39 (eds. T. Funaki and H. Osada) (2004), Math. Soc. Japan, pp. 213–232.
130. Y. Giga, Singular diffusivity-facets, shocks and more, ICIAM 2003, Applied Mathematics Entering the 21st Century, (eds. M. Hill and R. Moore), SIAM (2004) pp.121–138.
131. Y. Giga, Y. Kashima* and N. Yamazaki, Local solvability of a constrained gradient system of total variation, Abstract and Applied Analysis **8** (2004), 651–682. DOI: 10.1155/S1085337504311048
132. Y. Giga and H. Kuroda*, On breakdown of solutions of a constrained gradient system of total variation, Proc. Symp. of PDEs—Foz do Iguacu, Brazil (2003), Boletim da Sociedade Paranaense de Matemática **22** (2004), 9–20.
133. Y. Giga, S. Matsui and S. Sasayama*, Blow up rate for semilinear heat equations with subcritical nonlinearity, Indiana Univ. Math. J. **53** (2004), 483–514. DOI: 10.1512/iumj.2004.53.2401
134. Y. Giga, S. Matsui and S. Sasayama*, On blow up rate for sign-changing solutions in a convex domain, Math. Methods. in Appl. Sciences **27** (2004), 1771–1782. DOI: 10.1002/mma.562
135. Y. Giga and P. Rybka, Existence of self-similar evolution of crystals grown from super saturated vapor, Interfaces and Free Boundaries **6** (2004), 405–421. DOI: 10.4171/IFB/107
136. M.-H. Giga and Y. Giga, On singular vertical diffusion for some Hamilton-Jacobi equations, Viscosity solution theory of differential equations and its developments, RIMS Kôkyûroku (2005), Kôkyûroku 131–141.
137. Y. Giga, K. Inui*, A. Mahalov and S. Matsui, Uniform local solvability for the Navier-Stokes equations with the Coriolis force, Methods Appl. Anal. **12** (2005), 381–393. DOI: 10.4310/MAA.2005.v12.n4.a2
138. Y. Giga and S. Koike, A characterization of value functions by viscosity solutions (Nenseikai niyoru ataikansu no tokuchouzuke) ISCIE **49** (2005), 2–7 (in Japanese).
139. Y. Giga, H. Kuroda* and N. Yamazaki, An existence result for a discretized constrained gradient system of total variation flow in color image processing, Interdisciplinary Information Sciences **11** (2005), 199–204, Tohoku University.
140. Y. Giga and M. Ohnuma*, On strong comparison principle for semicontinuous viscosity solutions of some nonlinear elliptic equations, Int. J. Pure Appl. Math. **22** (2005), 165–184.
141. Y. Giga and P. Rybka, Stability of facets of self-similar motion of a crystal, Advances in Differential Equations **10** (2005), 601–634.
142. Y. Giga and N. Umeda♮, Blow-up directions at space infinity for solutions of semilinear heat equations, Bol. Soc. Parana. Mat. (3) **23** (2005), 9–28. [Correction: Bol. Soc. Parana. Mat. (3) **24** (2006), 19–24.]

143. H. Hontani, M.-H. Giga, Y. Giga and K. Deguchi, Expanding selfsimilar solutions of a crystalline flow with applications to contour figure analysis, Discrete Appl. Math. **147** (2005), 265–285. DOI: 10.1016/j.dam.2004.09.015
144. M.-H. Giga, Y. Giga and H. Hontani, Self-similar expanding solutions in a sector for a crystalline flow, SIAM J. Math. Anal. **37** (2006), 1207–1226. DOI: 10.1137/040614372
145. Y. Giga, K. Inui*, A. Mahalov and S. Matsui, Navier-Stokes equations in a rotating frame in $\mathbb{R}^3$ with initial data nondecreasing at infinity, Hokkaido Math. J. **35** (2006), 321–364.
146. Y. Giga, T. Ohtsuka* and R. Schätzle, On a uniform approximation of motion by anisotropic curvature by the Allen-Cahn equations, Interfaces and Free Boundaries **8** (2006), 317–348. DOI: 10.4171/IFB/146
147. Y. Giga and P. Rybka, Stability of facets of crystals growing from vapor, Discrete Contin. Dyn. Syst. **14** (2006), 689–706. DOI: 10.3934/dcds.2006.14.689
148. Y. Giga and N. Umeda♮, On blow-up at space infinity for semilinear heat equations, J. Math. Anal. Appl. **316** (2006), 538–555. DOI: 10.1016/j.jmaa.2005.05.007
149. Y. Giga, PDE-Based Image Processing (3), (PDE wo riyoushita Gazou Shori (3)), Medical Imaging Technology **25** (2007), 56–59 (in Japanese).
150. Y. Giga, K. Inui*, A. Mahalov and S. Matsui, Uniform local solvability for the Navier-Stokes equations with the Coriolis force, Kyoto Conference on the Navier-Stokes equations and their applications, RIMS Kôkyûroku Bessatsu **B1** (eds. Y. Giga et al), (2007), pp.187–198.
151. Y. Giga, K. Inui*, A. Mahalov and J. Saal♮, Global solvabiliy of the Navier-Stokes equations in spaces based on sum-closed frequency sets, Advances in Differential Equations **12** (2007), 721–736.
152. Y. Giga, K. Inui*, A. Mahalov, S. Matsui and J. Saal♮, Rotating Navier-Stokes equations in $\mathbb{R}^3_+$ with initial data nondecreasing at infinity : The Ekman boundary layer problem, Arch. Rational Mech. Anal. **186** (2007), 177–224. DOI: 10.1007/s00205-007-0053-9
153. Y. Giga, M. Kubo* and Y. Tonegawa, Magnetic clusters and fold energies, Proc. Royal Soc. Edinburgh Sect. A **137** (2007), 23–40. DOI: 10.1017/S0308210505000156
154. Y. Giga, H. Kuroda* and N. Yamazaki, Global solvability of constrained singular diffusion equation associated with essential variation, Proc. Free boundary problem 2005, International series of Numerical Math. **154** (2007), Birkhäuser, 209–218.
155. Y. Giga, A. Mahalov and B. Nicolaenko, The Cauchy problem for the Navier-Stokes Equations with spatially almost periodic initial data, Mathematical Aspects of Nonlinear Dispersive Equations (eds. J. Bourgain et al), Annal of Math. Stud. **163**, Princeton Press (2007), 213–222.

156. M. Geissert and Y. Giga, On the Stokes resolvent equations in locally uniformly L^p spaces in exterior domains, Functional analysis and evolution equations, 307–314, Birkhäuser, Basel (2008).
157. Y. Giga, Evolution equations with almost periodic initial data, International Conference for the 25th Anniversary of Viscosity Solutions, (eds. Y. Giga, K. Ishii, S. Koike, T. Ozawa and N. Yamada), Gakuto International Series, Mathematical Scienes and Applications **30**, Gakkotosho, Tokyo (2008), pp. 97–106.
158. Y. Giga, K. Inui*, A. Mahalov and J. Saal♮, Uniform global solvability of the rotating Navier-Stokes equations for nondecaying initial data, Indiana Univ. Math. J. **57** (2008), 2775–2791. DOI: 10.1512/iumj.2008.57.3795
159. Y. Giga, H. Jo*, A. Mahalov and T. Yoneda*, On time analyticity of the Navier-Stokes equations in a rotating frame with spatially almost periodic data, Physica D **237** (2008), 1422–1428. DOI: 10.1016/j.physd.2008.03.007
160. Y. Giga and Q. Liu*, A remark on the discrete deterministic game approach for curvature flow equations, Proc. Interantional Conference on Nonlinear Phenomena with Energy Dissipation, (eds. A. Damlamian, N. Kenmochi, M. Mimura and J. Sprekels), Gakuto International Series, Mathematical Sciences and Applications **29**, Gakkotosho, Tokyo (2008), pp.103–115.
161. Y. Giga and P. Rybka, Faceted crystal grown from solution—a Stefan type problem with a singular interfacial energy, Proceedings of the 4th JSAM-SIMAI Seminar on Industrial and Applied Mathematics, (eds. H. Fujita and M. Nakamura) Gakuto International Series, Mathematical Sciences and Applications **28**, Gakkotosho, Tokyo (2008), pp.31–41.
162. Y. Giga and P. Rybka, Facet bending in the driven crystalline curvature flow in the plane, J. Geometric Analysis **18** (2008), 109–147. DOI: 10.1007/s12220-007-9004-9
163. Y. Giga, Y. Seki* and N. Umeda♮, Blow up at space infinity for nonlinear heat equations, Recent Advances in Nonlinear Analysis, (eds. M. Chipot et al), Proceedings of the International Conference on Nonlinear Analysis, Hsinchu, Taiwan (2006), World Scientific, (2008), pp. 77–94.
164. Y. Giga and N. Umeda♮, On instant blow-up for semilinear heat equations with growing initial data, Methods and Applications of Analysis **15** (2008), 185–196. DOI: 10.4310/MAA.2008.v15.n2.a5
165. E. Yokoyama, Y. Giga and P. Rybka, A microscopic time scale approximation to the behavior of the local slope on the faceted surface under a nonuniformity in supersaturation, Physica D **237** (2008), 2845–2855. DOI: 10.1016/j.physd.2008.05.009
166. Y. Giga and Q. Liu*, A billiard-based game interpretation of the Neumann problem for the curve shortening equation, Adv. Differential Equations **14** (2009), 201–240.
167. Y. Giga and P. Rybka, Facet bending driven by the planar crystalline curvature with a generic nonuniform forcing term, J. Differential Equations **246** (2009), 2264–2303. DOI: 10.1016/j.jde.2009.01.009

168. Y. Giga, Y. Seki* and N. Umeda♮, Mean curvature flow closes open ends of noncompact surface of rotation, Comm. Partial Differential Equations **34** (2009), 1508–1529. DOI: 10.1080/03605300903296926
169. Y. Giga and J. Zhai*, Uniqueness of constant weakly anisotropic mean curvature immersion of sphere S^2 in $\mathbb{R}^3$, Advances in Differential Equations **14** (2009), 601–619.
170. M.-H. Giga and Y. Giga, Very singular diffusion equations—second and fourth order problems, Japanese J. Ind. Appl. Math. **27** (2010), 323–345. DOI: 10.1007/s13160-010-0020-y
171. Y. Giga, Interface dynamics—Effect of curvature (Kaimen Dynamics—Kyokuritsu no Kouka), Handbook in Applied Analysis, (ed. K. Masuda) Springer, (2010), 375–418 (in Japanese).
172. Y. Giga, P. Górka and P. Rybka, Nonlocal spatially inhomogeneous Hamilton-Jacobi equation with unusual free boundary, Discrete and Continuous Dynamical Systems **26** (2010), 493–519. DOI: 10.3934/dcds.2010.26.493
173. Y. Giga, P. Górka and P. Rybka, A comparison principle for Hamilton-Jacobi equations with discontinuous Hamiltonian, Proc. American Mathematical Society **139** (2011), 1777–1785. DOI: 10.1090/S0002-9939-2010-10630-5
174. Y. Giga and R. V. Kohn, Scale-invariant extinction time estimates for some singular diffusion equations, Discrete and Continuous Dynamical Systems **30** (2011), 509–535. DOI: 10.3934/dcds.2011.30.509
175. Y. Giga, A. Mahalov and T. Yoneda*, On a bound for amplitudes of Navier-Stokes flow with almost periodic initial data, J. Math. Fluid Mech. **13** (2011), 459–467. DOI: 10.1007/s00021-010-0030-1
176. Y. Giga and H. Miura♮, On vorticity directions near singularities for the Navier-Stokes flows with infinite energy, Comm. Math. Phys. **303** (2011), 289–300. DOI: 10.1007/s00220-011-1197-x
177. Y. Giga, N. Mizoguchi and T. Senba, Asymptotic behavior of type I blowup solutions to a parabolic-elliptic system of drift-diffusion type, Arch. Rational Mech. Anal. **201** (2011), 549–573. DOI: 10.1007/s00205-010-0394-7
178. Y. Giga and J. Saal♮, L^1-maximal regularity for the Laplacian and applications, Discrete and Continuous Dynamical Systems, supplement **2011** (2011), 495–504. DOI: 10.3934/proc.2011.2011.495
179. Y. Giga, Y. Seki* and N. Umeda♮, On a decay rate of quenching profile at space infinity for axisymmetric mean curvature flow, Discrete and Continuous Dynamical Systems **29** (2011), 1463–1470. DOI: 10.3934/dcds.2011.29.1463
180. T. Eto*, Y. Giga and K. Ishii, An area minimizing scheme for anisotropic mean curvature flow, Proc. Japan Acad. **88** (2012), 7–10. DOI: 10.3792/pjaa.88.7
181. T. Eto*, Y. Giga and K. Ishii, An area minimizing scheme for anisotropic mean-curvature flow, Adv. Differential Equations **17** (2012), 1031–1084.
182. Y. Giga, Q. Liu* and H. Mitake, Large-time asymptotics for one-dimensional Dirichlet problems for Hamilton-Jacobi equations with noncoercive

Hamiltonians, J. Differential Equations **252** (2012), 1263–1282. DOI: 10.1090/S0002-9947-2013-05905-3

183. K. Abe[*] and Y. Giga, Analyticity of the Stokes semigroup in spaces of bounded functions, Acta Mathematica **211** (2013), 1–46. DOI: 10.1007/s11511-013-0098-6
184. M.-H. Giga and Y. Giga, On the role of kinetic and interfacial anisotropy in the crystal growth theory, Interface and Free Boundaries **15** (2013), 429–450. DOI: 10.4171/IFB/309
185. M.-H. Giga and Y. Giga, Ketshoseicho mondai to Kyotokui Kakusan Hoteishiki, Ijokakusan no Suri, RIMS Kôkyûroku (in Japanese) **1857** (2013), 33–56.
186. M.-H. Giga, Y. Giga and A. Nakayasu[*], On general existence results for one-dimensional singular diffusion equations with spatially inhomogeneous driving force, Proc. ERC Workshop on "Geometric Partial Differential Equations" (eds. M. Novaga et al.), Centro di Ricerca Matematica Ennio De Giorgi **15** (2013), 145–170. DOI: 10.1007/978-88-7642-473-1_8
187. M.-H. Giga, Y. Giga, T. Ohtsuka[*] and N. Umeda[♮], On behavior of signs for the heat equation and a diffusion method for data separation, Comm. Pure Appl. Anal. **12** (2013), 2277–2296. DOI: 10.3934/cpaa.2013.12.2277
188. M.-H. Giga, Y. Giga and N. Pozar[♮], Nenseikai riron ni yoru Kyotokui Kakusan Hoteishiki no sugaku kaiseki, Ijokakusan no Suri, RIMS Kôkyûroku (in Japanese) **1857** (2013), 57–77.
189. M.-H. Giga, Y. Giga and N. Pozar[♮], Anisotropic total variation flow of non-divergence type on a higher dimensional torus, Adv. Math. Sci. Appl. **23** (2013), 235–266.
190. Y. Giga, A remark on a Liouville problem with boundary for the Stokes and the Navier-Stokes equations, Discrete and Continuous Dynamical Systems, Series S **6** (2013), 1277–1289.
191. Y. Giga, P. Górka and P. Rybka, Evolution of regular bent rectangles by the driven crystalline curvature flow in the plane with a non-uniform forcing term, Adv. Differential Equations **18** Numbers 3–4 (2013), 201–242.
192. Y. Giga and N. Hamamuki[*], Hamilton-Jacobi equations with discontinuous source terms, Commun. Partial Differential Equations **38** (2013), 199–243. DOI: 10.1080/03605302.2012.739671
193. Y. Giga and G. Pisante[♮], On representation of boundary integrals involving the mean curvature for mean-convex domains, Proc. ERC Workshop on "Geometric Partial Differential Equations" (eds. M. Novaga et al.), Centro di Ricerca Matematica Ennio De Giorgi **15** (2013), 171–187. DOI: 10.1007/978-88-7642-473-1_9194
194. Y. Giga and J. Saal[♮], An approach to rotating boundary layers based on vector Radon measures, J. Math. Fluid Mech. **15** (2013), 89–127. DOI: 10.1007/s00021-012-0094-1
195. K. Abe[*] and Y. Giga, The L^∞-Stokes semigroup in exterior domains, J. of Evolution Equations **14** (2014), 1–28. DOI: 10.1007/s00028-013-0197-z

196. T. Asai* and Y. Giga, On self-similar solutions to the surface diffusion flow equations with contact angle boundary conditions, Interfaces and Free Boundaries **16** (2014), 539–573. DOI: 10.4171/IFB/329
197. M.-H. Giga, Y. Giga and N. Pozar♮, Periodic total variation flow of non-divergence type in $\mathbb{R}^n$, J. Math. Pures Appl. **102** (2014), 203–233. DOI: 10.1016/j.matpur.2013.11.007
198. M.-H. Giga, Y. Giga and P. Rybka, A comparison principle for singular diffusion equations with spatially inhomogeneous driving force for graphs, Arch. Ration. Mech. Anal. **211** (2014), 419–453. Erratum: **212** (2014), 707. DOI: "10.1007/s00205-013-0676-y, Erratum: 10.1007/s00205-014-0720-6
199. Y. Giga, P.-Y. Hsu* and Y. Maekawa*, A Liouville theorem for the planer Navier-Stokes equations with the no-slip boundary condition and its application to a geometric regularity criterion, Comm. in Partial Differential Equations **39** (2014), 1906–1935. DOI: 10.1080/03605302.2014.912662
200. Y. Giga, H. Kuroda* and H. Matsuoka, Fourth-order total variation flow with Dirichlet condition: Characterization of evolution and extinction time estimates, Adv. Math. Sci. Appl. **24** (2014), 499–534.
201. Y. Giga, Q. Liu* and H. Mitake, Singular Neumann problems and large-time behavior of solutions of noncoercive Hamilton-Jacobi equations, Trans. Amer. Math. Soc. **366** (2014), 1905–1941. DOI: 10.1090/S0002-9947-2013-05905-3
202. K. Abe*, Y. Giga and M. Hieber, Stokes resolvent estimates in spaces of bounded functions, Annales scientifiques de l'É.N.S. **48** (2015), 537–559.
203. K. Abe*, Y. Giga, K. Schade and T. Suzuki*, On the Stokes semigroup in some non-Helmholtz domains, Arch. der Math. **104** (2015), 177–187. DOI: 10.1007/s00013-015-0729-6
204. Y. Giga, P. Górka and P. Rybka, Bent rectangles as viscosity solutions over a circle, Nonlinear Analysis **125** (2015), 518–549. DOI: 10.1016/j.na.2015.05.033
205. Y. Giga, N. Hamamuki* and A. Nakayasu*, Eikonal equations in metric spaces, Trans. Amer. Math. Soc. **367** (2015), 49–66. DOI: 10.1090/S0002-9947-2014-05893-5
206. Y. Giga and H. Kuroda*, A counterexample to finite time stopping property for one-harmonic map flow, Comm. Pure Appl. Anal. **14** (2015), 121–125. DOI: 10.3934/cpaa.2015.14.121
207. Y. Giga and J. Saal♮, Uniform exponential stability of the Ekman spiral, Arkiv för Matematik **53** (2015), 105–126. DOI: 10.1007/s11512-014-0203-x208
208. T. Ohtsuka*, T.-Y. R. Tsai and Y. Giga, A level set approach reflecting sheet structure with single auxiliary function for evolving spirals on crystal surfaces, Journal of Scientific Computing **62** (2015), 831–874. DOI: 10.1007/s10915-014-9877-2
209. M. Bolkart and Y. Giga, On L^∞-BMO estimates for derivatives of the Stokes semigroup, Math. Z. **284** (2016), 1163–1183. DOI: 10.1007/s00209-016-1693-y

210. R. Farwig, Y. Giga and P.-Y. Hsu*, Initial values for the Navier-Stokes equations in spaces with weights in time, Funkcialaj Ekvacioj **59** (2016), 199–216. DOI: 10.1619/fesi.59.199
211. Y. Giga and N. Kajiwara*, On a resolvent estimate for bidomain operators and its applications. [preprint: HUPSM #1096 (2016)]
212. Y. Giga, H. Mitake and H. V. Tran, On asymptotic speed of solutions to level-set mean curvature flow equations with driving and source terms, SIAM J. Math. Anal. **48** (2016), 3515–3546. DOI: 10.1137/15M1052755
213. Y. Giga and N. Pozar♮, A level set crystalline mean curvature flow of surfaces, Adv. Differential Equations, **21** (2016), 631–698.
214. Y. Giga and X. Xiang, Lorentz space estimates for vector fields with divergence and curl in Hardy spaces, Tamkang Journal of Mathematics **47** (2016), 249–260. DOI: 10.5556/j.tkjm.47.2016.1932
215. K. Abe*, Y. Giga, K. Schade and T. Suzuki*, On the Stokes resolvent estimates for cylindrical domains, J. Evol. Eq., **17** (2017) 17–49. DOI: 10.1007/s00028-016-0350-6
216. M. Endo*, Y. Giga, D. Götz and C. Liu, Stability of a two-dimensional Poiseuille-type flow for a viscoelastic fluid, J. Math. Fluid Mech. **19** (2017), 17–45. DOI: 10.1007/s00021-016-0294-1
217. Y. Giga, M. Gries, M. Hieber, A. Hussein and T. Kashiwabara, Bounded H^∞-calculus for the hydrostatic Stokes operator on L^p spaces and applications, Proc. Amer. Math. Soc., electronically published (2017). DOI: 10.1090/proc/13676
218. Y. Giga and T. Namba*, Well-posedness of Hamilton-Jacobi equations with Caputo's time-fractional derivative, Comm. in Partial Differential Equations (2017), 1532–4133 (Online).
219. H. Koba*, C. Liu and Y. Giga, Energetic variational approaches for incompressible fluid systems on an evolving surface, Quart. Appl. Math. **75** (2017), 359–389. DOI: 10.1090/qam/1452
220. M. Bolkart, Y. Giga, T.-H. Miura*, T. Suzuki* and Y. Tsutsui♮, On analyticity of the L^p-Stokes semigroup for some non-Helmholtz domains, Math. Nachr., to appear. DOI: 10.1002/mana.201600016
221. M. Bolkart, Y. Giga and T. Suzuki*, Analyticity of the Stokes semigroup in *BMO*-type spaces, Journal of Math. Soc. Japan, to appear.
222. R. Farwig and Y. Giga, Well-chosen weak solutions of the instationary Navier-Stokes system and their uniqueness, Hokkaido Mathematical Journal, to appear.
223. R. Farwig, Y. Giga and P.-Y. Hsu*, The Navier-Stokes equations with initial values in Besov spaces of type $B^{-1+3/q}_{q,\infty}$, J. Korean Math. Soc., to appear.
224. Y. Giga, S. Ibrahim, S. Shen and T. Yoneda*, Global well-posedness for a two-fluid model, Differential Integral Equations, to appear.

II. Recent preprints

1. Y. Giga, J. Sauer and K. Schade, Strong stability of 2D viscoelastic Poiseuille-type flows. [preprint: HUPSM #1046 (2014)] (Revised version available in 2017)
2. M. Bolkart, Y. Giga, T. Suzuki* and Y. Tsutsui♮, Equivalence of *BMO*-type norms with applications to the heat and Stokes semigroup. [preprint: HUPSM #1090 (2016)]
3. R. Farwig, Y. Giga and P.-Y. Hsu*, On the continuity of the solutions to the Navier-Stokes equations with initial data in critical Besov spaces. [preprint: HUPSM #1093 (2016)]
4. Y. Giga, M. Muszkieta and P. Rybka, A duality based approach to total variation flow in H^{-s}. [preprint: HUPSM #1102 (2017)]
5. Y. Giga and N. Pozar♮, Approximation of general facets by regular facets with respect to anisotropic total variation energies and its application to the crystalline mean curvature flow. [preprint: HUPSM #1100 (2017)]
6. T.-H. Miura*, Y. Giga and C. Liu, An energetic variational approach for nonlinear diffusion equations in moving thin domains. [preprint: HUPSM #1101 (2017)]
7. T. Ohtsuka*, Y.-H. R. Tsai and Y. Giga, Growth rate of crystal surfaces with several dislocation centers. [preprint: HUPSM #1099 (2017)]

III. Books

1. Y.-G. Chen♮ and Y. Giga, "Tracking evolving surfaces" (Ugoku Kyokumen wo oikakete), Nippon Hyoron Sha (1996) (in Japanese) 128pp.
2. R. Agemi, Y. Giga and T. Ozawa eds, "Nonlinear Waves", Gakuto International Series, Mathematical Sciences and Applications **10** (1997) 542pp.
3. Y. Giga and M.-H. Giga, "Nonlinear Partial Differential Equations, Asymptotic behaviour of solutions and self-similar solutions" (Hisenkei Henbibun Houteishiki—Kai no Zenkinkyodo to Jiko Soji Kai), Kyoritsu Shuppan, 1999 (in Japanese) 288pp.
4. Y. Giga, "Surface Evolution Equations—a level set method", Lipschitz Lecture Notes **44**, University of Bonn, 2002, 231pp. (Also Hokkaido Univ. Technical Report Series in Math. #71, 2002).
5. Y. Giga, "Surface Evolution Equations. A Level Set Approach", Monographs in Mathematics **99**, Birkhäuser, Basel-Boston-Berlin, xii+264pp. (2006)
6. Y. Giga, H. Kozono, H. Okamoto and Y. Shibata eds, "Kyoto Conference on the Navier-Stokes equations and their applications", 2007, March, RIMS Kôkyûroku, Bessatsu **B1**, (2007)
7. Y. Giga, K. Ishii, S. Koike, T. Ozawa and N. Yamada, eds, "International Conference for the 25th Anniversary of Viscosity Solutions", Gakkotosho International Series, Mathematical Sciences and Applications **30**, Gakkotosho, Tokyo (2008).
8. M.-H. Giga, Y. Giga and J. Saal♮, "Nonlinear Partial Differential Equations. Asymptotic Behavior of Solutions and Self-Similar Solutions", Progress in

Nonlinear Differential Equations and their Applications **79**, Birkhäuser, Boston-Basel-Berlin, xviii+294pp (2010). DOI: 10.1007/978-0-8176-4651-6

9. T. Adachi, Y. Giga, T. Hishida, Y. Kagei, H. Kozono and T. Ogawa eds, "Mathematical Analysis on the Navier-Stokes Equations and Related Topics", Gakuto International Series, Mathematical Sciences and Applications **35** (2011), 241pp.
10. Y. Giga and T. Kobayashi eds, "What Mathematics Can Do for You: Essays and Tips from Japanese Industry Leaders", Springer Japan (2013), viii+146pp.
11. L. Ambrosio, Y. Giga, P. Rybka and Y. Tonegawa eds, "Variational Methods for Evolving Objects", Advanced Studies in Pure Mathematics **67**, Mathematical Society of Japan, Tokyo (2015), 298pp.
12. H. Amann, Y. Giga, H. Kozono, H. Okamoto and M. Yamazaki eds, "Recent Developments of Mathematical Fluid Mechanics", Birkhäuser Basel (2016), ix +482pp.

Contents

An Implicit Boundary Integral Method for Interfaces Evolving by Mullins-Sekerka Dynamics

Chieh Chen, Catherine Kublik and Richard Tsai

Abstract We present an algorithm for computing the nonlinear interface dynamics of the Mullins-Sekerka model for interfaces that are defined implicitly (e.g. by a level set function) using integral equations. The computation of the dynamics involves solving Laplace's equation with Dirichlet boundary conditions on multiply connected and unbounded domains and propagating the interface using a normal velocity obtained from the solution of the PDE at each time step. Our method is based on a simple formulation for implicit interfaces, which rewrites boundary integrals as volume integrals over the entire space. The resulting algorithm thus inherits the benefits of both level set methods and boundary integral methods to simulate the nonlocal front propagation problem with possible topological changes. We present numerical results in both two and three dimensions to demonstrate the effectiveness of the algorithm.

Keywords Mullins-Sekerka · Boundary integral methods · Implicit interface
Level set methods · Free boundary problem · Dendritic growth simulation

C. Chen (✉)
Goldman Sachs, 6011 Connection Drive, Irving, TX 75039, USA
e-mail: Chieh.Chen@gs.com

C. Kublik
Department of Mathematics, University of Dayton, 300 College Park,
Dayton, OH 45469, USA
e-mail: ckublik1@udayton.edu

R. Tsai
Department of Mathematics and Institute for Computational Engineering and Sciences,
University of Texas at Austin, 2515 Speedway, 78712, Austin, USA
e-mail: tsai@kth.se

R. Tsai
Department of Mathematics, KTH Royal Institute of Technology,
SE-100 44 Stockholm, Sweden

Y. Maekawa and S. Jimbo (eds.), *Mathematics for Nonlinear Phenomena—Analysis and Computation*, Springer Proceedings in Mathematics & Statistics 215,
DOI 10.1007/978-3-319-66764-5_1

1 Introduction

This paper proposes an algorithm for simulating interfacial motions in Mullins-Sekerka dynamics using a boundary integral equations formulated on implicit surfaces. The proposed method is able to simulate the dynamics on unbounded domains, and as such will provide a tool to better understand the dynamics obtained in these situations. As pointed out in Gurtin's paper [23], the behavior of the dynamics are far more interesting on domains that are unbounded. However, it seems that there is no other computational method that can simulate the nonlinear interfacial motion of Mullins-Sekerka dynamics in three dimensions on unbounded domains, allowing the interfaces to merge, break up, without the need of finding explicit representations.

The Mullins-Sekerka flow is a Stefan-type free boundary problem involving a nonlocal interfacial motion dynamically controlled by the solution of Laplace's equation with appropriate boundary conditions obtained on both sides of the interface. As a result, it is reasonable to consider boundary integral methods, especially for exterior domains, combined with level set methods [20, 40], for easy tracking and the capability of handling topological changes.

Various numerical methods have been proposed to solve elliptic problems on irregular domains. We mention here some of the few "usual suspects" that use finite element methods [4, 15, 24–26], finite difference techniques [7, 11, 18, 27, 28, 33], and boundary integral methods [3, 46]. Finite element methods require an explicit representation (e.g. triangulation) of the domain which makes them less tractable for an evolving domain. There is also a wide range of finite difference methods for solving elliptic equations, some of them using the level set method [18, 19, 35]. The Immersed Interface Method (IIM) [33, 34] is a popular technique for solving elliptic equations, particularly if the coefficients in the equation are discontinuous. This technique uses an adaptive finite difference scheme with a locally adaptive stencil. Unlike finite element method, the IIM can be used with an implicit representation of the interface. Nevertheless, both finite element and finite difference based techniques discretize the whole domain and therefore cannot be applied directly to "exterior" problems without imposing additional artificial boundary conditions.

In contrast, boundary integral methods (BIMs) use an integral representation of the solution, namely the solution is defined by an integral of a suitable potential over the interface. If the domain boundaries are fixed and suitable parameterizations are available, BIMs can be a natural choice to solve exterior problems.

In this paper, we present an implicit boundary integral algorithm for solving the following free-boundary problem

$$\begin{cases} \Delta u(\mathbf{x},t) = 0, & \mathbf{x} \in \mathbb{R}^m \setminus \Gamma_t,\ m = 2 \text{ or } 3, \\ u(\mathbf{x}) = -\kappa(\mathbf{x}), & \mathbf{x} \in \Gamma_t, t \geq 0, \\ v_n = -\left[\frac{\partial u}{\partial n}\right]_{\Gamma_t}, & \mathbf{x} \in \Gamma_t, t \geq 0, \\ \lim_{|\mathbf{x}|\to\infty} |u(\mathbf{x},t)| = u_\infty(t), & \\ \Gamma_0 = \partial\Omega_0^-, & \end{cases} \tag{1}$$

where $\kappa(\mathbf{x})$ is the mean curvature (except in 1D where it is just the curvature) of the interface Γ_t at $\mathbf{x}$ and $v_n(\mathbf{x})$ is the normal velocity of the interface Γ_t at $\mathbf{x}$. This problem is sometimes referred to as the Mullins-Sekerka problem [38] which was first studied as a solidification process. For example, Ω_t^+ may represent the solid phase of a material at time $t \geq 0$ and $\Omega_t^- = \Omega_t \setminus \Omega_t^+$ may represent the liquid phase of the material. The interface between the two phases is denoted Γ_t. In general, this problem models the solidification or liquidation of materials of negligible specific heat, where the function u represents the scaled temperature and Γ_t is the boundary of the bounded domain Ω_t, namely $\Gamma_t = \partial \Omega_t$ at time t. The boundary condition $u = -\kappa$ comes from the Gibbs-Thomson relation and is related to the surface tension effect.

In this paper, we take $u_\infty = 0$ in three dimensions. The case $u_\infty \neq 0$ can be done but requires a modification of the integral equations described in Section 2. To obtain the so-called Mullins-Sekerka instability in 3D, it is necessary, however, that $u_\infty < 0$: we simulate such a case in Section 6.2 and explain briefly how the integral equations need to be modified. In two dimensions, the far field environment u_∞ is determined by the compatibility condition

$$\int_{\Gamma_t} \frac{\partial u(\mathbf{x}, t)}{\partial n} dS(\mathbf{x}) = 0, \ \forall t > 0, \tag{2}$$

(see [30]). We note that this additional condition is only necessary in two dimensions. In three dimensions, u_∞ can be taken to be any non zero value. This difference between the two and three-dimensional case is the direct consequence of the asymptotic behavior of the fundamental solution of Laplace's equation in two and three dimensions. For a detailed derivation, we refer the reader to Theorem 6.11 and Section 6.2 of Kress [30]. In two dimensions, the boundary condition at infinity (assuming that the temperature is uniformly bounded) produces an isolated boundary in which the flux $\frac{\partial u}{\partial n}$ is zero at infinity. In three dimensions, the boundary condition at infinity with $u_\infty = 0$ is an isothermal boundary condition which models the liquidation of material.

In this setup, Ω_t is the solid region and Ω_t^c is the liquid region. $\left[\frac{\partial u}{\partial n}\right]_{\Gamma_t}$ is the jump in the normal derivative of the function u across the interface Γ_t defined as the sum of the outward normal derivatives of u from each side of Γ_t. More precisely,

$$\left[\frac{\partial u}{\partial n}\right] := \left(\frac{\partial u}{\partial n^+}\right)_- - \left(\frac{\partial u}{\partial n^+}\right)_+, \tag{3}$$

where $+$ and $-$ refer to outside and inside respectively.

We observe that the two-dimensional Mullins-Sekerka problem has nontrivial stationary solutions. For example, a collection of disjoint circles with the same radius R will be stationary under the Mullins-Sekerka dynamics. To see this, we note that in this case, both the solution inside and outside will be constant equal to $-\frac{1}{R}$, and thus $v_n = 0$.

This free boundary problem has interesting properties concerning volume and surface area. See for example [13], where coarsening rates in Mullins-Sekerka evolution in bounded domains with periodic conditions are studied.

In the following, we describe some notable differences in evolutions defined in two and three dimensions. Let $A(t)$ denote the surface area in 3D or length in 2D of Γ_t at time t and $V(t)$ denote the volume in 3D or area in 2D of Ω_t at time t. Then we have

$$\begin{aligned}\frac{dV(t)}{dt} &= -\int_{\Gamma_t} v_n ds \\ &= \int_{\Gamma_t} \left[\frac{\partial u}{\partial n}\right] ds \\ &= \int_{\Omega_t} \Delta u dx - \int_{\Gamma_t} \left(\frac{\partial u}{\partial n^+}\right)_+ ds.\end{aligned}$$

The first term is zero because u is harmonic inside Ω_t. In two dimensions, the second term turns out to be zero because of the compatibility condition (2) necessary in the two-dimensional case (see [30]). Thus the total area enclosed by Γ_t is conserved. This compatibility condition is not valid in three dimensions and hence the motion is not volume preserving in three dimensions.

For the area, we have the following:

$$\begin{aligned}\frac{dA(t)}{dt} &= -\int_{\Gamma_t} \kappa v_n ds \\ &= -\int_{\Gamma_t} u \left[\frac{\partial u}{\partial n}\right] ds \\ &= \int_{\Gamma_t} u\,(\nabla u)_+ \cdot \mathbf{n}^+ ds - \int_{\Omega_t} |\nabla u|^2 d\mathbf{x} - \underbrace{\int_{\Omega_t} \Delta u d\mathbf{x}}_{=0}.\end{aligned}$$

We now show that in two dimensions

$$\int_{\Gamma_t} u \nabla u \cdot \mathbf{n}^+ ds = -\int_{\mathbb{R}^2 \setminus \bar{\Omega}_t} |\nabla u|^2 d\mathbf{x}.$$

Let $R > 0$ be sufficiently large so that $\bar{\Omega}_t \subset B_R(\mathbf{x_0})$ where $B_R(\mathbf{x_0})$ is the ball centered at $\mathbf{x_0} \in \Omega_t$ with radius R. Applying the divergence theorem in the "annulus" $B_R(\mathbf{x_0}) \setminus \bar{\Omega}_t$ for a fixed $t > 0$, we obtain

$$\begin{aligned}\int_{\partial B_R(\mathbf{x_0})} u \nabla u \cdot \mathbf{n}^+ ds - \int_{\Gamma_t} u \nabla u \cdot \mathbf{n}^+ ds &= \int_{B_R(\mathbf{x_0}) \setminus \bar{\Omega}_t} \nabla \cdot (u \nabla u)\, d\mathbf{x} \\ &= \int_{B_R(\mathbf{x_0}) \setminus \bar{\Omega}_t} |\nabla u|^2 d\mathbf{x} + \int_{B_R(\mathbf{x_0}) \setminus \bar{\Omega}_t} u \Delta u d\mathbf{x}\end{aligned}$$

$$= \int_{B_R(\mathbf{x_0})\setminus\bar{\Omega}_t} |\nabla u|^2 d\mathbf{x}, \tag{4}$$

since u is harmonic in $\mathbb{R}^m \setminus \bar{\Omega}_t$. Standard results in potential theory show that the two-dimensional bounded exterior solution u satisfy

$$u(\mathbf{x}, t) = u_\infty + O\left(\frac{1}{|\mathbf{x}|}\right), \quad \text{and} \quad |\nabla u(\mathbf{x}, t)| = O\left(\frac{1}{|\mathbf{x}|^2}\right),$$

as $|\mathbf{x}| \to \infty$ uniformly for all directions. Therefore

$$\lim_{R\to\infty} \int_{\partial B_R(\mathbf{x_0})} u\nabla u \cdot \mathbf{n}^+ ds = \lim_{R\to\infty} \int_{\partial B_R(\mathbf{x_0})} u\frac{\partial u}{\partial r} ds = 0.$$

Finally, taking R to infinity in equation (4), we obtain

$$\int_{\Gamma_t} u\nabla u \cdot \mathbf{n}^+ ds = -\int_{\mathbb{R}^2\setminus\bar{\Omega}_t} |\nabla u|^2 d\mathbf{x},$$

thus giving

$$\frac{dA(t)}{dt} = -\int_{\mathbb{R}^2\setminus\bar{\Omega}_t} |\nabla u|^2 d\mathbf{x} - \int_{\Omega_t} |\nabla u|^2 d\mathbf{x} = -\int_{\mathbb{R}^2\setminus\Gamma_t} |\nabla u|^2 d\mathbf{x}. \tag{5}$$

It follows that the evolution in two dimensions is area preserving while decreasing the length of the boundary.

Another interesting property of the Mullins-Sekerka dynamics is that the two-dimensional motion induced by (1) does not preserve convexity, see e.g. [6].

In this paper, we solve (1) using an implicit boundary interface method. In other words, we define Γ_t implicitly through a level set function φ, such that for $t \geq 0$, we have

$$\Gamma_t = \left\{\mathbf{x} \in \mathbb{R}^m : \varphi(\mathbf{x}, t) = 0\right\}.$$

The algorithm operates by alternatively solving for the solution u while maintaining the interface fixed, and then propagating the interface according to the normal velocity obtained from the jump in the normal derivative of u across the interface In the present work, the solution of the boundary value problem given by the first three equations in (1) is solved using an implicit boundary integral formulation [31, 32] that allows the boundary integral to be rewritten as an integral over $\mathbb{R}^m$.

2 Integral Equations for Laplace's Equation

We present below the boundary integral equation formulations most relevant to this paper and some of their useful properties. Throughout the paper, we consider $\Omega \subset \mathbb{R}^m$ to be a bounded set, $m = 2$ or 3 and the boundary $\partial\Omega = \Gamma$ to be a disjoint collection of closed compact C^2 interfaces. We also denote by $\mathbf{n}^+$ and $\mathbf{n}^-$ the outward and the inward unit normal vectors respectively. The exterior is defined to be the unbounded domain.

We consider the Dirichlet problem (interior and exterior) in Ω, an $L+1$ connected region (bounded or unbounded) with L holes labeled $\{S_i\}_{i=1}^L$ each having boundary $\Gamma_i = \partial S_i$:

$$\begin{cases} \Delta u(\mathbf{x}) = 0 & \mathbf{x} \in \Omega \\ u(\mathbf{x}) = f(\mathbf{x}), & \mathbf{x} \in \Gamma, \end{cases} \tag{6}$$

and

$$\begin{cases} \Delta u(\mathbf{x}) = 0 & \mathbf{x} \in \mathbb{R}^m \setminus \bar{\Omega}, \\ u(\mathbf{x}) = f(\mathbf{x}), & \mathbf{x} \in \Gamma, \\ \lim_{|\mathbf{x}|\to\infty} |u(\mathbf{x})| = u_\infty, & m = 2 \\ \lim_{|\mathbf{x}|\to\infty} |u(\mathbf{x})| = 0, & m = 3. \end{cases} \tag{7}$$

Since Dirichlet boundary conditions are imposed, we introduce an unknown density β defined on Γ and represent the solution of (6) using the double layer potential formulation

$$u(\mathbf{x}) = \int_\Gamma \beta(\mathbf{y}) \frac{\partial \Phi(\mathbf{x}, \mathbf{y})}{\partial \mathbf{n}_\mathbf{y}^+} dS(\mathbf{y}), \quad \mathbf{x} \in \Omega,$$

where Φ is the fundamental solution of Laplace's equation defined as

$$\Phi(\mathbf{x}, \mathbf{y}) = \begin{cases} \frac{1}{2\pi} \ln |\mathbf{x} - \mathbf{y}| & \text{for } m = 2, \\ -\dfrac{1}{m(m-2)\rho_m |\mathbf{x} - \mathbf{y}|^{m-2}} & \text{for } m \geq 3, \end{cases} \tag{8}$$

where ρ_m is the volume of the unit ball in $\mathbb{R}^m$.

In this context, the naive integral equation is ill-posed, both for the interior and the exterior problem, since there are L nontrivial homogeneous solutions that span the nullspace, see e.g. [17]. To alleviate this issue, we follow the approach first suggested by Mikhlin [36] and developed by Greenbaum et al. [21]. The procedure amounts to adding a linear combination $\sum_{i=1}^L A_i \Phi(\mathbf{x} - \mathbf{z_i})$ to the integral equation, where $\mathbf{z_i} \in S_i$, and the coefficients A_i play the role of Lagrange multipliers. These L constants give L degrees of freedom that are used to make sure the constructed solution satisfies the appropriate equation and boundary conditions. For the full derivation of the modified integral equation, we refer the reader to the work of Mikhlin and Greenbaum et al. [21, 36].

The interior problem is therefore solved as follows:

1. Find the density β and the constants $\{A_i\}_{i=1}^{L}$ such that

$$\int_{\Gamma} \beta(\mathbf{y}) \frac{\partial \Phi(\mathbf{x},\mathbf{y})}{\partial n_y^+} dS(\mathbf{y}) + \frac{1}{2}\beta(\mathbf{x}) + \sum_{i=1}^{L} A_i \Phi(\mathbf{x}-\mathbf{z_i}) = f(\mathbf{x}), \text{ for } \mathbf{x} \in \Gamma,$$
$$\int_{\Gamma_i} \beta(\mathbf{y}) dS(\mathbf{y}) = 0, \text{ for } 1 \le i \le L. \tag{9}$$

2. Reconstruct the solution u in Ω using the double layer potential formulation

$$u(\mathbf{x}) = \int_{\Gamma} \beta(\mathbf{y}) \frac{\partial \Phi(\mathbf{x},\mathbf{y})}{\partial n_y^+} dS(\mathbf{y}) + \sum_{i=1}^{L} A_i \Phi(\mathbf{x}-\mathbf{z_i}), \qquad \text{for } \mathbf{x} \in \Omega. \tag{10}$$

Note that for a simply connected domain Ω (i.e. a domain without any hole), (9) should be replaced by

$$\int_{\Gamma} \beta(\mathbf{y}) \frac{\partial \Phi(\mathbf{x},\mathbf{y})}{\partial n_y^+} dS(\mathbf{y}) + \frac{1}{2}\beta(\mathbf{x}) = f(\mathbf{x}), \text{ for } \mathbf{x} \in \Gamma.$$

For the exterior problem it is also necessary to modify the integral equation in order to obtain an invertible system. It can be shown, e.g. [30], that if the kernel is modified using $\frac{1}{|\mathbf{x}-\mathbf{y}|^{m-2}}$, the system becomes well-posed. The exterior problem is therefore solved as follows:

1. Find the density β and the constants $\{A_i\}_{i=1}^{L}$ such that

$$\int_{\Gamma} \beta(\mathbf{y}) \left(\frac{\partial \Phi(\mathbf{x},\mathbf{y})}{\partial n_y^+} - \frac{1}{|\mathbf{x}-\mathbf{y}|^{m-2}} \right) dS(\mathbf{y}) - \frac{1}{2}\beta(\mathbf{x}) + \sum_{i=1}^{L} A_i \Phi(\mathbf{x}-\mathbf{z_i}) = f(\mathbf{x}),$$
$$\int_{\Gamma_i} \beta(\mathbf{y}) dS(\mathbf{y}) = 0, \text{ for } 1 \le i \le L-1, \quad \text{for } \mathbf{x} \in \Gamma,$$
$$m = 2: \quad \sum_{i=1}^{L} A_i = 0,$$
$$m = 3: \quad \int_{\Gamma_L} \beta(\mathbf{y}) dS(\mathbf{y}) = 0. \tag{11}$$

2. Reconstruct the solution u in $\mathbb{R}^m \setminus \bar{\Omega}$ using the double layer potential formulation

$$u(\mathbf{x}) = \int_{\Gamma} \beta(\mathbf{y}) \left(\frac{\partial \Phi(\mathbf{x}, \mathbf{y})}{\partial n_y^+} - \frac{1}{|\mathbf{x} - \mathbf{y}|^{m-2}} \right) dS(\mathbf{y}) + \sum_{i=1}^{L} A_i \Phi(\mathbf{x} - \mathbf{z_i}), \qquad \text{for } \mathbf{x} \in R^m \setminus \bar{\Omega}. \tag{12}$$

The condition $\sum_{i=1}^{L} A_i = 0$, necessary in the two dimensional case, ensures that the solution is bounded. Indeed, if Γ is a compact set and β is uniformly bounded on Γ, then the first integral in the right-hand side of (12) will be bounded as well. Thus, the second term on the right-hand side of (12) will determine the asymptotic behavior of the solution u. For the cases that we consider here, Γ is compact and β is uniformly bounded on Γ. We note that the condition $\sum_{i=1}^{L} A_i = 0$ is not needed in three dimensions since $\Phi(r) = O\left(\frac{1}{r}\right)$ and thus vanishes at infinity.

3 Overview of Existing Numerical Methods

In this section, we give a brief overview of numerical methods for solving integral equations. Boundary integral methods typically provide highly accurate numerical solutions, however they require highly accurate parameterizations of the boundaries, high order quadratures and specialized dense matrix solvers.

For each of the boundary integral equations obtained in the previous section, we need to solve a Fredholm equation of the second kind. In other words, we need to find a function γ defined on Γ, such that

$$q(x) = \int_{\Gamma} \gamma(y(s)) K(x, y(s)) ds + C_0 \gamma(x),$$

where C_0 is a constant and K is the normal derivative of the fundamental solution of Laplace's equation to Γ. To solve these equations numerically it is necessary to discretize the above integrals. Three discretization methods are typically used: the Nyström method [2, 39], the collocation method [2] and the Galerkin method [2, 12]. Each of these discretization methods leads to a discrete system of the form

$$(I + \mathtt{K}\Lambda)\gamma = q,$$

where I is the identity matrix, $\mathtt{K}$ is a dense matrix, Λ is a diagonal matrix (for example containing the quadrature weights of the Nyström method), γ is the vector of unknowns, and q is a known vector obtained from the boundary conditions.

Since $\mathtt{K}$ is dense, this system is usually solved using an iterative procedure. In addition, low rank approximations may be constructed to improve the efficiency of the numerical solver. One very successful approach is the Fast Multipole Method introduced by Greengard and Rokhlin in 1987 [22]. The use of hierarchical matrices [8] to solve this dense system is also popular.

In this paper, we use an exact integral formulation that allows the boundary integral to be rewritten exactly as a volume integral over a thin tubular neighborhood of the boundary. This enables us to perform the computations on a fixed grid regardless of the location of the interface. Should the interface evolve in time, all computations will be performed on the mesh that is used by the level set function at each time step. This makes the algorithm easy to implement for evolving interfaces in two and three dimensions.

4 Exact Integral Formulations Using Signed Distance Functions

The exact integral formulation we use in this work was first proposed in [32] and extended in [31]. This formulation allows the computation of integrals of the form

$$\int_{\Gamma} v(\mathbf{x})dS, \tag{13}$$

in the level set framework, where Γ is represented implicitly by a level set function.

In [32], with the choice of $\varphi = d_\Gamma$, a signed distance function to Γ, the integral (13) is expressed as an average of integrals over nearby level sets of d_Γ, where these nearby level sets continuously sweep a thin tubular neighborhood around the boundary Γ of radius ε. Consequently, (13) is *equivalent* to the volume integral shown on the right-hand side below:

$$\int_{\Gamma} v(\mathbf{x})dS = \int_{\mathbb{R}^m} v(\mathbf{x}^*)J(\mathbf{x}; d_\Gamma)\delta_\varepsilon(d_\Gamma(\mathbf{x}))d\mathbf{x}, \tag{14}$$

where $\mathbf{x}^*$ is the closest point on Γ to $\mathbf{x}$, $J(\mathbf{x}; d_\Gamma)$ accounts for the change in curvature between the nearby level sets and the zero level set and δ_ε is an averaging kernel compactly supported in $[-\varepsilon, \varepsilon]$ with p ($p \in \mathbb{N}$) vanishing moments, defined as

$$\int_{-\infty}^{\infty} \delta_\varepsilon(\eta)\eta^k d\eta = \begin{cases} 1 & k = 0, \\ 0 & 0 < k \le p - 1, \end{cases}$$

We remark that the author in [37] used a similar formulation to study the heat equation defined in tubular neighborhoods of Γ.

We now briefly give a justification for (14). Suppose that Γ is a smooth closed hypersurface in $\mathbb{R}^m$ and assume that $\mathbf{x}$ is sufficiently close to Γ so that the closest point mapping

$$\mathbf{x}^* = P_\Gamma(\mathbf{x}) = \text{argmin}_{y \in \Gamma} |\mathbf{x} - \mathbf{y}|$$

is continuously differentiable. Then the restriction of P_Γ to Γ_η is a diffeormorphism between Γ_η and Γ, where $\Gamma_\eta := \{\mathbf{x} : d_\Gamma(\mathbf{x}) = \eta\}$. As a result, it is possible to write integrals over Γ using points on Γ_η as:

$$\int_\Gamma v(\mathbf{x})dS = \int_{\Gamma_\eta} v(P_\Gamma(\mathbf{x}))J(\mathbf{x};\eta)dS,$$

where the Jacobian $J(\mathbf{x},\eta)$ comes from the change of variable defined by P_Γ restricted on Γ_η. Averaging the above integrals respectively with a kernel δ_ε compactly supported in $[-\varepsilon,\varepsilon]$, we obtain

$$\int_\Gamma v(\mathbf{x})dS = \int_{-\varepsilon}^{\varepsilon} \delta_\varepsilon(\eta)\int_{\Gamma_\eta} v(P_\Gamma(\mathbf{x}))J(\mathbf{x};\eta)dS\, d\eta.$$

Formula (14) then follows from the coarea formula [16] applied to the integral on the right-hand side. Because the distance function is used, it is possible to obtain a closed form for the Jacobian $J(\mathbf{x},\eta)$. See [31, 32] for proofs.

We can now use this formulation to evaluate integral equations. Consider a general integral equation of the form

$$\int_\Gamma \beta(\mathbf{y})K(\mathbf{x},\mathbf{y})dS(\mathbf{y}) + \lambda\beta(\mathbf{x}) = f(\mathbf{x}), \quad \text{for } \mathbf{x}\in\Gamma,$$

where $\lambda\in\mathbb{R}$ and Γ is a closed C^2 hypersurface embedded in $\mathbb{R}^m$. We define for any function $u:\Gamma\mapsto\mathbb{R}$ its constant extension along the normals $\bar{u}$ as

$$\forall\mathbf{x}\in\mathbb{R}^m, \qquad \bar{u}(\mathbf{x}) = u(P_\Gamma(\mathbf{x})).$$

Analytically, the formulation in [32] gives us the following

$$\int_{|d_\Gamma|\le\varepsilon} \bar{\beta}(\mathbf{y})\tilde{K}(\mathbf{x},\mathbf{y})\delta_\varepsilon(d_\Gamma(\mathbf{y}))dy + \lambda\bar{\beta}(\mathbf{x}) = \bar{f}(\mathbf{x}),$$

where $\tilde{K}(\mathbf{x},\mathbf{y}) = K(\mathbf{x},\mathbf{y})J(\mathbf{x};\Gamma)$ can be thought of as a weighted restriction.

On a uniform cartesian grid with grid spacing h, the above expression can be discretized as: $\forall\mathbf{x}_j$ s.t. $|d_\Gamma(\mathbf{x}_j)|<\varepsilon$,

$$\sum_{k:|d_\Gamma(\mathbf{x}_k)|<\varepsilon} \bar{\beta}(\mathbf{x}_k)\tilde{K}(\mathbf{x}_j,\mathbf{x}_k)\delta_\varepsilon(d_\Gamma(\mathbf{x}_k))h^m + \lambda\bar{\beta}(\mathbf{x}_j) = \bar{f}(\mathbf{x}_j). \tag{15}$$

The number of equations in the corresponding system is the number of grid points trapped inside the tubular neighborhood $|d_\Gamma(\mathbf{x})|<\varepsilon$.

5 Algorithm for the Mullins-Sekerka Dynamics

In this section, we highlight the main ideas of the algorithm and provide a discussion of the interesting and specific features of the algorithm.

5.1 *Algorithm*

There is variety of work that has been done on the computation of Mullins-Sekerka flows. Zhu et al. [46] proposed a boundary integral method for computing the two space dimensional Mullins-Sekerka free boundary problem. In [29], Karali et al. used a set of ODE to simulate the interactions of circular bubbles under the Mullins-Sekerka flow both in two and three dimensions. In [9] the authors proposed a level set based approach using finite difference to compute the two dimensional Mullins-Sekerka dynamics. Finally, in [5] the authors proposed a parametric finite element approach for two and three dimensions. In this paper, we propose an implicit boundary algorithm for simulating the resulting interfacial motion from Mullins-Sekerka dynamics on unbounded domains in both two and three dimensions. Our algorithm is based on the technique described in [32]. However, because of the complexity of the Mullins-Sekerka dynamics, it is necessary to expand the basic method to more general configurations.

The overall structure of the proposed algorithm is defined by the general level set framework:

1. Given d_{Γ_n} on the grid at time $n\Delta t$ ($n \in \mathbb{N}$), solve the Laplace problems (6)–(7). Output: the density β in $\{d_{\Gamma_n}(\mathbf{x}_j) < \varepsilon\}$.
2. Compute a suitable extension $\tilde{v}_n$ of v_n in $\{d_{\Gamma_n}(\mathbf{x}_j) < \varepsilon\}$.
3. Solve the level set equation in $\{d_{\Gamma_n}(\mathbf{x}_j) < \varepsilon\}$ for Δt amount of time

$$\varphi_t + \tilde{v}_n|\nabla\varphi| = 0, \tag{16}$$

   using $\varphi(x, 0) = d_{\Gamma_n}$ as initial condition.
4. Redistance: Given $\varphi(\mathbf{x}, \Delta t)$, compute the signed distance function d_{Γ_n} to the zero level set of φ.
5. Repeat the above steps until the desired final time.

In the following, we shall describe what we think are important details that are specific to the Mullins-Sekerka simulations, leaving the other steps to the standard level set literature.

5.1.1 Dealing with Multiply Connected Regions in Step 1

One complexity is the fact that the regions can be multiply connected. Numerically, it means that we first need to identify the connected components and how multiply

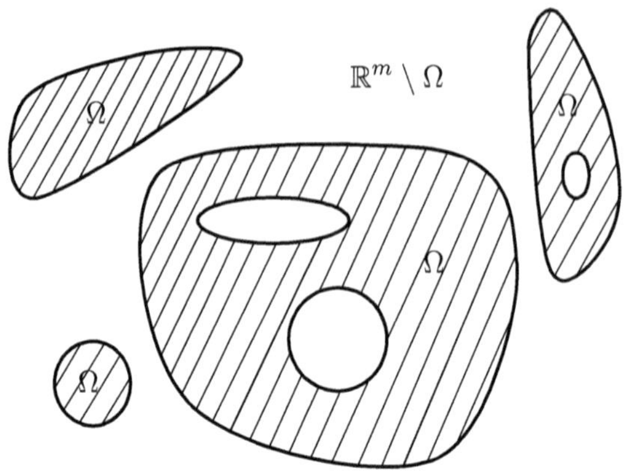

Fig. 1 Non-simply connected domain

connected each component is before solving the correct system. See Figure 1 for an illustration of a typical configuration with simply and multiply connected components. To perform this task, we use a connected component labeling algorithm which was first proposed in the context of computer graphics. See the appendix.

5.1.2 Velocity Extension in Step 2

Another difficulty in the context of level set methods is the velocity extension. In order to propagate an interface moving with normal velocity v_n, it is first necessary to extend $v_n(\mathbf{x}, t)$ defined on $\Gamma_t \times [0, \infty)$ to $\tilde{v}_n(\mathbf{x}, \tau)$ defined on $\mathbb{R}^m \times [0, \infty)$. This is typically done by solving the system

$$\begin{cases} \frac{\partial \tilde{v}_n}{\partial \tau} + sign(\varphi)\frac{\nabla\varphi}{|\nabla\varphi|} \cdot \nabla\tilde{v}_n = 0, \\ \tilde{v}_n(\mathbf{x}, 0) = v_n(\mathbf{x}, t), \end{cases} \tag{17}$$

for a sufficient amount of time so that $\nabla\tilde{v}_n \cdot \nabla\varphi = 0$ in a neighborhood of Γ_t. This extension extends v_n in a constant fashion along streamlines normal to the level sets of φ, which in practice is performed in a neighborhood of the interface, see [10] for a more extensive discussion on this topic. However, this extension requires the knowledge of the normal velocity on the interface which in this case necessitates the evaluation of a hypersingular integral

$$\frac{\partial u}{\partial n_{\mathbf{x}}}(P_\Gamma(\mathbf{x})) = \int_\Gamma \frac{\partial^2 \Phi}{\partial_{n_{\mathbf{x}}} \partial_{n_{\mathbf{y}}}}(P_\Gamma(\mathbf{x}), \mathbf{y})\beta(\mathbf{y})dS(\mathbf{y}).$$

We circumvent this issue by approximating the normal velocity v_n at the grid points $\mathbf{x}$ closest to the interface Γ_t (namely in a very narrow band around the interface taken to be a few grid points wide) using the equation

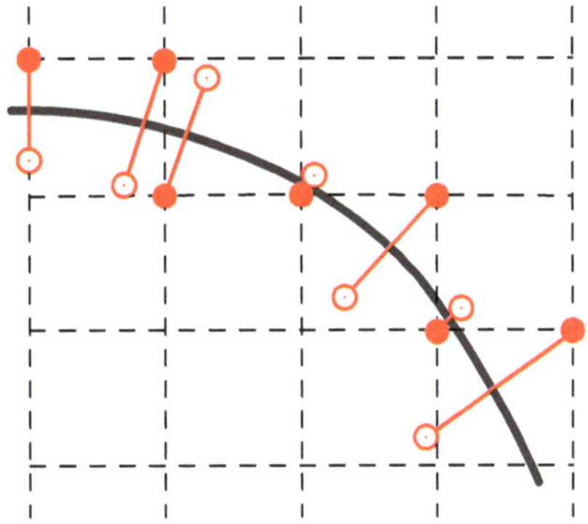

Fig. 2 Velocity extension. The solid dots represent the grid points $\mathbf{x}$ and the empty ones their corresponding mirror points across the interface

$$v_n(\mathbf{x}) := sign(d_{\Gamma_t}(\mathbf{x}))\left(\frac{\partial u}{\partial n^+}(\mathbf{x}) - \frac{\partial u}{\partial n^+}\left(\mathbf{x} - 2d_{\Gamma_t}(\mathbf{x})\nabla d_{\Gamma_t}(\mathbf{x})\right)\right), \tag{18}$$

where the normal derivatives $\frac{\partial u}{\partial \mathbf{n}^+}$ are computed using the interface integral

$$\frac{\partial u}{\partial n_{\mathbf{x}}}(\mathbf{x}) = \int_{\Gamma_t} \frac{\partial^2 \Phi}{\partial_{n_{\mathbf{x}}}\partial_{n_{\mathbf{y}}}}(\mathbf{x}, \mathbf{y})\beta(\mathbf{y})dS(\mathbf{y}),$$

where $\mathbf{x} \notin \Gamma_t$. The idea behind equation (18) is to compute the jump in the normal derivative of u at a point $\mathbf{x}$ close to Γ_t by using its symmetrical point (or mirror point) $\mathbf{x} - 2d_{\Gamma_t}(\mathbf{x})\nabla d_{\Gamma_t}(\mathbf{x})$ with respect to Γ_t. This is illustrated in Figure 2.

Note that the extension $\tilde{v}_n$ in (18) is continuous and agrees with v_n on Γ_t. We point out that this step can be interpreted as the initialization step for the PDE in equation (17). Thus, once the normal velocity $\tilde{v}_n$ is obtained in this very narrow band, we extend it to the next layer of grid points using the standard extension (17).

Finally the normal velocity may tend to ∞ as parts of Γ_t vanish in finite time. Numerically, we regularize the computed normal velocity by imposing a maximum normal velocity v_∞. This means we replace v_n by $\max(v_n, v_\infty)$.

5.1.3 Literature on Redistancing and Involving the Level Set Equations

To maintain the completeness of this exposition, we briefly mention some established literature on the level set equation (16) and the construction of signed distance functions (Step 4 in the algorithm).

The level set equation (16), first proposed by Osher and Sethian in their seminal paper [40], convects the values of the level set function φ with the velocity field $\tilde{v}_n$ which has been extended off of Γ_t. Typically, the extension is performed in a neighborhood of the interface since we are only interested in the location of the interface. Such techniques are referred to as local level set methods [1, 41].

The construction of signed distance functions (or distance reinitialization) is a procedure that replaces a general level set function by the signed distance function $d(\mathbf{x}, t)$ which is the value of the distance from $\mathbf{x}$ to Γ_t taken to be positive inside and negative outside (or vice versa). This assures that φ does not become too flat nor too steep near the interface. To construct the signed distance function, one needs to find a function d such that

$$|\nabla d| = 1 \text{ almost everywhere,} \qquad \text{subject to } \{\mathbf{x} : \varphi(\mathbf{x}, t) = 0\} = \{\mathbf{x} : d(\mathbf{x}, t) = 0\}\,.$$

This construction can be performed very efficiently, e.g. fast sweeping, fast marching, etc. [10, 42–45].

6 Numerical Simulations

In this section, we show the results of some numerical simulations using the algorithm. The aim of the following simulations is to show the computational properties of the algorithm. We discretize the integral equations (9) and (11) using finite differences and compute them using simple Riemann sums over uniform grids. After discretization, we obtain a linear system of the form $Ax = b$ where the vector x consists of the unknown density β and the constants A_i. See (9) and (11).

To solve equation (17), we use an upwind scheme with a WENO-3 scheme for the spatial derivatives, and a third order TVD-RK3 method for time. We propagate enough to ensure the velocity information is transmitted throughout an ε-tubular neighborhood of the boundary Γ_t. For the level set equation, we use a Godunov scheme with WENO-3 discretization for the eikonal term, and a TVD-RK3 method for time.

6.1 Two Dimensions

We start by illustrating that the simulated two-dimensional Mullins-Sekerka does indeed preserve area and decrease length. In this test case, we computed the evolution of the perimeter and area of an ellipse under the Mullins-Sekerka dynamics. The plots for the perimeter and area versus time are shown in Figure 3.

As a second test problem, we simulated the evolution of an elongated tube to corroborate the fact that the two-dimensional Mullins-Sekerka does not preserve convexity at all times during the evolution. This is illustrated in Figure 4.

Finally, we demonstrate the capability of our algorithm to handle mergings of connected components. In this particular test example, we started with two ellipses that were close enough to finally merge into one connected component. Figure 5 shows the evolution of these two ellipses and Table 1 shows the numerical errors in the conservation of total area before and after merging.

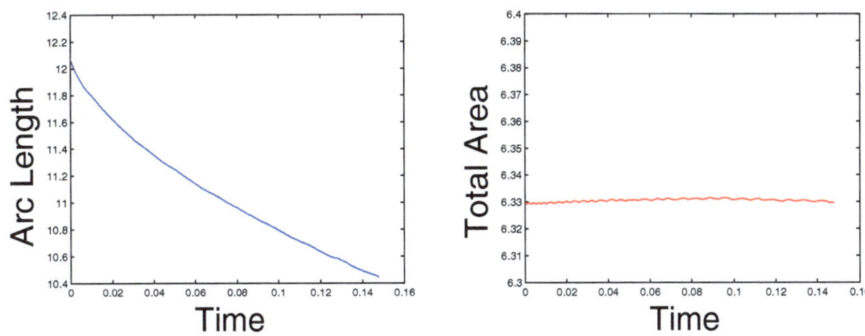

Fig. 3 Evolution of the perimeter and area of an ellipse under the Mullins-Sekerka model

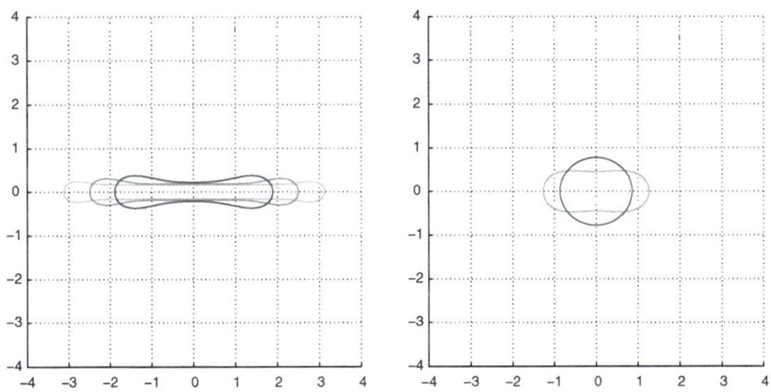

Fig. 4 Evolution of a thin tube under the two-dimensional Mullins-Sekerka dynamics

Table 1 Merging time and area jump for the merging of two ellipses evolving under the two-dimensional Mullins-Sekerka dynamics

h	4/128	4/256	4/512
Initial area	2.6965	2.69611	2.695495
Start merge area	2.696655	2.691792	2.686751
End merge area	2.804205	2.724255	2.694007
Area jump	0.107550	0.032463	0.007256
Relative area error	0.03988	0.01206	0.00270

6.2 *Three Dimensions*

In these simulations, we assume that the far-field condition is $\lim_{|\mathbf{x}|\to\infty} |u(\mathbf{x})| = u_\infty$ which allows the solution to converge to a non-zero constant. This u_∞ is referred to as the far-field environment which corresponds to the temperature at infinity. The dynamics of the Mullins-Sekerka model depend on this far-field environment and give rise to interesting behaviors that are illustrated in the next simulations.

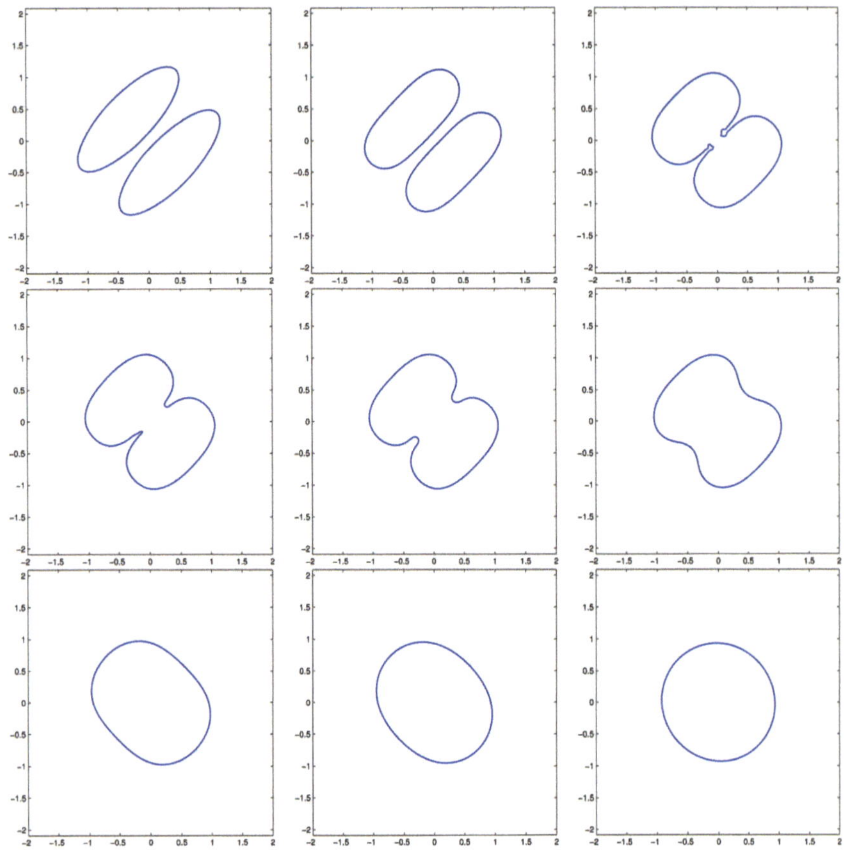

Fig. 5 Two ellipses merging in a two-dimensional Mullins-Sekerka flow simulation

While it is difficult to obtain rigorous analytical results on the general 3D Mullins-Sekerka model on unbounded domains with a general far-field environment u_∞, we can simulate these different situations. We note that to obtain the so-called Mullins-Sekerka instability (or dendritic growth) it is necessary that $u_\infty < 0$ (Figure 6).

We observe that in Section 2, we only provided a method for solving the three-dimensional exterior problem with $u_\infty = 0$. We thus briefly provide the equations for solving the **general 3D exterior problem**:

1. Find the density β and the constants $\{A_i\}_{i=1}^{L}$ such that

$$\begin{cases} \int_\Gamma \beta(\mathbf{y}) \left(\frac{\partial \Phi(\mathbf{x},\mathbf{y})}{\partial n_y^+} - \frac{1}{|\mathbf{x}-\mathbf{y}|^{m-2}} \right) dS(\mathbf{y}) - \frac{1}{2}\beta(\mathbf{x}) + \sum_{i=1}^{L} A_i \Phi(\mathbf{x} - \mathbf{z_i}) = f(\mathbf{x}) - u_\infty, \\ \int_{\Gamma_i} \beta(\mathbf{y}) dS(\mathbf{y}) = 0, \quad 1 \le i \le L, \quad \mathbf{x} \in \Gamma. \end{cases}$$

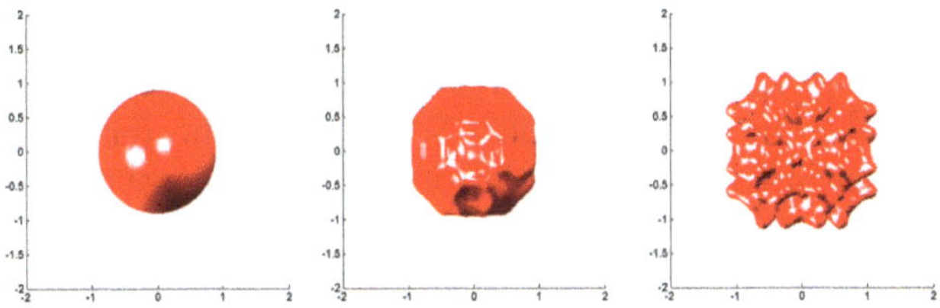

Fig. 6 Instability of dendritic growth in the 3D Mullins-Sekerka while maintaining crystal symmetry

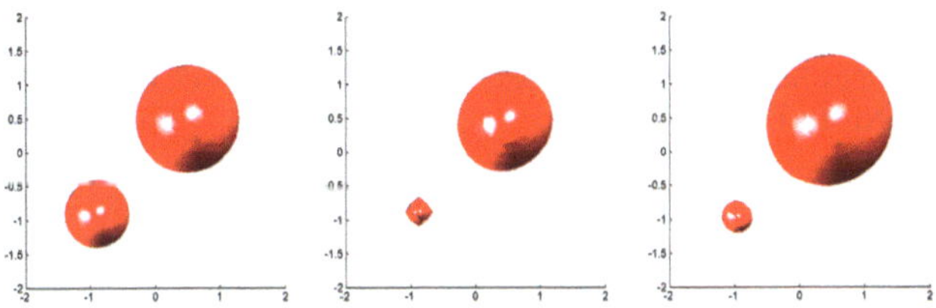

Fig. 7 Effect of the far-field value u_∞ on the dynamics of two spheres. On the left plot, both spheres grow in a low-temperature environment. In the middle, both shrink in a high- temperature environment. On the right, the larger sphere grows and absorbs the mass of the smaller sphere

2. Reconstruct the solution u in $\mathbb{R}^m \setminus \bar{\Omega}$ using the double layer potential formulation

$$u(\mathbf{x}) = \int_\Gamma \beta(\mathbf{y}) \left(\frac{\partial \Phi(\mathbf{x}, \mathbf{y})}{\partial n_y^+} - \frac{1}{|\mathbf{x} - \mathbf{y}|^{m-2}} \right) dS(\mathbf{y}) + \sum_{i=1}^{L} A_i \Phi(\mathbf{x} - \mathbf{z_i}) + u_\infty, \quad \text{for } \mathbf{x} \in R^m \setminus \bar{\Omega}.$$

We focus on two features of the three-dimensional Mullins-Sekerka dynamics. First, we look at the isothermal process and show that it depends on the far-field environment u_∞. Second, we simulate a case where instabilities develop: see Figure 6.

In Figure 7 we simulate the dynamics of two initial spheres with different radii using various far-field values u_∞. If u_∞ is small (i.e. low temperature), both spheres grow (solidification). If u_∞ is large (high-temperature), both spheres shrink (liquidation). If u_∞ is set at phase transition point, we see that the larger sphere grows at the expense of the smaller one, as expected in the two-dimensional case.

7 Conclusion

We described an implicit boundary integral algorithm for computing the dynamics of the Mullins-Sekerka model on unbounded domains in two and three dimensions. We used a formulation described in [32] and further investigated in [31] to compute integral equations using a signed distance function. We showed that our algorithm was able to handle topological changes, such as mergings of connected components, and was able to simulate the so-called Mullins-Sekerka instability (or dendritic growth).

This paper demonstrates the versatility of our computational method and in particular its capability to simulate on unbounded domains in three dimensions. Coupled with fast multipole methods, this algorithm has the potential to be a powerful tool to better understand the dynamics of the Mullins-Sekerka model on unbounded domains.

Acknowledgements Kublik's research was partially funded by a University of Dayton Research Council Seed Grant and a Dr. Schraut Faculty Research Award. Chen's and Tsai's research is partially supported by Simons Foundation, NSF Grants DMS-1318975, DMS-1217203, and ARO Grant No. W911NF-12-1-0519.

Appendix

Connected Component Labeling

As explained in Section 2, the system of equations to solve depends on many properties including the number of connected components of the region, whether the region is bounded or not (exterior vs interior), and the orientation of the region. In our numerical simulations, we adapted a technique called connected component labeling (CCL), see e.g. [14], to find necessary topological information needed to obtain the correct set of equations. These include:

1. The boundedness of the connected component C_i. This decides which formulation (interior vs exterior) to use.
2. The orientation of C_i which determines the normal direction.
3. The total number of connected boundary components of the boundary of C_i, denoted by $\Gamma^i = \partial C_i$.
4. The selection of $\mathbf{z}_i$ in (9) and (11) for each component Γ^j. This involves
    - the separation of Γ^i into boundary components Γ^i_j, each bounding a hole in the region except for the most exterior one. We denote the exterior boundary as Γ^i_0.
    - For each hole delimited by Γ^i_j , $j \neq 0$, find a point $\mathbf{z}_i$ inside the hole that gives the least singular value of $\Phi(\mathbf{x} - \mathbf{z}_i)$. This means that $\mathbf{z}_\mathbf{i}$ should be as far from the boundary as possible.

Since we deal with closed interfaces, every grid point belongs to a unique connected component and has a well-defined component label. Note that a connected component C_i may belong to Ω or $\mathbf{R}^m \setminus \Omega$.

We use the algorithm as follows:

- We label the unbounded component as C_0. This is the only true exterior component for the boundary integral formulations.
- Each interior component ($d_\Gamma > 0$, i.e. the solid phase) will have positive label i and each exterior component ($d_\Gamma < 0$ i.e. the liquid phase) will take negative label $-i$ (except for the unbounded one).
- Each boundary piece of each component will have label j.
- The point $\mathbf{z_i}$ in C_i is chosen to be the point with largest signed distance function in absolute value.

We adopt the two-pass CCL algorithm with $2m$-connectivity in $\mathbb{R}^m$. This algorithm uses equivalence classes for labels: after the first pass, points in the same connected component may not have the same label, but the labels of points in the same connected component will be assigned to the same equivalent class in the second pass. The root of the equivalence class denotes the smallest label (in absolute value) in the equivalence class. The scanning process works as follows:

1. First Pass

   a. Begins with label 0 which denotes the unbounded component C_0.
   b. At the current point scanned, we look at the sign of the signed distance function of its $2m$ neighbors already visited (in 2D, west and north of the current point)
      - If none of the neighbors have the same sign as the current point, we create a new label (positive or negative, based on the sign of the distance function at the current point) and a new equivalence class.
      - If only one neighbor has the same sign, we pick the label for the current point to be the root of that neighboring point's equivalence class.
      - If more than one neighbor has the same sign as the current point, we pick the label for the current point to be the smallest root of the neighbors that have the same sign's equivalence classes. Furthermore, we combine the equivalence classes of the neighboring points that have the same sign since they are connected through the current point.
   c. The largest root of all equivalence classes with a given sign (interior or exterior) will give the total number of connected components that have that sign. Thus, the sum is the total number of connected components.

2. Second Pass

   a. At each point, we assign its label to be the root of its equivalence class.
   b. We update and store the points with largest absolute distance within each equivalence class. These are the points $\mathbf{z_i}$.
   c. For each point $\mathbf{x}$ within the ε tubular neighborhood of the boundary ($|d_\Gamma| < \varepsilon$), we look for the root of its equivalence class (say i) and the root of its projection point's ($P_\Gamma(\mathbf{x})$) equivalence class (say j) that takes the opposite sign. To obtain j, we look at the vertices of the cell $P_\Gamma(\mathbf{x})$ falls in and scan

for the label with opposite sign. This step identifies the boundary piece Γ_j^i and collects points within the ε neighborhood of the boundary Γ_j^i.

d. We store the total number of connected boundary pieces.

References

1. D. Adalsteinsson, J.A. Sethian, A fast level set method for propagating interfaces. J. Comput. Phys. **118**(2), 269–277 (1995)
2. K.E. Atkinson, in *The Numerical Solution of Integral Equations of the Second Kind* (Cambridge University Press, Cambridge 1997)
3. K.E. Atkinson, G. Chandler, Boundary integral equation methods for solving Laplace's equation with nonlinear boundary conditions: the smooth boundary case. Math. Comput. **55**(191), 451–472 (1990)
4. I. Babuška, The finite element method for elliptic equations with discontinuous coefficients. Computing **5**, 207–213 (1970)
5. J.W. Barrett, H. Garcke, R. Nürnberg, On stable parametric finite element methods for the Stefan problem and the Mullins-Sekerka problem with applications to dendritic growth. J. Comput. Phys. **229**(18), 6270–6299 (2010)
6. P.W. Bates, X. Chen, X. Deng, A numerical scheme for the two phase Mullins-Sekerka problem. Electron. J. Differ. Equs. (1995)
7. J. Bedrossian, J.J. von Brecht, S. Zhu, E. Sifakis, J. Teran, A second order virtual node method for elliptic problems with interfaces and irregular domains. J. Comput. Phys. **229**, 6405–6426 (2010)
8. S. Börn, L. Grasedyck, W. Hackbusch, *Hierarchical Matrices* (Technical report, Max-Planck Institut fur Mathematik in den Naturwissenschaften, Leipzig, 2003)
9. S. Chen, B. Merriman, S. Osher, P. Smereka, A simple level set method for solving Stefan problem. J. Comput. Phys. **135**(1), 8–29 (1997)
10. L.-T. Cheng, Y.-H. Tsai, Redistancing by flow time dependent Eikonal equation. J. Comput. Phys. **227**(2), 4002–4017 (2008)
11. I.-L. Chern, Y.-C. Shu, A coupling interface method for elliptic interface problems. J. Comput. Phys. **225**, 2138–2174 (2007)
12. P.G. Ciarlet, in *The Finite Element Method for Elliptic Problems*. SIAM. North-Holland, Amsterdam (1978)
13. S. Conti, B. Niethammer, F. Otto, Coarsening rates in off-critical mixtures. SIAM J. Math. Anal. **37**(6), 1732–1741 (2006)
14. M. Dillencourt, H. Samet, M. Tamminen, A general approach to connected-component labeling for arbitrary image representations. J. ACM **39**(2), 253–280 (1992)
15. J. Dolbow, I. Harari, An efficient finite element method for embedded interface problems. J. Numer. Methods Eng. **78**, 229–252 (2009)
16. H. Federer, in *Geometric Measure Theory* (Springer, Berlin, 1969)
17. G.B. Folland, in *Introduction to Partial Differential Equations* (Princeton University Press, Princeton, 1976)
18. F. Gibou, R. Fedkiw, A fourth order accurate discretization for the Laplace and heat equations on arbitrary domains, with applications to the Stefan problem. J. Comput. Phys. **202**, 577–601 (2005)
19. F. Gibou, R. Fedkiw, L. Cheng, M. Kang, A second order accurate symmetric discretization of the Poisson equation on irregular domains. J. Comput. Phys. **176**, 1–23 (2002)
20. Y. Giga, in *Surface Evolution Equations: A Level Set Approach*. Monographs in Mathematics, vol. 99 (Birkhäuser Verlag, Basel, 2006)
21. A. Greenbaum, L. Greengard, G.B. McFadden, Laplace's equation and the Dirichlet-Neumann map in multiply connected domains. J. Comput. Phys. **105**, 267–278 (1993)

22. L. Greengard, V. Rokhlin, A fast algorithm for particle simulations. J. Comput. Phys. **73**(2), 325–348 (1987)
23. M. Gurtin, On the two-phase Stefan problem with interfacial energy and entropy. Arch. Rational Mech. Anal. **96**, 199–241 (1986)
24. A. Hansbo, P. Hansbo, An unfitted element method, based on Nitsche's method for elliptic interface problems. Comput. Methods Appl. Mech. Eng. **191**, 5537–5552 (2002)
25. A. Hansbo, P. Hansbo, A finite element method for the simulation of strong and weak discontinuities in solid mechanics. Comput. Methods Appl. Mech. Eng. **193**, 3523–3540 (2004)
26. J. Huang, J. Zou, A mortar element method for elliptic problems with discontinuous coefficients. IMA J. Numer. Anal. **22**, 549–576 (2002)
27. H. Johansen, Cartesian grid embedded boundary finite difference methods for elliptic and parabolic differential equations on irregular domains. Ph.D. thesis, University of California, Berkeley, 1997
28. H. Johansen, P. Colella, A cartesian grid embedded boundary method for Poisson's equation on irregular domains. J. Comput. Phys. **147**, 60–85 (1998)
29. G. Karali, P. Kevrekidis, Bubble interactions for the Mullins-Sekerka problem: some case examples. Math. Comput. Simul. **80**(4), 707–720 (2009)
30. R. Kress, *Linear Integral Equations*, 2nd edn. (Springer, New York, 1999)
31. C. Kublik, R. Tsai, Integration over curves and surfaces defined by the closest point mapping. Res. Math.Sci. **3**, 1–17 (2016)
32. C. Kublik, N.M. Tanushev, R. Tsai, An implicit interface boundary integral method for Poisson's equation on arbitrary domains. J. Comput. Phys. **247**, 269–311 (2013)
33. R. Leveque, Z. Li, The immersed interface method for elliptic equations with discontinuous coefficients and singular sources. SIAM J. Numer. Anal. **31**, 1019–1044 (1994)
34. Z. Li , K. Ito, The immersed interface method: numerical solutions of PDEs involving interfaces and irregular domains (frontiers in applied mathematics), *Society for Industrial and Applied Mathematics* (2006)
35. X. Liu, R. Fedkiw, M. Kang, A boundary condition capturing method for Poisson's equation on irregular domains. J. Comput. Phys. **160**(1), 151–178 (2000)
36. S.G. Mikhlin, in *Integral Equations* (Pergamon, London, 1957)
37. T.-H. Miura, Zero width limit of the heat equation on moving thin domains. UTMS Preprint Series, the University of Tokyo, 2015
38. W.W. Mullins, R.F. Sekerka, Morphological stability of a particle growing by diffusion or heat flow. J. Appl. Phys. **34**, 323–329 (1963)
39. E. Nyström, Über die praktische Auflösung von Integralgleichungen mit Anwendungen auf Randwertaufgaben. Acta Math. **54**, 185–204 (1930)
40. S. Osher, J.A. Sethian, Fronts propagating with curvature dependent speed: algorithms based on Hamilton-Jacobi formulations. J. Comput. Phys. **79**, 12–49 (1988)
41. D. Peng, B. Merriman, S. Osher, H.-K. Zhao, M. Kang, A PDE-based fast local level set method. J. Comput. Phys. **155**(2), 410–438 (1999)
42. G. Russo, P. Smereka, A remark on computing distance functions. J. Comput. Phys. **163**, 51–67 (2000)
43. J. Sethian, A fast marching level set method for monotonically advancing fronts. Proc. Natl. Acad. Sci. **93**(4), 1591–1595 (1996)
44. Y.-H. Tsai, L. Cheng, S. Osher, H.-K. Zhao, Fast sweeping methods for a class of Hamilton-Jacobi equations. SIAM J. Numer. Anal. **41**(2), 673–694 (2003)
45. J. Tsitsiklis, Efficient algorithms for globally optimal trajectories. IEEE Trans. Autom. Control **40**, 1528–1538 (1995)
46. J. Zhu, X. Chen, T.Y. Hou, An efficient boundary integral method for the Mullins-Sekerka problem. J. Comput. Phys. **127**, 246–267 (1996)

Geometric Interfacial Motion: Coupling Surface Diffusion and Mean Curvature Motion

V. Derkach, J. McCuan, A. Novick-Cohen and A. Vilenkin

Abstract Mean curvature motion as well as surface diffusion both constitute geometric interfacial motions which have received considerable attention. However in many applications a complex coupling of surfaces occurs whose evolution may be described using both these types of motion. A variety of such physical problems are described. While sometimes an anisotropic formulation might seem to be preferable, often an isotropic formulation is helpful to consider. Some analytic and numerical results are being presented, with some supporting experimental evidence. Many open questions remain.

Keywords Surface diffusion · Mean curvature motion · Microstructure Numerical simulations · Coupled geometric motions

While arguably motion by mean curvature and motion by surface diffusion, as well as systems describable by coupling these motions, may all be traced back to a series of papers by W. Mullins from 1956–1958 [35–37], our interest in these directions began some 40 years later and is reflected in a series of papers, in part in collaboration with J. Cahn, R. Dal Passo, C. Elliott, L. Giacomelli, J. Kanel, L. Peres-Hari, E. Rabkin, O. Zelekman-Smirin, and A. Zigelman, [12–19, 29–33, 38–41, 43, 51–53].

V. Derkach (✉) · A. Novick-Cohen
Department of Mathematics, Technion IIT, 32000 Haifa, Israel
e-mail: vadder@technion.ac.il

A. Novick-Cohen
e-mail: amync@technion.ac.il

J. McCuan
School of Mathematics, Georgia Tech, Atlanta, GA 30332, USA
e-mail: mccuan@math.gatech.edu

A. Vilenkin
Racah Institute of Physics, The Hebrew University, 91904 Jerusalem, Israel
e-mail: vilenkin@mail.huji.ac.il

Y. Maekawa and S. Jimbo (eds.), *Mathematics for Nonlinear Phenomena—Analysis and Computation*, Springer Proceedings in Mathematics & Statistics 215,
DOI 10.1007/978-3-319-66764-5_2

Below is a partial list of some of relevant physical problems.

1. Polycrystalline films are material layers composed of a number of individual crystals, or grains. When these films are thin, the effects of their exterior surface can affect the stability of the film. Since thin films are in common use in numerous technological applications, the interest in their stability is considerable.
2. Sintering is a technology with a long history, which is still in active use to this day. In this process, a collection of metallic particles, produced previously by evaporation and condensation, are placed into a mold. Under pressure, the collections of particles can be made to combine to yield one solid metallic block with some specific shape.
3. Under appropriate compositional conditions, ordering can be induced in a previously homogenous material. If the composition differs slightly from these conditions, the excess composition can emerge as droplets along the boundaries between the ordered regions. This phenomena can be modeled with some success by a coupled Allen-Cahn/Cahn-Hilliard system of equations with degenerate mobilities.

In all of the above applications, the coupled motion of surface diffusion with motion by mean curvature appears quite naturally. In each of these applications, there are additional effects which are often neglected and which arguably should be included. However, the coupled motion, in and of itself, is not overly well understood, so it is reasonable to isolate it and study it, even given its limitations. So we shall first define motion by mean curvature and motion by surface diffusion, then return to consider some systems coupling these motions.

1 Motion by Mean Curvature

Initially let us take

$$\Gamma(t) := \{X(\alpha, t) : [0, 1] \times [0, T] \to R^2\}, \tag{1}$$

to be a smooth (and smoothly parameterized) embedded curve in R^2. For $\alpha \in (0, 1)$, we may define $\hat{n}$, the unit normal to $\Gamma(t)$ at α, and we may define κ, the mean curvature of $\Gamma(t)$ at α, via the inscribed circle at α. Then setting $\hat{V} := \frac{\partial}{\partial t} X$ for $(\alpha, t) \in (0, 1) \times (0, T)$, the motion known as motion by mean curvature is given by

$$V_n = \mathcal{A}\kappa, \tag{2}$$

where $V_n := \hat{n} \cdot \hat{V}$ is the normal velocity, and $\mathcal{A}$ is a dimensional parameter, known as the reduced mobility whose dimensions are given by $[\mathcal{A}] = L^2/T$. It is straightforward to later extend the above definition to smooth (and smoothly parameterized) embedded hypersurfaces in R^3, taking κ to be the sum of the principle curvatures. An early reference, if not the earliest reference, to motion by mean curvature is

Mullins 1956 [35]. In the mathematical literature, early references include Brakke 1978 [11], Huisken 1984 [27], Grayson 1987 [25]. For surveys, see Giga 2006 [22] and Bellettini 2013 [9].

2 Motion by Surface Diffusion

To define motion by surface diffusion, we may proceed as above and let us initially consider $\Gamma(t) := X(\alpha, t)$ to be a smooth and smoothly parameterized embedded curve in R^2 prescribed as in (1), with $\hat{n}$, κ, $\hat{V}$, V_n defined as above for $(\alpha, t) \in (0, 1) \times (0, T)$. Taking $s\ (= s(t))$ to be an arc-length parametrization of $\Gamma(t)$, e.g. $s := \int_0^\alpha |\frac{\partial}{\partial\tilde{\alpha}} X(\tilde{\alpha}, t)|\, d\tilde{\alpha}$, the motion known as motion by surface diffusion is given by $V_n = -\mathcal{B}\frac{\partial^2 \kappa}{\partial s^2}$, where $\mathcal{B}$ is a dimensional kinetic coefficient with dimensions $[\mathcal{B}] = L^4/T$. The above definition of surface diffusion may be extended to smooth (and smoothly parameterized) embedded hypersurfaces in R^3, by defining κ as the sum of the principle curvatures and by replacing the operator $\frac{\partial^2}{\partial s^2}$ with the Laplace-Beltrami operator Δ_s, yielding that

$$V_n = -\mathcal{B}\Delta_s \kappa. \tag{3}$$

The Laplace-Beltrami operator on a smooth hypersurface may be defined as $\Delta_s := \sum_{i=1}^{2} \frac{\partial^2}{\partial s_i^2}$, where s_i $i = 1, 2$ are locally defined arc-length parameterizations in the directions of the principle curvatures; in this sense Δ_s indeed generalizes $\frac{\partial^2}{\partial s^2}$. An early reference, if not the earliest reference, to surface diffusion in the literature is Mullins 1957 [36]. Unlike motion by mean curvature, convex curves evolving by surface diffusion can later lose convexity [24]. Pinch off of embedded curves may also occur [23]. Some recent references include [3, 34].

3 Coupling the Two Motions

Why couple the two motions discussed above? Let us focus on the first problem in the list given earlier, that of polycrystalline films. Because the microstructure of polycrystalline materials, namely the way in which the individual crystals or grains are arranged within the specimen, strongly influences the material properties, it has attracted many studies since Mullins' 1956 [35] work on the dynamics of grain boundaries, and earlier static discussions by Smith [45] and Herring [26]. Much emphasis has been placed on considering grain boundaries within the material, especially in large systems and on the distribution of grain boundary properties, see e.g. [7]. Many additional effects have been included, such anisotropy, elastic effects, and defects [44].

When a polycrystalline specimen is thin enough to be considered a film, or more specifically a thin film, the exterior surface is nowhere too far from the grain boundaries within, and the effects of the dynamics of the exterior surface on the structural stability of the specimen need to be taken into account. A classical though somewhat simplistic model for the evolution of the exterior surface, due to Mullins 1957 [36], predicts that it evolves by surface diffusion. While many of the grain boundaries lie totally within the interior of the material, some continue on up until they intersect the exterior surface. This gives rise a system in which numerous grain boundaries are coupled with the exterior surface, which we assume for simplicity to evolve by mean curvature and by surface diffusion, respectively. As there is much to be studied in regard to these coupled motion, a possible simplification is to consider some special simple geometries.

4 Some Special Geometries

In polycrystalline specimens with *bamboo structure*, the specimen is one grain thick throughout. Accordingly all of the grains intersect the exterior surface at least twice, once above and once below. Within the interior of the specimen, triple junctions form where three-grain boundaries intersect. On the exterior surface, where three intersecting grain boundaries intersect with the exterior surface, quadruple junctions form. Although assuming bamboo structure simplifies the analysis somewhat, it still is quite complicated. A further possible simplification is to consider the 2D symmetric thermal grooving geometry proposed by Mullins [36], described below. Other special geometries are possible to consider, some of which will be described a bit further on.

5 A Phase Field Approach

One approach to modeling this type of coupled motions is to embed the problem description within a phase field approach, and one way to accomplish this is to consider certain types of Allen-Cahn/Cahn-Hilliard system with degenerate mobility. Such a system was derived to describe the dynamics of simultaneous order-disorder and phase transitions, as the formal continuum limit of discrete lattice dynamics, in the context of the Krzanowski instability [12]. For the resultant system, it was possible to formally derive the coupled motion by mean curvature and surface diffusion by considering an appropriately defined long time asymptotic limit [39, 40].

6 The Mullins' Sharp Interface Approach

Mullins considered a geometrically simple exterior surface-grain boundary configuration with important experimental implications and outlined a dynamic problem for its evolution [36], following the general thermodynamic framework outlined by Gibbs. The geometry which he considered consisted of an initially flat exterior surface of infinite extent, coupled with an initially flat grain boundary of semi-infinite extent which intersected the exterior surface perpendicularly. The exterior surface was assumed to evolve by surface diffusion, and the grain boundary, which in theory could evolve by motion by mean curvature, could be expected to remain flat due to symmetry considerations. Although this reflects a 3D problem, its dynamics reduce to a 2D problem due to the uniform cross-section of its geometry. To complete the problem formulation, boundary conditions are required at the triple junction, in addition to far-field and initial conditions. The following conditions were proposed by Mullins: 1) persistence, meaning that the grain boundary remained attached to the exterior surface, 2) balance of mechanical forces, known also as Herring's law or Young's law for isotropic systems, 3) continuity of the chemical potential, which, based on thermodynamic considerations of Gibbs, implies continuity of the mean curvature, 4) balance of mass flux along the exterior surface, assuming that there is no mass flux along the grain boundary. While Mullins formulated the conditions above within the framework of a linearized approximation for surface diffusion, the same physical conditions can be readily formulated within a fully nonlinear framework [31]. Assuming asymptotic planarity and fixing the initial location of the exterior surface along the $x-$axis and the initial location of the grain boundary to lie along the negative $y-$axis, yields the following problem formulation for the evolution of the exterior surface,

$$\begin{cases} y_t = -\mathcal{B}\left[\frac{\kappa_x}{\sqrt{1+y_x^2}}\right]_x, \quad \kappa = \frac{y_{xx}}{(1+y_x^2)^{3/2}}, \quad x \in (-\infty, 0) \cup (0, \infty), \ t \in (0, \infty), \\ y(0^+, t) = y(0^-, t), \quad t \in (0, \infty), \\ \arctan(y_x(0^+, t)) = \arctan(y_x(0^-, t)) + 2 \arcsin(m/2), \quad t \in (0, \infty), \\ \kappa(0^+, t) = \kappa(0^-, t), \quad t \in (0, \infty), \\ \frac{\kappa_x}{\sqrt{1+y_x^2}}|_{(0^+,t)} = \frac{\kappa_x}{\sqrt{1+y_x^2}}|_{(0^-,t)}, \quad t \in (0, \infty), \\ y(\pm\infty, t) = y_x(\pm\infty, t) = 0, \quad t \in (0, \infty), \\ y(x, 0) = 0, \quad x \in (-\infty, 0) \cap (0, \infty), \end{cases} \tag{4}$$

where

$$m = \gamma_{exterior}/\gamma_{gb}. \tag{5}$$

Here $\gamma_{exterior}$, γ_{gb} denote the surface free energies of the exterior surfaces and of the grain boundaries, respectively. The physically natural energy for the coupled systems is $E = \gamma_{exterior} A_{exterior} + \gamma_{gb} A_{gb}$, where $A_{exterior}$, A_{gb} denote the total area of the exterior surfaces and of the grain boundaries, respectively. It is convenient to normalize the energy by setting $E_{exterior} := A_{exterior}$, $E_{gb} := A_{gb}$, and defining

$$E_w = E_{exterior} + m E_{gb}, \tag{6}$$

as the *weighted energy*. For the coupled systems a natural restriction on m is that

$$m \in [0, 2), \tag{7}$$

since when $m = 2$, the energy of two exterior surface equals that of one-grain boundary, so that two exterior surfaces coupling with a grain boundary *wet* the grain boundary.

Mullins assumed that $0 < m \ll 1$, which is physically reasonable, and linearized the system in (4) about $y \equiv 0$. For the resultant linear system, known in the literature as *Mullins' small slope approximation*, he found a self-similar solution of the form $y(x,t) = \frac{m}{\sqrt{4-m^2}}(\mathcal{B}t)^{1/4}Y(\xi)$ where $\xi = x/(\mathcal{B}t)^{1/4}$, and derived a series formula for the dimensionless function $Y(\xi)$ which allowed him to predict that $y(0,t) = a_0\, m(\mathcal{B}t)^{1/4}$ with $a_0 \approx -.39$. The quantity $y(0,t)$ is known as the *thermal groove depth*, since the kinetic coefficient, $\mathcal{B}$, increases with temperature. Based on the linearization of the equation in (4), he also identified a traveling wave [37] describing a groove which steadily translates along the exterior surface with speed $V_0 = \mathcal{B}(m/3d)^3$, where d is the thermal groove depth.

Despite the years which have passed, many questions remain regarding the nonlinear problem formulation given in (4). For example, it is not yet known whether it possesses self-similar solutions, although a positive answer has been demonstrated for (4) with partially linearized boundary conditions [3]. In the next two sections we outline some of what is known regarding traveling wave solutions for problem formulations related to (4).

7 Traveling Wave Solutions

In [29–32], together with J. Kanel we considered a configuration in which a grain boundary is attached at a thermal groove to an exterior surface in a *quarter loop geometry*. The quarter loop geometry refers to a two grain system which may be assumed to be of semi-infinite extent, in which the exterior surface is asymptotically flat, and where there is one internal grain boundary, which intersects the exterior surface and partitions the specimen into two grains, then asymptotes to one side, say to the right, to some specific depth below the exterior surface, say $-H$. If we assume that the profiles of both the exterior surface and the grain boundary can be described by graphs of functions, say by $h = h(x,t)$ and by $u = u(x,t)$, then we may obtain a problem formulation, much as in (4) above, except that now we need to consider $x \in (-\infty, s(t)) \cup (s(t), \infty)$, where $s = s(t)$ with $s(0) = 0$ since the groove can move, and we need to amplify the resultant equations with equations governing the motion for the grain boundary, such as

$$\begin{cases} u_t = \mathcal{A}\kappa, \quad \kappa = u_{xx}(1+u_x^2)^{-3/2}, \quad x \in (-\infty, s(t)) \cup (s(t), \infty), \ t \in (0,\infty), \\ y(s(t)^-, t) = u(s(t)^+, t), \quad t \in (0,\infty), \\ \arctan(u_x(0^+, t)) = -\frac{\pi}{2} + \frac{1}{2}[\arctan(y_x(0^+, t)) + \arctan y_x(0^-, t))], \quad t \in (0,\infty), \\ u(\infty, t) = -H, \quad t \in (0,\infty), \\ u(x, 0) = u_0(x), \quad x \in (0,\infty), \end{cases} \tag{8}$$

where $u_0(x)$ needs to be prescribed. In [30] it was demonstrated that there indeed exists a traveling wave solution to the resultant system if $0 < m$ is sufficiently small.

It turns out that the restriction on m resulted from looking for traveling wave solutions which could be described as graphs of functions. The restriction could be removed by formulating the traveling wave problem in terms of the angles, $\Phi(s_1)$ and $\Psi(s_2)$, where $\Phi(s_1)$ and $\Psi(s_2)$ denote respective the angles between the tangent to the grain boundary and the exterior surface and the x−axis, and where s_1, s_2 are respective arc-length parameterizations for $\Phi(s_1)$ and $\Psi(s_2)$, with $s_1 \in (0, \infty)$, $s_2 \in (-\infty, 0) \cup (0, \infty)$, and with the thermal groove as the location where $s_1 = s_2 = 0$. Setting $s = (V/\mathcal{B})^{1/3} s_2$, where V is the wave speed, the problem to be satisfied by Ψ could be formulated as:

$$\begin{cases} \Psi_{sss} = \sin\Psi, \quad s \in (-\infty, 0) \cup (0, \infty), \\ \Psi(0^+) - \Psi(0^-) + 2\arcsin(m/2), \\ \Psi_s(0^+) = \Psi_s(0^-), \\ \Psi_{ss}(0^+) = \Psi_{ss}(0^-), \\ \Psi(\pm\infty) = 0. \end{cases} \tag{9}$$

Based on this formulation and using a shooting argument, the following could be proven:

Theorem [31]. For all $m \in [0, 2)$, there exists a traveling wave solution to the coupled exterior surface and grain boundary problem (9) for the quarter loop geometry. For $0 < m \ll 1$, there exists a traveling wave solution which can be described via graphs of functions.

For all $0 \leq m < 2$, given a solution Ψ to (9), Φ can be prescribed in term of Ψ, and the wave speed is given by $V = \mathcal{B}(\lambda/H)^3$ where λ is the unique real solution to the following algebraic equation

$$\lambda^3 + \lambda^2 \Psi_{ss}(0+) - \frac{\mathcal{A}H^2}{\mathcal{B}}\Big[\frac{\pi}{2} - \frac{1}{2}(\Psi(0^+) + \Psi(0^-)\Big] = 0. \tag{10}$$

The existence of nonclassical (not graphs of functions) traveling wave solutions could be demonstrated numerically for $m \in (0, 2)$ sufficiently large. For the case of classical traveling wave solutions for $0 < m \ll 1$ sufficiently small, spectral stability of the 2D traveling wave solution in 3D was demonstrated numerically by Beck, Pan and Wetton [8]. Further stability results are possible [28].

8 Estimating the Reduced Mobility

The equation for motion by mean curvature, (2), contains a kinetic parameter, $\mathcal{A}$, the reduced mobility. In the context of evolution of microstructure, it is not so obvious how to estimate this parameter. Shvindlerman and colleagues developed a methodology for estimating $\mathcal{A}$ based on the half-loop geometry [2, 5]. The half-loop geometry refers to a bi-crystal with bamboo structure, consisting of a layer with finite height and with infinite lateral extent, containing two crystals. One crystal is embedded within the other, and the grain boundary which separates them is U-shaped, with ends which are asymptotically parallel. They analyzed grain boundaries in the laboratory fitting the description above, which appeared to move with constant speed. Neglecting the effects of the exterior surface and assuming the grain boundary within its interior to move by mean curvature motion, they looked for traveling wave solutions with this shape. Such mean curvature motion solutions, which they referred to as U-shape solutions, were already known to exist [35] and can be prescribed explicitly by $Y_{\pm}(x - V_0 t)$, where

$$V_0 = \frac{A\pi}{4Q}, \quad Y_{\pm}(x) = \pm\Big(Q - \frac{2Q}{\pi}\arcsin\left(e^{-\frac{\pi}{2Q}x}\right)\Big), \quad x \in \mathrm{R}^+, \tag{11}$$

and $2Q$ is the asymptotic width of the embedded crystal, far from the "nose" of the U-shaped profile. Later in the mathematical literature, these solutions came to known as grim reaper solutions [1]. These solutions could be used to estimate the reduced mobility $\mathcal{A}$, using the formula for V_0 in (11), [2, 5, 44].

Based on our remarks up to now, it is clear that if the exterior surfaces evolve by surface diffusion, then they should have some effect of the motion, and hence in the process of estimation or measurement of $\mathcal{A}$. Anna Zigelman set out to estimate this effect in her MSc and PhD theses [43, 52, 53]. To simplify the analysis, she considered the effect of surface diffusion only in the neighborhood of the thermal groove. She derived equations for the leading order corrections to the shape of the exterior surface, the grain boundary profile, the shape of the thermal groove, as well as to the traveling wave velocity, by formal asymptotic analysis based on the assumption that $0 < m \ll 1$ in two distinguished limits:

$$(i) \quad L/Q = m^{1/3}, \qquad (ii) \quad L/Q = O(1). \tag{12}$$

Limit (i) can be considered as a thin film (thin specimen) limit [43, 52], while limit (ii) corresponds better to the conditions of Shvindlerman's original experiments [52]. In both cases, the exterior surface could be shown at leading order to satisfy a fourth order parabolic problem which decouples from the remainder of the system describing the motion. The leading order perturbation to the shape of the thermal groove is described by a second order parabolic equation with force which depends on the angle of inclination of the exterior surface along the thermal groove, as well as on V_1, the perturbation to the velocity, V_0 at the nose of the profile. After some analysis [52, 53], V_1, the perturbation to the velocity, could be prescribed as

$$V_1(t) = \frac{d}{dt}(r * \mathrm{B})(t), \quad r(t) = \mathcal{L}^{-1}\Big[\frac{1}{\omega\,\mathcal{L}[G](\omega)}\Big](t), \tag{13}$$

where $\mathcal{L}$, $\mathcal{L}^{-1}$ denote, respectively, the Laplace transform and its inverse, $*$ denotes the convolution operator, $\mathrm{B}(t)$ depends on the inclination of the exterior surface along the thermal groove, the shape of the thermal groove, and on an appropriately defined heat kernel, and $\mathcal{G}$ is a weighted integral of the heat kernel.

In the experiments the motion was viewed from above, and measurements were made not of V_n and κ, but rather of $(V_n)_{\text{effective}}$, the projection of the normal velocity at the nose on the plane of observation, and $\kappa_{\text{effective}}$, the curvature at the nose of the projection of the groove root on the plane of observation. The measurements utilized the formula

$$(V_n)_{\text{effective}} = A_{\text{effective}}\ \kappa_{\text{effective}}, \tag{14}$$

and our results can be used to correct to the reduced mobility measurements by noting that

$$\frac{A_{\text{effective}}}{A} = \frac{(V_n)_{\text{effective}}}{V_n} \cdot \frac{\kappa}{\kappa_{\text{effective}}}, \tag{15}$$

where V_n, κ can be approximated via their zeroth order description and the leading order corrections.

9 Parametric Surfaces

Vadim Derkach developed and implemented a numerical code in his MSc and PhD Theses [15, 16, 18, 19] based on parametric surfaces, staggered grids, and finite differences designed to follow the evolution of certain special geometries containing a small number of grains, with the goal of gaining insight into various known physical phenomena such as pitting, grain annihilation, dewetting, and void formation (Figure 1). The small systems studied numerically by Vadim modeled thin or even nano-thin polycrystalline films with the bamboo structure mentioned earlier. Here "thin" and "nano-thin" refers to material layers ranging in thickness from a fraction of nanometer to several microns, and typically solid thin films are polycrystalline. They are highly important in many technological applications, such as electronic semiconductors, microchips, computer memory, optical coatings, and much more. Thermal grooving, pitting and hole formation influences thin film stability and robustness, and wetting/dewetting phenomena can lead to total film degradation. Let us now describe the numerical framework which was developed for studying small thin polycrystalline systems, which was later adapted for studying more specifically wetting and dewetting dynamics. Vadim's MSc Thesis [15, 18] treated the special five and three-grain systems portrayed in Fig. 2, which led the way to the study in his PhD Thesis [16] of three-grain systems bounded within a triangular domain, which upon reflection could yield a "tiling" of the plane with a thin material layer. In these three-grain triangle systems, the initial film height was approximately uniform, see Fig. 3, and projections of the initial resultant tilings are portrayed in Figs. 4–5.

Let us now consider these three-grain systems in the triangular domains in some detail. Looking at Fig. 3, we see that they contain three exterior surfaces, three-grain boundaries, three triple junctions along the exterior surface which develop into

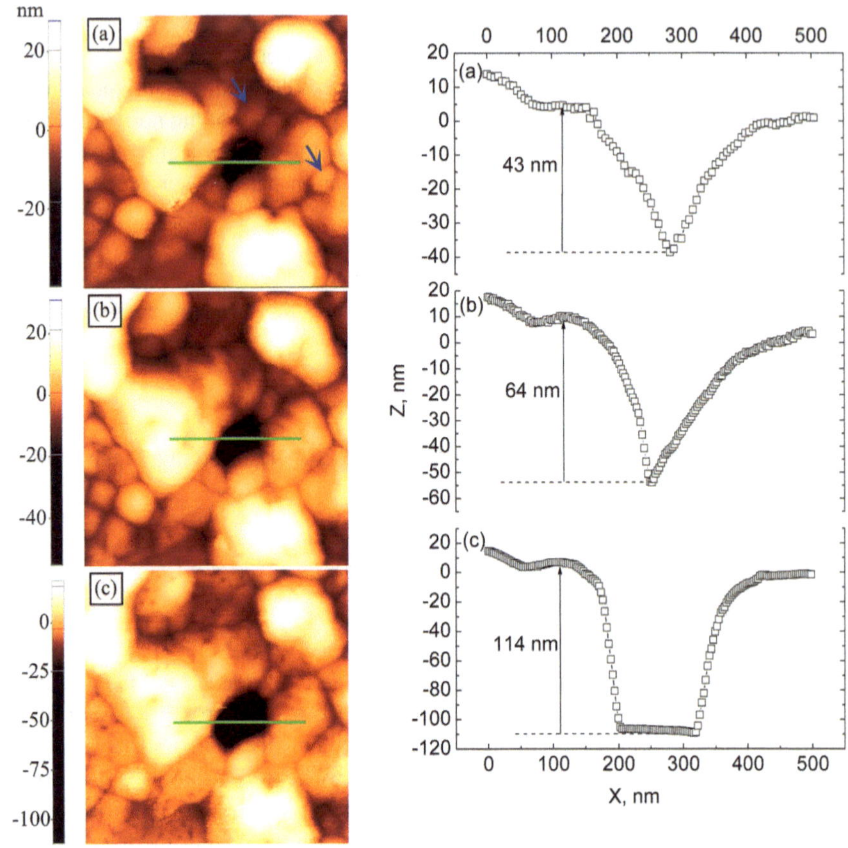

Fig. 1 (a–c). AFM images of a Cu thin film on sapphire substrate after the first (a), second (b), and third (c) consecutive anneal for 10 min at 300°C. The lateral dimensions of the image are 1×1 μm^2, and the color code labeled "nm" indicates the film height. A small grain visible in (a), sinks after the second anneal, leading to hole formation after the third anneal. Grains which disappeared after the second anneal without hole formation are marked in (a) by arrows which are blue in the online version. (d–f). The line topography profiles taken along the respective lines in (a–c). From Derkach, Novick-Cohen, Vilenkin and Rabkin [18], reproduced with permission

thermal grooves, as well as one internal triple junction line. There is a quadruple junction where the four triple junction lines meet. The triangular domain is bounded on three sides by vertical planes, as well as by a "mid-plane" located below at $z = 0$.

It is possible to prescribe a problem formulation for the evolution of these three-grain systems in accordance with the Mullins' framework discussed earlier. Namely, the exterior surfaces are assumed to evolve by surface diffusion, and the grain boundaries are assumed to evolve by mean curvature motion. Along the thermal grooves, the conditions imposed include: (i) persistence, (ii) balance of mechanical force (angle conditions), (iii) continuity of the chemical potential, (iv) balance of mass flux. Along the internal triple junction line, (I) persistence and (II) angle conditions

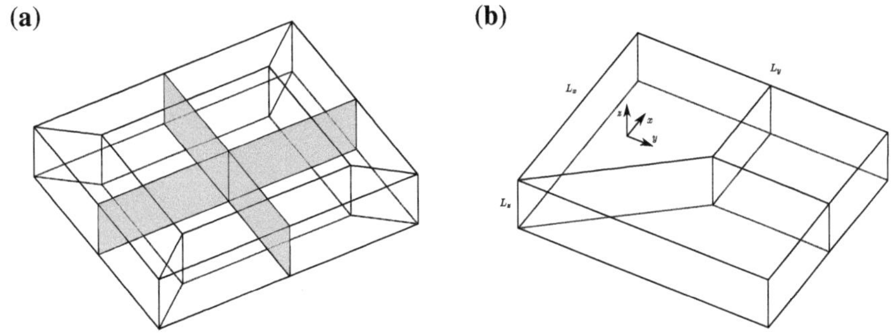

Fig. 2 Sketch of 5-grain and 3-grain geometries, see [19]

Fig. 3 Three grains in a triangular domain

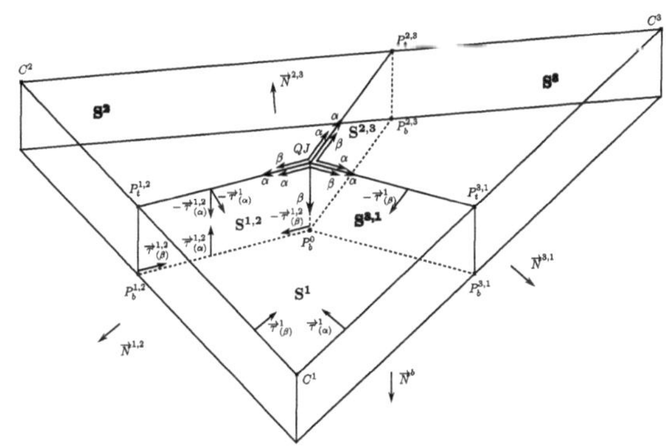

are imposed. Symmetry is imposed with respect to the three planes bounding the triangular domain along its sides as well as with respect to the "mid-plane" at $z = 0$. At the corners of the domain and at the intersection of the evolving surfaces with the bounding planes, conditions can be imposed in accordance with (i)–(iv), (I)–(II), and symmetry.

The six evolving surfaces can be described by extending the parametric representation for a regular evolving curve in R^2 given in (1) to the parametric representation of a regular evolving surface in R^3 by setting $\mathcal{S}(t) := \{X(\alpha, \beta, t) : [0, 1]^2 \times [0, T] \to R^3\}$, where

$$X(\alpha, \beta, t) = (x(\alpha, \beta, t), y(\alpha, \beta, t), z(\alpha, \beta, t)), \quad 0 \le \alpha, \beta \le 1, \; t \in [0, T]. \tag{16}$$

The coordinate orientations for the various surfaces are indicated in Fig. 3. The normal velocity V_n of X is given by $V_n = \langle X_t, \vec{n} \rangle$, where $\vec{n}$, the unit normal to X, may be expressed as $\vec{n} = \frac{X_\alpha \times X_\beta}{\|X_\alpha \times X_\beta\|}$. The mean curvature of X is given by

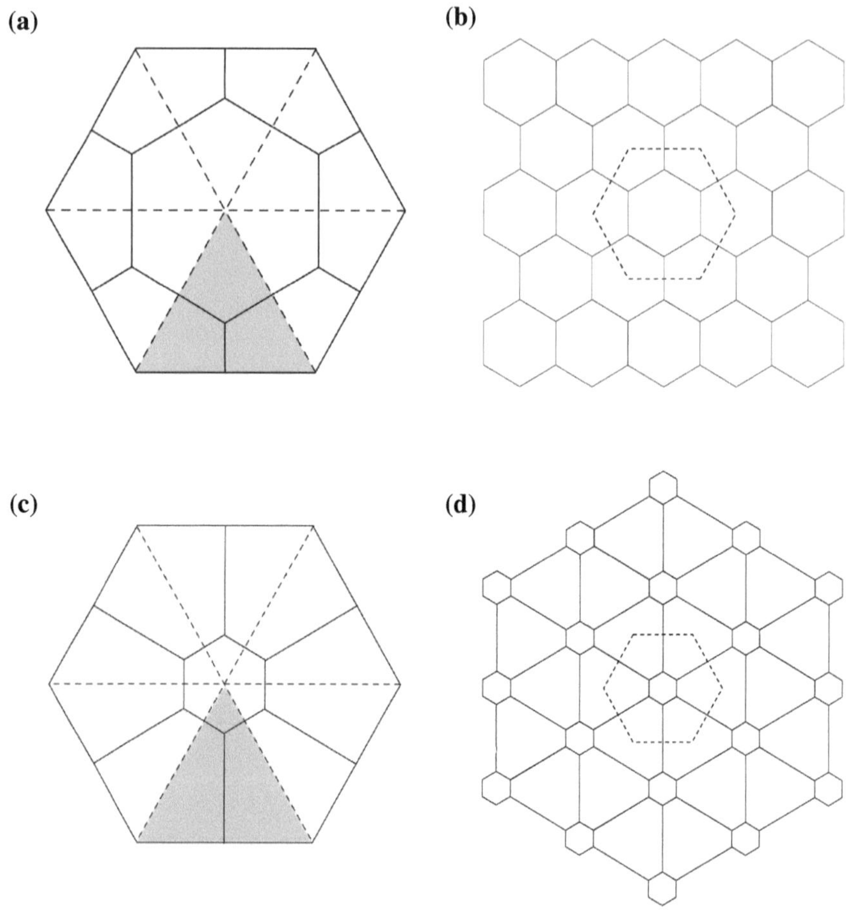

Fig. 4 The x-y projection of the three-grain system shown in the grey equilateral triangular domain in (a) can be extended by reflection to yield the hexagonal tiling of R^2 portrayed in (b). The x-y projection of the three-grain system shown in the grey equilateral triangular domain in (c) can be extended by reflection to yield the tiling of R^2 portrayed in (d) comprised of hexagons and truncated triangles

$$H = \frac{\left\langle \left\langle X_\beta, X_\beta \right\rangle X_{\alpha\alpha} - 2\left\langle X_\alpha, X_\beta \right\rangle X_{\alpha\beta} + \left\langle X_\alpha, X_\alpha \right\rangle X_{\beta\beta}, \vec{n} \right\rangle}{2\left[\left\langle X_\alpha, X_\alpha \right\rangle \left\langle X_\beta, X_\beta \right\rangle - \left\langle X_\alpha, X_\beta \right\rangle^2 \right].} \tag{17}$$

Rather than specifying tangential velocities, we impose the conditions

$$\langle X_\alpha, X_{\alpha\alpha} \rangle = 0, \quad \left\langle X_\beta, X_{\beta\beta} \right\rangle = 0, \tag{18}$$

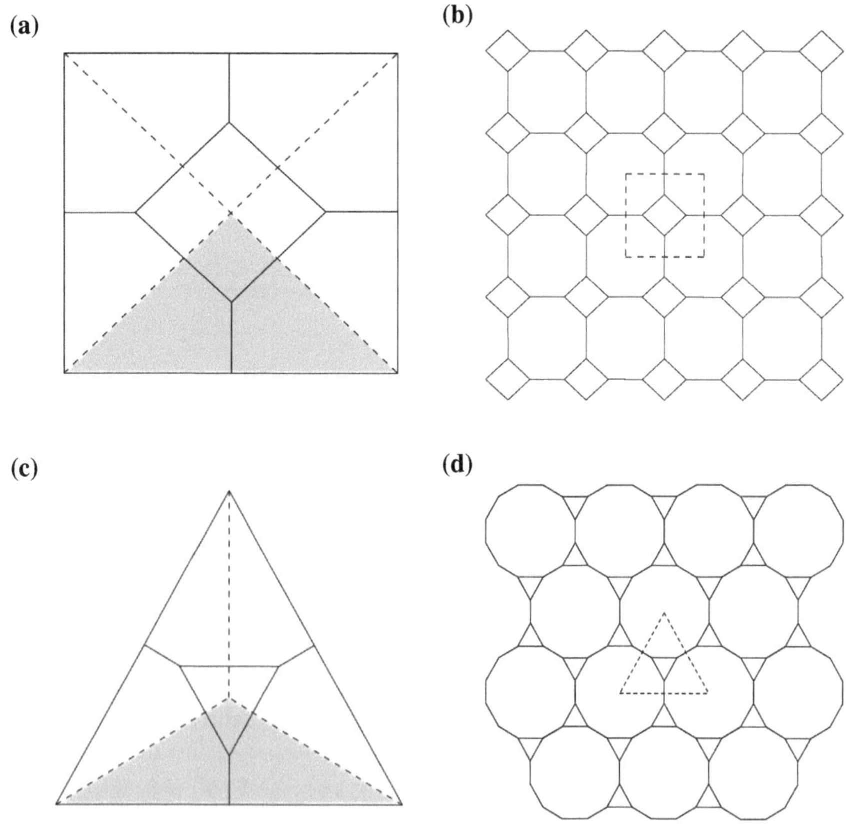

Fig. 5 The x y projection of the three-grain system shown in the grey equilateral triangular domain with angles $\pi/4$, $\pi/2$, $\pi/4$ in (a) can be extended by reflection to yield the tiling of R^2 portrayed in (b) comprised of hexagons and truncated triangles. The x-y projection of the three-grain system shown in the grey equilateral triangular domain with angles $\pi/6$, $2/3\,\pi$, $\pi/6$ in (c), yields upon reflection a truncated square tiling and a truncated hexagonal tiling, respectively, of R^2, portrayed in (d)

which guarantees uniform grid spacing in sense that $\|X_\alpha\|_\alpha = \|X_\beta\|_\beta = 0$; a similar approach was used in [42] for the coupled evolution of three curves in the plane. Let X^1, X^2, X^3 denote two exterior surfaces and a grain boundary which couple along a thermal groove. Along the thermal groove (i) persistence can be expressed as $X^1 = X^2 = X^3$, and (ii) balance of mechanical forces in the plane orthogonal to the thermal groove implies that $\overrightarrow{\tau}^1 + \overrightarrow{\tau}^2 + m\overrightarrow{\tau}^3 = 0$, where for $i = 1, 2, 3$, $\overrightarrow{\tau}^i$ is an outward unit tangent vector to X^i in the plane orthogonal to the thermal groove, see Fig. 6, and m was defined in (5). The angle θ indicated in Fig. 6 satisfies

$$\theta = \pi - 2\ \arcsin(m/2), \tag{19}$$

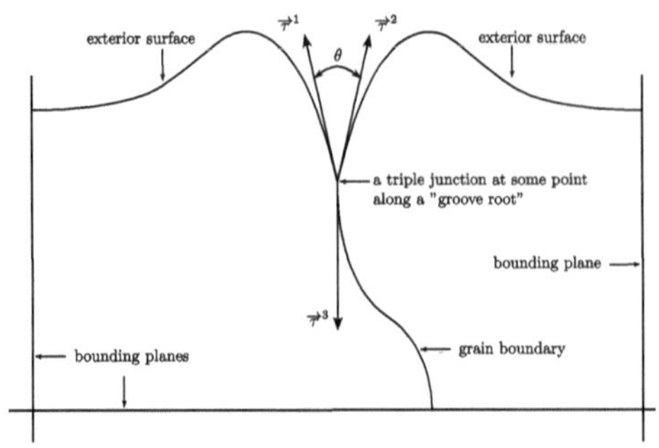

Fig. 6 The normal cross-section to a thermal groove

and is known as the *dihedral angle*. Moreover, (iii) continuity of the chemical potential can be expressed as $H^1 = H^2$, since the chemical potential can be identified with the mean curvature along the exterior surface, and (iv) balance of mass flux along the exterior surface may be expressed as $\langle \vec{\tau}^1, \nabla_s H^1 \rangle + \langle \vec{\tau}^2, \nabla_s H^2 \rangle = 0$, since along the exterior surface the mass flux is proportional to $\nabla_s H$.

Along the internal triple junction line, letting now X^1, X^2, X^3 denote the three-grain boundaries which couple along the internal triple junction line, (I) persistence may be expressed as $X^1 = X^2 = X^3$, and (II) Young's law may be expressed as $\vec{\tau}^1 + \vec{\tau}^2 + \vec{\tau}^3 = 0$, respectively.

Let Π denote one of the three vertical bounding planes, $\vec{N}$ denote a vector which is normal to Π, and r_0 denote a point on Π. If X denotes an evolving exterior surface which intersects Π, then persistence, symmetry, and zero mass flux through Π may be expressed as $\left\langle X - r_0, \vec{N} \right\rangle = 0$, $\left\langle \vec{n}, \vec{N} \right\rangle = 0$, and $\left\langle \vec{N}, \nabla_s H \right\rangle = 0$, respectively, and if X denotes an evolving grain boundary surface then persistence and symmetry may be expressed as $\left\langle X - r_0, \vec{N} \right\rangle = 0$ and $\left\langle \vec{n}, \vec{N} \right\rangle = 0$, respectively. Where two exterior surfaces intersect a bounding plane (see $P_t^{1,2}$, $P_t^{2,3}$, $P_t^{3,1}$ in Fig. 3) or an exterior surface intersects two bounding planes (see C^1, C^2, C^3 in Fig. 3), conditions are formulated in accordance with persistence, symmetry and zero mass flux. At the intersection of a grain boundary and a bounding plane with the mid-plane (see $P_b^{1,2}$, $P_b^{2,3}$, $P_b^{3,1}$ in Fig. 3) or at the intersection of the interior triple junction line with the mid-plane (see P_b^0 in Fig. 3), conditions are formulated in accordance with persistence and symmetry.

To follow the motion of the various coupled surfaces using regular parametric representations, it is necessary to prescribe boundary conditions at the quadruple junction. The condition which we impose is based on the assumptions that (i) Young's law holds along all triple junction lines up to the quadruple junction, and (ii) limiting regularity at the quadruple junction, implying that the triple junction lines become asymptotically straight as the quadruple junction is approached. Assumptions (i)–(ii) imply the unique construction, up to similarity, of a tetrahedron with prescribed angles at the quadruple junction, based on the tangents to the triple junctions lines meeting at the quadruple junction. See Fig. 7.

Assumptions (i)–(ii), together with a bit of trigonometry, imply that

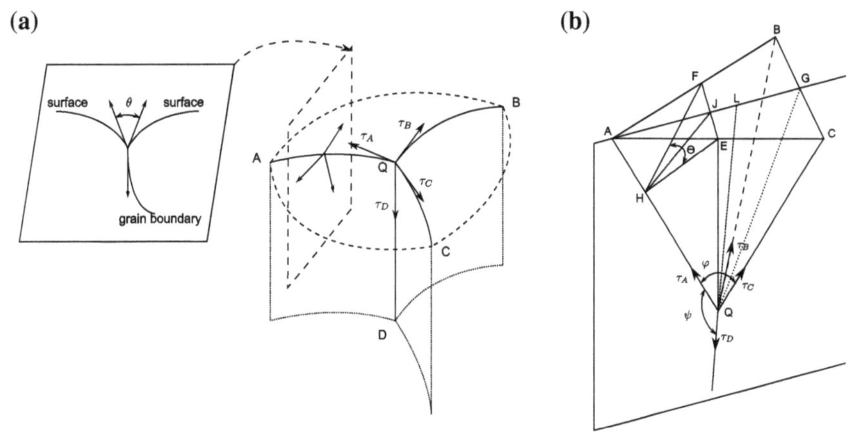

Fig. 7 (a) Sketch of the quadruple junction. Here τ_A, τ_B, τ_C denote tangents to the triple junction lines which meet at the quadruple junction. (b) The tetrahedral construction based on the tangents τ_A, τ_B, τ_C

$$\cos(\varphi) = -\frac{2-m^2}{4-m^2}, \quad 0 \le \varphi \le \frac{2\pi}{3}, \tag{20}$$

$$\cos(\psi) = -\frac{1}{\sqrt{3}}\frac{m}{\sqrt{4-m^2}}, \quad \frac{\pi}{2} \le \psi \le \pi, \tag{21}$$

where $\varphi = \cos^{-1}(\tau_A \cdot \tau_C)$ and $\psi = \cos^{-1}(\tau_A \cdot \tau_D)$, see Fig. 7b. Recalling from (see (7)) that $m \in [0, 2)$, it follows from (21) that the construction requires that $m \in [0, \sqrt{3}) \subset [0, 2)$. The breakdown of the construction when $m = \sqrt{3}$ can be seen as indicating $m = \sqrt{3}$ as a quadruple junction wetting criterion, Derkach, 2010 [15]. Smith [45] employed a similar construction in 1948 in analyzing the stability of triple junctions in the microstructure in two-phase alloys. Straumal, Kogtenkova and Zięba [47] found a similar criteria in 2007 in the context of wetting of triple junctions by melt. The (F, ϵ, δ) minimal surfaces approach for soap-bubbles (Taylor 1976 [49]) is relevant here also [48].

The problem formulation outlined above yields a PDE system which was solved numerically using finite difference methods to spatially discretize the system. Staggered grids were employed for each evolving α-β surface, using ghost points on each grid to accommodate the boundary conditions. The mean curvatures of the exterior surfaces were defined as additional variables [6]. The resultant DAE (differential and algebraic equation [4]) system was solved using a backward Euler (implicit) method. Newton iterations were used to solve the resultant nonlinear equations at each time step. Partial parallelization was used in calculating the Jacobian matrix and in solving the large system of linear equations which results at each Newton's method iteration. The approach here extends that of Pan and Wetton [42] to three spatial dimensions. For details, see [15, 16, 19].

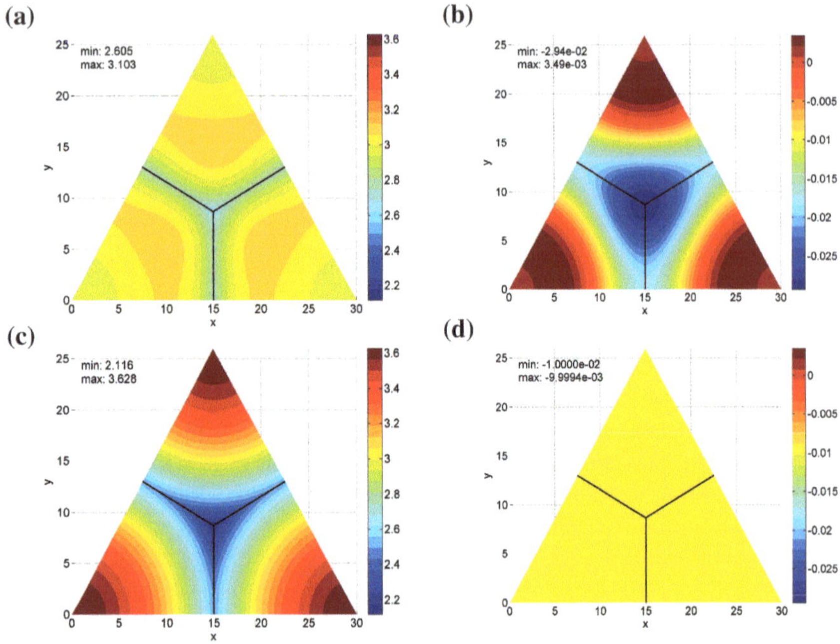

Fig. 8 Here initially the grain volume fractions are equal, $L_{C^1C^2} = 30$, $L_z = 3$, and $m = 0.3$. Exterior surfaces heights at times (a) $t = 50$, (c) $t = 6300$, and mean curvatures at times (b) $t = 50$, (d) $t = 6300$. Threefold symmetry is maintained. The mean curvatures converge to $\approx 10^{-2} = mL^{-1}_{C^1C^3}$ in accordance with the spherically capped hexagon configuration, despite earlier small slope predictions indicating hole formation at the quadruple junctions, Génin et al. [21]

The algorithm was implemented in [18] where good qualitative agreement with experiment could be seen with regard to physical features such as pitting, grain annihilation, and hole formation. Results from more recent simulations which were undertaken for initial conditions corresponding to the triangular configurations shown in Fig. 4 are portrayed in Figs. 8, 9 and 10. Note that in Fig. 8, where the initial configuration possesses three-fold symmetry, the three fold symmetry is maintained, the grain boundaries remain flat, and the exterior surfaces seem to approach constant mean curvature spherical caps. In Fig. 9, where the initial conditions have reflectional symmetry but not three-fold symmetry and one of the three grains is slightly smaller that the other, the system seems to be evolving towards the limiting spherically capped configuration seen in Fig. 8, indicating some measure of stability for this configuration. In Fig. 10, the initial conditions were similar to those portrayed in Fig. 9, but here the third smaller grain was taken to be smaller than before, and the system no longer seems to be evolving toward the spherically capped configuration.

To gain some comprehension and interpretation of these results, let us consider steady states with threefold symmetry with triangular bounding planes, whose x-y projection yield *hexagonal tilings* upon extension by reflectional symmetry. For

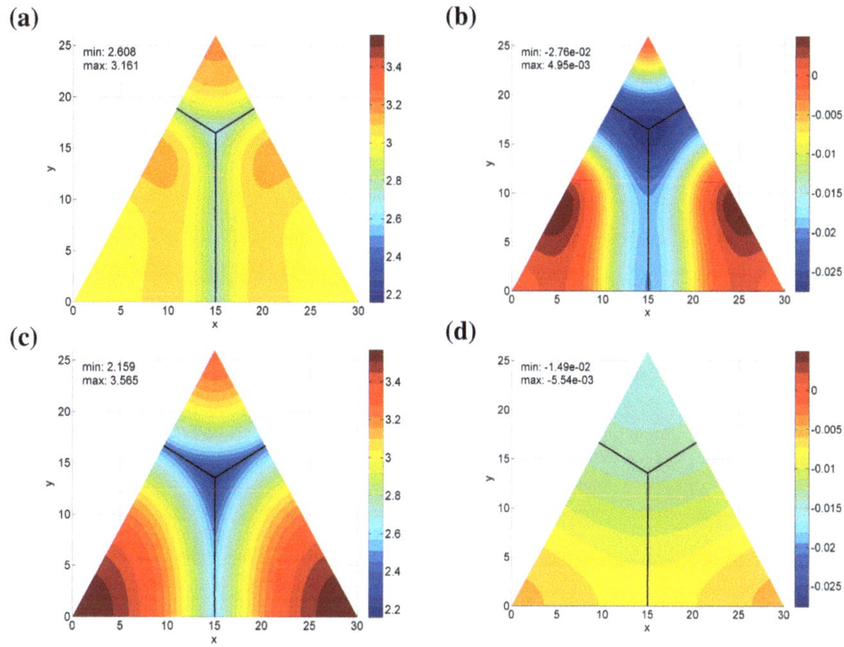

Fig. 9 Here the initial grain volume fractions are given by $\phi_{C^1} = \phi_{C^3} = 0.45, \phi_{C^2} = 0.1, L_{C^1C^2} = 30$, $L_z = 3$ and $m = 0.3$. Exterior surface heights at times (a) $t = 50$, (c) $t = 2310$, and mean curvatures at times (b) $t = 50$, (d) $t = 2310$. The smaller grain is seen to grow, and the system appears to approach the steady state seen in Fig. 8

these configurations there exist steady states for our dynamic problem formulation if the initial volume is sufficiently large and if the parameter m is sufficiently small. Some details follow below. Further details will appear in [17]. By considering the reflectional extension to hexagons, these configurations can be viewed as spherically capped hexagonal cylinders. Some discussions in these directions were given some 30 years by Finn [20] and Srolovitz and Safran [46].

To understand the restrictions, let us consider some details of the geometry, see Fig. 11. The intersections of the spherical caps with the flat vertical grain boundaries define circular sections, whose maximal height relative to the height of the spheres' centers is given by $z_{\max} = (L_{C^3C^1}\sqrt{4 - m^2})/(2m)$, and whose minimal height, which is attained at the quadruple junctions, is given by $z_{\min} = (L_{C^3C^1}\sqrt{3 - m^2})/(\sqrt{3}m)$. In particular, the construction requires $0 < m < \sqrt{3}$, and hence $m = \sqrt{3}$ appears again as a quadruple junction wetting criterion. This restriction may be seen as being sharp by considering the following *nonexistence result.*

Theorem. Given a domain Ω containing a corner with interior angle α. If $\alpha + \theta < \pi$, then there exist no constant mean curvature surfaces defined over Ω which are bounded from below which meet $\partial\Omega$ with contact angle $\theta_c = \frac{1}{2}(\pi - \theta)$.

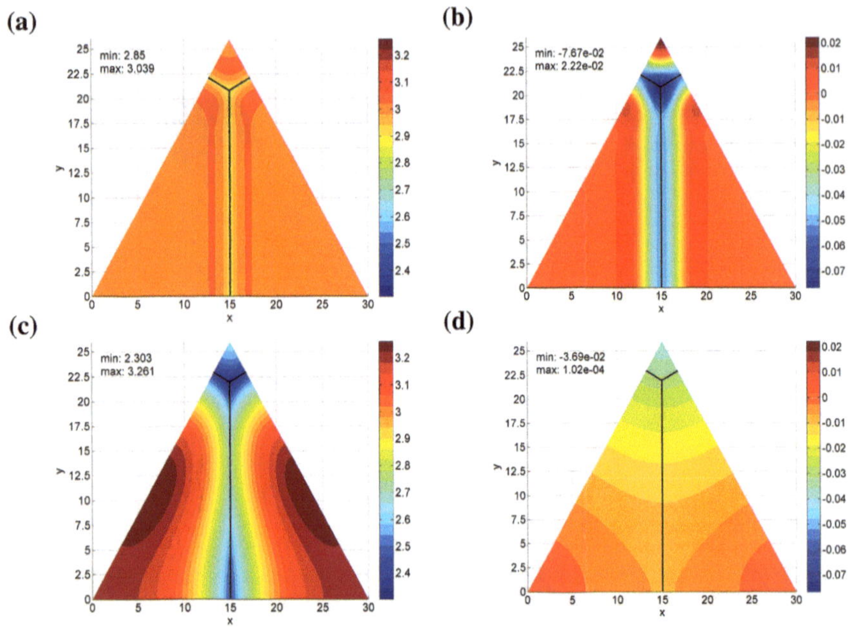

Fig. 10 Here the initial grain volume fractions are given by $\phi_{C^1} = \phi_{C^3} = 0.485$, $\phi_{C^2} = 0.03$, $L_{C^1C^2} = 30$, $L_z = 3$ and $m = 0.3$. Exterior surface heights at times (a) $t = 1$, (c) $t = 703$, and mean curvatures at times (b) $t = 1$, (d) $t = 703$. We observe shrinkage and sinking of the smaller grain. Apparently if initially one grain is much smaller than the others, the steady state seen in Fig. 8 is not approached

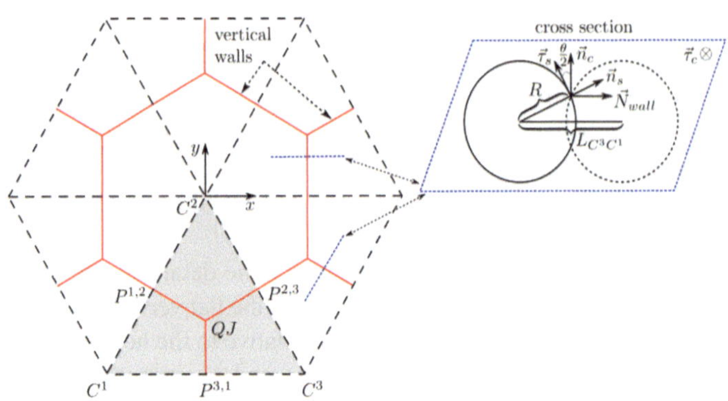

Fig. 11 The x-y projection of the three-grain system shown in the grey equilateral triangle extends by reflection to yield a tiling by hexagonal grains. The exterior surface above each hexagon grain is a spherical cap. The spherical caps intersect each other with dihedral angle $\theta = \pi - 2\arcsin(m/2)$ along thermal grooves lying above the sides of the hexagon. Vertical planar grain boundaries surround each grain, intersecting with three fold symmetry along internal triple junction lines which lie above the corners of each hexagon. See Finn, 1985 [20], Srolovitz and Safran, 1986 [46], Derkach, 2016 [16]

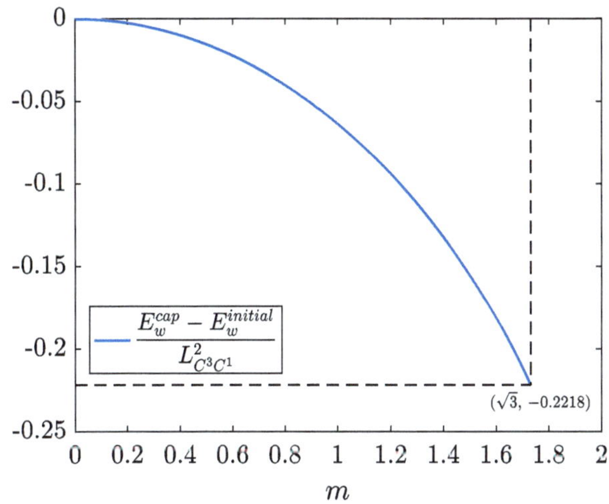

Fig. 12 A plot of $(E_w^{cap} - E_w^{initial})L^{-2}_{C^3C^1}$ for $m \in (0, \sqrt{3})$

Proof. The proof follows from Finn [20, Corollary 5.5] and the definitions given there.

In the context of the equilateral triangular system, $\alpha = \frac{2\pi}{3}$ and it follows from (19) that $\theta = 2\cos^{-1}(m/2)$. The restriction $\alpha + \theta < \pi$ implies that the nonexistence of solutions which are bounded from below if $m > \sqrt{3}$. The non-boundedness from below can be shown [20] to occur at the vertex of the interior corner. In terms of the geometry of the spherical capped hexagon, this implies that any such steady states for $m \in [\sqrt{3}, 2)$ must necessarily penetrate the film at the quadruple junction.

The results above were concerned with equilibrium capillary surfaces in unbounded cylinder with unbounded mass. In the context of thin films, the film thickness required to guarantee film stability is a technologically critical issue. This leads us to consider volume restrictions for the realization of these configurations. The volume of the spherical cap over the hexagon and above height $z_{\min}$ is given by

$$V^{cap} = L^3_{C^3C^1}\left[\mathcal{F}(m) - \frac{\sqrt{3-m^2}}{2m}\right],$$

where

$$\mathcal{F}(m) = \frac{3}{2}\int_0^1 \left[\frac{\xi\sqrt{3-m^2\xi^2}}{3m} + \frac{4-m^2\xi^2}{2m^2}\arcsin\left(\frac{m\xi}{\sqrt{3}\sqrt{4-m^2\xi^2}}\right)\right]d\xi.$$

Letting L_z^* denote the height of a flat configuration with equivalent volume, namely the height L_z^* such that $V^{cap} = \frac{\sqrt{3}}{2}L_z^* L^2_{C^3C^1}$, we obtain that realization of the capped hexagon configuration requires that

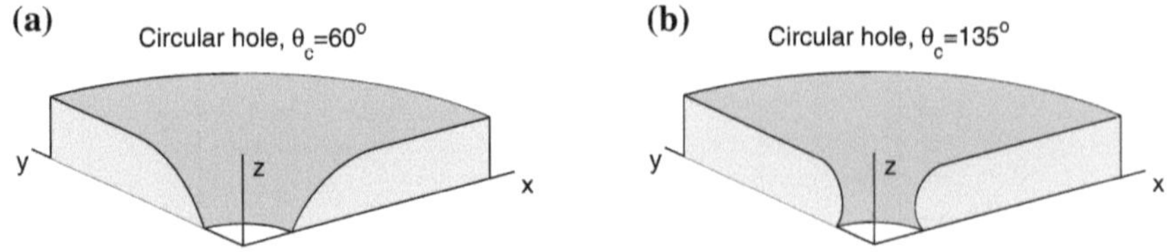

Fig. 13 Sketches of the initial circular 3D-domains which used to model hole growth and hole shrinkage by wetting and dewetting in mono-crystals. (a), (b), circular holes in mono-crystals with contact angles $\theta_c = 60^o$ and $\theta_c = 135^o$, respectively

$$L_z \geq L_z^* = \frac{2}{\sqrt{3}}\Big[\mathcal{F}(m) - \frac{\sqrt{3-m^2}}{2m}\Big] L_{C^3C^1}.$$

Note in particular that if $m = 0.3$, then $L_z^* \approx 0.02946 L_{C^3C^1}$.

If the conditions are such that the spherically capped hexagon configuration can be realized, is it energetically preferable relative to say the equivalent volume flat configuration? The relevant energy to consider in this regard is the weighted energy, E_w, (see (6)) which for the capped hexagons is

$$E_w^{cap} = L_{C^3C^1}^2 \mathcal{G}(m) + \sqrt{3}\, m L_z L_{C^3C^1}, \tag{22}$$

where the function $\mathcal{G}(m)$ may be explicitly prescribed, and which for the equivalent volume flat configuration with flat grain boundaries is given by

$$E_w^{initial} = \frac{\sqrt{3}}{2} L_{C^3C^1}^2 + \sqrt{3}\, m L_z L_{C^3C^1}. \tag{23}$$

If follows from (22)–(23) that $(E_w^{cap} - E_w^{initial})/L_{C^3C^1}^2$ is a function of m only. This function was calculated numerically using Matlab and is portrayed in Fig. 12; it can be clearly seen in Fig. 12 that $E_w^{cap} \leq E_w^{initial}$ for $m \in (0, \sqrt{3})$.

In the discussions above, weighted energies were compared for configurations with the same volume and where the length, $L_{C^3C^1}$, of side of the bounding triangle was fixed. Note that by virtue of the geometry, $L_{hexagon} = \frac{1}{\sqrt{3}} L_{triangle} = \frac{1}{\sqrt{3}} L_{C^3C^1}$. Let us compare the weighted energies for spherically capped hexagon configurations with the same volume, but allowing now the length of the triangle (hexagon) side to vary. Taking V, the volume of the system, to be fixed, implies that, L_z, the height of the equivalent volume flat triangular system satisfies $L_z = \frac{2}{\sqrt{3}} V L_{C^3C^1}^{-2}$. It then follows from (22) that E_w^{cap} may be expressed as

$$E_w^{cap} = L_{C^3C^1}^2 \mathcal{G}(m) + \frac{2mV}{L_{C^3C^1}}, \tag{24}$$

which attains a minimum at

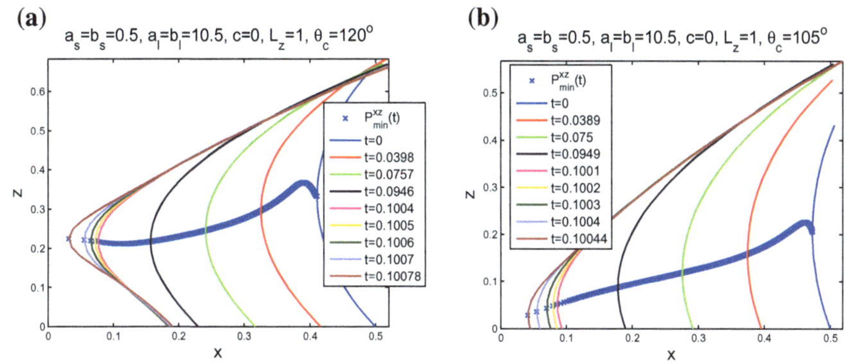

Fig. 14 Diagram showing the void creation in the wetting case when $\theta_c > \dfrac{\pi}{2}$, where the initial hole radius was 0.5 and the initial film height was $L_z = 1$. Profiles of the free boundary along xz bounding plane for: (a) $\theta_c = 120^o$; (b) $\theta_c = 105^o$. Evolution in (a) appears to be asymptotically self-similar, see Bernoff, Bertozzi and Witelski [10], whereas what is happening in (b) is less clear

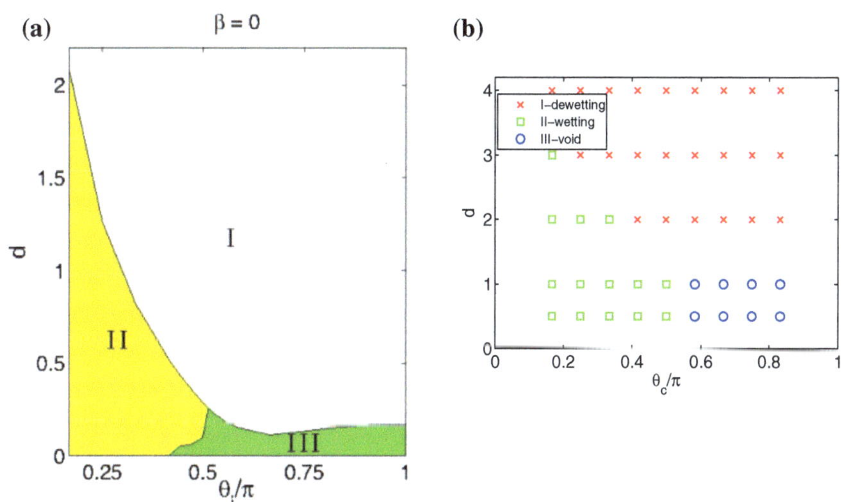

Fig. 15 Diagram indicating the dependence of I) dewetting, II) wetting, III) void creation on the initial hole diameter and on the contact angle θ_c. (a) Results from the simulations of Wang, Jiang, Bao and Srolovitz [50] (reproduced with permission) for holes in 2D films with initial diameter d. Here $\beta = 0$ indicates isotropy is being assumed. (b) Results from our simulations for 3D circular holes with initial diameter d, see Fig. 13. Comparing the 2D results with the 3D results, we conclude that the second curvature appearing in the 3D calculations acts effectively as a drag force, allowing holes with larger initial radius to shrink

$$L^{\min}_{C^3C^1} = \left[\frac{mV}{\mathcal{G}(m)}\right]^{\frac{1}{3}}. \tag{25}$$

We conjecture that (25) is indicative of an energetically preferred length scale for grains with prescribed volume in a bamboo structured system.

Should coarsening be expected to ensue in the spherically capped system? It follows from (22) that the weighted energy per unit area of the capped hexagon system is given by

$$\frac{2}{\sqrt{3}}\frac{E^{cap}_w}{L^2_{C^3C^1}} = \frac{2}{\sqrt{3}}\mathcal{G}(m) + \frac{2mL_z}{L_{C^3C^1}}. \tag{26}$$

Since (26) decreases as $L_{C^3C^1}$ increases, there is an energetic preference for coarsening. However coarsening may be hindered by dynamic considerations such as local energy barriers.

In concluding we remark that while many open questions remain regarding the system of capped hexagons, we see in Figs. 13, 14 and 15 that the algorithmic approach developed for studying systems of grains can be adapted for studying wetting and dewetting phenomena, [16].

Acknowledgements The authors would like to acknowledge support from the Israel Science Foundation (Grants #1200/16). This work was supported by the Russell Berrie Nanotechnology Institute at the Technion through the Nevet Program and Capital Equipment Usage Grants.

References

1. S. Angenent, On the formation of singularities in the curve shortening flow. J. Differ. Geom. **33**, 601–633 (1991)
2. A.V. Antonov, Ch.V. Kopetskii, Ya.M. Mukovskii, L.S. Shvindlerman, The shape of a moving grain boundary. Phys. Status Solidi **9**, 45–50 (1972)
3. T. Asai, Y. Giga, On self-similar solutions to the surface diffusion flow equations with contact angle boundary conditions. Interfaces Free Boundaries **16**, 539–573 (2014)
4. U.M. Asher, L.R. Petzold, *Computer Methods for Ordinary Differential Equations and Differential-Algebraic Equations* (SIAM, Philadelphia, 1998)
5. V.Yu. Aristov, Ye.M. Fridman, L.S. Shvindlerman, Boundary migration in aluminium bicrystals. Phys. Met. Metallogr. **35**, 188–191 (1973)
6. A. Averbuch, M. Israeli, I. Ravve, Electromigration of intergranular voids in metal films for microelectronic interconnects. J. Comput. Phys. **189**, 481–502 (2003)
7. K. Barmak, E. Eggeling, D. Kinderlehrer, R. Sharp, S. Ta'asan, A.D. Rollett, K.R. Coffey, Grain growth and the puzzle of its stagnation in thin films: the curious tale of a tail and an ear. Prog. Mater. Sci. **58**, 987–1055 (2013)
8. M. Beck, Z. Pan, B. Wetton, Stability of travelling wave solutions for coupled surface and grain boundary motion. Phys. D **239**, 1730–1740 (2010)
9. G. Bellettini, *Lecture notes on mean curvature flow, barriers, and singular perturbations.* Appunti. Scuola Normale Superiore di Pisa (Nuova Serie) **12**, Pisa, 2013
10. A.J. Bernoff, A.L. Bertozzi, T.P. Witelski, Axisymmetric surface diffusion: dynamics and stability of self-similar pinchoff. J. Stat. Phys. **93**, 725–776 (1998)
11. K. Brakke, The Motion of a Surface by its Mean Curvature. Mathematical Notes, vol. 20 (Princeton University Press, Princeton, 1978)

12. J.W. Cahn, A. Novick-Cohen, Evolution equations for phase separation and ordering in binary alloys. J. Stat. Phys. **76**, 877–909 (1994)
13. J.W. Cahn, C.M. Elliott, A. Novick-Cohen, The Cahn-Hilliard equation: motion by minus the Laplacian of the mean curvature. Eur. J. Appl. Math. **7**, 287–301 (1996)
14. R. Dal Passo, L. Giacomelli, A. Novick-Cohen, Existence for an Allen-Cahn/Cahn-Hilliard system with degenerate mobility. Interfaces Free Boundaries **1**, 199–226 (1999)
15. V. Derkach, Surface evolution and grain boundary migration in various 3-D systems of 3 grains. M.Sc. Thesis, Department of Mathematics, Technion-IIT, Israel, 2010
16. V. Derkach, Surface and grain boundary evolution in thin single- and poly-crystalline films. PhD. thesis, Department of Mathematics, Technion-IIT, Israel (2017)
17. V. Derkach, J. McCuan, A. Novick-Cohen, in preparation
18. V. Derkach, A. Novick-Cohen, A. Vilenkin, E. Rabkin, Grain boundary migration and grooving in thin 3D systems. Acta Mat. **65**, 194–206 (2014)
19. V. Derkach, A. Novick-Cohen, A. Vilenkin, Grain boundary migration with thermal grooving effects: a numerical approach. J. Elliptic Parabol. Equ. **2**, 389–413 (2016)
20. R. Finn, *Equilibrium Capillary Surfaces* (Springer, Berlin, 1985)
21. F.Y. Génin, W.W. Mullins, P. Wynblatt, Capilary instabilities in thin films: a model of thermal pitting at grain boundary vertices. Acta. Metall. Mater. **40**, 3239–3248 (1992)
22. Y. Giga, *Surface Evolution Equations. A Level Set Approach*. Monographs in Mathematics, vol. 99 (Birkhäuser Verlag, Basel, 2006)
23. Y. Giga, K. Ito, On pinching of curves moved by surface diffusion. Commun. Appl. Anal. **2**, 393–405 (1998)
24. Y. Giga, K. Ito, Loss of convexity of simple closed curves moved by surface diffusion, in *Topics in Nonlinear Analysis*. Progress in Nonlinear Differential Equations and Their Applications, vol. 35 (Birkhäuser, Basel, 1999), pp. 305–320
25. M. Grayson, The heat equation shrinks embedded plane curves to round points. J. Differ. Geom. **26**, 285–314 (1987)
26. C. Herring, The use of classical macroscopic concepts in surface energy problems, in *Structure and Properties of Solid Surfaces*, ed. by R. Gomer, C.S. Smith (University of Chicago, Chicago, 1953), pp. 5–81
27. G. Huisken, Flow by mean curvature of convex surfaces into spheres. J. Differ. Geom. **20**, 237–266 (1984)
28. J. Kanel, A. Novick-Cohen, In preparation
29. J. Kanel, A. Novick-Cohen, A. Vilenkin, Coupled surface and grain boundary motion: non-classical traveling wave solutions. Adv. Differ. Eqns. **9**, 299–327 (2004)
30. J. Kanel, A. Novick-Cohen, A. Vilenkin, Coupled surface and grain boundary motion: a traveling wave solution. Nonlinear Anal. TMA **59**, 1267–1292 (2004)
31. J. Kanel, A. Novick-Cohen, A. Vilenkin, A traveling wave solution for coupled surface and grain boundary motion. Acta. Mater. **51**, 1981–1989 (2003)
32. J. Kanel, A. Novick-Cohen, A. Vilenkin, A numerical study of grain boundary motion in bicrystals. Acta Mater. **53**, 227–235 (2005)
33. J. Kanel, A. Novick-Cohen, A. Vilenkin, Numerical analysis of a 3D radially symmetric grain attached to a free crystal surface. Acta Mater. **54**, 2589–2595 (2006)
34. J. LeCrone, G. Simonett, On well-posedness, stability, and bifurcation for the axisymmetric surface diffusion flow. SIAM J. Math. Anal. **45**, 2834–2869 (2013)
35. W.W. Mullins, Two dimensional motion of idealized grain boundaries. J. Appl. Phys. **27**, 900–904 (1956)
36. W.W. Mullins, Theory of thermal grooving. J. Appl. Phys. **28**, 333–339 (1957)
37. W.W. Mullins, The effect of thermal grooving on grain boundary motion. Acta Metall. **6**, 414–427 (1958)
38. A. Novick-Cohen, Order-disorder and phase separation: modeling grain sintering, in *Proceedings of the 9th International Symposium (CMDS9), Istanbul, Turkey*, ed. by E. Inan, K.Z. Markov (World Scientific, 1999), pp. 56-68

39. A. Novick-Cohen, Triple-junction motion for Allen-Cahn/Cahn-Hilliard systems. Phys. D **137**, 1–24 (2000)
40. A. Novick-Cohen, L. Peres Hari, Geometric motion for a degenerate Allen-Cahn/Cahn-Hilliard system: the partial wetting case. Phys. D **209**, 205–235 (2005)
41. A. Novick-Cohen, A. Vilenkin, O. Zelekman-Smirin, The effects of grain grooves on grain boundary migration in nano-films. Acta Mater. **53**, 813–822 (2010)
42. Z. Pan, B. Wetton, A numerical method for coupled surface and grain boundary motion. Eur. J. Appl. Math. **19**, 311–327 (2008)
43. A. Rotman (A. Zigelman), Verification of the influence of surface energies on the effective mobility. M.Sc. Thesis, Department of Mathematics, Technion-IIT, Israel, 2009
44. G. Gottstein, L.S. Shvindlerman, *Grain Boundary Migration in Metals: Thermodynamics, Kinetics, Applications*, 2nd edn. (CRC Press, Boca Raton, 2010)
45. C.S. Smith, Grains, phases, and interfaces: an interpretation of microstructure. Trans. AIME **175**, 15–51 (1948)
46. D.J. Srolovitz, S.A. Safran, Capillarity instabilities in thin film. J. Appl. Phys. **60**, 247–254 (1986)
47. B.B. Straumal, O. Kogtenkova, P. Zięba, Wetting transition of grain-boundary triple junctions. Acta Mat. **56**, 925–933 (2008)
48. J.E. Taylor, private communication
49. J.E. Taylor, The structure of singularities in soap-bubble-like and soap-film-like minimal surfaces. Ann. Math. **103**, 489–539 (1976)
50. Y. Wang, W. Jiang, W. Bao, D.J. Srolovitz, Sharp interface model for solid-state dewetting problems with weakly anisotropic surface energies. Phys. Rev. B **91**, 045303 (2015)
51. A. Vilenkin, A. Novick–Cohen, Grain boundary migration in thin films: the effects of grain groves in nano films, in *AES–ATEMA' 2007 International Conference* (2007), pp. 127–134. Accessed 6–10 Aug 2007
52. A. Zigelman, Coupled grain boundary and exterior surface motion. Ph.D. thesis, Department of Mathematics, Technion-IIT, Israel, 2015
53. A. Zigelman, A. Novick-Cohen, A. Vilenkin, The influence of the exterior surface on grain boundary mobility measurements. SIAM J. Appl. Math. **74**, 819–843 (2014)

A Note on Regularity Criteria for Navier-Stokes System

Note on Navier-Stokes

Jishan Fan and Tohru Ozawa

Abstract We use some interpolation inequalities on Besov spaces to show a regularity criterion for n-dimensional Navier-Stokes system.

Keywords Regularity criterion · Navier-Stokes · Incompressible

Mathematics Subject Classifications (2010) 35Q30 · 76D03 · 76D09

1 Introduction

We consider the Navier-Stokes equations:

$$\partial_t u + u \cdot \nabla u + \nabla \pi = \Delta u, \tag{1.1}$$

$$\operatorname{div} u = 0, \tag{1.2}$$

$$u(\cdot, 0) = u_0, \ \operatorname{div} u_0 = 0 \ \text{ in } \mathbb{R}^n, \quad n \geq 3. \tag{1.3}$$

Here, u is the velocity field and π is the scalar pressure.

The global well-posedness for (1.1)–(1.3) was studied first by Leray and Hopf [6, 9]. They proved that a weak solution exists globally in time. However, the regularity of weak solutions is still a big open problem. On the other hand, the development of regularity criteria is of major importance for both theoretical and practical purpose. There has been a lot of progress on studies on the Navier-Stokes equations in this

Dedicated to Professor Yoshikazu Giga on the occasion of his 60th birthday.

J. Fan
Department of Applied Mathematics, Nanjing Forestry University, Nanjing 210037, People's Republic of China
e-mail: fanjishan@njfu.com.cn

T. Ozawa (✉)
Department of Applied Physics, Waseda University, Tokyo 169-8555, Japan
e-mail: txozawa@waseda.jp

Y. Maekawa and S. Jimbo (eds.), *Mathematics for Nonlinear Phenomena—Analysis and Computation*, Springer Proceedings in Mathematics & Statistics 215,
DOI 10.1007/978-3-319-66764-5_3

direction. The first result is obtained independently by Serrin [13] and Ohyama [12], which states that if weak solution u satisfies

$$u \in L^r(0, T; L^m) \text{ with } \frac{2}{r} + \frac{n}{m} \leq 1,\ n < m \leq \infty, \tag{1.4}$$

then u is smooth. After that, there are further developments and refinements by Fabes, Jones and Riviere [2], Giga [4], Sohr and von Wahl [14] and Struwe [16]. The limiting case $r = \infty$ and $m = n$ was proved by Escauriaza, Seregin, and Šverák [1]. Kozono and Shimada [8] refined (1.4) by the following condition:

$$u \in L^{\frac{2}{1-\alpha}}(0, T; \dot{B}^{-\alpha}_{\infty,\infty}) \text{ for } 0 < \alpha < 1. \tag{1.5}$$

Fan-Ozawa [3] used the method in [8] to prove the regularity criterion

$$\omega := \operatorname{curl} u \in L^{\frac{2}{2-\alpha}}(0, T; \dot{B}^{-\alpha}_{\infty,\infty}) \text{ for } 0 < \alpha < 1, \tag{1.6}$$

when $n = 3$.

The aim of this paper is to prove (1.6) for any $n \geq 3$ and give a new proof of (1.5). We will prove

Theorem 1.1 *Let $n \geq 3$ and $s > \frac{n}{2}$ and let $u_0 \in H^s(\mathbb{R}^n)$ with $\operatorname{div} u_0 = 0$ in $\mathbb{R}^n$. Let (u, π) be a local strong solution to the problem (1.1)–(1.3). If (1.6) or (1.5) holds true with $0 < T < \infty$, then the solution (u, π) can be extended beyond $T > 0$.*

To prove Theorem 1.1, we will use the following commutator estimate due to Kato-Ponce [7]:

$$\|\Lambda^s(fg)\|_{L^p} \leq C(\|f\|_{L^{p_1}}\|\Lambda^s g\|_{L^{q_1}} + \|\Lambda^s f\|_{L^{p_2}}\|g\|_{L^{q_2}}), \tag{1.7}$$

$$\|\Lambda^s(fg) - f\Lambda^s g\|_{L^p} \leq C(\|\nabla f\|_{L^{p_1}}\|\Lambda^{s-1} g\|_{L^{q_1}} + \|\Lambda^s f\|_{L^{p_2}}\|g\|_{L^{q_2}}) \tag{1.8}$$

with $s > 0$, $\Lambda := (-\Delta)^{\frac{1}{2}}$ and $\frac{1}{p} = \frac{1}{p_1} + \frac{1}{q_1} = \frac{1}{p_2} + \frac{1}{q_2}$ with $1 < p_1, p_2, q_1, q_2 < \infty$.

We will also use the following interpolation inequalities [5, 10, 11]:

$$\|\nabla u\|_{L^q} \leq C\|\nabla u\|^{1-\theta}_{\dot{B}^{-\alpha}_{\infty,\infty}}\|u\|^{\theta}_{\dot{H}^{s+\alpha}}, \tag{1.9}$$

$$\|\Lambda^s u\|_{L^{\frac{2q}{q-2}}} \leq C\|\nabla u\|^{\theta}_{\dot{B}^{-\alpha}_{\infty,\infty}}\|u\|^{1-\theta}_{\dot{H}^{s+\alpha}}, \tag{1.10}$$

where $s > 0,\ 0 < \alpha < 1,\ 0 < \theta = \frac{2}{q} < 1$ and $q = \frac{2(s-1+2\alpha)}{\alpha}$, and

$$\|u\|_{L^p} \leq C\|u\|^{1-\tilde{\theta}}_{\dot{B}^{-\alpha}_{\infty,\infty}}\|u\|^{\tilde{\theta}}_{\dot{H}^{s+\alpha}}, \tag{1.11}$$

$$\|\Lambda^s u\|_{L^{\frac{2p}{p-2}}} \leq C\|u\|^{\tilde{\theta}}_{\dot{B}^{-\alpha}_{\infty,\infty}}\|u\|^{1-\tilde{\theta}}_{\dot{H}^{s+\alpha}}, \tag{1.12}$$

where $s > 0, 0 < \alpha < 1, 0 < \tilde{\theta} = \frac{2}{p} < 1$ and $p = \frac{2(s+2\alpha)}{\alpha}$.

2 Proof of Theorem 1.1

This section is devoted to the proof of Theorem 1.1, we only need to establish a priori estimates.

First, testing (1.1) by u and using (1.2), we see that

$$\frac{1}{2}\frac{d}{dt}\int |u|^2 dx + \int |\nabla u|^2 dx = 0. \tag{2.1}$$

Case 1. Let (1.6) hold true.

Taking Λ^s to (1.1), testing by $\Lambda^s u$, and using (1.2), (1.8), (1.9) and (1.10), we find that

$$\begin{aligned}
&\frac{1}{2}\frac{d}{dt}\int |\Lambda^s u|^2 dx + \int |\Lambda^{s+1} u|^2 dx \\
&= -\int (\Lambda^s(u\cdot\nabla u) - u\cdot\nabla\Lambda^s u)\Lambda^s u dx \\
&\le C\|\nabla u\|_{L^q}\|\Lambda^s u\|_{L^{\frac{2q}{q-2}}}\|\Lambda^s u\|_{L^2} \\
&\le C\|\nabla u\|_{\dot B^{-\alpha}_{\infty,\infty}}\|u\|_{\dot H^{s+\alpha}}\|\Lambda^s u\|_{L^2} \\
&\le C\|\omega\|_{\dot B^{-\alpha}_{\infty,\infty}}\|u\|_{\dot H^{s+\alpha}}\|\Lambda^s u\|_{L^2} \\
&\le C\|\omega\|_{\dot B^{-\alpha}_{\infty,\infty}}\|\Lambda^s u\|_{L^2}^{2-\alpha}\|\Lambda^{s+1} u\|_{L^2}^{\alpha} \\
&\le \frac{1}{2}\|\Lambda^{s+1}u\|_{L^2}^2 + C\|\omega\|_{\dot B^{-\alpha}_{\infty,\infty}}^{\frac{2}{2-\alpha}}\|\Lambda^s u\|_{L^2}^2,
\end{aligned}$$

which gives

$$\|\Lambda^s u\|_{L^\infty(0,T;L^2)} \le C. \tag{2.2}$$

Here, we have used the following inequality [15]:

$$\|\nabla u\|_{\dot B^{-\alpha}_{\infty,\infty}} \le C\|\omega\|_{\dot B^{-\alpha}_{\infty,\infty}}. \tag{2.3}$$

Case 2. Let (1.5) hold true.

Taking Λ^s to (1.1), testing by $\Lambda^s u$, and using (1.2), (1.7), (1.11) and (1.12), we deduce that

$$\begin{aligned}
&\frac{1}{2}\frac{d}{dt}\int |\Lambda^s u|^2 dx + \int |\Lambda^{s+1} u|^2 dx \\
&= -\int \Lambda^s \operatorname{div}(u\otimes u)\cdot\Lambda^s u dx \\
&= \int \Lambda^s(u\otimes u) : \nabla\Lambda^s u dx \\
&\le \|\Lambda^s(u\otimes u)\|_{L^2}\|\nabla\Lambda^s u\|_{L^2} \\
&\le C\|u\|_{L^p}\|\Lambda^s u\|_{L^{\frac{2p}{p-2}}}\|\Lambda^{s+1}u\|_{L^2}
\end{aligned}$$

$$\le C\|u\|_{\dot{B}^{-\alpha}_{\infty,\infty}}\|u\|_{\dot{H}^{s+\alpha}}\|\Lambda^{s+1}u\|_{L^2} \le C\|u\|_{\dot{B}^{-\alpha}_{\infty,\infty}}\|\Lambda^s u\|_{L^2}^{1-\alpha}\|\Lambda^{s+1}u\|_{L^2}^{1+\alpha}$$
$$\le \frac{1}{2}\|\Lambda^{s+1}u\|_{L^2}^2 + C\|u\|_{\dot{B}^{-\alpha}_{\infty,\infty}}^{\frac{2}{1-\alpha}}\|\Lambda^s u\|_{L^2}^2,$$

which leads to (2.2).

This completes the proof. □

Acknowledgements Fan is partially supported by NSFC (No. 11171154).

References

1. L. Escauriaza, G.A. Seregin, V. Šverák, $L^{3,\infty}$-Solutions of the Navier-Stokes equations and backward uniqueness. Rus. Math. Surv. **58**(2), 211–248 (2003)
2. E. Fabes, B. Jones, N.M. Riviere, The initial value provem for the Navier-Stokes equations with data in L^p. Arch. Rational Mech. Anal. **45**, 222–248 (1972)
3. J. Fan, T. Ozawa, On the regularity criteria for the generalized Navier-Stokes equations and Lagrangian averaged Euler equations. Differ. Integral Equs. **21**(5–6), 443–457 (2008)
4. Y. Giga, Solutions for semilinear parabolic equations in L^p and regularity of weak solutions of the Navier-Stokes equations. J. Differ. Equs. **62**, 186–212 (1986)
5. H. Hajaiej, L. Molinet, T. Ozawa, B. Wang, Necessary and sufficient conditions for the fractional Gagliardo-Nirenberg inequalities and applications to Navier-Stokes and generalized Boson equations. RIMS Kokyuroku Bessatsu **26**, 159–175 (2011)
6. E. Hopf, Über die Anfangswertaufgabe für die hydrodynamischen Grundgleichungen. Math. Nachr. **4**, 213–231 (1951)
7. T. Kato, G. Ponce, Commutator estimates and the Euler and Navier-Stokes equations. Commun. Pure Appl. Math. **41**, 891–907 (1988)
8. H. Kozono, Y. Shimada, Bilinear estimates in homogeneous Triebel-Lizorkin spaces and the Navier-Stokes equations. Math. Nachr. **276**, 63–74 (2004)
9. J. Leray, Sur le mouvement d'un liquide visqueux emplissant l'espace. Acta Math. **63**, 193–248 (1934)
10. S. Machihara, T. Ozawa, Interpolation inequalities in Besov spaces. Proc. Am. Math. Soc. **131**, 1553–1556 (2002)
11. Y. Meyer, Oscillating patterns in some nonlinear evolution equations, in *Mathematical Foundation of Turbulent Viscous Flows*. Lecture Notes in Mathematics, vol. 1871, ed. by M. Cannone, T. Miyakawa (Springer, Berlin, 2006), pp. 101–187
12. T. Ohyama, Interior regularity of weak solutions to the Navier-Stokes equation. Proc. Jpn. Acad. **36**, 273–277 (1960)
13. J. Serrin, On the interior regularity of weak solutions of the Navier-Stokes equations. Arch. Rational Mech. Anal. **9**, 187–195 (1962)
14. H. Sohr, W. von Wahl, On the regularity of the pressure of weak solutions of Navier-Stokes equations. Arch. Math. **46**, 428–439 (1986)
15. E.M. Stein, *Singular Integrals and Differentiability Properties of Functions* (Princeton University Press, Princeton, 1970)
16. M. Struwe, On partial regularity results for the Navier-Stokes equations. Commun. Pure Appl. Math. **41**, 437–458 (1988)

On Periodic and Almost Periodic Solutions to Incompressible Viscous Fluid Flow Problems on the Whole Line

Matthias Hieber, Thieu Huy Nguyen and Anton Seyfert

Dedicated to Yoshikazu Giga on the occasion of his 60th Birthday.

Abstract It is shown that a large class of semilinear evolution equations on the whole line with periodic or almost periodic forces admit periodic or almost periodic mild solutions. The approach presented generalizes the method described in [28] to the case of the whole line and to forces which are almost periodic in the sense of H. Bohr. It relies on interpolation methods and on $L^p - L^q$-smoothing properties of the underlying linearized equation. Applied to incompressible fluid flow problems, the approach yields new results on (almost) periodic solutions to the Navier-Stokes-Oseen equations, to the flow past rotating obstacles, to the Navier-Stokes equations in the rotational setting as well as to Ornstein–Uhlenbeck type equations.

Anton Seyfert is supported by the DFG International Research Training Group IRTG 1529 on Mathematical Fluid Dynamics at TU Darmstadt.

M. Hieber (✉)
Fachbereich Mathematik, TU Darmstadt, Schlossgartenstr. 7, 64289 Darmstadt, Germany
e-mail: hieber@mathematik.tu-darmstadt.de

M. Hieber
607 Benedum Engineering Hall, University of Pittsburgh,
Pittsburgh, PA 15261, USA

T.H. Nguyen
School of Applied Mathematics and Informatics, Hanoi University of Science and Technology Vien Toan ung dung va Tin hoc, Dai hoc Bach khoa Hanoi,
1 Dai Co Viet, Hanoi, Vietnam
e-mail: huy.nguyenthieu@hust.edu.vn

A. Seyfert
Fachbereich Mathematik, TU Darmstadt, Schlossgartenstr. 7,
64289 Darmstadt, Germany
e-mail: seyfert@mathematik.tu-darmstadt.de

Y. Maekawa and S. Jimbo (eds.), *Mathematics for Nonlinear Phenomena—Analysis and Computation*, Springer Proceedings in Mathematics & Statistics 215,
DOI 10.1007/978-3-319-66764-5_4

Keywords Periodoc and almost periodic solutions
Semilinear evolution equations · Incompressible fluid flow
Navier-Stokes-Oseen equations · Flow past rotating obstacles
Navier-Stokes-Coriolis equations

2010 Mathematics Subject Classification 35Q30 · 35B10 · 76D07

1 Introduction

There is a long history of periodic or almost period solutions to abstract evolution equations as well as to ordinary and partial differential equations. While the question for periodic solutions is rather well studied and understood, the corresponding situation for almost periodic solutions is much less clear. Several approaches to almost periodic functions have been developed by Bochner, Stepanov, Besicovitch and Weyl after the fundamental definition given by H. Bohr [11] in 1925. These classes of functions play an important role in many areas of mathematics among them differential equations, dynamical systems and harmonic analysis. For general information about the theory of almost periodic functions we refer to [4, 13, 52], Chapter II. A central result from the abstract point of view is the following: a bounded and uniformly continuous function of $\mathbb{R}$ into a Banach space X with countable spectrum is almost periodic in the sense of Bohr whenever $c_0 \not\subset X$. Here c_0 is the space of all sequences converging to 0. The latter condition on X is related to Kadet's theorem on primitives of almost periodic functions, see e.g. [4].

When considering models arising from the theory of incompressible viscous fluid flow, first results on the existence of *periodic* solutions for the Navier-Stokes equations in *bounded* domains go back to Serrin [62]. His method for proving periodic solutions is based on their stability properties. Further results in this direction are due to Kaniel and Shinbrot [44] and Miyakawa and Teramoto [58]. Maremonti [55] obtained for the first time results on periodic solutions for the Navier-Stokes equations on the whole space $\mathbb{R}^d$. Kozono and Nakao introduced in [48] the notion of a mild solution to periodic problems and proved the existence of periodic solutions for the equations of Navier Stokes in $\mathbb{R}^d$.

Combining the technique of 'invading domains' [40] with methods due to Prodi [60], Prouse [61] and Yudovich [72], Maremonti and Padula [56] obtained results on periodic solutions for the Navier-Stokes equations on *fixed exterior domains*. In a different approach, Galdi and Sohr [24] extended Serrin's method, hereby relying on the decay of the solutions at spatial infinity.

A further approach to periodic solutions to the Navier-Stokes equations is due to Yamazaki [70]. He exploited interpolation properties of weak-L^d spaces and the iteration scheme method [31, 45] to prove the existence and uniqueness of periodic solutions on exterior domains.

For results concerning the Navier-Stokes flow past a moving or rotating obstacle, we refer to the work of Galdi and Silvestre [22], who assumed a prescribed motion of the body given by a time-periodic translation velocity and time-periodic angular velocity. The existence of strong, periodic solutions was also shown by them under a smallness condition. For further recent results concerning periodic solutions for the Navier-Stokes equations in the whole space or past moving cylinders, we refer to the work of Galdi [20, 21], Kyed [51], Maekawa and Sauer [53] and Nguyen [59].

For results on time-periodic solutions to degenerate quasilinear parabolic equations of the form

$$u_t = u^\gamma(\Delta u + u + f),$$

we refer to the work of Giga and Mizoguchi [36], where the existence of unique positive periodic solutions for positive right hand sides f was proved under the assumption that the first Dirichlet eigenvalue of $-\Delta$ satisfies certain conditions. Furthermore, results on periodic solutions to quasilinear parabolic equations of the form

$$u_t = u^2(u_{xx} + u - f),$$

describing curvatures u of evolving curves, were obtained by Giga and Mizoguchi in [37].

A *general approach to periodic solutions* to problems arising in incompressible fluid flows was recently developed by Geissert, Hieber and Nguyen in [28]. It is based on interpolation methods and $L^p - L^q$ smoothing properties of the semigroup associated with the linearized problem. In order to describe this approach more specifically, consider first the linear equation

$$u'(t) + Au(t) = BF(t), \quad t \geq 0, \quad u(0) = u_0, \tag{1}$$

where $-A$ is the generator of a C_0-semigroup $(e^{-tA})_{t\geq 0}$ on some Banach space Y and B plays the role of a 'connection operator' between various spaces involved. In the problems described above related to incompressible fluid flow, a typical choice for A and the so called 'connection operator' B would be the Stokes operator and $B = \mathbb{P}\mathrm{div}$. Here $\mathbb{P}$ denotes the Helmholtz projection. Note, however, that in other cases $B = Id$ is allowed, too. The approach allows to treat the important situation where $0 \in \sigma(A)$. Of central importance are smoothing properties of the semigroup e^{-tA} or its gradient, which are valid e.g. for the Stokes semigroup on exterior domains.

It was shown in [28], that equation (1) admits a bounded mild solution u in a suitable interpolation space Y for given bounded F and the norm of u can be controlled by that of F provided one has the $L^p - L^q$-smoothing properties for the linear semigroups involved. This leads to the fact that the solution operator $S_T : L^\infty(\mathbb{R}_+; X) \to BC(\mathbb{R}_+; (Y_1, Y_2)_{\theta,\infty})$, mapping a given T-periodic right-hand side f to a T-periodic mild solution $\hat{u}$ of (1) is a bounded, linear operator. The operator S_T was defined as

$$S_T f(t) := \mathrm{e}^{-At}x + \int_0^t \mathrm{e}^{-A(t-s)} Bf(s)\,\mathrm{d}s, \quad t > 0,$$

where

$$x := a + \int_0^T e^{-(T-s)A} Bf(s)ds \text{ and } a := w^* - \lim_{k\to\infty} \int_0^{(k-1)T} e^{-(kT-s)A} Bf(s)ds. \tag{2}$$

However, the solution may not be unique. The above T-periodic solution is unique if and only if $\frac{2n\pi i}{T} \notin \sigma_p(-A)$ for all $n \in \mathbb{Z}$. The latter condition may be verified in many situations, e.g., for the Stokes operator A on $L^q_\sigma(\Omega)$, where $\Omega \subset \mathbb{R}^n$ denotes a bounded or exterior domain with smooth boundary. The choice of these interpolation spaces is inspired by the work of Yamazaki [70] on the Navier-Stokes equations with time dependent external forces. Using a fixed point argument, the linear result was extended to the semilinear situation given by

$$u'(t) + Au(t) = G(u)(t), \quad t \geq 0, \tag{3}$$

where the Nemytskii operator G is a locally Lipschitz continuous map acting on some function space of bounded functions such as $BC(\mathbb{R}_+, X)$ and maps periodic functions into periodic functions.

In this paper we extend the approach described above to semilinear evolution equations on the *whole line* and to the setting of *almost periodic functions*. We show the existence of a unique, almost periodic solution to a large class of semilinear evolution equations of the form

$$u'(t) + Au(t) = BG(u)(t), \quad t \in \mathbb{R}, \tag{4}$$

in small balls around given almost periodic functions $u_{ap} \in BC(\mathbb{R}; Y)$. Here we understand almost periodic functions as uniformly almost periodic functions in the sense of H. Bohr, [11]. It is well known that the definitions of uniformly almost periodic functions given in terms of a relative density of the set of almost periods, i.e. the Bohr criterion, is equivalent to the definitions in terms of compactness of the set of translates (Bochner's criterion) or as the closure of the set of trigonometric polynomials in sup-norm, are equivalent, see e.g. [13]. This does not hold true for Stepanov's class of almost periodic functions and also not for the classes introduced by Besicovitch and Weyl. These classes of almost periodic functions are described in more detail in the following section. For results on Leray's problem for almost periodic flows in the sense of Stepanov and Besicovitch we refer to the work of Berselli and Romito [7]. Results on almost periodic solutions to parabolic systems can be found also in [18].

In order to describe our approach, we consider first the linear problem of the form

$$u'(t) + Au(t) = Bf(t), \quad t \in \mathbb{R}, \tag{5}$$

where $-A$ is again the generator of a C_0-semigroup e^{-tA} on a suitable interpolation space Y, $f : \mathbb{R} \to X$ an almost periodic function and where B is the 'connection' operator between X and Y. We then consider the solution operator $\mathbf{S} : BC(\mathbb{R}; X) \to BC(\mathbb{R}; Y)$ mapping a given almost periodic right-hand side f to an almost periodic mild solution $\hat{u}$ of (5) defined by

$$(\mathbf{S}f)(t) := \int_{-\infty}^{t} e^{-(t-\tau)A} Bf(\tau)d\tau, \quad t \in \mathbb{R},$$

and show that $\mathbf{S}$ is a bounded linear operator from $BC(\mathbb{R}; X)$ to $BC(\mathbb{R}; Y)$. The linear result is then extended to the semilinear situation described by (4), where the Nemytskii operator G is a locally Lipschitz continuous map acting on $BC(\mathbb{R}; X)$ and maps almost periodic functions to almost periodic functions.

Our approach yields new results on the existence of almost periodic solutions for various types of fluid flow problems, e.g. for the Navier-Stokes-Oseen flow in exterior domains, the flow past rotating obstacles, the Navier-Stokes equations in the rotational setting and also for Ornstein–Uhlenbeck or parabolic equations with rough coefficients. We hereby make use of the fact that the linearized problem is given by a C_0-semigroup, which, in the fluid flow problems we are considering in this article, is the Stokes, the Stokes–Oseen and the Stokes–Coriolis semigroup in the L^p-setting. Fundamental contributions regarding these semigroups are due to Yoshikazu Giga; see e.g. [29–33, 38] and also [1, 2].

Besides time-periodic solutions to fluid flow problems it is also very interesting to consider the corresponding initial value problem for *spatially* almost periodic initial data. For the case of the Navier-Stokes equations we refer here to the work of Giga, Mahalov, Nicolaenko and Yoneda; see e.g. [34, 35].

The general question of *uniqueness* of almost periodic solutions for semilinear evolution equations is a rather delicate one. Concerning the equations of Navier-Stokes, Yamazaki [70] proved the uniqueness of an almost periodic solution in exterior domains for small almost periodic L^3_w-solutions within the class of solutions having sufficiently small $L^\infty(L^3_w)$-norm. Different uniqueness criteria in this context were developed by Farwig, Nakatsuka and Taniuchi in [17] and [16] as well as by Lions and Masmoudi in [54].

We prove in this article that the periodic and almost periodic solutions to the fluid flow problems as well as to the semilinear diffusion equations described in Section 5 and 6 are uniquely determined within the class of small $L^\infty(L^3_w)$ and $L^\infty(L^{d(r-1)/2})$ solutions, respectively.

This paper is organized as follows. In Section 2 we collect basic facts on almost periodic functions and develop then in Section 3 the abstract framework for our approach to obtain periodic and almost periodic solutions to linear evolution equations on the line in certain interpolation spaces. Section 4 deals with the situation of general semilinear problems. Our main abstract results are contained in Theorems 3.4 and 4.2. In Section 5 we apply our method to various equations arising in incompressible fluid flows. In Section 6 we show that our approach also yields the existence and uniqueness results for periodic and almost periodic solutions to semilinear Ornstein–Uhlenbeck and diffusion equations.

2 Preliminaries on Almost Periodic Functions

In the following we are interested in almost periodic solutions to equation (4) and define the class of almost periodic functions following Bohr's approach, see [11]. To this end, given a Banach space X, we denote by $BC(\mathbb{R}; X)$ the space of all bounded continuous functions from $\mathbb{R}$ to X and by $BUC(\mathbb{R}; X)$ the space of all uniformly continuous functions from $\mathbb{R}$ to X, both equipped with the sup-norm $\|f\|_\infty = \sup_{t\in\mathbb{R}} \|f(t)\|_X$.

We then call a function $f \in C(\mathbb{R}; X)$ *almost periodic* if for each $\varepsilon > 0$ there exists $L_\epsilon > 0$ such that for every $a \in \mathbb{R}$ there exists $T \in [a, a + L_\epsilon)$ such that

$$\sup_{t\in\mathbb{R}} \|f(t+T) - f(t)\|_X < \varepsilon.$$

We denote the class of all almost periodic functions $f : \mathbb{R} \to X$ by $AP(\mathbb{R}; X)$.

Furthermore, a subset $D \subset \mathbb{R}$ is called relatively dense in $\mathbb{R}$ if there is an $L > 0$ such that for each $a \in \mathbb{R}$ we have $D \cap [a, a + L) \neq \emptyset$. Setting $E(f, \varepsilon) := \{\tau \in \mathbb{R} : \|f(\cdot + \tau) - f(\cdot)\|_\infty < \varepsilon\}$, we may reformulate the above definition as follows: $f \in BC(\mathbb{R}; X)$ is almost periodic if for each $\varepsilon > 0$ the set $E(f, \varepsilon)$ is relatively dense in $\mathbb{R}$.

In the following, we collect well-known properties of almost periodic functions. For further information and proofs of the following assertions we refer to the books by Corduneanu [13] and Levitan and Zhikov [52].

Given $T > 0$ let $P_T(\mathbb{R}; X)$ denote the space of all T-periodic continuous functions from $\mathbb{R} \to X$ and let $TP(\mathbb{R}; X)$ be the space of all trigonometric polynomials from $\mathbb{R} \to X$. Here, linear combinations of functions of the form $e_{i\eta} \otimes x$ with $\eta \in \mathbb{R}$ and $x \in X$ are called trigonometric polynomials and

$$(e_{i\eta} \otimes x)(t) := e^{i\eta t}x, \quad t \in \mathbb{R}.$$

Proposition 2.1 *Let X be a Banach space.*

(a) For each $T > 0$, $P_T(\mathbb{R}; X) \subset AP(\mathbb{R}; X)$ and $TP(\mathbb{R}; X) \subset AP(\mathbb{R}; X)$.
(b) If $f \in AP(\mathbb{R}; X)$, then the set $\{f(t) : t \in \mathbb{R}\}$ is relatively compact in X.
(c) $AP(\mathbb{R}; X) \subset BUC(\mathbb{R}; X)$.
(d) Equipped with $\|f\|_{AP} = \|f\|_{L^\infty(X)}$, the space $AP(\mathbb{R}; X)$ is a Banach space. It is even a commutative Banach algebra under pointwise multiplication if X is a commutative Banach algebra.
(e) If $f \in AP(\mathbb{R}; X)$, $g \in AP(\mathbb{R}; Y)$ and $\cdot : X \times Y \to Z$ is continuous, then $g \cdot f \in AP(\mathbb{R}; Z)$.

(f) (Bochner's criterion [10]): Let $f \in BC(\mathbb{R}; X)$. Then $f \in AP(\mathbb{R}; X)$ if and only if the set

$$\tau(f) = \{\tau_r(f) = f(\cdot + r), r \in \mathbb{R}\}$$

is relatively compact in $(BC(\mathbb{R}; X), \|\cdot\|_\infty)$.
(g) (Maak's criterion): Let $f \in C(\mathbb{R}; X)$. Then $f \in AP(\mathbb{R}; X)$ if and only if given $\varepsilon > 0$ there exists a partition $\mathbb{R} = \cup_{i=1}^M E_i$ such that for all $\xi, \eta \in E_i$, $1 \le i \le M$

$$\sup_{t \in \mathbb{R}} \| f(t + \xi) - f(t + \eta)\| < \varepsilon.$$

(h) $AP(\mathbb{R}; X)$ coincides with the closure of $TP(\mathbb{R}; X)$ with respect to the sup-norm, i.e.

$$AP(\mathbb{R}; X) = \overline{span}\{e_{i\eta} \otimes x : \eta \in \mathbb{R}, x \in X\}.$$

(i) Let $(S(t))_{t \ge 0}$ be the shift semigroup on $BUC(\mathbb{R}; X)$ and $f \in BUC(\mathbb{R}; X)$. Then $f \in AP(\mathbb{R}; X)$ if and only if the orbit $\{S(t)f : t \in \mathbb{R}\}$ is relatively compact in $BUC(\mathbb{R}; X)$.
(j) Let $f \in BUC(\mathbb{R}; X)$ have countable Carleman spectrum. Then $f \in AP(\mathbb{R}; X)$ provided $c_0 \not\subset X$.
k) Let $f \in BUC(\mathbb{R}; X)$ and let $u \in BUC(\mathbb{R}; X)$ be a mild solution of

$$u'(t) + Au(t) = f(t), \quad t \in \mathbb{R},$$

where A is a closed, linear operator on a Banach space X such that $\sigma(A) \cap i\mathbb{R}$ is countable. Then $u \in AP(\mathbb{R}; X)$ provided $c_0 \not\subset X$ or $u(\mathbb{R})$ is relatively weakly compact.
(l) Let $f \in AP(\mathbb{R}; X)$ and $T > 0$. Then there exists a unique T-periodic function $f^T \in C_T(\mathbb{R}; X)$ such that

$$\lim_{k\to\infty} \|\frac{1}{k}\sum_{j=1}^{k-1} f(\cdot + jT) - f^T(\cdot)\|_\infty = 0.$$

(m) Let $f \in AP(\mathbb{R}; X)$. Then the mean value of f exists in X, i.e.

$$M(f) = \lim_{T\to\infty} \frac{1}{2T}\int_{-T}^{T} f(t)dt \in X.$$

(n) If $f \in AP(\mathbb{R}; X)$ and $\gamma > 0$, then $|f|^\gamma \in AP(\mathbb{R}; X)$.

More general definitions of almost periodicity were introduced by Stepanov, Weyl and Besicovitch for real-valued functions. In fact, for $p \in (1, \infty)$ the norm

$$\|f\|_{S^{p,r}} := \sup_{t\in\mathbb{R}} \left(\frac{1}{r}\int_t^{t+r} |f(s)|^p ds\right)^{1/p}$$

was introduced by Stepanov [64] in 1925. Then the space $\mathcal{S}^p(\mathbb{R})$ of almost periodic functions *in the sense of Stepanov* is defined as the closure of $TP(\mathbb{R};\mathbb{R})$ with respect to the norm $\|\cdot\|_{\mathcal{S}^{p,r}}$. Note that although the definition of the above norm depends on r, the topology induced is independent of the value of r. *Weyl* [69] extended this definition by considering the closure of $TP(\mathbb{R};\mathbb{R})$ with respect to the seminorm

$$\|f\|_{\mathcal{W}^p} := \lim_{r\to\infty} \|f\|_{\mathcal{S}^{p,r}},$$

and we denote this closure by $\mathcal{W}^p(\mathbb{R};\mathbb{R})$. Finally, *Besicovitch* [8, 9] introduced the space $\mathcal{B}^p(\mathbb{R})$ as the closure of $TP(\mathbb{R};\mathbb{R})$ with respect to the seminorm

$$\|f\|_{\mathcal{B}^p} := \lim \sup_{R\to\infty} \left(\frac{1}{2R}\int_R^R |f(s)|^p ds\right)^{1/p}.$$

Observe that their examples of functions $f \in L^\infty(\mathbb{R};\mathbb{R})$ with $\|f\|_{\mathcal{B}^p} = 0$ even though $f \not\equiv 0$. For $1 < p < \infty$ we then have

$$AP(\mathbb{R};\mathbb{R}) \subset \mathcal{S}^p(\mathbb{R};\mathbb{R}) \subset \mathcal{W}^p(\mathbb{R};\mathbb{R}) \subset \mathcal{B}^p(\mathbb{R};\mathbb{R}),$$

with strict inclusions.

3 Almost Periodic Solutions to Linear Evolution Equations

Let X be a Banach space, (Y_1, Y_2) be an interpolation couple of Banach spaces Y_1 and Y_2 denoted by $(Y_1, Y_2)_{\theta,\infty}$ the real interpolation space between Y_1 and Y_2 for some $0 < \theta < 1$. Consider the inhomogeneous linear Cauchy problem of the form

$$u'(t) + Au(t) = Bf(t), \quad t \in \mathbb{R}, \tag{6}$$

where $-A$ is a generator of a C_0-semigroup e^{-tA} on Y_1 and Y_2, $f(t) \in X$ for all $t \in \mathbb{R}$ and where B is the 'connection' operator between X and $Y := (Y_1, Y_2)_{\theta,\infty}$ such that $e^{-tA}B \in \mathcal{L}(X, Y_i)$ for $i = 1, 2$ and $t \geq 0$. We call a function $u \in C(\mathbb{R}; (Y_1, Y_2)_{\theta,\infty})$ a *mild solution* to (6) if u satisfies the integral equation

$$u(t) = \int_{-\infty}^t e^{-(t-\tau)A} Bf(\tau)d\tau, \quad t \in \mathbb{R}. \tag{7}$$

The following assumption is central to our approach.

Assumption (A).

Let (Y_1, Y_2) be an interpolation couple. Assume that Y_i has a Banach pre-dual Z_i for $i = 1, 2$ and such that $Z_1 \cap Z_2$ is dense in Z_i for $i = 1, 2$. Let $-A$ be the generator of a bounded C_0-semigroup $(e^{-tA})_{t\geq 0}$ on Y_1 and Y_2. Furthermore, suppose that there exists constants $\alpha_1, \alpha_2 \in \mathbb{R}$ with $0 < \alpha_2 < 1 < \alpha_1$ and $L > 0$ such that

$$\begin{aligned} \|e^{-tA}Bv\|_{Y_1} &\leq Lt^{-\alpha_1}\|v\|_X, \quad t > 0, \\ \|e^{-tA}Bv\|_{Y_2} &\leq Lt^{-\alpha_2}\|v\|_X, \quad t > 0, \end{aligned} \tag{8}$$

where B is given as above. We then have the following result which is a variant of Lemma 2.1 in [28].

Lemma 3.1 *Assume (A). Let $\theta \in (0, 1)$ such that $1 = (1-\theta)\alpha_1 + \theta\alpha_2$, let $f \in L^\infty(\mathbb{R}; X)$. Then equation (6) admits a mild solution u satisfying*

$$\|u(t)\|_Y \leq \tilde{M}\|f\|_{L^\infty(\mathbb{R};X)}, \quad t \in \mathbb{R}. \tag{9}$$

for some constant $\tilde{M} \geq 1$.

Proof We will show by a duality argument that the integral (7) is well defined in $Y = (Y_1, Y_2)_{\theta,\infty}$. By the duality theorem of real interpolation spaces we have that $(Z_1, Z_2)'_{\theta,1} = Y$. Hence, for a given $\psi \in (Z_1, Z_2)_{\theta,1}$, we may estimate

$$\begin{aligned} |\langle u(t), \psi\rangle| &= |\int_{-\infty}^t \langle e^{-(t-s)A}Bf(s), \psi\rangle ds| = |\int_{-\infty}^t \langle f(s), B'(e^{(t-s)A})'\psi\rangle ds| \\ &\leq \int_{-\infty}^t \|f(s)\|_X\|B'(e^{(t-s)A})'\psi\|_{X'}ds \leq \|f\|_{L^\infty(\mathbb{R};X)}\int_{-\infty}^t \|B'(e^{(t-s)A})'\psi\|_{X'}ds \\ &= \|f\|_{L^\infty(\mathbb{R};X)}\int_0^\infty \|B'(e^{rA})'\psi\|_{X'}dr. \end{aligned} \tag{10}$$

In the following, we show that the above right-hand side is finite. In fact, due to (8), we have

$$\|B'(e^{rA})'\phi\|_{X'} \leq Cr^{-\alpha_1}\|\phi\|_{Y_1} \quad \text{and} \quad \|B'(e^{rA})'\phi\|_{X'} \leq Cr^{-\alpha_2}\|\phi\|_{Y_2}, \quad r > 0.$$

Hence, the sublinear operator

$$\begin{aligned} T\colon Z_1 + Z_2 &\to L^{1/\alpha_1,\infty}(\mathbb{R}_+; \mathbb{R}) + L^{1/\alpha_2,\infty}(\mathbb{R}_+; \mathbb{R}), \\ \phi_1 + \phi_2 &\mapsto \|B'(e^{\cdot A})'\phi_1\|_{X'} + \|B'(e^{\cdot A})'\phi_2\|_{X'} \end{aligned}$$

is well defined with some bounds

$$\|T\phi_1\|_{L^{1/\alpha_1,\infty}} \leq C\|\phi_1\|_{Z_1} \quad \text{and} \quad \|T\phi_2\|_{L^{1/\alpha_2,\infty}} \leq C\|\phi_2\|_{Z_2}. \tag{11}$$

By the interpolation theorem of Marcinkiewicz, T maps $(Z_1, Z_2)_{\theta,1}$ into

$$(L^{1/\alpha_1,\infty}(0,\infty), L^{1/\alpha_2,\infty}(0,\infty))_{\theta,1} = L^1(0,\infty),$$

and there exists a constant $\tilde{M}$ such that

$$\|T\psi\|_{L^1} \le \tilde{M}\|\psi\|_Z. \tag{12}$$

Plugging this inequality into (10) yields the assertion. □

The above Lemma 3.1 implies that we may define the solution operator $\mathbf{S} : L^\infty(\mathbb{R}; X) \to BC(\mathbb{R}; Y)$ by

$$\mathbf{S}(f)(t) := \int_{-\infty}^{t} e^{-(t-\tau)A} Bf(\tau)d\tau, \quad t \in \mathbb{R}. \tag{13}$$

Applying the inequality (9), we immediately obtain the following corollary.

Corollary 3.2 *The solution operator* $\mathbf{S} : L^\infty(\mathbb{R}; X) \to BC(\mathbb{R}; (Y_1, Y_2)_{\theta,\infty})$ *defined as in* (13) *is a bounded linear operator satisfying* $\|\mathbf{S}\| \le \tilde{M}$ *where* $\tilde{M}$ *is defined as in* (9).

Let us note that in many applications, e.g. incompressible fluid flow on exterior domains in $\mathbb{R}^d$, one may assume that X has a Banach pre-dual V; i.e. $X = V'$ for some space V. An important example is the space $X = L^{\frac{d}{2},\infty}(\mathbb{R}^d) = (L^{\frac{d}{d-2},1}(\mathbb{R}^d))'$. Therefore, in this situation, the assumption (8) may be replaced by estimates for the dual operators restricted on V considered as a subspace of X' through the canonical embedding. The importance of this observation lies in the fact that these dual estimates may be verified quite easily by using $L^q - L^r$ smoothing estimates for the underlying semigroup. More precisely, we have the following corollary.

Corollary 3.3 *Assume that X has a Banach pre-dual V, considered as a subspace of X' through the canonical embedding. Then the assertion of Lemma 3.1 remains true provided assumptions* (8) *are replaced by*

$$\begin{aligned} \|B'e^{-tA'}\varphi\|_V \le & Lt^{-\alpha_1}\|\varphi\|_{Z_1}, \quad t > 0 \\ \|B'e^{-tA'}\varphi\|_V \le & Lt^{-\alpha_2}\|\varphi\|_{Z_2}, \quad t > 0. \end{aligned} \tag{14}$$

Our first result on the unique existence of periodic or almost periodic solutions to the linear equation (6) reads as follows.

Theorem 3.4 *Assume that Assumption (A) holds and let* $\theta \in (0, 1)$ *be given by* $1 = (1-\theta)\alpha_1 + \theta\alpha_2$. *Assume that* $f \in BC(\mathbb{R}; X)$ *and let* $Y = (Y_1, Y_2)_{\theta,\infty}$.
(a) Then equation (6) *admits a unique mild solution* $\hat{u} \in BC(\mathbb{R}; Y)$ *satisfying*

$$\|\hat{u}\|_{BC(\mathbb{R};Y)} \le \tilde{M}\|f\|_{L^\infty(\mathbb{R};X)}, \tag{15}$$

where $\tilde{M}$ *is given as in Lemma 3.1.*
(b) If $f \in P_T(\mathbb{R}; X)$ *for some* $T > 0$, *then* $\hat{u} \in P_T(\mathbb{R}; Y)$.
(c) If $f \in AP(\mathbb{R}; X)$, *then* $\hat{u} \in AP(\mathbb{R}; Y)$.

Proof The assertion (a) follows from Lemma 3.1. In order to proof (c), let $\varepsilon > 0$ and set $\delta := \varepsilon/\tilde{M}$. As f is almost periodic, there is an $L_\delta > 0$, such that for each $a \in \mathbb{R}$ there is a $T \in [a, a + L_\delta]$ with $\|f(\cdot + T) - f(\cdot)\|_{L^\infty(\mathbb{R},X)} < \delta$. Since

$$(\mathbf{S}f)(t + T) = (Sf(\cdot + T))(t), \quad t \in \mathbb{R},$$

it follows from Lemma 3.1 that

$$\|\hat{u}(\cdot + T) - \hat{u}(\cdot)\|_{BC(\mathbb{R};Y)} = \|\mathbf{S}f(\cdot + T) - \mathbf{S}f\|_{BC(\mathbb{R};Y)} \leq \tilde{M}\|f(\cdot + T) - f(\cdot)\|_{L^\infty(\mathbb{R};X)} \leq \tilde{M}\delta \leq \varepsilon,$$

which means that $\hat{u}$ is an almost periodic solution to (6). Assertion (b) is proved similarly. □

Remark 3.5 Note that the proof of Theorem 3.4 shows that the solution operator $\mathbf{S} \in \mathcal{L}(L^\infty(\mathbb{R}; X), BC(\mathbb{R}; Y))$ defined as in (13) with $Y := (Y_1, Y_2)_{\theta,\infty}$ maps a given (almost)-periodic right-hand side f to an (almost)-periodic mild solution $\hat{u}$ of (6) satisfying (15).

4 Almost Periodic Solutions to Semilinear Evolution Equations

In this section we consider the semilinear evolution equation

$$u'(t) + Au(t) = BG(u)(t), \quad t \in \mathbb{R}, \tag{16}$$

where A and B satisfy Assumption (A) and G maps functions in $BC(\mathbb{R}; (Y_1, Y_2)_{\theta,\infty})$ into $BC(\mathbb{R}; X)$. Similarly to equation (6), by a *mild solution to* (16) we understand a function $u \in C(\mathbb{R}; Y)$ satisfying the equation

$$u(t) = \int_{-\infty}^{t} e^{-(t-\tau)A} BG(u)(\tau)d\tau. \tag{17}$$

For $u_0 \in BC(\mathbb{R}; Y)$, $u_p \in P_T(\mathbb{R}; Y)$, $u_{ap} \in AP(\mathbb{R}; Y)$ and $R > 0$ we set

$$\begin{aligned}
B^0_{R,Y}(u_0) &:= \{w \in BC(\mathbb{R}; Y) : \|w - u_0\|_{BC(\mathbb{R};Y)} \leq R\},\\
B^p_{R,Y}(u_p) &:= \{w \in P_T(\mathbb{R}; Y) : \|w - u_p\|_{BC(\mathbb{R};Y)} \leq R\},\\
B^{ap}_{R,Y}(u_{ap}) &:= \{w \in AP(\mathbb{R}; Y) : \|w - u_{ap}\|_{BC(\mathbb{R};Y)} \leq R\},
\end{aligned}$$

and introduce the following assumptions on the function G.

Assumption (G).
Assume that $G : BC(\mathbb{R}; Y) \to BC(\mathbb{R}; X)$ and that there exists $u_0 \in BC(\mathbb{R}; Y)$ such that

$$\|G(u) - G(v)\|_{L^\infty(\mathbb{R};X)} \leq L\|u - v\|_{BC(\mathbb{R};Y)}, \quad u, v \in B^0_{R,Y}(u_0),$$

for $L, R > 0$ satisfying

$$\tilde{M}L < 1 \text{ and } R(1 - L\tilde{M}) > \tilde{M}\|G(u_0)\|_{L^\infty(\mathbb{R},X)} + \|u_0\|_{BC(\mathbb{R};Y)}.$$

Assumption (G^p).
Assume that $G : BC(\mathbb{R}; Y) \to BC(\mathbb{R}; X)$ maps T-periodic functions to T-periodic functions and that there exists an $u_p \in P_T(\mathbb{R}; Y)$ such that

$$\|G(u) - G(v)\|_{L^\infty(\mathbb{R};X)} \leq L\|u - v\|_{BC(\mathbb{R};Y)}, \quad u, v \in B^p_{R,Y}(u_p),$$

for $L, R > 0$ satisfying

$$\tilde{M}L < 1 \text{ and } R(1 - L\tilde{M}) > \tilde{M}\|G(u_p)\|_{L^\infty(\mathbb{R},X)} + \|u_p\|_{BC(\mathbb{R};Y)}.$$

Assumption (G^{ap}).
Assume that $G : BC(\mathbb{R}; Y) \to BC(\mathbb{R}; X)$ maps almost periodic functions to almost periodic functions and that there exists an $u_{ap} \in AP(\mathbb{R}; Y)$ such that

$$\|G(u) - G(v)\|_{L^\infty(\mathbb{R};X)} \leq L\|u - v\|_{BC(\mathbb{R};Y)}, \quad u, v \in B^{ap}_{R,Y}(u_{ap}),$$

for $L, R > 0$ satisfying

$$\tilde{M}L < 1 \text{ and } R(1 - L\tilde{M}) > \tilde{M}\|G(u_{ap})\|_{L^\infty(\mathbb{R},X)} + \|u_{ap}\|_{BC(\mathbb{R};Y)}.$$

Remark 4.1 It follows from Proposition 2.1 (d) and (n) that the above assumption that G is mapping almost periodic functions to almost periodic functions is fulfilled in particular for G given by $G(u) = u^2$ and $G(u) = |u|^\gamma$ for $\gamma > 0$.

The following result gives a criteria for the existence of a unique periodic or almost periodic mild solution to equation (16) in $B^p_R(u_p)$ and $B^{ap}_R(u_{ap})$, respectively.

Theorem 4.2 *Assume Assumptions (A) and let $\theta \in (0, 1)$ be given by $1 = (1 - \theta)\alpha_1 + \theta\alpha_2$.*

(a) Let $u_0 \in BC(\mathbb{R}; Y)$ and assume (G). Then there exists a unique mild solution $\hat{u}$ to equation (16) in $B^0_{R,Y}(u_0)$.
(b) Let $u_p \in P_T(\mathbb{R}; Y)$ and assume (G^p). Then there exists a unique periodic mild solution $\hat{u}$ to equation (16) in $B^p_{R,Y}(u_p)$.
(c) Let $u_{ap} \in AP(\mathbb{R}; Y)$ and assume (G^{ap}). Then there exists a unique almost periodic mild solution $\hat{u}$ to equation (16) in $B^{ap}_{R,Y}(u_{ap})$.

Proof The proof of the above assertions is similar and we hence give a proof only of assertion (c). To this end, choose $L, R > 0$ according to assumption (G^{ap}). We define the mapping Φ by

$$\Phi : BC(\mathbb{R}; (Y_1, Y_2)_{\theta,\infty}) \to BC(\mathbb{R}; (Y_1, Y_2)_{\theta,\infty}), \quad v \mapsto \mathbf{S}(G(v)),$$

where $\mathbf{S}$ is the solution operator of

$$u'(t) + Au(t) = BG(v)(t), \quad t \in \mathbb{R}, \tag{18}$$

given in (13). In the following, we prove that Φ maps $B^{ap}_{R,Y}(u_{ap})$ into itself and that ϕ is a contraction on $B^{ap}_{R,Y}(u_{ap})$. To this end, note that by Theorem 3.4 and by assumption we obtain

$$\begin{aligned}\|\Phi(u) - u_{ap}\|_{BC(\mathbb{R};Y)} \le& \|\Phi(u) - \Phi(u_{ap})\|_{BC(\mathbb{R};Y)} + \|\Phi(u_{ap}) - u_{ap}\|_{BC(\mathbb{R};Y)} \\ \le& \tilde{M}L\|u - u_{ap}\|_{BC(\mathbb{R};Y)} + \tilde{M}\|G(u_{ap})\|_{L^\infty(\mathbb{R};X)} + \|u_{ap}\|_{BC(\mathbb{R};Y)} \\ \le& R, \quad u \in B^{ap}_{R,Y}(u_{ap}).\end{aligned}$$

Moreover, it follows from Theorem 3.4 and from assumption (G) that

$$\|\Phi(u) - \Phi(v)\|_{BC(\mathbb{R};Y)} \le \tilde{M}L\|u - v\|_{BC(\mathbb{R};Y)}, \quad u, v \in B^{ap}_{R,Y}(u_{ap}).$$

The contraction principle yields the existence of a unique fixed point $\hat{u} \in B^{ap}_R(u_{ap})$ of Φ. Clearly, $\hat{u}$ is the unique almost periodic mild solution of (16) in $B^{ap}_{R,Y}(u_{ap})$.

In order to prove the uniqueness assertions, let $u_{ap} \in AP(\mathbb{R}; Y)$ and assume that $u, v \in B^{ap}_R(u_{ap})$ are two almost periodic mild solutions of (16). Then

$$\begin{aligned}\|u - v\|_{BC(\mathbb{R};Y)} =& \|\mathbf{S}(G(u) - G(v))\|_{BC(\mathbb{R};Y)} \le \tilde{M}\|G(u) - G(v)\|_{BC(\mathbb{R};Y)} \\ \le& \tilde{M}L\|u - v\|_{BC(\mathbb{R};Y)}, \quad u, v \in B^{ap}_{R,Y}(u_{ap}).\end{aligned}$$

Since by assumption $\tilde{M}L < 1$, we proved the uniqueness assertion for the case of almost periodic functions. The other cases are proved similarly. □

5 Almost Periodic Solutions to Incompressible Fluid Flow Problems on Exterior Domains

In this section, we apply the approach described above to periodic and almost periodic solutions to various models arising in incompressible fluid flow, such as the flows induced by the equations of Navier-Stokes, the Navier-Stokes-Oseen equations on fixed exterior domains and past rotating obstacles,

Throughout this section, $\Omega \subset \mathbb{R}^d$, $d \ge 3$ denotes an exterior domain with smooth boundary. We start by considering the spaces

$$C^\infty_{c,\sigma}(\Omega) := \{v \in C^\infty_c(\Omega) : \operatorname{div} v = 0 \text{ in } \Omega\} \text{ and } L^p_\sigma(\Omega) := \overline{C^\infty_{c,\sigma}(\Omega)}^{\|\cdot\|_{L^p}}. \tag{19}$$

Further, for $1 < r < \infty$ and $1 \le q \le \infty$, we denote by $L^{r,q}(\Omega)$ the Lorentz spaces. It is well known (see e.g. [6, 67]) that $L^{r,r}(\Omega) = L^r(\Omega)$. For $q = \infty$ the space $L^{r,\infty}(\Omega)$ is also called the weak-L^r space and denoted by $L^r_w(\Omega) := L^{r,\infty}(\Omega)$.

For $1 < r < \infty$ let $\mathbb{P} = \mathbb{P}_r$ be the Helmholtz projection on $L^r(\Omega)$, i.e., the projection onto $L^r_\sigma(\Omega)$ corresponding to the Helmholtz decomposition of L^r-vector fields (see [19]) as

$$L^r(\Omega) = L^r_\sigma(\Omega) \oplus \{\nabla p \in L^r : p \in L^r_{loc}(\Omega)\}.$$

By interpolation, $\mathbb{P} = \mathbb{P}_{r,q}$ defines a bounded projection on $L^{r,q}(\Omega)$ whenever $1 < r < \infty$ and $1 \le q \le \infty$. We also make use of solenoidal Lorentz spaces $L^{r,q}_\sigma(\Omega)$, which are defined by

$$L^{r,q}_\sigma(\Omega) := Rg(\mathbb{P}_{r,q}).$$

We then have (see e.g. [12, Theorem 5.2])

$$L^{r,q}(\Omega) = L^{r,q}_\sigma(\Omega) \oplus G^{r,q}(\Omega),$$

where $G^{r,q}(\Omega) := \{\nabla p \in L^{r,q}(\Omega) : p \in L^{r,q}_{loc}(\Omega)\}$. Furthermore, by interpolation

$$L^{r,q}_\sigma(\Omega) = (L^{r_1}_\sigma(\Omega), L^{r_2}_\sigma(\Omega))_{\theta,q} \tag{20}$$

provided $1 < r_0 < r < r_1 < \infty$, $1 \le q \le \infty$ and $\frac{1}{r} = \frac{1-\theta}{r_0} + \frac{\theta}{r_1}$. In addition, if $1 \le q < \infty$, then

$$(L^{r,q}_\sigma(\Omega))' = L^{r',q'}_\sigma(\Omega), \quad r' = \frac{r}{r-1}, \; q' = \frac{q}{q-1} \text{ and } q' = \infty \text{ if } q = 1. \tag{21}$$

Finally, we denote by $\|\cdot\|_{s,w}$ the norm of $L^s_{\sigma,w}(\Omega) := L^{s,\infty}_\sigma(\Omega)$ and by $\|\cdot\|_{\infty,s,w}$ the norm in $BC(\mathbb{R}_+, L^s_{\sigma,w})$, i.e., $\|v\|_{\infty,s,w} := \sup_{t\in\mathbb{R}_+} \|v(t)\|_{s,w}$ for $v \in BC(\mathbb{R}_+, L^s_{\sigma,w})$.

5.1 Almost Periodic Solutions to the Oseen and Navier-Stokes-Oseen Equations

Given $u_\infty \in \mathbb{R}^3$, consider the Navier-Stokes-Oseen equations in an exterior domain $\Omega \subset \mathbb{R}^3$ with smooth boundary

$$\begin{cases} u_t + (u\cdot\nabla)u - \Delta u + \nabla p = \operatorname{div} F & \text{in } \Omega\times\mathbb{R}, \\ \nabla\cdot u = 0 & \text{in } \Omega\times\mathbb{R}, \\ u = 0 & \text{on } \partial\Omega\times\mathbb{R}, \\ \lim_{|x|\to\infty} u(t,x) = u_\infty. \end{cases} \tag{22}$$

We are now rewriting equation (22) by means of the *Oseen operator*. In order to do so, let $\varphi \in C_c^\infty(\mathbb{R}^3)$ such that $\varphi \geq 0$, $\varphi \equiv 1$ on a neighborhood of Ω^c and $\operatorname{supp} \varphi \Subset B(0, r)$ for some $r > 0$. We define

$$b_\infty := \operatorname{Bog}_D((\nabla \varphi) u_\infty),$$

where Bog_D denotes Bogovskiĭ's operator on $D = \{x \in \Omega : |x| < r\}$; see e.g. [19] or [25]. Note that $b_\infty \in C_c^\infty(\overline{\Omega})$ satisfies $\operatorname{div} b_\infty = 0$ and $b_\infty = u_\infty$ on $\partial\Omega$. Moreover, for any $\alpha \in \mathbb{N}_0^n$ there exists $C > 0$ such that $\|\nabla^\alpha b_\infty\|_{L^\infty(\Omega)} \leq C|u_\infty|$. Defining the Oseen operator A_O by

$$A_O u := Au + \mathbb{P}(u_\infty \nabla v), \quad D(A_O) := W_0^{1,p}(\Omega) \cap W^{2,p}(\Omega) \cap L_\sigma^p(\Omega)$$

and setting $v := u - u_\infty + b_\infty$, we see that u is a solution of (22) if and only if v solves

$$v'(t) + A_O v(t) = \mathbb{P} \operatorname{div} G(v)(t), \quad t \in \mathbb{R}, \tag{23}$$

where

$$G(v) = F - \nabla b_\infty - (u_\infty - b_\infty) \otimes b_\infty - b_\infty \otimes v + v \otimes v - v \otimes b_\infty.$$

It is well known (see [46, 57]) that $-A_O$ generates a bounded analytic semigroup e^{-tA_O} on $L_\sigma^p(\Omega)$, called the Oseen semigroup. The $L^r - L^p$ smoothing properties of the Oseen semigroup (see [46], Theorem 2.1) combined with interpolating methods yield the following $L^{r,q} - L^{p,q}$ smoothing properties of Oseen semigroup, see also [28], Lemma 4.3.

Lemma 5.1 *(a) Let $p \in (1, \infty)$, $1 \leq q \leq \infty$. Then $-A_O$ generates a bounded analytic semigroup e^{-tA_O} on $L_\sigma^p(\Omega)$ and on $L_\sigma^{p,q}(\Omega)$.*

(b) Let $1 < p \leq q < \infty$. Then there exists $C > 0$ such that

$$\|e^{-tA_O} f\|_{q,1} \leq C t^{-\frac{3}{2}(\frac{1}{p} - \frac{1}{q})} \|f\|_{p,1}, \quad t > 0. \tag{24}$$

(c) Let $1 < p \leq q \leq 3$. Then there exists $C > 0$ such that

$$\|\nabla e^{-tA_O} f\|_{q,1}, \ \|\nabla e^{-tA_O'} f\|_{q,1} \leq C t^{-\frac{1}{2} - \frac{3}{2}(\frac{1}{p} - \frac{1}{q})} \|f\|_{p,1}, \quad t > 0. \tag{25}$$

We then obtain the following theorem on the existence of periodic and almost periodic solutions to the Navier-Stokes-Oseen equations (23).

Theorem 5.2 *Suppose that $F \in BC(\mathbb{R}; L^{3/2,\infty}(\Omega)^{d^2})$. Let $Y = L_\sigma^{3,\infty}(\Omega)$ and $T > 0$.*

(a) If F is T-periodic (almost periodic), then there exists a T-periodic (almost periodic) mild solution $\hat{u} \in P_T(\mathbb{R}; Y)$ ($\hat{u} \in AP(\mathbb{R}; Y)$) of

$$v'(t) + A_O v(t) = \mathbb{P}\operatorname{div} F(t), \quad t \in \mathbb{R},$$

and there exists a constant $\tilde{M} > 0$, independent of F, such that

$$\|\hat{u}\|_{BC(\mathbb{R};Y)} \leq \tilde{M}\|F\|_{L^\infty(\mathbb{R}; L^{\frac{3}{2},\infty}(\Omega)^{3\times 3})}$$

(b) If $F \in P_T(\mathbb{R}; L^{3/2,\infty}(\Omega)^{d^2})$ and $\|G(0)\|_{L^\infty(\mathbb{R}; L^{3/2,\infty}_\sigma(\Omega))} < \frac{1}{2}(\frac{1}{2\tilde{M}} - \|b_\infty\|_{L^{3,\infty}(\Omega)})^2$, then there exists a unique mild solution $\hat{v} \in P_T(\mathbb{R}; Y)$ of (23) in $B^p_{R,Y}(0)$ provided $R = (4\tilde{M})^{-1} - \frac{1}{2}\|b_\infty\|_{L^{3,\infty}(\Omega)}$.
(c) If $F \in AP(\mathbb{R}; L^{3/2,\infty}(\Omega)^{d^2})$ and $\|G(0)\|_{L^\infty(\mathbb{R}; L^{3/2,\infty}_\sigma(\Omega))} < \frac{1}{2}(\frac{1}{2\tilde{M}} - \|b_\infty\|_{L^{3,\infty}(\Omega)})^2$, then there exists a unique mild solution $\hat{v} \in AP(\mathbb{R}; Y)$ of (23) in $B^{ap}_{R,Y}(0)$ provided $R = (4\tilde{M})^{-1} - \frac{1}{2}\|b_\infty\|_{L^{3,\infty}(\Omega)}$.

Proof We show that Assumption (A) is fulfilled for the Oseen semigroup e^{-tA_O} and choose

$$Y_1 := L^{6,\infty}_\sigma(\Omega), \quad Y_2 := L^{2,\infty}_\sigma(\Omega), \quad X := L^{\frac{3}{2},\infty}_\sigma(\Omega)$$

and let B be given by $B = \mathbb{P}\operatorname{div}$. Note that the pre-dual spaces of Y_1 and Y_2 are given by $Z_1 = L^{\frac{6}{5},1}_\sigma(\Omega)$ and $Z_2 = L^{2,1}_\sigma(\Omega)$, respectively, and that the pre-dual of X is the space $L^{3,1}_\sigma(\Omega)$. It thus follows from Lemma 5.1 that

$$\begin{aligned}
\|e^{-At}B\Psi\|_{Y_i} &\leq \sup_{\varphi\in Z_i, \|\varphi\|_{Z_i}=1} |\langle e^{-At}B\Psi, \varphi\rangle| = \sup_{\varphi\in Z_i, \|\varphi\|_{Z_i}=1} |\langle \Psi, \nabla e^{-A't}\varphi\rangle| \\
&\leq \sup_{\varphi\in Z_i, \|\varphi\|_{Z_i}=1} \|\Psi\|_X \|\nabla e^{-A't}\varphi\|_{L^{3,1}(\Omega)} \\
&\leq Ct^{-\alpha_i}\|\Psi\|_X, \quad t > 0,
\end{aligned}$$

where $\alpha_1 = \frac{5}{4}$ and $\alpha_2 = \frac{3}{4}$. Choosing $\theta = \frac{1}{2}$ we have $1 = (1-\theta)\alpha_1 + \theta\alpha_2$ and by interpolation

$$(Y_1, Y_2)_{\theta,\infty} = (L^{6,\infty}_\sigma(\Omega), L^{2,\infty}_\sigma(\Omega))_{\theta,\infty} = L^{3,\infty}_\sigma(\Omega).$$

The assertion (a) follows now from Theorem 3.4.

In order to prove (b) note that

$$\|G(0)\|_{L^\infty(\mathbb{R};X)} \leq \|F - \nabla b_\infty - (u_\infty - b_\infty)\otimes b_\infty\|_{L^\infty(\mathbb{R};X)},$$

and that by Hölder's inequality,

$$\|G(v) - G(w)\|_{L^\infty(\mathbb{R};X)} = \| - b_\infty \otimes (v - w) + v\otimes v - w\otimes w - (v-w)\otimes b_\infty\|_{L^\infty(\mathbb{R};X)}$$

$$\leq 2(\|b_\infty\|_{L^{3,\infty}(\Omega)} + 2R)\|v - w\|_{C_b(\mathbb{R};L^{3,\infty}_\sigma(\Omega))}, \quad u, v \in B_R^{ap}(0).$$

In particular, the assumption of Theorem 4.2 are fulfilled with $u_{ap} = 0$, $L = 2(\|b_\infty\|_{L^{3,\infty}(\Omega)} + 2R)$ and $R = (4\tilde{M})^{-1} - \frac{1}{2}\|b_\infty\|_{L^{3,\infty}(\Omega)}$, which by Theorem 4.2 yields the assertion for almost periodic functions. The case of periodic solutions is proved similarly. The proof of Theorem 4.2 is complete. □

Remarks 5.3 (a) Since $\|G(0)\|_{L^\infty(\mathbb{R};X)} \leq \|F\|_{L^\infty(\mathbb{R};X)} + C|u_\infty|$, we see that the assumptions of Theorem 5.2 (b) are fulfilled in particular provided F and u_∞ are sufficiently small.
(b) Setting $u_\infty = 0$ in Theorem 5.2, we are in the situation of the Navier-Stokes equation and rediscover a result by Yamazaki [70] on almost periodic solutions to the Navier-Stokes equations.
(c) If u_∞ is of the form $u_\infty = (0, 0, k)$, we may consider the case of almost periodic motions of flows along a translating and rotating body. This will be done in one of the following sections.

5.2 *Periodic and Almost Periodic Solutions to the Navier Stokes Equations in Besov Spaces*

In this section we consider the Navier-Stokes equations on $\mathbb{R}^d$, i.e.

$$\begin{aligned} u_t + (u \cdot \nabla)u - \Delta u + \nabla p &= F \quad \text{in } \mathbb{R}^d \times \mathbb{R}, \\ \nabla \cdot u &= 0 \quad \text{in } \mathbb{R}^d \times \mathbb{R}. \end{aligned} \tag{26}$$

within the setting of homogeneous Besov spaces. For the definition and properties of these spaces, we refer to [3] or [5].

Let $\chi \in C^\infty(\mathbb{R}^d; \mathbb{R})$ be a smooth cut-off function such that $\operatorname{supp}\chi \subseteq B(0, 4/3)$, $0 \leq \chi \leq 1$ and $\chi \equiv 1$ in $B(0, 3/4)$. Set $\phi(\xi) := \chi(\xi/2) - \chi(\xi)$ and $h := \mathcal{F}^{-1}\phi$, where $\mathcal{F}$ denotes the Fourier transform. The homogeneous Littlewood-Paley decomposition $(\dot{\Delta}_j)_{j\in\mathbb{Z}}$ is then defined by

$$\dot{\Delta}_j u(x) = 2^{jd} \int_{\mathbb{R}^d} h(2^j y)u(x - y)dy = (\mathcal{F}^{-1}\phi(2^{-j}\cdot)\mathcal{F}u)(x).$$

We will make use of the operator $\dot{S}_j u = \sum_{j' \leq j-1} \dot{\Delta}_j u$. Given $s \in \mathbb{R}$ and $p, q \in [1, \infty]$, the homogeneous Besov spaces are defined as $\dot{B}^s_{p,q} = \{u \in \mathcal{S}'_h : \|u\|_{\dot{B}^s_{p,q}} < \infty\}$, where $\mathcal{S}'_h$ is the set of all tempered distributions such that $\lim_{j\to-\infty} \dot{S}_j u = 0$ in the topology of tempered distributions and

$$\|u\|_{\dot{B}^s_{p,q}} = (\sum_{j\in\mathbb{Z}} 2^{sqj}\|\dot{\Delta}_j u\|^q_{L^p})^{1/q} \quad \text{for } q < \infty, \quad \text{and} \quad \|u\|_{\dot{B}^s_{p,q}} = \sup_{j\in\mathbb{Z}} 2^{sj}\|\dot{\Delta}_j u\|_{L^p} \quad \text{for } q = \infty.$$

The following smoothing property of the heat semigroup will be important.

Lemma 5.4 *Let* $1 < p < \infty$, $1 \le q \le \infty$ *and* $\sigma, \tau \in \mathbb{R}$ *with* $\sigma \le \tau$. *Then there is a* $C > 0$ *such that*

$$\|e^{-t\Delta} f\|_{\dot{B}^{\tau}_{p,q}(\mathbb{R}^d)} \le C t^{-\frac{\tau-\sigma}{2}} \|f\|_{\dot{B}^{\sigma}_{p,q}(\mathbb{R}^d)}, \quad t > 0. \tag{27}$$

Proof By Lemma 2.4. in [5] we know that for each annulus $\mathcal{C}$, there exist constants $c > 0$ and $C > 0$ such that for each $t, \lambda > 0$ and $u \in L^p(\mathbb{R}^d)$ with $\operatorname{supp} \mathcal{F}u \subset \lambda\mathcal{C}$

$$\|e^{t\Delta} u\|_{L^p} \le C e^{-ct\lambda^2} \|u\|_{L^p}.$$

Now choose $\mathcal{C}$ such that $\operatorname{supp} \phi \subseteq \mathcal{C}$, where ϕ is the same as in the Littlewood-Paley decomposition. This implies that $\operatorname{supp} \mathcal{F}\dot{\Delta}_j u \subseteq 2^j\mathcal{C}$. We then have

$$2^{j\tau} \|\dot{\Delta}_j e^{-t\Delta} u\|_{L^p} \le C 2^{j\tau} e^{-ct2^{2j}} \|\dot{\Delta}_j u\|_{L^p} = C e^{-ct2^{2j}} t^{\frac{\tau-\sigma}{2}} 2^{j(\tau-\sigma)} t^{-\frac{\tau-\sigma}{2}} 2^{j\sigma} \|\dot{\Delta}_j u\|_{L^p}.$$

Due to [5], Lemma 2.35, the term $e^{-ct2^{2j}} t^{\frac{\tau-\sigma}{2}} 2^{j(\tau-\sigma)}$ can be bounded from above by some constant $K > 0$ independently of t and j and we hence obtain

$$2^{j\tau} \|\dot{\Delta}_j e^{-t\Delta} u\|_{L^p} \le K t^{-\frac{\tau-\sigma}{2}} 2^{j\sigma} \|\dot{\Delta}_j u\|_{L^p}.$$

Taking the $\ell^q(\mathbb{Z})$-norm on both sides yields the desired result.

We will further make use of the following embedding result, see [5], Proposition 2.20.

Let $s \in \mathbb{R}$, $q \in [1, \infty]$ and $p_1, p_2 \in [1, \infty]$ such that $p_1 \le p_2$. Then there is a constant $C > 0$ such that

$$\|u\|_{\dot{B}^{s-d(\frac{1}{p_1}-\frac{1}{p_2})}_{p_2,q}} \le C \|u\|_{\dot{B}^{s}_{p_1,q}}. \tag{28}$$

Furthermore, in order to estimate the nonlinear term, we need to estimates products of functions in homogeneous Besov spaces. To this end, consider the product of u and v which is given formally by

$$uv = \sum_{j,j' \in \mathbb{Z}} \dot{\Delta}_j u \dot{\Delta}_{j'} v.$$

By employing the Bony decomposition

$$\dot{T}_u v = \sum_{j' \le j-1} \dot{S}_{j-1} u \dot{\Delta}_{j'} v, \quad \dot{R}(u, v) = \sum_{|j-j'| \le 1} \dot{\Delta}_j u \dot{\Delta}_{j'} v,$$

the product may be reformulated as

$$uv = \dot{T}_u v + \dot{T}_v u + \dot{R}(u, v). \tag{29}$$

The operators $\dot{T}$ and $\dot{R}$ may be estimated as follows, see [5], Theorem 2.47, and Theorem 2.52:
(a) Given $s \in \mathbb{R}$, $t \in (\infty, 0)$ and $p, r_1, r_2 \in [1, \infty]$, there exists a constant $C > 0$ such that

$$\|\dot{T}_u v\|_{\dot{B}^{s+t}_{p,r}} \leq C\|u\|_{\dot{B}^t_{\infty,r_1}} \|v\|_{\dot{B}^s_{p,r_2}}, \quad u \in \dot{B}^t_{\infty,r_1}, v \in \dot{B}^s_{p,r_2}, \text{ and where } r^{-1} = \min(1, \frac{1}{r_1} + \frac{1}{r_2}). \tag{30}$$

(b) Given $s_1, s_2 \in \mathbb{R}$, $p, p_1, p_2, r, r_1, r_2 \in [1, \infty]$ such that $s_1 + s_2 > 0$, $\frac{1}{p} = \frac{1}{p_1} + \frac{1}{p_2}$ and $\frac{1}{r} = \frac{1}{r_1} + \frac{1}{r_2}$, there exists a constant $C > 0$ such that

$$\|\dot{R}(u, v)\|_{\dot{B}^{s_1+s_2}_{p,r}} \leq C\|u\|_{\dot{B}^{s_1}_{p_1,r_1}} \|v\|_{\dot{B}^{s_2}_{p_2,r_2}}, \quad u \in \dot{B}^{s_1}_{p_1,r_1}, v \in \dot{B}^{s_2}_{p_2,r_2}. \tag{31}$$

Combining the above results we arrive at the following Lemma.

Lemma 5.5 *Let $p \in [2, d)$, and $s = d/p - 1$. Then there is a constant $C > 0$ such that*

$$\|uv\|_{\dot{B}^{s-1}_{p,\infty}} \leq C\|u\|_{\dot{B}^s_{p,\infty}} \|v\|_{\dot{B}^s_{p,\infty}}, \quad u \in \dot{B}^s_{p,\infty}, v \in \dot{B}^s_{p,\infty}.$$

Proof Due to (29) it suffices to prove the above estimate for $\dot{T}_u v$, $\dot{T}_v u$ and $\dot{R}(u, v)$. The latter one follows from (28) and (31) as

$$\|\dot{R}(u, v)\|_{\dot{B}^{s-1}_{p,\infty}} \leq C\|\dot{R}(u, v)\|_{\dot{B}^{2s}_{p/2,\infty}} \leq C\|u\|_{\dot{B}^s_{p,\infty}} \|v\|_{\dot{B}^s_{p,\infty}}.$$

Similarly, the other two estimates follow from (28) and (30) by

$$\|\dot{T}_u v\|_{\dot{B}^{s-1}_{p,\infty}} \leq C\|u\|_{\dot{B}^{-1}_{\infty,\infty}} \|v\|_{\dot{B}^s_{p,\infty}} \leq C\|u\|_{\dot{B}^s_{p,\infty}} \|v\|_{\dot{B}^s_{p,\infty}}.$$

Let us further note that the Helmholtz projection $\mathbb{P}$ is well defined on $\dot{B}^s_{p,q}$ for any $s \in \mathbb{R}$, $q \in [1, \infty]$ and $1 < p < \infty$. Hence equation (26) may be rewritten equivalently as

$$u'(t) - \Delta u(t) = \mathbb{P}G(u)(t), \quad t \in \mathbb{R}, \tag{32}$$

where $G(u)(t) = F(t) - \operatorname{div}(u(t) \otimes u(t))$.

We are now in the position to formulate our result on the existence of almost periodic solutions to equation (32). For $0 < s \in \mathbb{R}$, $p \in [2, d)$ and $3 \leq d \in \mathbb{N}$ satisfying $s = \frac{d}{p} - 1$ we set

$$Y = \dot{B}^{s}_{p,\infty}(\mathbb{R}^d) \text{ and } X = \dot{B}^{s-2}_{p,\infty}(\mathbb{R}^d).$$

Theorem 5.6 *Let $0 < s \in \mathbb{R}$, $p \in [2, d)$ and $3 \leq d \in \mathbb{N}$ such that $s = \frac{d}{p} - 1$. Let $F \in BC(\mathbb{R}; X)$.*

(a) If F is T-periodic (almost periodic), then there is a unique T-periodic (almost periodic) mild solution $\hat{u} \in BC(\mathbb{R}; Y)$ to

$$u'(t) - \Delta u(t) = F,$$

and there exists a constant $\tilde{M} > 0$, independent of F, such that $\|\hat{u}\|_{BC(\mathbb{R};Y)} \leq \tilde{M}\|F\|_{BC(\mathbb{R};X)}$.
(b) There exists a constant $C(d, p) \geq 1$, such that if $F \in P_T(\mathbb{R}; X)$ and $\|F\|_{BC(\mathbb{R};X)} < \frac{1}{(4C(d,p)\tilde{M})^2}$, then (32) admits a unique mild solution $\hat{u} \in P_T(\mathbb{R}; Y)$ in $B^{p}_{R,Y}(0)$ provided $R = (4C(d, p)\tilde{M})^{-1}$.
(c) There is a constant $C(d, p) \geq 1$, such that if $F \in AP(\mathbb{R}; X)$ and $\|F\|_{BC(\mathbb{R};X)} < \frac{1}{(4C(d,p)\tilde{M})^2}$, then (32) admits a unique mild solution $\hat{u} \in AP(\mathbb{R}; Y)$ in $B^{ap}_{R,Y}(0)$ provided $R = (4C(d, p)\tilde{M})^{-1}$.

Proof. For $\delta \in (0, 1)$ set $Y_1 := \dot{B}^{s+\delta}_{p,\infty}(\mathbb{R}^d)$ and $Y_2 := \dot{B}^{s-\delta}_{p,\infty}(\mathbb{R}^d)$. By Lemma 5.4,

$$\|e^{-t\Delta}\psi\|_{Y_i} \leq Ct^{-\alpha_i}\|\psi\|_X, \quad i \in \{1, 2\}, \tag{33}$$

where $\alpha_1 = 1 + \delta/2$ and $\alpha_2 = 1 - \delta/2$. For $\theta = 1/2$, we obtain $1 = (1 - \theta)\alpha_1 + \theta\alpha_2$ and

$$(Y_1, Y_2)_{\theta,\infty} = (\dot{B}^{s+\delta}_{p,\infty}(\mathbb{R}^d), \dot{B}^{s-\delta}_{p,\infty}(\mathbb{R}^d))_{\theta,\infty} = \dot{B}^{s}_{p,\infty}(\mathbb{R}^d). \tag{34}$$

Note that Y_1 and Y_2 are dual spaces, as $\dot{B}^{\sigma}_{p,\infty}(\mathbb{R}^d) = (\dot{B}^{-\sigma}_{p',1}(\mathbb{R}^d))'$. Hence, the assumptions of Theorem 3.4 are fulfilled and assertion a) follows.

For the proof of assertion (c) note that $\|G(0)\|_{BC(\mathbb{R},X)} = \|F\|_{BC(\mathbb{R},X)}$. Lemma 5.5 and the fact that div maps $\dot{B}^{s}_{p,q}$ continuously to $\dot{B}^{s-1}_{p,q}$ yields

$$\|G(u) - G(v)\|_{BC(\mathbb{R},X)} \leq C(d, p)(\|u\|_{BC(\mathbb{R},Y)} + \|v\|_{BC(\mathbb{R},Y)})\|u - v\|_{BC(\mathbb{R},Y)}.$$

Hence, for $u_{ap} = 0$, $R = (4C(d, p)\tilde{M})^{-1}$ and $L = 2RC(d, p)$, the assumption (G^{ap}) is satisfied and Theorem 4.2 yields the desired result. The proof of assertion b) can be carried out in a similar way. □

5.3 Almost Periodic Navier-Stokes Flow along a Translating and Rotating Obstacle

In this subsection we consider the Navier-Stokes flows along a translating and rotating obstacle D with the translation velocity $u_\infty = (0, 0, k)$ and the angular velocity $\omega = (0, 0, a)$ being both constant. Then, the equations for the flows along the obstacle read as

$$\begin{cases} u_t + (u \cdot \nabla)u - \Delta u + k\partial_3 u + \nabla p = (\omega \times x) \cdot \nabla u - \omega \times u + \operatorname{div} F & \text{in } \Omega \times \mathbb{R}, \\ \nabla \cdot u = 0 & \text{in } \Omega \times \mathbb{R}, \\ u = \omega \times x - k\mathbf{e}_3 & \text{on } \partial\Omega \times \mathbb{R}, \end{cases} \tag{35}$$

where $\omega = (0, 0, a)$ is a constant vector representing the angular velocity around the x_3-axis of a given obstacle $D \subset \mathbb{R}^3$ with smooth boundary. Here $\Omega := \mathbb{R}^3 \backslash D(0)$ with $D(0)$ being the position of D at $t = 0$, and $\mathbf{e}_3 = (0, 0, 1)$.

We define $b_{\omega,k} := -\frac{1}{2}\nabla \times (\varphi(x)|x|^2\omega) + \frac{1}{4}\nabla \times \nabla \times (\varphi(x)k|x|^2\mathbf{e}_3)$, where φ is the cut-off function defined as in Section 5.1. Then $b_{\omega,k} \in C_c^\infty(\Omega)$, $\operatorname{div} b_{\omega,k} = 0$ and $b_{\omega,k} = \omega \times x - k\mathbf{e}_3$ on $\partial\Omega$. Moreover, an elementary calculation shows that

$$b_{\omega,k} = \operatorname{div} F_w + \operatorname{div} F_k$$

for $F_w := \begin{pmatrix} 0 & \frac{a}{2}\varphi(x)|x|^2 & 0 \\ -\frac{a}{2}\varphi(x)|x|^2 & 0 & 0 \\ 0 & 0 & 0 \end{pmatrix}$ and

$$F_k := \frac{k}{4} \begin{pmatrix} |x|^2\partial_3\varphi(x) + 2x_3\varphi(x) & 0 & -|x|^2\partial_1\varphi(x) - 2x_1\varphi(x) \\ 0 & |x|^2\partial_3\varphi(x) + 2x_3\varphi(x) & -|x|^2\partial_2\varphi(x) - 2x_2\varphi(x) \\ 0 & 0 & 0 \end{pmatrix}.$$

Introducing for $1 < p < \infty$ the operator

$$\begin{aligned} Lu &:= -\mathbb{P}\left[\Delta u - k\partial_3 u + (\omega \times x) \cdot \nabla u - \omega \times u\right], \\ D(L) &:= \left\{u \in L^p_\sigma(\Omega) \cap W^{2,p}(\Omega) : u|_{\partial\Omega} = 0 \text{ and } (\omega \times x) \cdot \nabla u \in L^p(\Omega)\right\}, \end{aligned} \tag{36}$$

we may rewrite (35) as an equation for $v := u - b_{\omega,k}$ of the form

$$v'(t) + Lv(t) = \mathbb{P} \operatorname{div} G(v)(t), \quad t \in \mathbb{R}, \tag{37}$$

where

$$G(v) = F + F_\omega + F_k + (\omega \times x) \otimes b_{\omega,k} + \nabla b_{\omega,k} + b_{\omega,k} \otimes v + v \otimes v + b_{\omega,k} \otimes v + b_{\omega,k} \otimes b_{\omega,k}.$$

It is known that $-L$ generates a bounded C_0-semigroup e^{-tL} on $L^p_\sigma(\Omega)$ for $p \in (1, \infty)$, which is, however, not analytic; see [26]. Moreover, the semigroup e^{-tL}

satisfies the $L^p - L^q$ smoothing estimates, see [26, 43] and [63, Theorem 1.2]. Hence, by interpolation we obtain the following lemma; see also [28], Lemma 4.6.

Lemma 5.7. *a) Let $1 < p < \infty$ and $1 \le q \le \infty$. Then, $-L$ generates a bounded semigroup e^{-tL} on $L^p_\sigma(\Omega)$ and on $L^{p,q}_\sigma(\Omega)$.*
b) Let $1 < p \le q < \infty$. Then there exists $C > 0$ such that

$$\|e^{-tL} f\|_{q,1} \le C t^{-\frac{3}{2}(\frac{1}{p}-\frac{1}{q})} \|f\|_{p,1}, \quad t > 0. \tag{38}$$

Let $1 < p \le q \le 3$. Then there exists $C > 0$ such that

$$\|\nabla e^{-tL} f\|_{q,1},\ \|\nabla e^{-tL'} f\|_{q,1} \le C t^{-\frac{1}{2}-\frac{3}{2}(\frac{1}{p}-\frac{1}{q})} \|f\|_{p,1}, \quad t > 0. \tag{39}$$

We then obtain the following result concerning periodic and almost periodic solutions to equation (35).

Theorem 5.8. *Suppose that $F \in BC(\mathbb{R}; L^{3/2,\infty}(\Omega)^{3\times 3})$. Let $Y = L^{3,\infty}_\sigma(\Omega)$ and $T > 0$.*

(a) If F is $T-$periodic (almost periodic), then there exists a T-periodic (almost periodic) mild solution $\hat{u} \in P_T(\mathbb{R}; Y)$ ($\hat{u} \in AP(\mathbb{R}; Y)$) of

$$v'(t) + Lv(t) = \mathbb{P} \operatorname{div} F(t), \quad t \in \mathbb{R},$$

and there exists a constant $\tilde{M} > 0$, independent of F, such that

$$\|\hat{u}\|_{BC(\mathbb{R};Y)} \le \tilde{M} \|F\|_{L^\infty(\mathbb{R}; L^{\frac{3}{2},\infty}(\Omega)^{3\times 3})}$$

(b) If $F \in P_T(\mathbb{R}; L^{3/2,\infty}(\Omega)^{d^2})$ and $\|G(0)\|_{L^\infty(\mathbb{R}; L^{3/2}_\sigma(\Omega))} < \frac{1}{2}(\frac{1}{2\tilde{M}} - \|b_{\omega,k}\|_Y)^2$, then there exists a unique mild solution $\hat{v} \in P_T(\mathbb{R}; Y)$ of (37) in $B^P_{R,Y}(0)$ provided $R = (4\tilde{M})^{-1} - \frac{1}{2}\|b_{\omega,k}\|_{L^{3,\infty}(\Omega)}$.
(c) If $F \in AP(\mathbb{R}; L^{3/2,\infty}(\Omega)^{d^2})$ and $\|G(0)\|_{L^\infty(\mathbb{R}; L^{3/2}_\sigma(\Omega))} < \frac{1}{2}(\frac{1}{2\tilde{M}} - \|b_{\omega,k}\|_Y)^2$, then there exists a unique mild solution $\hat{v} \in AP(\mathbb{R}; Y)$ of (37) in $B^{ap}_{R,Y}(0)$ provided $R = (4\tilde{M})^{-1} - \frac{1}{2}\|b_{\omega,k}\|_{L^{3,\infty}(\Omega)}$.

Proof. Similarly as above we verify that Assumption (A) is fulfilled for e^{-tL} provided we set $Y_1 := L^{6,\infty}_\sigma(\Omega)$, $Y_2 := L^{2,\infty}(\Omega)$, $X := L^{\frac{3}{2},\infty}_\sigma(\Omega)$ and $B = \mathbb{P}\operatorname{div}$. Hence, assertion a) follows from Theorem 3.4. Moreover,

$$\|G(0)\|_{L^\infty(\mathbb{R};X)} \le \|F + F_\omega + F_k + (\omega \times x) \otimes b_{\omega,k} + \nabla b_{\omega,k} + b_{\omega,k} \otimes b_{\omega,k}\|_{L^\infty(\mathbb{R};X)}.$$

Hölder's inequality implies

$$\begin{aligned}\|G(v) - G(u)\|_{L^\infty(\mathbb{R};X)} &= \|b_{\omega,k} \otimes (v-u) + v \otimes v - u \otimes v + (v-u) \otimes b_{\omega,k}\|_{L^\infty(\mathbb{R};X)} \\ &\le 2(\|b_{\omega,k}\|_{L^{3,\infty}(\Omega)} + 2R)\|v-u\|_{BC(\mathbb{R};L^{3,\infty}_\sigma(\Omega))}, \quad u, v \in B_R(0).\end{aligned}$$

Hence, the assumptions of Theorem 4.2 are fulfilled for $u_{ap} = 0, L = 2(\|b_{\omega,k}\|_{L^{3,\infty}(\Omega)} + 2R)$ and $R = (4\tilde{M})^{-1} - \frac{1}{2}\|b_{\omega,k}\|_{L^{3,\infty}(\Omega)}$.

Remark 5.9. Note that

$$\|G(0)\|_{L^\infty(\mathbb{R};X)} \leq \|F\|_{L^\infty(\mathbb{R};X)} + C_1|a| + C_2|k|.$$

Hence, the assumptions of Theorem 5.2 b) are in particular fulfilled whenever F, a and k are sufficiently small.

5.4 *Almost Periodic Solutions to the Navier-Stokes Equations in Aperture Domains*

Consider an aperture domain $\Omega \subset \mathbb{R}^d$, $d \geq 3$, of class C^3, i.e. there exists a ball $B_R(0)$ such that $\Omega \cup B_R(0) = \mathbb{R}^d_{+k} \cup \mathbb{R}^d_{-k} \cup B_R(0)$ where for fixed $k > 0$, $\mathbb{R}^d_{\pm k} := \{x \in \mathbb{R}^d : \pm x_n > k\}$ denote half spaces. Since Ω is assumed to be connected, we may consider two disjoint subdomains Ω_+ and Ω_- of Ω and a smooth $(d-1)$-dimensional manifold M, called the aperture of Ω, with the following properties:

$$\Omega_\pm \cup B_R(0) = \mathbb{R}^d_{\pm k} \cup B_R(0) \text{ and } M \cup \partial M = \partial\Omega_+ \cap \partial\Omega_- \subset \overline{B_R(0)} \text{ and } \Omega = \Omega_+ \cup \Omega_- \cup M.$$

In the following, we investigate almost periodic solutions for the Navier-Stokes equations on an aperture domain $\Omega \subset \mathbb{R}^d$ with zero-flux through the aperture M, i.e. we consider

$$\begin{cases} u_t + (u \cdot \nabla)u - \Delta u + \nabla p = \operatorname{div} F & \text{in } \Omega \times \mathbb{R}, \\ \nabla \cdot u = 0 & \text{in } \Omega \times \mathbb{R}, \\ u = 0 & \text{on } \partial\Omega \times \mathbb{R}, \\ \int\limits_M u(t,x) \cdot n d\sigma = 0. & \end{cases} \tag{40}$$

Here $u_0 \in L^{d,\infty}_\sigma(\Omega)$ and $F \in L^\infty(\mathbb{R}; L^{d/2,\infty}_\sigma(\Omega))$ is assumed to be almost periodic.

For $1 < p < \infty$ we consider the Stokes operator $A = -\mathbb{P}\Delta$ on $L^p_\sigma(\Omega)$ with

$$D(A) = W^{2,p}(\Omega) \cap W^{1,p}_0(\Omega) \cap L^p_\sigma(\Omega). \tag{41}$$

For an aperture domain, it is well known that $-A$ generates a bounded analytic semigroup $(e^{-tA})_{t\geq 0}$ on $L^p_\sigma(\Omega)$, see [42], which satisfies $L^p - L^q$ smoothing estimates for $d \geq 3$, see [42, Theorem 2.1]. By interpolation, the semigroup e^{-tA} extends to a bounded analytic semigroup on $L^{p,q}_\sigma(\Omega)$ for $1 \leq q \leq \infty$ and satisfies the following $L^p - L^q$ smoothing estimates.

Lemma 5.10. *(a) Let $1 < p < \infty$, $1 \leq q \leq \infty$ and $d \geq 3$. Then, $-A$ defined as in (41) generates a bounded analytic semigroup e^{-tA} on $L^p_\sigma(\Omega)$ and on $L^{p,q}_\sigma(\Omega)$.*
(b) Let $1 < p \leq q < \infty$. Then there exists a constant $C > 0$ such that

$$\|e^{-tA} f\|_{q,1} \leq Ct^{-\frac{d}{2}(\frac{1}{p}-\frac{1}{q})}\|f\|_{p,1}, \quad t > 0. \tag{42}$$

(c) Let $1 < p \leq q \leq d$. Then there exists a constant $C > 0$ such that

$$\|\nabla e^{-tA} f\|_{q,1},\ \|\nabla e^{-tA'} f\|_{q,1} \leq Ct^{-\frac{1}{2}-\frac{d}{2}(\frac{1}{p}-\frac{1}{q})}\|f\|_{p,1}, \quad t > 0. \tag{43}$$

Applying the Helmholtz projection, equation (40) can be rewritten on $L^{r,q}_\sigma(\Omega)$ as

$$u'(t) + Au(t) = \mathbb{P}\operatorname{div} G(u)(t), \quad t \in \mathbb{R}, \tag{44}$$

where

$$G(u)(t) = -u(t) \otimes u(t) + F.$$

Applying our method yields the following result on periodic and almost periodic solutions for the Navier-Stokes equations on an aperture domain. The first result in this direction goes back to Kubo [50] and Crispo and Maremonti [14]. The latter considered the case of the periodic flux without external force $\operatorname{div} F$. One may transform the case of periodic fluxes to the case of zero-fluxes by setting $u(t,x) := \alpha(t)b(x) + U(t,x)$ where $\alpha(t)$ is the periodic flux, $b(x)$ is the solution of a stationary Stokes problem with flux 1 and U solves an instationary Navier-Stokes problem with suitable right-hand side. For details see [14].

Theorem 5.11. *Suppose that $F \in BC(\mathbb{R}; L^{d/2,\infty}(\Omega)^{d^2})$. Let $Y = L^{d,\infty}_\sigma(\Omega)$ and $T > 0$. (a) If F is $T-$periodic (almost periodic), then there exists a T-periodic (almost periodic) mild solution $\hat{u} \in P_T(\mathbb{R}; Y)$ ($\hat{u} \in AP(\mathbb{R}; Y)$) of*

$$v'(t) + Av(t) = \mathbb{P}\operatorname{div} F(t), \quad t \in \mathbb{R},$$

and there exists a constant $\tilde{M} > 0$, independent of F, such that

$$\|\hat{u}\|_{BC(\mathbb{R};Y)} \leq \tilde{M}\|F\|_{L^\infty(\mathbb{R}; L^{\frac{d}{2},\infty}(\Omega))}. \tag{45}$$

(b) If $F \in P_T(\mathbb{R}; L^{d/2,\infty}(\Omega)^{d^2})$ and $\|F\|_{L^\infty(\mathbb{R}; L^{d/2,\infty}_\sigma(\Omega))} < \frac{1}{8(\tilde{M})^2}$, then equation (44) admits a unique mild solution $\hat{v} \in P_T(\mathbb{R}; Y)$ in $B^p_{R,Y}(0)$ provided $R = (4\tilde{M})^{-1}$.
(c) If $F \in AP(\mathbb{R}; L^{d/2,\infty}(\Omega)^{d^2})$ and $\|F\|_{L^\infty(\mathbb{R}; L^{d/2,\infty}_\sigma(\Omega))} < \frac{1}{8(\tilde{M})^2}$, then equation (44) admits a unique mild solution $\hat{v} \in AP(\mathbb{R}; Y)$ in $B^{ap}_{R,Y}(0)$ provided $R = (4\tilde{M})^{-1}$.

5.5 Almost Periodic Solutions to the Navier-Stokes Flow in the Rotational Framework

Consider the Navier-Stokes equations in the rotational framework on $\mathbb{R}^3$ which are given as

$$\begin{cases} u_t + (u \cdot \nabla)u - \Delta u + \nabla p = \omega e_3 \times u + \operatorname{div} F & \text{in } \mathbb{R}^3 \times \mathbb{R}, \\ \nabla \cdot u = 0 & \text{in } \mathbb{R}^3 \times \mathbb{R}, \end{cases} \tag{46}$$

where ω is the speed of rotation and e_3-the unit vector in x_3-direction. Applying the Helmholtz projection $\mathbb{P}$ we obtain the equation

$$\left\{ u_t + A_{SC}u = \mathbb{P}\operatorname{div}(-u \otimes u + F) \ \text{ in } \mathbb{R}^3 \times \mathbb{R}, \right. \tag{47}$$

where $A_{SC} := -\mathbb{P}\Delta - \mathbb{P}(\omega e_3 \times \nabla u)$ denotes the *Stokes-Coriolis operator* on $L^3_{\sigma,w}(\mathbb{R}^3)$. It was shown by Giga, Inui, Mahalov and Matsui [32], as well as by Hieber and Shibata [41], that this operator generates a C_0-semigroup T_p on $L^p_\sigma(\mathbb{R}^3)$ admitting an explicit representation formula of the form

$$T_p(t)f := \mathcal{F}^{-1}[\cos(\Omega\frac{\xi_3}{|\xi|}t)e^{-|\xi|^2 t} I\widehat{f}(\xi) + \sin(\Omega\frac{\xi_3}{|\xi|}t)e^{-|\xi|^2 t} R(\xi)\widehat{f}(\xi)], \quad t \geq 0,\ f \in L^p_\sigma(\mathbb{R}^3)^3,$$

where

$$R = R(\xi) = \begin{pmatrix} 0 & \frac{\xi_3}{|\xi|} & -\frac{\xi_2}{|\xi|} \\ -\frac{\xi_3}{|\xi|} & 0 & \frac{\xi_1}{|\xi|} \\ \frac{\xi_2}{|\xi|} & -\frac{\xi_1}{|\xi|} & 0 \end{pmatrix}.$$

This semigroup is called the *Stokes-Coriolis semigroup*. Since the $L^r - L^p$ estimates are valid for $e^{-tA_{SC}}$ for $1 < r \leq 2 \leq p < \infty$, see [41, Prop. 2.4], the above interpolating procedure yields $L^{r,q} - L^{p,q}$ smoothing properties for $(e^{-tA_{SC}})_{t\geq 0}$ and its gradient as in Lemma 5.1. For an almost periodic function $F \in BC(\mathbb{R}, L^{3/2}_w(\mathbb{R}^3)^{3\times 3})$ consider the linear problem

$$u_t + A_{SC}u = \mathbb{P}\operatorname{div} F \ \text{ on } \mathbb{R}^3 \times \mathbb{R}, \tag{48}$$

as well as the nonlinear problem

$$u_t + A_{SC}u = \mathbb{P}\operatorname{div} G(u)(t) \ \text{ on } \mathbb{R}^3 \times \mathbb{R}, \tag{49}$$

where $G(u)(t) = -u(t) \otimes u(t) + F(t)$. Applying Theorem 4.2 we obtain the following result.

Theorem 5.12. *Let $F \in BC(\mathbb{R}, L^{3/2}_w(\mathbb{R}^3)^{3\times 3})$, $Y = L^{3,\infty}_\sigma(\mathbb{R}^3)$ and $T > 0$.*

(a) If F is $T-$periodic (almost periodic), then there exists a T-periodic (almost periodic) mild solution $\hat{u} \in P_T(\mathbb{R}; Y)$ ($\hat{u} \in AP(\mathbb{R}; Y)$) of (48) *and there exists a constant $\tilde{M} > 0$ such*

$$\|\hat{u}\|_{\infty,3,w} \leq \tilde{M}\|F\|_{\infty,\frac{3}{2},w}.$$

(b) If $F \in P_T(\mathbb{R}; L^{3/2,\infty}(\Omega)^{d^2})$ and $\|F\|_{\infty,\frac{3}{2},w}$ is small enough, then equation (49) *admits a unique mild solution $\hat{v} \in P_T(\mathbb{R}; Y)$ in $B^p_{R,Y}(0)$ provided R is small enough.*
(c) If $F \in AP(\mathbb{R}; L^{3/2,\infty}(\Omega)^{d^2})$ and $\|F\|_{\infty,\frac{3}{2},w}$ is small enough, then equation (49) *admits a unique mild solution $\hat{v} \in AP(\mathbb{R}; Y)$ in $B^{ap}_{R,Y}(0)$ provided R is small enough.*

Let us remark that periodic solutions for the Navier-Stokes equations in the rotational setting were recently investigated by Kozono, Mashiko and Takada in [49]. They proved the existence of a unique mild periodic solution $u \in BC(\mathbb{R}; L^r_\sigma(\mathbb{R}^3)) \cap BC(\mathbb{R}; W^{1,q}(\mathbb{R}))$ to (48) under suitable smallness assumptions on F.

6 Almost Periodic Solutions to Semilinear Ornstein–Uhlenbeck and Diffusion Equations with Rough Coefficients

6.1 Almost Periodic Solutions to Semilinear Ornstein–Uhlenbeck Equations on Exterior Domains

For an exterior domain $\Omega \subset \mathbb{R}^d$ with $C^{1,1}$ boundary consider the Ornstein–Uhlenbeck equation

$$\begin{aligned} u_t - \Delta u - Mx\cdot\nabla u &= G(u)(t), \quad t\in\mathbb{R},\ x\in\Omega,\\ u &= 0, \quad t\in\mathbb{R}, \quad x\in\partial\Omega, \end{aligned}$$

where $G(u) = |u|^{r-1}u + F$ for some fixed $r \in \mathbb{N}$, $M \in \mathbb{R}^{d\times d}$ and a given almost periodic function F. For operators of the form

$$\mathcal{L}u(x) := \Delta u(x) + Mx\cdot\nabla u(x),\ x\in\Omega,$$

we define Ornstein–Uhlenbeck operators L on $L^p(\Omega)$ by

$$Lu := \mathcal{L}u, \quad D(L) := \{u \in W^{2,p}(\Omega)\cap W^{1,p}_0(\Omega) : Mx\cdot\nabla u \in L^p(\Omega)\}.$$

The above equation (50) may be hence rewritten as

$$u'(t) - Lu(t) = G(u)(t), \quad t\in\mathbb{R}. \tag{50}$$

In this case, the right-hand side doesn't have a divergent form and hence, in order to apply our method, we choose $B = Id$. Note that $L^p - L^q$ smoothing estimates for e^{tL} are known due to the results given in [27] and [39] and that the gradient estimates aren't needed in the present situation.

In the following, the choice of the underlying spaces will depend on the exponent r of the nonlinear term. More precisely, we define the spaces involved as

$$X := L^{\frac{d(r-1)}{2r},\infty}(\Omega) = L^{\frac{d(r-1)}{d(r-1)-2r},1}(\Omega)', \quad Y_1 := L^{\frac{2d(r-1)}{5-r},\infty}(\Omega), \quad Y_2 := L^{\frac{2d(r-1)}{r+3},\infty}(\Omega) \tag{51}$$

and choose $\theta = 1/2$. Then, $Y = (Y_1, Y_2)_{\frac{1}{2},\infty} = L^{\frac{d(r-1)}{2},\infty}(\Omega)$ and the preduals of Y_1 and Y_2 are given by

$$Z_1 = L^{\frac{2d(r-1)}{(2d+1)(r-1)-4},1}(\Omega) \text{ and } Z_2 = L^{\frac{2d(r-1)}{(2d-1)(r-1)-4},1}(\Omega).$$

We then have the following result.

Theorem 6.1. *Let $1 < r < 5$ such that $d(r-1) > 2r$. Assume that $M \in \mathbb{R}^{d\times d}$ satisfies $tr M = 0$ and suppose that $F \in BC(\mathbb{R}; X)$.*

(a) If F is $T-$periodic (almost periodic), then there exists a T-periodic (almost periodic) mild solution $\hat{u} \in P_T(\mathbb{R}; Y)$ ($\hat{u} \in AP(\mathbb{R}; Y)$) of

$$u'(t) - Lu(t) = F(t), \quad t \in \mathbb{R},$$

and there exists a constant $\tilde{M} > 0$ such $\|u\|_{BC(\mathbb{R};Y)} \le \tilde{M}\|F\|_{L^\infty(\mathbb{R};X)}$.
(b) If $F \in P_T(\mathbb{R}; X)$ and $\|F\|_{L^\infty(\mathbb{R};X)} < [(r(r-1)\tilde{M})^{1/(r-2)}\tilde{M}]^{-1}$, then equation (50) admits a unique mild solution $\hat{v} \in P_T(\mathbb{R}; Y)$ in $B^p_{R,Y}(0)$ provided $R = [r(r-1)\tilde{M}]^{1/(2-r)}$.
(c) If $F \in AP(\mathbb{R}; X)$ and $\|F\|_{L^\infty(\mathbb{R};X)} < [(r(r-1)\tilde{M})^{1/(r-2)}\tilde{M}]^{-1}$, then equation (50) admits a unique mild solution $\hat{v} \in AP(\mathbb{R}; Y)$ in $B^{ap}_{R,Y}(0)$ provided $R = [r(r-1)\tilde{M}]^{1/(2-r)}$.

Proof. Choosing $\alpha_1 = \frac{5}{4}$ and $\alpha_2 = \frac{3}{4}$ we have $\alpha_1 > 1 > \alpha_2 > 0$ and $1 = \frac{\alpha_1}{2} + \frac{\alpha_2}{2}$. Then the $L^p - L^q$ estimates for Ornstein–Uhlenbeck semigroup on exterior domains, see [27], Thm.3.1, yield

$$\|e^{tA}u\|_{\frac{d(r-1)}{d(r-1)-2r},1} \le Ct^{-\frac{5}{4}}\|u\|_{\frac{2d(r-1)}{(2d+1)(r-1)-4},1}$$

$$\|e^{tA}u\|_{\frac{d(r-1)}{d(r-1)-2r},1} \le Ct^{-\frac{3}{4}}\|u\|_{\frac{2d(r-1)}{(2d-1)(r-1)-4},1}.$$

Theorem 3.4 now yields Assertion (a).

Next, we consider the estimate for $G(u)$. Using the weak Hölder inequality we arrive at

$$\|G(u) - G(v)\|_{L^{\frac{d(r-1)}{2r},\infty}(\Omega)} \le \sum_{j=0}^{r-1} \left\| |u - v||u|^j |v|^{r-1-j} \right\|_{L^{\frac{d(r-1)}{2r},\infty}(\Omega)}$$
$$\le r\|u - v\|_{L^{\frac{d(r-1)}{2},\infty}(\Omega)} R^{r-1}, \quad u, v \in B^p_{R,Y}(0).$$

Applying Theorem 4.2 with $u_p = 0$ and $L = rR^{r-1}$ yields assertion (b). Assertion c) is proved similarly.

6.2 Almost Periodic Solutions to Diffusion Equations with Rough Coefficients

Consider a measurable function $b : \mathbb{R}^d \to \mathbb{R}$ satisfying

$$b \in L^\infty(\mathbb{R}^d) \text{ and such that } b \ge \delta > 0 \text{ for some } \delta > 0. \tag{52}$$

In this section we look for almost periodic solutions to the equation

$$u_t - b\Delta u = G(u)(t), \quad t \in \mathbb{R}, \; x \in \mathbb{R}^d,$$

where as in Section 5, the function G is assumed to be of the form $G(u) = |u|^{r-1}u + F$ for some fixed $r \in \mathbb{N}$ and a given almost periodic function F. Note that by a result due to Duong and Ouhabaz [15] the operator $-L$ defined on $L^p(\mathbb{R}^d)$ by $Lu := -b\Delta u$ generates a bounded analytic semigroup T on $L^p(\mathbb{R}^d)$ for all $1 < p < \infty$ such that $T(t)f(x) = \int_{\mathbb{R}^d} K(t, x, y)f(y)dy$ for $t > 0$ and a.a. $x, y \in \mathbb{R}^n$ and where

$$|K(t, x, y)| \le \frac{M}{t^{d/2}} e^{\frac{-a|x-y|^2}{ct}}, \quad t > 0, x, y \in \mathbb{R}^d,$$

for some constants $M, a, c > 0$. The above equation (51) may be hence rewritten as

$$u'(t) + Lu(t) = G(u)(t), \quad t \in \mathbb{R}. \tag{53}$$

The above heat kernel estimate allows us to verify Assumption (A) and to follow the above lines in order to obtain the following result. In particular, let the spaces X and Y be defined as in (51).

Theorem 6.2. *Let $1 < r < 5$ such that $d(r-1) > 2r$. Let b satisfy (52) and suppose that $F \in BC(\mathbb{R}; X)$.*

(a) If F is $T-$periodic (almost periodic), then there exists a T-periodic (almost periodic) mild solution to the equation

$$u'(t) + Lu(t) = F(t), \quad t \in \mathbb{R},$$

and a constant $\tilde{M} > 0$*, independent of* F*, such that* $\|u\|_{BC(\mathbb{R};Y)} \leq \tilde{M}\|F\|_{L^\infty(\mathbb{R};X)}$.
(b) If $F \in P_T(\mathbb{R}; X)$ *and* $\|F\|_{L^\infty(\mathbb{R};X)} < [(r(r-1)\tilde{M})^{1/(r-2)}\tilde{M}]^{-1}$*, then equation* (53) *admits a unique mild solution* $u \in P_T(\mathbb{R}; Y)$ *in* $B^p_{R,Y}(0)$ *where* $R = [r(r-1)\tilde{M}]^{1/(2-r)}$.
(c) If $F \in AP(\mathbb{R}; X)$ *and* $\|F\|_{L^\infty(\mathbb{R};X)} < [(r(r-1)\tilde{M})^{1/(r-2)}\tilde{M}]^{-1}$*, then equation* (53) *admits a unique mild solution* $u \in AP(\mathbb{R}; Y)$ *in* $B^{ap}_{R,Y}(0)$ *where* $R = [r(r-1)\tilde{M}]^{1/(2-r)}$.

Acknowledgements This work is financially supported by Vietnam National Foundation for Science and Technology Development (NAFOSTED). The work of the second author is also supported by Vietnam Institute for Advanced Study in Mathematics (VIASM).

References

1. K. Abe, Y. Giga, Analyticity of the Stokes semigroup in spaces of bounded functions. Acta Math. **211**, 1–46 (2013)
2. K. Abe, Y. Giga, M. Hieber, Stokes resolvent estimates in spaces of bounded functions. Ann. Éc. Norm. Supér. **48**, 537–559 (2015)
3. H. Amann, Operator valued Fourier multipliers, vector-valued Besov spaces, and applications. Math. Nachr. **186**, 5–56 (1997)
4. W. Arendt, C.J.K. Batty, M. Hieber, F. Neubrander, *Vector-Valued Laplace Transform and Cauchy Problems*. Monographs in Mathematics, vol. 96, 2nd edn. (Birkhäuser, Basel, 2011)
5. H. Bahouri, J.-Y. Chemin, R. Danchin, *Fourier Analysis and Nonlinear Partial Differential Equations* (Springer, Berlin, 2011)
6. J. Bergh, J. Löfström, *Interpolation Spaces* (Springer, Berlin, 1976)
7. L. Berselli, M. Romito, On Leray's problem for almost periodic flows. J. Math. Sci. Univ. Tokyo **19**, 69–130 (2012)
8. A.S. Besicovitch, Analysis of conditions of almost periodicity. Acta Math. **58**, 217–230 (1932)
9. A.S. Besicovitch, *Almost Periodic Functions* (Dover Publications, New York, 1955)
10. S. Bochner, Abstrakte fastperiodische Funktionen. Acta Math. **61**, 149–184 (1933)
11. H. Bohr, Zur Theorie der fastperiodische Funktionen, I and II. Eine Verallgemeinerung der Theorie der Fourierreihen. *Acta Math.*, **45**, (1925), 29-127 and **46**, (1925), 101-214
12. W. Borchers, T. Miyakawa, On stability of exterior stationary Navier-Stokes flows. Acta Math. **174**, 311–382 (1995)
13. C. Corduneanu, *Almost Periodic Functions* (Wiley, New York, 1968)
14. F. Crispo, P. Maremonti, Navier-Stokes equations in aperture domains: Global existence with bounded flux and time-periodic solutions. Math. Meth. Appl. Sci. **31**, 249–277 (2008)
15. X. Duong, E. Ouhabaz, Complex multiplicative perturbations of elliptic operators: heat kernel bounds and holomorphic functional calculus. Diff. Integral Equ. **12**, 395–418 (1999)
16. R. Farwig, T. Nakatsuka, Y. Taniuchi, Uniqueness of solutions on the whole time axis to the Navier-Stokes equations in unbounded domains. Comm. Partial. Differ. Equ. **40**, 1884–1904 (2015)
17. R. Farwig, Y. Taniuchi, Uniqueness of almost periodic-in-time solutions to the Navier-Stokes equations in unbounded domains. J. Evol. Equ. **11**, 485–508 (2011)
18. C. Foias, S. Zaidman, Almost periodic solutions of parabolic systems. Ann. Scoula Norm. Sup. Pisa **15**, 247–262 (1961)
19. G. P. Galdi, *An Introduction to the Mathematical Theory of the Navier-Stokes Equations, Steady State Problems*. Springer Monographs in Math. 2nd edn. (Springer, Berlin, 2011)

20. G.P. Galdi, Existence and uniqueness of time-periodic solutions to the Navier-Stokes equations in the whole plane. Discret. Contin. Dyn. Syst. **6**, 1237–1257 (2013)
21. G.P. Galdi, On the time-periodic flow of a viscous liquid past a moving cylinder. Arch. Ration. Mech. Anal. **210**, 451–498 (2013)
22. G.P. Galdi, A.L. Silvestre, Existence of time-periodic solutions to the Navier-Stokes equations around a moving body. Pac. J. Math. **223**, 251–267 (2006)
23. G.P. Galdi, A.L. Silvestre, On the motion of a rigid body in a Navier-Stokes liquid under the action of a time-periodic force. Indiana Univ. Math. J. **58**, 2805–2842 (2009)
24. G.P. Galdi, H. Sohr, Existence and uniqueness of time-periodic physically reasonable Navier-Stokes flows past a body. Arch. Ration. Mech. Anal. **172**, 363–406 (2004)
25. M. Geissert, H. Heck, M. Hieber, *On the equation* $\operatorname{div} u = g$ *and Bogovskii's operator in Sobolev spaces of negative order* vol. 168. Partial differential equations and functional analysis, Operator Theory: Advances and Applications, (Birkhäuser, Basel, 2006), pp. 113–121
26. M. Geissert, H. Heck, M. Hieber, L_p-Theory of the Navier-Stokes flow in the exterior of a moving or rotating obstacle. J. Reine Angew. Math. **596**, 45–62 (2006)
27. M. Geissert, H. Heck, M. Hieber, I. Wood, The Ornstein-Uhlenbeck semigroup in exterior domains. Arch. Math. **85**, 554–562 (2005)
28. M. Geissert, M. Hieber, T.H. Nguyen, A general approach to time periodic incompressible viscous fluid flow problems. Arch. Rational Mech. Anal. **220**, 1095–1118 (2016)
29. Y. Giga, Analyticity of the semigroup generated by the Stokes operator on L_r-spaces. Math. Z. **178**, 297–329 (1981)
30. Y. Giga, Domains of fractional powers of the Stokes operator in L_r spaces. Arch. Ration. Mech. Anal. **89**, 251–265 (1985)
31. Y. Giga, Solutions for semilinear parabolic equations in L^p and regularity of weak solutions of the Navier-Stokes system. J. Differ. Equ. **61**, 186–212 (1986)
32. Y. Giga, K. Inui, A. Mahalov, S. Matsui, Uniform local solvability of the Navier-Stokes equations with the Coriolis force. Methods Appl. Anal. **12**, 381–394 (2005)
33. Y. Giga, K. Inui, A. Mahalov, J. Saal, Uniform global solvabiliy of the Navier-Stokes equations for nondecaying initial data. Indiana Univ. Math. J. **57**, 2775–2792 (2008)
34. Y. Giga, A. Mahalov, B. Nicolaenko, in *The Cauchy problem for the Navier-Stokes equations with spatially almost periodic initial data*. Mathematical Aspects of nonlinear dispersive Equations, Annals of Mathematics Studies (Princeton University Press, Princeton, 2007) **163** 213–222
35. Y. Giga, A. Mahalov, T. Yoneda, On a bound for amplitudes of the Navier-Stokes flow with almost periodic initial data. J. Math. Fluid Math. **13**, 459–467 (2011)
36. Y. Giga, N. Mizoguchi, On time periodic solutions of the Dirchilet problem for degenerate parabolic equations of nondivergence type. J. Math. Anal. Appl. **201**, 396–416 (1996)
37. Y. Giga, N. Mizoguchi, Existence of periodic solutions to equations of evolving curves. SIAM J. Math. Anal. **27**, 5–39 (1996)
38. Y. Giga, H. Sohr, On the Stokes operator in exterior domains. J. Fac. Sci. Univ. Tokyo Sect. IA Math., **36**, 103–130, (1989)
39. R. Haller-Dintelmann, J. Wiedl, Kolmogorov kernel estimates for the Ornstein-Uhlenback operator. Ann. Sc. Norm. Super. Cl. Sci **4**, 729–748 (2005)
40. J.G. Heywood, The Navier-Stokes equations: On the existence, regularity and decay of solutions. Indiana Univ. Math. J. **29**, 639–681 (1980)
41. M. Hieber, Y. Shibata, The Fujita-Kato approach to the Navier-Stokes equations in the rotational framework. Math. Z. **265**, 481–491 (2010)
42. T. Hishida, The nonstationary Stokes and Navier-Stokes flows through an aperture, in *Contributions to Current Challenges in Mathematical Fluid Mechanics*. Advances in Mathematical Fluid Mechanics, ed. by G.P. Galdi, et al. (Birkhäuser, Basel, 2004), pp. 79–123
43. T. Hishida, Y. Shibata, $L_p - L_q$ Estimate of the stokes operator and Navier-Stokes flows in the exterior of a rotating obstacle. Arch. Ration. Mech. Anal. **193**, 339–421 (2009)
44. S. Kaniel, M. Shinbrot, A reproductive property of the Navier-Stokes equations. Arch. Rational Mech. Anal. **24**, 363–369 (1967)

45. T. Kato, Strong L^p-solutions of Navier-Stokes equations in $\mathbb{R}^n$ with applications to weak solutions. Math. Z. **187**, 471–480 (1984)
46. T. Kobayashi, Y. Shibata, On the Oseen equation in the three dimensional exterior domains. Math. Ann. **310**, 1–45 (1998)
47. H. Kozono, M. Yamazaki, Exterior problem for the stationary Navier-Stokes equations in the Lorentz space. Math. Ann. **310**, 279–305 (1998)
48. H. Kozono, M. Nakao, Periodic solutions solutions to the Navier-Stokes equations in unbounded domains. Tohoku Math. J. **48**, 33–50 (1996)
49. H. Kozono, Y. Mashiko, R. Takada, Existence of periodic solutions and their asymptotic stability to the Navier-Stokes equations with Coriolis force. J. Evol. Equ. **14**, 565–601 (2014)
50. T. Kubo, Periodic solutions to the Navier-Stokes equations in a perturbed half space and an aperture domain. Math. Methods Appl. Sci. **28**, 1341–1357 (2005)
51. M. Kyed, The existence and regularity of time-periodic solutions to the three dimensional Navier-Stokes equations in the whole space. Nonlinearity **27**, 2909–2935 (2014)
52. B.M. Levitan, V.V. Zhikov, *Almost Periodic Functions and Differential Equations* (Cambridge University Press, Cambridge, 1982)
53. Y. Maekawa, J. Sauer, Maximal regularity for the time-periodic Stokes operator on unbounded and bounded domains. J. Math. Soc. Japan. (to appear)
54. P.-L. Lions, N. Masmoudi, Uniqueness of mild solutions of Navier-Stokes system in L^N. Comm. Partial Differ. Equ. **26**, 2211–2226 (2001)
55. P. Maremonti, Existence and stability of time periodic solutions to the Navier-Stokes equations in the whole space. Nonlinearity **4**, 503–529 (1991)
56. P. Maremonti, M. Padula, Existence, uniqueness, and attainability of periodic solutions of the Navier-Stokes equations in exterior domains. J. Math. Sci. **93**, 719–746 (1999)
57. T. Miyakawa, On nonstationary solutions of the Navier-Stokes equations in an exterior domain. Hiroshima Math. J. **12**, 115–140 (1982)
58. T. Miyakawa, Y. Teramoto, Existence and periodicity of weak solutions to the Navier-Stokes equations in a time dependent domain. Hiroshima Math. J. **12**, 513–528 (1982)
59. T.H. Nguyen, Periodic motions of Stokes and Navier-Stokes flows around a rotating obstacle. Arch. Ration. Mech. Anal. **213**, 689–703 (2014)
60. G. Prodi, Qualche risultato riguardo alle equazioni di Navier-Stokes nel caso bidimensionale. Rend. Sem. Mat. Univ. Padova **30**, 1–15 (1960)
61. G. Prouse, Soluzioni periodiche dell'equazione di Navier-Stokes. Atti Accad. Naz. Lincei Rend. Cl. Sci. Fis. Mat. Natur., **35**, 443-447, (1963)
62. J. Serrin, A note on the existence of periodic solutions of the Navier-Stokes equations. Arch. Rational Mech. Anal. **3**, 120–122 (1959)
63. Y. Shibata, *On the Oseen semigroup with rotating effect*, Functional Analysis and Evolution Equations (Birkhäuser, Basel, 2008), pp. 595–611
64. V. Stepanov, Sue quelques généralisations des fonctions presque périodiques. C.R. Acad. Sci. Paris, **181**, 90–92, (1925)
65. Y. Taniuchi, On stability solutions of periodic solutions in unbounded domains. Hokkaido Math. J. **28**, 147–173 (1999)
66. Y. Taniuchi, On the uniqueness of time-periodic solutions to the Navier-Stokes equations in unbounded domains. Math. Z. **261**, 597–615 (2009)
67. H. Triebel, *Interpolation Theory, Function Spaces, Differential Operators* (North-Holland, Amsterdam, 1978)
68. G. Van Baalen, P. Wittwer, Time periodic solutions of the Navier-Stokes equations with nonzero constant boundary conditions at infinity. SIAM J. Math. Anal. **43**, 1787–1809 (2011)
69. H. Weyl, Integralgleichungen und fastperiodische Funktionen. Math. Ann. **97**, 338–356 (1926)
70. M. Yamazaki, The Navier-Stokes equations in the weak-L^n space with time-dependent external force. Math. Ann. **317**, 635–675 (2000)
71. T. Yoshizawa, *Stability theory and the Existence of Periodic Solutions and Almost Periodic Solutions*, Applied Mathematical Sciences (Springer, Berlin, 1975)
72. V. Yudovich, Periodic motions of a viscous incompressible fluid. Sov. Math., Dokl., **1**, 168-172, (1960)

Remarks on Viscosity Solutions for Mean Curvature Flow with Obstacles

K. Ishii, H. Kamata and S. Koike

Abstract Obstacle problems for mean curvature flow equations are concerned. Existence of Lipschitz continuous viscosity solutions are obtained under several hypotheses. Comparison principle globally in time is also discussed.

Keywords Mean curvature flow with obstacles · Viscosity solutions

1 Introduction

In 1991, Chen, Giga, and Goto in [2], and Evans and Spruck in [5] started to study the motion of a hypersurface moving by its mean curvature via the level set method in order to analyze its behavior even after singularities occur. When an initial hypersurface is bounded, in the study of mean curvature flow equations, it is natural to suppose solutions to be constant outside a compact set in the space variables. Under this boundary condition, in [2, 5], the existence and uniqueness of solutions were obtained. We refer to Giga's survey [6] on the mean curvature flow equations via the level set approach.

On the other hand, if the initial hypersurface is unbounded, it is impossible to require solutions to be constant outside compact sets. To deal with this situation,

K. Ishii—KI was supported in part by Grant-in-Aid for Scientific Research (No. 23340028) of Japan Society for the Promotion of Science.
S. Koike—SK was supported in part by Grant-in-Aid for Scientific Research (No. 23340028) of Japan Society for the Promotion of Science.

K. Ishii (✉)
Graduate School of Maritime Science, Kobe University, Higashinada, Kobe 658-0022, Japan
e-mail: ishii@maritime.kobe-u.ac.jp

H. Kamata · S. Koike
Mathematical Institute, Graduate School of Sciences, Tohoku University, Aoba, Sendai 980-8578, Japan
e-mail: koike@math.tohoku.ac.jp

Y. Maekawa and S. Jimbo (eds.), *Mathematics for Nonlinear Phenomena—Analysis and Computation*, Springer Proceedings in Mathematics & Statistics 215,
DOI 10.1007/978-3-319-66764-5_5

Giga, Goto, Ishii, and Sato in [7] established the comparison principle even when solutions are allowed to grow linearly in x. Later, Ishii and Souganidis in [8] extended it to more general mean curvature-type equations with no hypothesis on the growth rate in x. We refer to related works [1, 11].

We are interested in the case when hypersurfaces move under some restrictions. For instance, if an ellipsoid moves via its mean curvature required not to exceed a fixed ball which is inside of the initial ellipsoid, then the final shape would be the ball even if their centers do not coincide. It is natural to ask how hypersurfaces behave in general when there are regions called obstacles which they cannot exceed. Moreover, if such obstacles move in time, then the hypersurface may become larger instead of shrinking to a fixed shape. Thus, to deal with various situations, we will discuss the uniqueness of unbounded viscosity solutions for obstacle problems.

In order to treat the motion of hypersurfaces under these restrictions, recently, Mercier in [9] proposed to study mean curvature flows with obstacles. More precisely, he obtained an existence result via Perron's method, and the comparison principle to show the uniqueness of bounded viscosity solutions. We refer to [10] for a related topics.

Our aim here is to study existence, uniqueness, and regularity for viscosity solutions of mean curvature flow with obstacles. More precisely, we obtain the comparison principle for viscosity solutions of the mean curvature flow equation with obstacles and of its approximate equations. Particularly, for approximate equations, we allow solutions to grow quadratically in the space variable. Furthermore, we show the existence of Lipschitz continuous viscosity solutions via Bernstein method for approximate equations with penalization and regularization.

This paper is organized as follows: In Section 2, we establish the comparison principle for unbounded viscosity solutions of obstacle problems for mean curvature flow. Section 3 is devoted to the comparison principle for approximate equations of obstacle problems. More precisely, we show the comparison principle for bounded viscosity solutions, and for viscosity solutions with quadratic growth in x. We obtain the existence of Lipschitz continuous viscosity solutions of obstacle problems for the mean curvature flow in a periodic setting in Section 4. In Appendix, we show the existence of periodic solutions for penalized problems.

2 Preliminaries

Setting $\mathbb{R}^n_0 := \mathbb{R}^n \setminus \{0\}$, and $\mathbb{S}^n := \{X \mid n \times n \text{ symmetric matrices}\}$, we define $F : \mathbb{R}^n_0 \times \mathbb{S}^n \to \mathbb{R}$ by

$$F(\xi, X) := -\mathrm{Tr}\left(X - \frac{\xi \otimes \xi}{|\xi|^2} X\right).$$

We denote by $B_r(x)$ the open ball in $\mathbb{R}^n$ with its center x and its radius $r > 0$. We shall simply write B_r for $B_r(0)$. For $\varepsilon \in (0, 1]$, we define

$$\langle x \rangle_\varepsilon := \sqrt{|x|^2 + \varepsilon} \quad \text{for } x \in \mathbb{R}^n.$$

We will also write $\langle x \rangle := \langle x \rangle_1$.

For $(\xi, X) \in \mathbb{R}^n \times \mathbb{S}^n$, we denote $F_*(\xi, X)$ and $F^*(\xi, X)$, respectively, by

$$F_*(\xi, X) := \liminf_{\varepsilon \to 0} \{F(\eta, Y) \mid (\eta, Y) \in \mathbb{R}^n_0 \times \mathbb{S}^n,\ \eta \in B_\varepsilon(\xi), |X - Y| < \varepsilon\},$$

and $F^*(\xi, X) := -(-F)_*(\xi, X)$. We notice that $F_*(0, O) = F^*(0, O) = 0$.

Given obstacles $\psi, \phi \in C(\mathbb{R}^n \times [0, \infty))$, we will define the notion of viscosity solutions of

$$\max\{\min\{u_t + F(Du, D^2u), u - \psi\}, u - \phi\} = 0 \quad \text{in } \mathbb{R}^n \times (0, \infty). \tag{1}$$

In general, we shall consider general second-order partial differential equations of parabolic type:

$$G(x, t, u, u_t, Du, D^2u) = 0 \quad \text{in } \mathbb{R}^n \times (0, \infty), \tag{2}$$

where $G : \mathbb{R}^n \times (0, \infty) \times \mathbb{R} \times \mathbb{R} \times \mathbb{R}^n_0 \times \mathbb{S}^n \to \mathbb{R}$ is a given continuous function. Note that, to deal with (1), it is sufficient to set

$$G(x, t, r, s, p, X) = \max\{\min\{s + F(p, X), r - \psi(x, t)\}, r - \phi(x, t)\}. \tag{3}$$

For $(x, t, r, s, p, X) \in \mathbb{R}^n \times (0, \infty) \times \mathbb{R} \times \mathbb{R} \times \mathbb{R}^n \times \mathbb{S}^n$, we define

$$G_*(x, t, r, s, p, X) := \liminf_{\varepsilon \to 0} \{G(\bar{x}, \bar{t}, \bar{r}, \bar{s}, \bar{p}, \bar{X}) \mid (\bar{x}, \bar{t}, \bar{r}, \bar{s}, \bar{p}, \bar{X}) \in U_\varepsilon(x, t, r, s, p, X)\},$$

and $G^*(x, t, r, s, p, X) := -(-G)_*(x, t, r, s, p, X)$, where $U_\varepsilon(x, t, r, s, p, X)$ is the set of points $(\bar{x}, \bar{t}, \bar{r}, \bar{s}, \bar{p}, \bar{X}) \in \mathbb{R}^n \times (0, \infty) \times \mathbb{R} \times \mathbb{R} \times \mathbb{R}^n_0 \times \mathbb{S}^n$ such that

$$\max\{|x - \bar{x}|, |t - \bar{t}|, |r - \bar{r}|, |s - \bar{s}|, |p - \bar{p}|, |X - \bar{X}|\} < \varepsilon.$$

Thus, if we take G as in (3), then $G_*(x, t, r, s, p, X)$ (resp., $G^*(x, t, r, s, p, X)$) becomes

$$\max\{\min\{s + F_*(p, X), r - \psi(x, t)\}, r - \phi(x, t)\}$$

$$\left(\text{resp., } \max\{\min\{s + F^*(p, X), r - \psi(x, t)\}, r - \phi(x, t)\}\right).$$

In order to present our definition, we introduce parabolic semijets $\mathscr{P}^{2,\pm}$ and their "closure" $\overline{\mathscr{P}}^{2,\pm}$ in the following manner: For $u : \mathbb{R}^n \times (0, \infty) \to \mathbb{R}$ and $(x, t) \in \mathbb{R}^n \times (0, \infty)$,

$$\mathscr{P}^{2,+}u(x,t) := \left\{ \begin{array}{l|l} (\tau, p, X) \in & u(x+h, t+r) \le u(x,t) + \tau r + \langle p, h\rangle \\ \mathbb{R} \times \mathbb{R}^n \times \mathbb{S}^n & +\frac{1}{2}\langle Xh, h\rangle + o(|h|^2 + |r|) \text{ as } |h| + |r| \to 0 \end{array} \right\},$$

$\mathscr{P}^{2,-}u(x,t) := -\mathscr{P}^{2,+}(-u)(x,t)$, and $\overline{\mathscr{P}}^{2,\pm}u(x,t)$ are given by

$$\left\{ \begin{array}{l|l} (\tau, p, X) \in & \exists (x_k, t_k) \in \mathbb{R}^n \times (0,\infty) \text{ and } (\tau_k, p_k, X_k) \in \mathscr{P}^{2,\pm}u(x_k, t_k) \\ \mathbb{R} \times \mathbb{R}^n \times \mathbb{S}^n & \text{such that } \lim_{k\to\infty} (x_k, t_k, u(x_k, t_k), p_k, X_k) = (x, t, u(x,t), p, X) \end{array} \right\}.$$

We adapt the following definition. We refer to [4, 6] for equivalent definitions.

Definition 2.1 *We call $u : \mathbb{R}^n \times (0,\infty) \to \mathbb{R}$ a viscosity subsolution (resp., supersolution) of (2) if u is locally bounded and,*

$$G_*(x, t, u^*(x,t), \tau, p, X) \le 0 \quad \big(\textit{resp., } G^*(x, t, u_*(x,t), \tau, p, X) \ge 0\big)$$

for any $(x,t) \in \mathbb{R}^n \times (0,\infty)$ and $(\tau, p, X) \in \overline{\mathscr{P}}^{2,+}u^*(x,t)$ (resp., $\overline{\mathscr{P}}^{2,-}u_*(x,t)$). We also call $u : \mathbb{R}^n \times (0,\infty) \to \mathbb{R}$ a viscosity solution of (2) if it is a viscosity sub- and supersolution of (2).

Although our proof of the comparison principle follows the argument in [8], we give it for the reader's convenience.

Theorem 2.2 *Assume that $\psi, \phi \in C(\mathbb{R}^n \times [0,\infty))$ satisfy that*

$$\psi \le \phi \quad \textit{in } \mathbb{R}^n \times [0,\infty), \tag{4}$$

and for each $T > 0$, there is $\omega_T \in C([0,\infty))$ such that $\omega_T(0) = 0$ and

$$|\psi(x,t) - \psi(y,t)| \vee |\phi(x,t) - \phi(y,t)| \le \omega_T(|x-y|) \quad \textit{for } x, y \in \mathbb{R}^n, t \in [0,T]. \tag{5}$$

Let u and $v : \mathbb{R}^n \times [0,\infty) \to \mathbb{R}$ be, respectively, a viscosity subsolution and a viscosity supersolution of (1). If we assume

$$\limsup_{r\to 0} \left\{ u^*(x,0) - v_*(y,0) \,\middle|\, x, y \in \mathbb{R}^n, |x-y| \le r \right\} \le 0, \tag{6}$$

then we have

$$u^* \le v_* \quad \textit{in } \mathbb{R}^n \times [0,\infty). \tag{7}$$

Proof. In what follows, we shall write u and v for u^* and v_*, respectively. Note that F is geometric, i.e.,

$$F(\lambda p, \lambda X + \nu p \otimes p) = \lambda F(p, X) \quad \text{for } (p, X, \lambda, \nu) \in \mathbb{R}^n_0 \times \mathbb{S}^n \times (0,\infty) \times \mathbb{R}.$$

For any bounded increasing function $\xi \in C^1(\mathbb{R})$ such that $0 < \xi'(r)$, it is easy to see that $\xi(u)$ (resp., $\xi(v)$) is a bounded viscosity subsolution (resp., supersolution) of

$$\max\{\min\{u_t + F(Du, D^2u), u - \xi(\psi)\}, u - \xi(\phi)\} = 0 \quad \text{in } \mathbb{R}^n \times (0, \infty).$$

Moreover, we can verify (6) replacing u and v by $\xi(u)$ and $\xi(v)$.

Fix any $T > 0$. Now, setting $U := \xi(u)$, $V := \xi(v)$, and

$$\theta := \lim_{r \to 0} \sup \left\{ U(x,t) - V(y,t) \;\middle|\; x, y \in \mathbb{R}^n, |x - y| \leq r, t \in [0, T] \right\}, \quad (8)$$

we suppose $\theta > 0$. Then, we will get a contradiction, which yields $U \leq V$ in $\mathbb{R}^n \times [0, T]$. Hence, we obtain $u \leq v$ in $\mathbb{R}^n \times [0, T]$ for each $T > 0$.

Let us suppose that u and v are bounded, and $\theta > 0$ holds with $(U, V) := (u, v)$. For $\alpha, \delta, \gamma > 0$, and $(x, t), (y, s) \in \mathbb{R}^n \times [0, T]$, we set

$$\Phi_\delta^{\alpha,\gamma}(x, y, t) := u(x, t) - v(y, t) - \frac{\alpha}{4}|x - y|^4 - \delta(\langle x \rangle + \langle y \rangle) - \frac{\gamma}{T - t}.$$

We claim that for $k \in \mathbb{N}$, there are $\alpha_k > k$ and $\gamma_k, \delta_k > 0$ such that for $\delta \in (0, \delta_k)$,

$$\begin{aligned} &\sup \left\{ \Phi_\delta^{\alpha_k,\gamma_k}(x, y, t) \;\middle|\; x, y \in \mathbb{R}^n, t \in [0, T) \right\} \\ &= \sup \left\{ \Phi_\delta^{\alpha_k,\gamma_k}(x, y, t) \;\middle|\; x, y \in \mathbb{R}^n, t \in [0, T), \delta(\langle x \rangle + \langle y \rangle) \leq M + 1 \right\}, \end{aligned} \quad (9)$$

where $M := \sup\{u(x, t) - v(y, t) \mid x, y \in \mathbb{R}^n, 0 \leq t < \infty\}$. Indeed, letting $r_k \in (0, \frac{1}{k})$, we find $x_k, y_k \in \mathbb{R}^n$ and $t_k \in [0, T]$ such that

$$\theta - \frac{1}{3k} \leq u(x_k, t_k) - v(y_k, t_k), \quad \text{and} \quad |x_k - y_k| < r_k.$$

Setting $\alpha_k := r_k^{-3}$, we have

$$u(x_k, t_k) - v(y_k, t_k) - \frac{\alpha_k}{4}|x_k - y_k|^4 \geq \theta - \frac{2}{3k}.$$

We can select $\delta_k, \gamma_k \in (0, 1]$ such that

$$\theta - \frac{1}{k} \leq \Phi_\delta^{\alpha_k,\gamma_k}(x_k, y_k, t_k) \quad \text{for } \delta \in (0, \delta_k),$$

which implies that $M - \theta + 1 > 0$. We note that if $\delta(\langle x \rangle + \langle y \rangle) > M - \theta + 1$, then $\Phi_\delta^{\alpha_k,\gamma_k}(x, y, t) < \theta - \frac{1}{k}$. Therefore, we have shown the claim (9).

Now, let $(x_\delta^k, y_\delta^k, t_\delta^k) \in \mathbb{R}^n \times \mathbb{R}^n \times [0, T)$ be the point such that the mapping $(x, y, t) \in \mathbb{R}^n \times \mathbb{R}^n \times [0, T) \to \Phi_\delta^{\alpha_k,\gamma_k}(x, y, t)$ attains its maximum for $\delta \in (0, \delta_k)$.

We shall simply write x, y, t, and Φ, respectively, for $x_\delta^k, y_\delta^k, t_\delta^k$, and $\Phi_\delta^{\alpha_k,\gamma_k}$. Since it follows that

$$0 < \theta - \frac{1}{k} \leq \Phi(x, y, t) \leq M - \frac{\alpha_k}{4}|x - y|^4, \quad (10)$$

we have $|x - y| \leq 2\left(4\alpha_k^{-1}(M - \theta + 1)\right)^{1/4} =: R_k$. Hence, we have

$$\lim_{k\to\infty}\sup_{\delta\in(0,\delta_k)}|x-y|=0. \tag{11}$$

Setting $\hat{u}(x,t) := u(x,t) - \delta\langle x\rangle$, $\hat{v}(y,t) := v(y,t) + \delta\langle y\rangle$, and $p := x - y$, in view of Ishii's lemma (c.f. Theorem 7 in [3]), we can find $X = X_\delta^k$, $Y = Y_\delta^k \in \mathbb{S}^n$, and $\tau = \tau_\delta^k \in \mathbb{R}$ such that

$$\left(\tau + \frac{\gamma_k}{(T-t)^2}, \alpha_k|p|^2 p, X\right) \in \overline{\mathscr{P}}^{2,+}\hat{u}(x,t), \quad (\tau, \alpha_k|p|^2 p, Y) \in \overline{\mathscr{P}}^{2,-}\hat{v}(y,t), \tag{12}$$

$$\text{and} \quad -(\alpha_k + \|A\|)I_{2n} \le \begin{pmatrix} X & O \\ O & -Y \end{pmatrix} \le A + \frac{1}{\alpha_k}A^2. \tag{13}$$

Here and later, I_ℓ is the identity matrix of order $\ell \in \mathbb{N}$. Since $A \in \mathbb{S}^{2n}$ is given by

$$A = \alpha_k\left\{|p|^2\begin{pmatrix} I_n & -I_n \\ -I_n & I_n \end{pmatrix} + 2\begin{pmatrix} p\otimes p & -p\otimes p \\ -p\otimes p & p\otimes p \end{pmatrix}\right\},$$

we easily have

$$A^2 = 2\alpha_k^2\left\{|p|^4\begin{pmatrix} I_n & -I_n \\ -I_n & I_n \end{pmatrix} + 8|p|^2\begin{pmatrix} p\otimes p & -p\otimes p \\ -p\otimes p & p\otimes p \end{pmatrix}\right\}.$$

By (13), we find $K_1 > 0$, independent of k and $\delta \in (0, \delta_k)$, such that

$$-K_1\alpha_k(1+|p|^2)I_n \le Z \le K_1\alpha_k|p|^2(1+|p|^2)I_n, \quad \text{for } Z = X, -Y. \tag{14}$$

On the other hand, in view of (12), it follows that

$$\left(\tau + \frac{\gamma_k}{(T-t)^2}, \alpha_k|p|^2 p + \frac{\delta x}{\langle x\rangle}, X + \frac{\delta}{\langle x\rangle}I_n - \frac{\delta x\otimes x}{\langle x\rangle^3}\right) \in \overline{\mathscr{P}}^{2,+}u(x,t),$$

and

$$\left(\tau, \alpha_k|p|^2 p - \frac{\delta y}{\langle y\rangle}, Y - \frac{\delta}{\langle y\rangle}I_n + \frac{\delta y\otimes y}{\langle y\rangle^3}\right) \in \overline{\mathscr{P}}^{2,-}v(y,t).$$

Now, we shall first suppose that for $k \in \mathbb{N}$ and $\delta \in (0, \delta_k)$,

$$u(x,t) > \psi(x,t) \quad \text{and} \quad v(y,t) < \phi(y,t).$$

Hence, from the definition of viscosity solutions, we have

$$\tau + \frac{\gamma_k}{(T-t)^2} + F_*\left(\alpha_k|p|^2 p + \frac{\delta x}{\langle x\rangle}, X + \frac{\delta}{\langle x\rangle}I_n - \frac{\delta x \otimes x}{\langle x\rangle^3}\right)$$
$$\leq \tau + F^*\left(\alpha_k|p|^2 p - \frac{\delta y}{\langle y\rangle}, Y - \frac{\delta}{\langle y\rangle}I_n + \frac{\delta y \otimes y}{\langle y\rangle^3}\right). \quad (15)$$

Since (14) together with (11) implies that X_δ^k and Y_δ^k are bounded in $\delta \in (0, \delta_k)$, we may suppose that there are $\hat{X} := X^k, \hat{Y} := Y^k \in \mathbb{S}^n$ such that

$$\lim_{\delta\to 0}(X, Y) = \lim_{\delta\to 0}(X_\delta^k, Y_\delta^k) = (\hat{X}, \hat{Y}).$$

First, we consider the case of $\liminf_{\delta\to 0}|p| = 0$. Hence, (14) implies

$$\hat{X} \leq O \leq \hat{Y}.$$

Therefore, by taking $\delta \to 0$ in (15) along a subsequence if necessary, there is $t_k \in [0, T)$ such that

$$\frac{\gamma_k}{(T-t_k)^2} \leq F^*(0, \hat{Y}) - F_*(0, \hat{X}) \leq F^*(0, O) - F_*(0, O) = 0,$$

which gives a contradiction.

On the contrary, if we suppose $\inf_{\delta\in(0,\delta_k)}|p| > 0$, then we may suppose $\lim_{\delta\to 0} p = p_0$ for some $p_0 \in \mathbb{R}_0^n$. In this case, sending $\delta \to 0$ in (15) again, for some $t_k \in [0, T)$, we have

$$\frac{\gamma_k}{(T-t_k)^2} \leq F(\alpha_k|p_0|^2 p_0, \hat{Y}) - F(\alpha_k|p_0|^2 p_0, \hat{X}), \quad (16)$$

For $\xi = (\xi_1, \ldots, \xi_n) \in \mathbb{R}^n$, setting $\hat{\xi} = (\xi_1, \ldots, \xi_n, \xi_1, \ldots, \xi_n) \in \mathbb{R}^{2n}$, we have $A\hat{\xi} = 0$. Hence, since (13) implies $X_\delta^k \leq Y_\delta^k$, we have $\hat{X} \leq \hat{Y}$, which contradicts to (16)

Therefore, we may suppose that

$$(u - \psi)(x, t) \leq 0 \quad \text{or} \quad (v - \phi)(y, t) \geq 0. \quad (17)$$

We shall only show a contradiction when the second inequality in (17) holds for $\delta \in (0, \delta_k)$ since we can argue in a similar way for the other case. Noting $(u - \phi)(x, t) \leq 0$ by the definition of viscosity subsolutions of (1), we have

$$\theta - \frac{1}{k} \leq u(x,t) - v(y,t) \leq \phi(x,t) - \phi(y,t) \leq \omega_T(|x-y|) \leq \omega_T\left(\sup_{0<\delta<\delta_k}|x_\delta^k - y_\delta^k|\right),$$

which yields a contradiction for large $k \in \mathbb{N}$ by (11). □

Remark 2.3 *If we know* $\sup_{\mathbb{R}^n\times(0,T)} u + \sup_{\mathbb{R}^n\times(0,T)} (-v) < \infty$ *a priori for each* $T > 0$, *then we can show a stronger conclusion than (7),*

$$\lim_{r\to 0} \sup\{u(x,t) - v(y,t) \mid |x-y| \le r,\ t \in [0,T]\} \le 0,$$

because we can take $\xi(r) = r$ *in this case.*

3 Penalization and Regularization

Our approximate equations for (1) are as follows: for $\varepsilon \in (0, 1]$,

$$u_t + F_\varepsilon(Du, D^2u) + \beta_\varepsilon(u-\phi) - \beta_\varepsilon(\psi - u) = 0 \quad \text{in } \mathbb{R}^n \times (0,\infty). \tag{18}$$

Here, letting $\beta \in C^\infty(\mathbb{R})$ satisfy that

$$\beta'(r) > 0,\ \beta''(r) \ge 0 \text{ for } r > 0,\ \beta(r) = 0 \text{ for } r \le 0, \text{ and } \beta(r) = 2r - 1 \text{ for } r \ge 1,$$

we define $\beta_\varepsilon := \varepsilon^{-1}\beta$. We also define

$$F_\varepsilon(\xi, X) := -\mathrm{Tr}\left(\left\{(1+\varepsilon)I - \frac{\xi\otimes\xi}{\langle\xi\rangle_\varepsilon^2}\right\}X\right) \quad \text{for } (\xi, X) \in \mathbb{R}^n \times \mathbb{S}^n.$$

We first show the case when u and $-v$ are bounded from above because our argument is almost same as in the proof of Theorem 2.2.

Theorem 3.1 *Assume that* $\psi, \phi \in C(\mathbb{R}^n \times [0,\infty))$ *satisfy (5). Let* u *and* $v : \mathbb{R}^n \times [0,\infty) \to \mathbb{R}$ *be, respectively, a viscosity subsolution and a viscosity supersolution of (18). If we assume that for each* $T > 0$,

$$M_T := \sup\{u^*(x,t) - v_*(y,s) \mid (x,t), (y,s) \in \mathbb{R}^n \times [0,T]\} < \infty, \tag{19}$$

and (6), then we have (7).

Remark 3.2 *The reason why we have to assume (19) is that we cannot use the composition with monotone function because the left hand side of (18) is not geometric.*

Proof. We shall simply write u and v for u^* and v_*, respectively.

We use exactly the same argument and notations as in the proof of Theorem 2.2 to arrive at (12) and (13). Thus, we have

$$\begin{aligned} &\tau + \frac{\gamma_k}{(T-t)^2} + F_\varepsilon\left(\alpha_k|p|^2 p + \frac{\delta x}{\langle x\rangle}, X + \frac{\delta}{\langle x\rangle} I_n - \frac{\delta x \otimes x}{\langle x\rangle^3}\right) \\ &\quad \le \beta_\varepsilon((\psi - u)(x,t)) - \beta_\varepsilon((u-\phi)(x,t)), \end{aligned} \tag{20}$$

and

$$\tau + F_\varepsilon\left(\alpha_k|p|^2 p - \frac{\delta y}{\langle y\rangle}, Y - \frac{\delta}{\langle y\rangle}I_n + \frac{\delta y \otimes y}{\langle y\rangle^3}\right) \geq \beta_\varepsilon((\psi - v)(y,t)) - \beta_\varepsilon((v-\phi)(y,t)). \tag{21}$$

For the right hand side of (20) and (21), we can find $\rho_j \in \mathbb{R}$ $(j = 1, 2)$ such that

$$(\beta_\varepsilon(\psi - u) - \beta_\varepsilon(u-\phi))(x,t) - (\beta_\varepsilon(\psi - v) - \beta_\varepsilon(v-\phi))(y,t) \leq \{\beta_\varepsilon'(\rho_1) + \beta_\varepsilon'(\rho_2)\}\{v(y,t) - u(x,t) + \omega_T(|x-y|)\}. \tag{22}$$

Recalling $(x, y, t) = (x_\delta^k, y_\delta^k, t_\delta^k)$, we have

$$\theta - \frac{1}{k} \leq u(x_\delta^k, t_\delta^k) - v(y_\delta^k, t_\delta^k),$$

which implies that the right hand side of (22) is nonpositive by (11) for large $k \in \mathbb{N}$. Hence, for such $k \in \mathbb{N}$, (20) and (21) yield

$$\frac{\gamma_k}{(T-t)^2} \leq F_\varepsilon\left(\alpha_k|p|^2 p - \frac{\delta y}{\langle y\rangle}, Y - \frac{\delta}{\langle y\rangle}I_n + \frac{\delta y \otimes y}{\langle y\rangle^3}\right) - F_\varepsilon\left(\alpha_k|p|^2 p + \frac{\delta x}{\langle x\rangle}, X + \frac{\delta}{\langle x\rangle}I_n - \frac{\delta x \otimes x}{\langle x\rangle^3}\right).$$

Therefore, sending $\delta \to 0$ along a subsequence if necessary, we have

$$\frac{\gamma_k}{(T-t_k)^2} \leq F_\varepsilon(p_k, \hat{Y}) - F_\varepsilon(p_k, \hat{X})$$

for some $p_k \in \mathbb{R}^n$ and $t_k \in [0, T)$, which contradicts to $\hat{X} \leq \hat{Y}$. □

It is known in [7] that, under suitable hypotheses, the comparison principle holds for viscosity solutions growing at most linearly in x of the mean curvature flow equation in unbounded domains with no obstacles. We will modify the argument in [7] to allow the case when viscosity solutions may have quadratic growth in x. To this end, we first observe the following result.

Lemma 3.3 *(cf. Proposition 2.3 in [7]) Assume that $\psi, \phi : \mathbb{R}^n \times [0,\infty) \to \mathbb{R}$ satisfy (5). Let u and $v : \mathbb{R}^n \times [0,\infty) \to \mathbb{R}$ be, respectively, a viscosity subsolution and a viscosity supersolution of (18) such that for any $T > 0$, there exists $A_T > 0$ satisfying the following properties:*

$$\max\{u^*(x,t), -v_*(x,t)\} \leq A_T(1 + |x|^2) \quad \textit{for } (x,t) \in \mathbb{R}^n \times [0,T], \tag{23}$$

$$u^*(x,0) - v_*(y,0) \leq A_T\langle x - y\rangle \quad \textit{for } x, y \in \mathbb{R}^n. \tag{24}$$

Then, there exist $L_T > 0$ and $M_T > 0$ such that

$$u^*(x,t) - v_*(y,t) \le L_T\langle x-y\rangle + \frac{M_T}{T-t} \quad \textit{for } x, y \in \mathbb{R}^n \textit{ and } t \in [0,T).$$

Remark 3.4 *Since the argument below works for mean curvature flow equations with no obstacles, the conclusion in Theorem 3.5 below is stronger than that of Theorem 2.2.*

Proof. Again, we shall write u and v for u^* and v_*, respectively. Fix $T > 0$.

We choose $L_T > 2A_T$ such that

$$L_T(r^2+1)^{1/2} \ge \omega_T(r) \quad \text{for } r \ge 0,$$

where ω_T comes from (5). We shall delete T from $A_T, L_T > 0$ in what follows.

We next introduce nonnegative nondecreasing functions $\zeta_R \in C^2([0,\infty))$ for $R > 1$ (cf. p.447 of [7]) such that

$$\zeta_R(r) = \begin{cases} 0 & \text{for } 0 \le r \le R, \\ r^2 & \text{for } r \ge 2R, \end{cases} \quad \text{and} \quad B_0 := \sup_{R\ge 1}\sup_{r \ge 1}\left\{\frac{1}{r}\zeta_R'(r) + |\zeta_R''(r)|\right\} < \infty.$$

It is easy to verify

$$\gamma_1 := \sup\{|F_\varepsilon(p,X)| \mid p \in \mathbb{R}^n, |X| \le L + 2AB_0, 0 < \varepsilon < 1\} < \infty.$$

For $R, M > 0$, setting

$$\xi_M^R(x,y,t) := L\langle x-y\rangle + A\{\zeta_R(|x|) + \zeta_R(|y|)\} + \frac{M}{T-t} \quad \text{for } (x,y,t) \in \mathbb{R}^n \times \mathbb{R}^n \times [0,T),$$

we define

$$\Theta_M^R := \sup\left\{u(x,t) - v(y,t) - \xi_M^R(x,y,t) \;\middle|\; (x,y,t) \in \mathbb{R}^n \times \mathbb{R}^n \times [0,T)\right\}.$$

We claim that if $M_T = 2T^2\gamma_1$, then $\Theta_{M_T}^R \le 0$ for $R > 1$, which yields

$$u(x,t) - v(y,t) \le L\langle x-y\rangle + 2A\{\zeta_R(|x|) + \zeta_R(|y|)\} + \frac{M_T}{T-t}$$

for $x, y \in \mathbb{R}^n, 0 \le t < T$. Thus, sending $R \to \infty$, we conclude the proof.

In order to verify our claim, suppose $\Theta_{M_T}^R > 0$ for $R > 1$ and, we will get a contradiction. Since M_T and $R > 1$ are fixed, we shall simply write ξ and Θ for $\xi_{M_T}^R$ and $\Theta_{M_T}^R$, respectively.

For $\delta \in (0,1)$, setting $\eta_\delta(x,y,t,s) = \xi(x,y,t) + \frac{1}{2\delta}(t-s)^2$, we note

$$\Theta_\delta := \sup_{(x,t),(y,s)\in\mathbb{R}^n\times[0,T)} \{u(x,t) - v(y,s) - \eta_\delta(x,y,t,s)\} \ge \Theta > 0.$$

We first note that if $|x| \vee |y| \geq 2R$, then (23) and $L > 2A$ imply that for $\delta > 0$,

$$u(x,t) - v(y,s) - \eta_\delta(x,y,t,s) < 0 \quad \text{for } t, s \in [0,T). \tag{25}$$

Hence, there are $(x,y,t,s) \in B_{2R} \times B_{2R} \times [0,T) \times [0,T]$ such that

$$\Theta_\delta = u(x,t) - v(y,s) - \eta_\delta(x,y,t,s) > 0.$$

In view of (23), we have

$$\frac{(t-s)^2}{2\delta} + \frac{M_T}{T-t} \leq 2A(1+4R^2).$$

Assume next that $t = 0$, then for small $\delta > 0$, by (24), we have $u(x,0) - v(y,s) \leq A\langle x-y\rangle + v(y,0) - v(y,s) < 2A\langle x-y\rangle$, which yields $\Theta_\delta < 0$. Thus, we may suppose $t > 0$. We may also suppose $s > 0$ by the same reason. Hence, we verify that

$$\left(\frac{t-s}{\delta} + \frac{M_T}{(T-t)^2}, D_x\xi^R(x,y,t), D_x^2\xi^R(x,y,t)\right) \in \mathscr{P}^{2,+}u(x,t),$$

and

$$\left(\frac{t-s}{\delta}, -D_y\xi^R(x,y,t), -D_y^2\xi^R(x,y,t)\right) \in \mathscr{P}^{2,-}v(y,t).$$

Writing $U = u(x,t)$ and $V = v(y,t)$ for simplicity, we shall omit writing (x,y,t) for $D_x\xi^R$, $D_y\xi^R$, $D_x^2\xi^R$, $D_y^2\xi^R$. From the definition, we have

$$\begin{aligned}&\frac{M_T}{(T-t)^2} + F_\varepsilon(D_x\xi^R, D_x^2\xi^R) + \beta_\varepsilon(U - \phi(x,t)) - \beta_\varepsilon(\psi(x,t) - U)\\ &\quad \leq F_\varepsilon(-D_y\xi^R, -D_y^2\xi^R) + \beta_\varepsilon(V - \psi(y,t)) - \beta_\varepsilon(\phi(y,t) - V).\end{aligned}$$

Thus, we find $\rho_j \in \mathbb{R}$ ($j = 1,2$) such that

$$\begin{aligned}\frac{M_T}{(T-t)^2} &\leq F_\varepsilon(-D_y\xi^R, -D_y^2\xi^R) - F_\varepsilon(D_x\xi^R, D_x^2\xi^R)\\ &\quad -\{\beta_\varepsilon'(\rho_1) + \beta_\varepsilon'(\rho_2)\}\{U - V - \omega_T(|x-y|)\}\end{aligned} \tag{26}$$

However, by the choice of η_δ, we have $U - V \geq L\langle x-y\rangle$. Hence, (26) yields

$$\frac{M_T}{T^2} < \frac{M_T}{(T-t)^2} \leq F_\varepsilon(-D_y\xi^R, -D_y^2\xi^R) - F_\varepsilon(D_x\xi^R, D_x^2\xi^R). \tag{27}$$

At (y,t), we thus have

$$D_y^2\xi^R = L\left(\frac{I_n}{\langle p\rangle} - \frac{p\otimes p}{\langle p\rangle^3}\right) + 2A\left\{\zeta_R'(|y|)\left(\frac{I_n}{|y|} - \frac{y\otimes y}{|y|^3}\right) + \zeta_R''(|y|)\frac{y\otimes y}{|y|^2}\right\}, \tag{28}$$

where $p := x - y$. We note that when $y = 0$, the last term of the right hand side denotes 0. Since (28) implies $-(L + 2AB_0)I_n \le -D_y^2\xi^R \le (L + 2AB_0)I_n$, it follows $F_\varepsilon(-D_y\xi^R, -D_y^2\xi^R) \le \gamma_1$. Similarly, we obtain $-F_\varepsilon(D_x\xi^R, D_x^2\xi^R) \le \gamma_1$. Therefore, we obtain a contradiction by (27). □

Our comparison principle for unbounded viscosity solutions of (18) is as follows:

Theorem 3.5 *(cf. Theorem 2.1 in [7]) Assume $\psi, \phi \in C(\mathbb{R}^n \times [0,\infty))$ satisfy (5). Let u and $v : \mathbb{R}^n \times [0,\infty) \to \mathbb{R}$ be, respectively, a viscosity subsolution and a viscosity supersolution of (18) satisfying (23). If there is $\hat{\omega} \in C([0,\infty))$ such that*

$$\hat{\omega}(0) = 0, \quad \text{and} \quad u^*(x,0) - v_*(y,0) \le \hat{\omega}(|x-y|) \quad \text{for } x, y \in \mathbb{R}^n, \tag{29}$$

then we obtain that for each $T > 0$,

$$\limsup_{r\to 0}\{u^*(x,t) - v_*(y,t) \mid x, y \in \mathbb{R}^n, |x-y| \le r, t \in [0,T]\} \le 0.$$

Remark 3.6 *The above conclusion immediately yields (7).*

Proof. We shall simply write u and v as before. For fixed $T > 0$, suppose

$$\theta := \limsup_{r\to 0}\{u(x,t) - v(y,t) \mid x, y \in \mathbb{R}^n, |x-y| \le r, t \in [0,T)\} > 0,$$

and then we will get a contradiction. By Lemma 3.3, we notice $\theta < \infty$ since

$$u(x,t) - v(y,t) \le L_{2T}\langle x-y\rangle + \frac{M_{2T}}{2T-t} \quad \text{for } x, y \in \mathbb{R}^n \text{ and } 0 \le t \le T. \tag{30}$$

For any $k \in \mathbb{N}$, we can choose $r_k \in (0, \frac{1}{k})$, $x_k, y_k \in \mathbb{R}^n$, and $0 \le t_k < T$ such that

$$u(x_k,t_k) - v(y_k,t_k) > \theta - \frac{1}{3k}, \quad \text{and} \quad |x_k - y_k| < r_k.$$

Setting $\alpha_k := r_k^{-3} > k^3$, we find $\delta_k > 0$ and $\gamma_k \in (0,1)$ such that

$$u(x_k,t_k) - v(y_k,t_k) - \frac{\alpha_k}{4}|x_k - y_k|^4 - \frac{\delta}{3}(\langle x_k\rangle^3 + \langle y_k\rangle^3) - \frac{\gamma_k}{T - t_k} > \theta - \frac{1}{k}$$

for $\delta \in (0, \delta_k)$. For $\delta \in (0, \delta_k)$, setting

$$\Phi_\delta(x,y,t) := \frac{\alpha_k}{4}|x-y|^4 + \frac{\delta}{3}(\langle x\rangle^3 + \langle y\rangle^3) + \frac{\gamma_k}{T-t},$$

we note $\Theta_\delta := \sup\{u(x,t) - v(y,t) - \Phi_\delta(x,y,t) \mid x, y \in \mathbb{R}^n, t \in [0,T)\} \geq \theta - \frac{1}{k} > 0$ for large $k \in \mathbb{N}$. By (23), we can choose $x_\delta, y_\delta \in \mathbb{R}^n$ and $t_\delta \in [0,T)$ such that

$$\Phi_\delta(x_\delta, y_\delta, t_\delta) = \Theta_\delta > 0,$$

In view of (30) again, we find $\hat{L} > 0$, independent of $\delta \in (0, \delta_k)$, such that

$$\frac{\alpha_k}{4}|x_\delta - y_\delta|^4 \leq \hat{L}(1 + |x_\delta - y_\delta|),$$

which implies $\hat{A} := \sup\{|x_\delta - y_\delta| \mid \delta \in (0, \delta_k)\} < \infty$. Hence, we have

$$\frac{\alpha_k}{4}|x_\delta - y_\delta|^4 \leq \hat{L}(1 + \hat{A}). \tag{31}$$

On the other hand, since $\Theta_\delta > 0$ yields

$$\frac{\delta}{3}(|x_\delta|^3 + |y_\delta|^3) \leq \hat{L}(1 + \hat{A}),$$

we find $\hat{C} > 0$ such that

$$|x_\delta| + |y_\delta| \leq \hat{C}\delta^{-\frac{1}{3}}. \tag{32}$$

In view of (29), we may suppose $t > 0$. In fact, if $t = 0$, then there is $C' > 0$ such that

$$\theta - \frac{1}{k} \leq \hat{\omega}(|x_\delta - y_\delta|) \leq \hat{\omega}\left(C'k^{-\frac{3}{4}}\right),$$

which is a contradiction for large $k \in \mathbb{N}$ by (31). In what follows, we shall omit writing $\delta \in (0, \delta_k)$. Thanks to Theorem 7 in [3], we can find $X = X_\delta, Y = Y_\delta \in S^n$ and $\tau = \tau_\delta \in \mathbb{R}$ such that

$$\left(\tau + \frac{\gamma_k}{(T-t)^2}, \alpha_k|p|^2 p + \delta\langle x\rangle x, X + \delta\langle x\rangle I_n + \frac{\delta x \otimes x}{\langle x\rangle}\right) \in \overline{\mathscr{P}}^{2,+}u(x,t),$$

$$\left(\tau, \alpha_k|p|^2 p - \delta\langle y\rangle y, Y - \delta\langle y\rangle I_n - \frac{\delta y \otimes y}{\langle y\rangle}\right) \in \overline{\mathscr{P}}^{2,-}v(y,t),$$

and

$$-K_1\alpha_k|p|^2 I_{2n} \leq \begin{pmatrix} X & O \\ O & -Y \end{pmatrix} \leq A + \frac{1}{\alpha_k}A^2,$$

where $K_1 > 0$ is independent of $\delta \in (0, \delta_k)$. Here, we set $p = x - y$, and

$$A := \alpha_k|p|^2 \begin{pmatrix} I_n & -I_n \\ -I_n & I_n \end{pmatrix} + 2\alpha_k \begin{pmatrix} p \otimes p & -p \otimes p \\ -p \otimes p & p \otimes p \end{pmatrix}.$$

As in the proof of Theorem 2.2, we observe X, Y are bounded in $\delta \in (0, \delta_k)$, and $X \le Y$. Hence, we may suppose $\hat{X} := \lim_{\delta\to 0} X_\delta$ and $\hat{Y} := \lim_{\delta\to 0} Y_\delta$ along a subsequence of δ if necessary. Thus, we have

$$\hat{X} \le \hat{Y}.$$

By the definition, setting $q = q_\delta =: \alpha_k |x-y|^2(x-y)$, we have

$$\tau + \frac{\gamma_k}{(T-t)^2} + F_\varepsilon\left(q + \delta\langle x\rangle x, X + \delta\langle x\rangle I_n + \frac{\delta x\otimes x}{\langle x\rangle}\right)$$
$$\le -\beta_\varepsilon((u-\phi)(x,t)) + \beta_\varepsilon((\psi-u)(x,t)),$$

and

$$\tau + F_\varepsilon\left(q - \delta\langle y\rangle y, Y - \delta\langle y\rangle I_n - \frac{\delta y\otimes y}{\langle y\rangle}\right) \ge \beta_\varepsilon((v-\phi)(y,t)) - \beta_\varepsilon((\psi-v)(y,t)).$$

Hence, using these inequalities, we can find $\rho_j \in \mathbb{R}$ $(j=1,2)$ such that

$$\begin{aligned}\frac{\gamma}{(T-t)^2} \le & -F_\varepsilon\left(q + \delta\langle x\rangle x, X + \delta\langle x\rangle I_n + \frac{\delta x\otimes x}{\langle x\rangle}\right)\\ & +F_\varepsilon\left(q - \delta\langle y\rangle y, Y - \delta\langle y\rangle I_n - \frac{\delta y\otimes y}{\langle y\rangle}\right)\\ & -\{\beta_\varepsilon'(\rho_1) + \beta_\varepsilon'(\rho_2)\}\{u(x,t) - v(y,t) - \omega_T(|x-y|)\},\end{aligned}$$

The third term of the right hand side of this inequality is nonpositive for large $k \in \mathbb{N}$ because $\beta_\varepsilon' \ge 0$ holds, and (31) implies

$$u(x,t) - v(y,t) - \omega_T(|x-y|) \ge \theta - \omega_T(|x-y|) \ge 0.$$

Sending $\delta \to 0$ in the above with (32), we find $(\hat{t}, \hat{q}) \in [0,T) \times \mathbb{R}^n$ such that $\frac{\gamma_k}{(T-\hat{t})^2} \le F_\varepsilon(\hat{q}, \hat{Y}) - F_\varepsilon(\hat{q}, \hat{X})$, which is a contradiction. □

4 Existence of Lipschitz Continuous Solutions

Since the comparison principle for (18) in Section 3 yields the uniqueness of viscosity solutions, when ϕ, ψ, and g are periodic in $x \in \mathbb{R}^n$, the viscosity solution is periodic in x. Keeping this fact in mind, in this section, we establish the existence of Lipschitz continuous (periodic) viscosity solutions of (1).

For notational simplicity, we define $Q_T := \mathbb{T}^n \times [0,T]$ for $T \in (0,\infty)$, and $Q_\infty := \mathbb{T}^n \times [0,\infty)$. For $0 < T \le \infty$, we denote by $W^{2,\infty}(\mathbb{T}^n)$ and $W^{2,1}_\infty(Q_T)$, respectively, the sets of functions $f \in L^\infty(\mathbb{T}^n)$ such that $\frac{\partial f}{\partial x_k}, \frac{\partial^2 f}{\partial x_k \partial x_\ell} \in L^\infty(\mathbb{T}^n)$ $(1 \le k, \ell \le n)$, and the set of those $u \in L^\infty(Q_T)$ such that $\frac{\partial u}{\partial x_k}, \frac{\partial^2 u}{\partial x_k \partial x_\ell}, \frac{\partial u}{\partial t} \in L^\infty(Q_T)$, $(1 \le k, \ell \le n)$.

Lemma 4.1 *Let $g \in W^{2,\infty}(\mathbb{T}^n)$, and $\psi, \phi \in C(Q_\infty) \cap W^{2,1}_\infty(Q_\infty)$. Assume that*

$$\psi(x,0) \le g(x) \le \phi(x,0) \text{ for } x \in \mathbb{T}^n. \tag{33}$$

Let $u_\varepsilon \in C^{2,1}(Q_\infty)$ be classical solutions of

$$u_t + F_\varepsilon(Du, D^2u) + \beta_\varepsilon(u - \phi) - \beta_\varepsilon(\psi - u) = 0 \quad \text{in } Q_\infty \tag{34}$$

such that $u_\varepsilon(x,0) = g(x)$ for $x \in \mathbb{T}^n$. Then, there exists $C_1 > 0$ such that for $0 < \varepsilon < 1$, $0 < T \le \infty$, and for $(x,t) \in Q_T$,

$$\begin{cases} (i) & \min_{\mathbb{T}^n} g \wedge \inf_{Q_T} \phi \le u_\varepsilon(x,t) \le \max_{\mathbb{T}^n} g \vee \sup_{Q_T} \psi \\ (ii) & |Du_\varepsilon(x,t)| \le \|D\phi\|_{L^\infty(Q_T)} \vee \|D\psi\|_{L^\infty(Q_T)} \vee \|Dg\|_{L^\infty(\mathbb{T}^n)}, \\ (iii) & \left|\dfrac{\partial u_\varepsilon}{\partial t}(x,t)\right| \le \|\phi_t\|_{L^\infty(Q_T)} \vee \|\psi_t\|_{L^\infty(Q_T)} \vee C_1\|D^2 g\|_{L^\infty(\mathbb{T}^n)}. \end{cases}$$

Moreover, if we suppose

$$\psi \le \phi \quad \text{in } Q_\infty, \tag{35}$$

then there exists $C_2 > 0$ such that for $0 < \varepsilon < 1$, $0 < T \le \infty$, and for $(x,t) \in Q_T$,

$$\begin{cases} (iv) \ 0 \le \beta_\varepsilon(u_\varepsilon - \phi)(x,t) \le C_2(\|\phi_t\|_{L^\infty(Q_T)} + \|D^2\phi\|_{L^\infty(Q_T)}), \\ (v) \ \ 0 \le \beta_\varepsilon(\psi - u_\varepsilon)(x,t) \le C_2(\|\psi_t\|_{L^\infty(Q_T)} + \|D^2\psi\|_{L^\infty(Q_T)}). \end{cases}$$

Proof. We introduce linear operators $\mathcal{L}_\varepsilon$ by

$$\mathcal{L}_\varepsilon w := -\mathrm{Tr}\left\{\left((1+\varepsilon)I_n - \frac{Du_\varepsilon \otimes Du_\varepsilon}{\langle Du_\varepsilon \rangle_\varepsilon^2}\right) D^2 w\right\}.$$

We shall omit writing ε for u_ε, β_ε, and $\mathcal{L}_\varepsilon$, and fix $T \in (0, \infty]$.

Proof of (i) Fix any $\lambda > 0$. Suppose

$$\Theta := \sup_{(x,t) \in Q_T} e^{-\lambda t}(u(x,t) - M) > 0,$$

where $M := \max_{\mathbb{T}^n} g \vee \sup_{Q_T} \psi$. If there is $(x_k, t_k) \in Q_T$ such that

$$e^{-\lambda t_k}(u(x_k, t_k) - M) > \Theta - \frac{1}{k},$$

then we may suppose that $\lim_{k\to\infty}(x_k, t_k) = (\hat{x}, \hat{t})$ for some $(\hat{x}, \hat{t}) \in Q_T$. Thus, we have

$$\Theta = e^{-\lambda\hat{t}}(u(\hat{x}, \hat{t}) - M).$$

We notice that $\hat{t} < \infty$ even when $T = \infty$ since $\Theta > 0$. We will not notice this remark in the argument below.

We first suppose that $\hat{t} > 0$. Then, we have

$$0 \leq \mathscr{L}(e^{-\lambda t}(u - M)) \quad \text{at } (\hat{x}, \hat{t}).$$

Multiplying $e^{\lambda\hat{t}}$, we have

$$0 \leq F_\varepsilon(Du, D^2u) \leq -u_t + \beta(\psi - u) \quad \text{at } (\hat{x}, \hat{t}).$$

Noting $\lambda(u - M) = u_t$ at $(\hat{x}, \hat{t})$, we have

$$0 < \lambda(u - M) \leq \beta(\psi - u) \quad \text{at } (\hat{x}, \hat{t}),$$

which contradicts to the positivity of β. Hence, we may suppose $\hat{t} = 0$. However, this yields $0 < u(\hat{x}, 0) - M$, which is again a contradiction. Therefore, we have

$$u_\varepsilon \leq M \quad \text{in } Q_T.$$

By a symmetric argument, it is easy to show the lower bound for u_ε.

Proof of (ii) For $\lambda > 0$, setting $V(x, t) = e^{-\lambda t}|Du(x, t)|^2$, we suppose

$$0 < \Theta := \sup_{Q_T} V < \infty.$$

As in the proof of (i), we may find $(\hat{x}, \hat{t}) \in Q_T$ such that

$$\Theta = V(\hat{x}, \hat{t}).$$

In what follows, for simplicity, we shall write u_k, u_{ij} for $\dfrac{\partial u}{\partial x_k}, \dfrac{\partial^2 u}{\partial x_i \partial x_j}$ etc. We will use the summation convention for repeated indices i, j, k, ℓ from 1 to n.

Assuming $\hat{t} > 0$, we have

$$0 \leq \mathscr{L}V = 2u_k\mathscr{L}u_k - 2(1 + \varepsilon)|D^2u|^2 + \frac{2|D^2u Du|^2}{\langle Du\rangle_\varepsilon^2} \leq 2u_k\mathscr{L}u_k \quad \text{at } (\hat{x}, \hat{t}).$$

Note that $V_k(\hat{x}, \hat{t}) = 0$, and $V_t(\hat{x}, \hat{t}) \geq 0$. (Notice that if $\hat{t} < T$, then $V_t(\hat{x}, \hat{t}) = 0$ holds true.) Hence, we have

$$u_i u_{ik} = 0, \quad 2u_i u_{it} \geq \lambda|Du|^2 \quad \text{at } (\hat{x}, \hat{t}). \tag{36}$$

On the other hand, differentiating (18) in x_k, we have

$$\mathscr{L}u_k = -u_{tk} - \frac{2u_iu_ju_\ell u_{ij}u_{\ell k}}{\langle Du\rangle_\varepsilon^2} + \frac{2u_iu_{jk}u_{ij}}{\langle Du\rangle_\varepsilon^2} - \beta'(u-\phi)(u_k-\phi_k) + \beta'(\psi-u)(\psi_k-u_k).$$

Hence, using (36), we have

$$\begin{aligned}2u_k\mathscr{L}u_k &= -\lambda|Du|^2 - 2\beta'(u-\phi)(|Du|^2 - u_k\phi_k) + 2\beta'(\psi-u)(-|Du|^2+u_k\phi_k)\\ &\le -\lambda|Du|^2 - \beta'(u-\phi)(|Du|^2-|D\phi|^2) + \beta'(\psi-u)(-|Du|^2+|D\psi|^2)\end{aligned}$$

at $(\hat{x},\hat{t})$. We thus obtain that

$$0 < \lambda|Du|^2 < -\beta'(u-\phi)(|Du|^2-|D\phi|^2) + \beta'(\psi-u)(-|Du|^2+|D\psi|^2) \quad \text{at } (\hat{x},\hat{t}).$$

If $|Du|^2 \ge |D\phi|^2 \vee |D\psi|^2$, then this inequality fails. Therefore, we have

$$|Du(\hat{x},\hat{t})|^2 \le \|D\phi\|_{L^\infty(Q_T)}^2 \vee \|D\psi\|_{L^\infty(Q_T)}^2.$$

For any fixed $(x,t)\in Q_T$, we thus have

$$|Du|^2(x,t) \le e^{\lambda t}V(\hat{x},\hat{t}) \le e^{\lambda t}\left(\|D\phi\|_{L^\infty(Q_T)}^2 \vee \|D\psi\|_{L^\infty(Q_T)}^2\right). \tag{37}$$

Sending $\lambda\to 0$ in the above, we obtain (ii).

Next, we consider the case of $\hat{t}=0$. In this case, it is easy to see

$$Du(x,0) = Dg(x) \quad \text{for } x\in\mathbb{T}^n.$$

This together with (37) concludes the assertion (ii).

Proof of (iii) Setting $V(x,t) := e^{-\lambda t}u_t^2(x,t)$, we proceed the same argument as for (ii) to obtain

$$0 < \lambda u_t^2 \le -\beta'(u-\phi)(u_t^2-\phi_t^2) + \beta'(\psi-u)(-u_t^2+\psi_t^2) \quad \text{at } (\hat{x},\hat{t}),$$

where $V(\hat{x},\hat{t}) = \max_{Q_\infty} V > 0$ and $\hat{t}>0$. This implies

$$u_t^2(\hat{x},\hat{t}) \le \|\phi_t\|_{L^\infty(Q_T)}^2 \vee \|\psi_t\|_{L^\infty(Q_T)}^2,$$

which yields $u_t^2 \le \|\phi_t\|_{L^\infty(Q_T)}^2 \vee \|\psi_t\|_{L^\infty(Q_T)}^2$ in Q_T.

On the other hand, in view of (33), we see

$$u_t(\hat{x},0) = -F_\varepsilon(Dg(\hat{x}), D^2g(\hat{x})).$$

Hence, we find $C_1>0$ such that

$$|u_t(\hat{x},0)| \le C_1\|D^2g\|_{L^\infty(Q_T)}.$$

Therefore, in this case, we conclude the proof.

Proof of (iv), (v) Assuming (35), we shall only prove (iv). Fix $\lambda > 0$. Suppose

$$0 < \Theta := \sup_{\mathbb{T}^n \times (0,\infty)} e^{-\lambda t}\beta(u-\phi) < \infty.$$

As in the proof of (i), we may find $(\hat{x},\hat{t}) \in Q_T$ such that

$$\Theta = e^{-\lambda \hat{t}}\beta(u(\hat{x},\hat{t}) - \phi(\hat{x},\hat{t})).$$

This together with (35) implies

$$\beta(\psi(\hat{x},\hat{t}) - u(\hat{x},\hat{t})) = 0.$$

When $\hat{t} = 0$, $\beta(g(\hat{x}) - \phi(\hat{x},0)) = 0$ by our hypothesis. Thus, we may suppose $\hat{t} > 0$. Then, the maximum principle yields

$$0 \le \mathscr{L}(e^{-\lambda t}\beta(u-\phi)) \quad \text{at } (\hat{x},\hat{t}).$$

Hence, we have

$$\begin{aligned} 0 &\le \beta'(u-\phi)\mathscr{L}(u-\phi) - \beta''(u-\phi)\left\{(1+\varepsilon)|D(u-\phi)|^2 - \frac{\langle Du, D(u-\phi)\rangle^2}{\langle Du\rangle_\varepsilon^2}\right\} \\ &\le \beta'(u-\phi)(-u_t - \mathscr{L}\phi - \beta(u-\phi)) \quad \text{at } (\hat{x},\hat{t}). \end{aligned}$$

Noting $\lambda\beta(u-\phi) \le \beta'(u-\phi)(u-\phi)_t$, we have

$$0 < \lambda\beta(u-\phi) \le \beta'(u-\phi)(-\phi_t - \mathscr{L}\phi - \beta(u-\phi)) \quad \text{at } (\hat{x},\hat{t}),$$

which yields $\beta((u-\phi)(\hat{x},\hat{t})) \le C_2(\|\phi_t\|_{L^\infty(Q_T)} + \|D^2\phi\|_{L^\infty(Q_T)})$ for some $C_2 > 0$. Therefore, for any fixed $(x,t) \in Q_T$, we have

$$\beta((u-\phi)(x,t)) \le e^{\lambda t}C_2\left(\|\phi_t\|_{L^\infty(Q_T)} + \|D^2\phi\|_{L^\infty(Q_T)}\right),$$

which gives our assertion by sending $\lambda \to 0$. □

We shall show that the existence of Lipschitz continuous viscosity solutions of (1) under the periodic setting.

Theorem 4.2 *Assume $g \in W^{2,\infty}(\mathbb{T}^n)$ and $\phi, \psi \in W^{2,1}_\infty(Q_\infty)$. If (33) and (35) hold, then there exists a unique viscosity solution $u \in C(Q_\infty)$ of (1) such that*

$$u(x,0) = g(x) \quad \textit{for } x \in \mathbb{T}^n.$$

Moreover, we have

$$\frac{\partial u}{\partial x_k}, \frac{\partial u}{\partial t} \in L^\infty(Q_\infty).$$

Proof. By an approximation argument, we may suppose $\psi, \phi \in C^{2,1}(Q_\infty)$ and $g \in C^2(\mathbb{T}^n)$. For $\varepsilon > 0$, it is standard to find classical bounded solutions $u_\varepsilon \in C^{2,1}(Q_\infty)$ of (18) satisfying $u_\varepsilon(\cdot, 0) = g$ in $\mathbb{T}^n$. (See the appendix for this fact.)

Since such u_ε is unique, it is periodic in x. Thus, in view of Lemma 4.1, we can find $u \in C(Q_\infty)$ satisfying that $u_{x_k}, u_t \in L^\infty(Q_T)$ $(1 \le k \le n)$, and u_ε converges to u uniformly on Q_T for each $T \in (0, \infty)$. Therefore, it is standard to show that u is a viscosity solution of (1). Finally, Theorem 2.2 directly gives the uniqueness. □

5 Appendix

For the reader's convenience, we briefly show the existence of classical solutions of penalized problems.

Proposition 5.1 *Let $g \in W^{2,\infty}(\mathbb{T}^n)$, and $\psi, \phi \in C(Q_\infty) \cap W^{2,1}_\infty(Q_\infty)$. If (33) and (35) hold, then there exist classical solutions $u \in C^{2,1}(Q_\infty)$ of (34) such that $u(x, 0) = g(x)$ for $x \in \mathbb{T}^n$.*

We first approximate ϕ, ψ, and g by smooth functions. Next, we introduce cut-off functions $\eta_R \in C^\infty(\mathbb{R}^n)$: For each $R > 2$,

$$\begin{cases} M_1 := \sup_{R>2} \|\eta_R\|_{W^{2,\infty}(\mathbb{R}^n)} < \infty, \quad 0 \le \eta_R \le 1 \text{ in } \mathbb{R}^n, \\ \eta_R(x) = 1 \text{ for } |x| \le R-2, \quad \eta_R(x) = 0 \text{ for } |x| \ge R-1. \end{cases} \tag{38}$$

Set $g_R := \eta_R g$, $\phi_R := \eta_R \phi$, and $\psi_R := \eta_R \psi$. Assuming ϕ, $\psi \in W^{2,1}_\infty(Q_\infty)$ and $g \in W^{2,\infty}(\mathbb{T}^n)$, we note that

$$\begin{cases} \sup_{R>2} \left\{ \|g_R\|_{W^2_\infty(Q_\infty)} + \|\phi_R\|_{W^{2,1}_\infty(Q_\infty)} + \|\psi_R\|_{W^{2,1}_\infty(Q_\infty)} \right\} < +\infty, \\ g_R = 0 \quad \text{in } B^c_{R-1}, \quad \phi_R = \psi_R = 0 \quad \text{in } B^c_{R-1} \times (0, \infty). \end{cases} \tag{39}$$

We shall consider approximate problems for (18) with $u(\cdot, 0) = g$ in $\mathbb{R}^n$:

$$\begin{cases} u_t + F_\varepsilon(Du, D^2u) + \beta_\varepsilon(u - \psi_R) - \beta_\varepsilon(\psi_R - u) = 0 \quad \text{in } B_R \times (0, \infty), \\ u(\cdot, 0) = g_R \text{ in } B_R, \\ u = 0 \quad \text{on } \partial B_R \times [0, \infty). \end{cases} \tag{40}$$

For fixed $\varepsilon > 0$, letting $u_{\varepsilon,R}$ be the unique classical solution of (40), we shall show the estimates on $u_{\varepsilon,R}$ independent of $R > 2$. In what follows, we drop writing $\varepsilon > 0$ from F_ε and β_ε.

First of all, by following the same argument as for (i) in Lemma 4.1, it is immediate to show

$$\inf_{\mathbb{T}^n}(-g^-) \wedge \inf_{Q_\infty}(-\phi^-) \le u_{\varepsilon,R} \le \sup_{\mathbb{T}^n} g^+ \vee \sup_{Q_\infty} \psi^+ \quad \text{in } B_R \times (0,\infty).$$

(We only notice here that $\sup_{\mathbb{T}^n} g_R \le \sup_{\mathbb{T}^n} g^+$.)

By the same argument as for (iv) and (v) in Lemma 4.1, we can find $C_1 = C_1(\varepsilon) > 0$, independent of $R > 2$, such that

$$\begin{cases} \|\beta(u_{\varepsilon,R} - \phi_R)\|_{L^\infty(B_R\times(0,\infty))} \le C_1 \|\phi\|_{W^{2,1}_\infty(Q_\infty)}, \\ \|\beta(\psi_R - u_{\varepsilon,R})\|_{L^\infty(B_R\times(0,\infty))} \le C_1 \|\psi\|_{W^{2,1}_\infty(Q_\infty)}. \end{cases}$$

Lemma 5.2 *There is $C_2 = C_2(\varepsilon) > 0$, independent of $R > 2$, such that*

$$|u_{\varepsilon,R}(x,t) - g_R(x)| \le C_2 t \quad \textit{for } (x,t) \in B_R \times (0,\infty).$$

Proof. Let $\eta(x,t) := g_R(x) + C_3 t$, where

$$C_3 := 1 + \sup_{R>2} \|F(Dg_R, D^2 g_R)\|_{L^\infty(\mathbb{R}^n\times(0,\infty))} \vee \|\phi_t\|_{L^\infty(Q_\infty)} \vee \|\psi_t\|_{L^\infty(Q_\infty)}.$$

Since we see

$$\psi_R(x,t) - \eta(x,t) \le \zeta_R(x)\{\psi(x,0) - g(x)\} \le 0,$$

it is easy to verify

$$\eta_t + F(D\eta, D^2\eta) + \beta(\eta - \phi_R) - \beta(\psi_R - \eta) > 0 \quad \text{in } B_R \times (0,\infty),$$

in view of the maximum principle, we have $u_{\varepsilon,R} \le g_R + C_3 t$ in $B_R \times (0,\infty)$. We are also able to verify that $u_{\varepsilon,R} \ge g_R - C_3 t$ on $B_R \times (0,\infty)$. □

Lemma 5.3 *There is $C_4 = C_4(\varepsilon) > 0$, independent of $R > 2$, such that*

$$|Du_{\varepsilon,R}(x,t)| \le C_4 \quad \textit{for } t \in \partial B_R \times [0,\infty).$$

Proof. Fix $x_0 \in \partial B_R$ and set $y_0 := (1 + R^{-1})x_0$. Using the standard barrier:

$$\rho(x,t) := e^{-\alpha} - e^{-\alpha|x-y_0|^2},$$

we observe

$$\rho \ge 0 \text{ in } B_R \times [0,\infty), \quad \text{and} \quad \rho(x_0,t) = 0 \text{ for } t \in [0,\infty).$$

It is easy to see

$$\begin{aligned} &\rho_t = 0, \quad D\rho = 2\alpha e^{-\alpha|x-y_0|^2}(x - y_0), \\ &\text{and} \quad D^2\rho = 2\alpha e^{-\alpha|x-y_0|^2}\{I - 2\alpha(x-y_0)\otimes(x-y_0)\}. \end{aligned}$$

Thus, we calculate as follows: for $\mu > 0$, which will be fixed later,

$$\begin{aligned}&\mu\rho_t + F(\mu D\rho, \mu D^2\rho) + \beta(\mu\rho - \phi_R) - \beta(\psi_R - \mu\rho)\\ \geq& -2\mu\alpha(1+\varepsilon)ne^{-\alpha|x-y_0|^2}\{(1+\varepsilon)n - 2\alpha\varepsilon|x-y_0|^2\} - \beta(\psi_R - \mu\rho)\\ \geq& 2\mu\alpha e^{-\alpha|x-y_0|^2}\{-(1+\varepsilon)n + 2\alpha\varepsilon\} - \beta(\psi_R - \mu\rho).\end{aligned}$$

Fixing $\alpha > (1+\varepsilon)n/(2\varepsilon)$, and $\mu > 0$ satisfying $\psi_R \vee g_R \leq \mu\rho$ in $B_R \times (0,\infty)$, we see that $\mu\rho$ is a classical strict supersolution of (40). Hence, the maximum principle implies $u_{\varepsilon,R} \leq \mu\rho$ on $B_R \times [0,\infty)$. Thus, we can find $C_4 > 0$ such that

$$\frac{1}{R}\langle Du_{\varepsilon,R}(x_0,t), x_0\rangle \leq C_4.$$

Using a symmetric argument, we have $|Du_{\varepsilon,R}(x_0,t)| \leq C_4$. □

Setting $V := e^{-\lambda t}|Du|^2$ for $\lambda > 0$, we suppose that $\sup_{B_R\times[0,\infty)} V =: \theta > 0$. In view of Lemma 5.3, we may suppose that $(\hat{x},\hat{t}) \in B_R \times (0,\infty)$. Hence, following the argument for (iii), we can show $\sup_{R>2} \|Du_{\varepsilon,R}\|_{L^\infty(B_R\times(0,\infty))} < \infty$.

Next, considering $V := e^{-\lambda t}u_t^2$ for $\lambda > 0$, by the same argument as above with Lemma 5.2, we obtain $\sup_{R>2} \|(u_{\varepsilon,R})_t\|_{L^\infty(B_R\times(0,\infty))} < \infty$.

It is now standard to obtain the local $C^{2,1}$ estimates on $u_{\varepsilon,R}$, independent of large R (for fixed $\varepsilon > 0$!). Therefore, we can obtain classical solutions of (18).

References

1. G. Barles, S. Biton, O. Ley, Uniqueness for parabolic equations without growth condition and applications to the mean curvature flow in $\mathbb{R}^2$. J. Differ. Equ. **187**, 456–472 (2003)
2. Y.-G. Chen, Y. Giga, S. Goto, Uniqueness and existence of viscosity solutions of generalized mean curvature flow equations. J. Differ. Geom. **33**, 749–786 (1991)
3. M.G. Crandall, H. Ishii, The maximum principle for semicontinuous functions. Differ. Integral Equ. **3**(6), 1001–1014 (1990)
4. M.G. Crandall, H. Ishii, P.-L. Lions, User's guide to viscosity solutions of second order partial differential equations. Bull. Am. Math. Soc. **27**, 1–67 (1992)
5. L.C. Evans, J. Spruck, Motion of level sets by mean curvature. I. J. Differ. Geom. **33**, 635–681 (1991)
6. Y. Giga, *Surface Evolution Equations, A Level Set Approach*, vol. 99, Monographs in Mathematics (Birkhäuser Verlag, Basel, 2006)
7. Y. Giga, S. Goto, H. Ishii, M.-H. Sato, Comparison principle and convexity preserving properties for singular degenerate parabolic equations on unbounded domains. Indiana Univ. Math. J. **40**(2), 443–470 (1991)
8. H. Ishii, P.E. Souganidis, Generalized motion of noncompact hypersurfaces with velocity having arbitrary growth of the curvature tensor. Tôhoku Math. J. **47**, 227–250 (1995)
9. G. Mercier, Mean curvature flow with obstacles: a viscosity approach, Preprint (2014)
10. G. Mercier, M. Novaga, Mean curvature flow with obstacles: existence, uniqueness and regularity of solutions, Preprint (2014)
11. M.-H. Sato, Comparison principle for singular degenerate elliptic equations on unbounded domains. Proc. Jpn. Acad. Ser. A **66**, 252–256 (1990)

Remark on Stability of Scale-Critical Stationary Flows in a Two-Dimensional Exterior Disk

Yasunori Maekawa

Abstract We prove the asymptotic stability of stationary flows in a two-dimensional exterior disk under suitable decay and smallness conditions on the stationary flows and the initial perturbations. The class of stationary flows considered in this paper includes some typical circular flows which decay in the scale-critical order $O(|x|^{-1})$ as $|x| \to \infty$.

1 Introduction

In this paper, we consider the perturbed Stokes equations for viscous incompressible flows in the exterior disk

$$\begin{cases} \partial_t u - \Delta u + V \cdot \nabla u + u \cdot \nabla V + \nabla p = 0\,, \quad \operatorname{div} u = 0\,, \quad t > 0\,,\ x \in \Omega\,, \\ u|_{\partial\Omega} = 0\,,\ t > 0\,, \qquad u|_{t=0} = a\,,\ x \in \Omega\,. \end{cases} \tag{S$_V$}$$

Here, $u = u(t,x) = (u_1(t,x), u_2(t,x))^\top$ and $p = p(t,x)$ are respectively unknown velocity field and pressure field, and $V(x) = (V_1(x), V_2(x))^\top$ and $a(x) = (a_1(x), a_2(x))^\top$ are given velocity fields. We will use the standard notation for derivatives: $\partial_t = \frac{\partial}{\partial t}$, $\partial_j = \frac{\partial}{\partial x_j}$, $\Delta = \sum_{j=1}^2 \partial_j^2$, $\operatorname{div} u = \sum_{j=1}^2 \partial_j u_j$, $V \cdot \nabla u + u \cdot \nabla V = \sum_{j=1}^2 V_j \partial_j u + u_j \cdot \partial_j V$. The domain Ω in $\mathbb{R}^2$ is assumed to be the exterior domain to the unit disk centered at the origin, i.e., $\Omega = \mathbb{R}^2 \setminus \overline{B_1(0)}$ with $B_r(0) = \{x \in \mathbb{R}^2 \mid |x| < r\}$. The given solenoidal velocity field V is assumed to be time-independent and decay in the scale-critical order: $V(x) = O(|x|^{-1})$ as $|x| \to \infty$. Further conditions on V will be stated later.

The perturbed Stokes problem as in (Sv) naturally appears when one considers the linearization around a stationary solution to the Navier–Stokes equations: $\partial_t u - \Delta u + u \cdot \nabla u + \nabla p = f, \operatorname{div} u = 0$ in $|x| > 1$ and $u = b$ on $|x| = 1$ with some

Y. Maekawa (✉)
Department of Mathematics, Kyoto University, Kyoto, Japan
e-mail: maekawa@math.kyoto-u.ac.jp

Y. Maekawa and S. Jimbo (eds.), *Mathematics for Nonlinear Phenomena—Analysis and Computation*, Springer Proceedings in Mathematics & Statistics 215,
DOI 10.1007/978-3-319-66764-5_6

given data f and b (thus, V is supposed to be the stationary solution). Indeed, the analysis of (Sv) is one of the most important steps in studying the asymptotic stability of the stationary solution. Since the fluid domain is unbounded, the stability property of stationary solutions highly relies on their spatial decay, and the order $O(|x|^{-1})$, which is a scale-critical order for the Navier–Stokes equations, is a natural candidate for the borderline in achieving the general stability result for small stationary solutions. This is indeed the case for the three-dimensional problem; see, e.g., Borchers [2], Borchers-Miyakawa [3]. However, in the two-dimensional case as in (Sv), the problem is more complicated and no general result is known so far. Note that the existence of stationary solutions to the two-dimensional Navier–Stokes equations in exterior domains has been proved so far under some symmetry conditions on domains and given data; see Galdi [4], Russo [9], Yamazaki [11], and Pileckas-Russo [8]. In particular, the class of stationary solutions obtained in Yamazaki [11] includes the solutions decaying in the scale-critical order. Recently, Hillairet-Wittwer [6] proved the existence of stationary solutions around αU for large $|\alpha|$, where U is a radial circular flow given by (2) below. The stability of these stationary solutions is still widely open even when the stationary solutions are small enough. The crucial difficulty, in the two-dimensional case, to obtain the stability of stationary solutions decaying like $O(|x|^{-1})$ is the absence of the Hardy inequality of the form

$$\|\frac{1}{|x|} f\|_{L^2(\Omega)} \leq C\|\nabla f\|_{L^2(\Omega)}, \quad f \in \dot{W}_0^{1,2}(\Omega) = \overline{C_0^\infty(\Omega)}^{\|\nabla f\|_{L^2(\Omega)}}. \tag{1}$$

Here, $C_0^\infty(\Omega)$ is the set of smooth compactly supported functions in Ω. This inequality is well known in the three or higher dimensions, but it fails in general for the two-dimensional case; indeed, one must replace the factor $\frac{1}{|x|}$ by $\frac{1}{|x|\log(e+|x|)}$ in the left-hand side of the above inequality. The breakdown of the inequality (1) indicates that, when V decays in the scale-critical order, the linearized term $V \cdot \nabla u + u \cdot \nabla V$ is no longer handled as a perturbation from the pure Stokes operator in the two-dimensional case, no matter how small the size of V is.

Recently, the L^2 linear/nonlinear stability of V has been established by the author [7] when V is a special flow given by

$$V(x) = \alpha U(x), \qquad U(x) = \frac{x^\perp}{|x|^2}, \qquad x^\perp = (-x_2, x_1)^\top, \tag{2}$$

where α is a real number such that $0 < |\alpha| \ll 1$. More recently, Guillod [5] showed that another typical stationary flow $\delta \frac{x}{|x|^2}$ is stable for arbitrary L^2 perturbations when the real number δ is small enough. Interestingly, the Hardy-type inequality is available for $\delta \frac{x}{|x|^2}$, which is the key in the proof of [5]; see also Remark 1 below for more precise descriptions.

The radial circular flow αU represents a stationary flow around a rotating disk with a constant angular velocity α. To state the result of [7] more precisely, let us introduce some notations. We denote by $L^q_\sigma(\Omega)$, $1 < q < \infty$, the closure of $C^\infty_{0,\sigma}(\Omega)$ in the topology of $L^q(\Omega)^2$. Here, $C^\infty_{0,\sigma}(\Omega) = \{f \in C_0^\infty(\Omega)^2 \mid \operatorname{div} f = 0\}$. We denote by $\mathbb{P}$ the Helmholtz projection $\mathbb{P} : L^q(\Omega)^2 \to L^q_\sigma(\Omega)$, which is an orthogonal projection when $q = 2$. The Helmholtz projection in the space $L^2(\Omega)^2 \cap L^q(\Omega)^2$ is also well defined for $1 < q < \infty$. Then under a suitable regularity condition on V, the perturbed Stokes operator $\mathbf{A}_V$ in $L^2_\sigma(\Omega) \cap L^q_\sigma(\Omega)$ with the domain $D_{L^2_\sigma \cap L^q_\sigma}(\mathbf{A}_V)$ is defined as

$$
\begin{aligned}
D_{L^2_\sigma \cap L^q_\sigma}(\mathbf{A}_V) &= D_{L^2_\sigma \cap L^q_\sigma}(\mathbf{A}) \,, \\
\mathbf{A}_V u &= \mathbf{A} u + \mathbb{P}\big(V \cdot \nabla u + u \cdot \nabla V\big) \,, \qquad u \in D_{L^2_\sigma \cap L^q_\sigma(\Omega)}(\mathbf{A}_V) \,.
\end{aligned}
\tag{3}
$$

Here, $\mathbf{A} = -\mathbb{P}\Delta$ is the Stokes operator and

$$
\begin{aligned}
D_{L^2_\sigma \cap L^q_\sigma}(\mathbf{A}) \;&=\; D_{L^2_\sigma}(\mathbf{A}) \cap D_{L^q_\sigma}(\mathbf{A}) \\
&= \big(L^2_\sigma(\Omega) \cap W^{1,2}_0(\Omega)^2 \cap W^{2,2}(\Omega)^2\big) \cap \big(L^q_\sigma(\Omega) \cap W^{1,q}_0(\Omega)^2 \cap W^{2,q}(\Omega)^2\big) \,.
\end{aligned}
$$

It is well known that $\mathbf{A}$ defines a nonnegative self-adjoint operator in $L^2_\sigma(\Omega)$, and $-\mathbf{A}$ generates a bounded C_0-analytic semigroup acting on $L^2_\sigma(\Omega)$ and on $L^q_\sigma(\Omega)$, $1 < q < \infty$. Thus, $-\mathbf{A}$ generates a bounded C_0-analytic semigroup on $L^2_\sigma(\Omega) \cap L^q_\sigma(\Omega)$, $1 < q < \infty$. In this paper, we will assume that V is smooth enough and decays at spatial infinity. Then, the general perturbation theory of sectorial operators shows that $-\mathbf{A}_V$ also generates a C_0-analytic semigroup acting on $L^2_\sigma(\Omega) \cap L^q_\sigma(\Omega)$. Therefore, the central issue is the estimate of the semigroup in large time. In the special case $V = \alpha U$, the following result is proved in [7].

Theorem 1 *([7, Theorem 1.3]). There exists a constant $\alpha_0 \in (0, 1)$ such that if $|\alpha| \le \alpha_0$, then the following statement holds. Let $1 < q \le 2 \le p < \infty$. Then it follows that*

$$
\|e^{-t\mathbf{A}_{\alpha U}} f\|_{L^p(\Omega)} \le C t^{-\frac{1}{q}+\frac{1}{p}} \|f\|_{L^q(\Omega)} \,, \qquad t > 0 \,, \tag{4}
$$

$$
\|\nabla e^{-t\mathbf{A}_{\alpha U}} f\|_{L^2(\Omega)} \le C t^{-\frac{1}{q}} \|f\|_{L^q(\Omega)} \,, \qquad t > 0 \,, \tag{5}
$$

for $f \in L^2_\sigma(\Omega) \cap L^q_\sigma(\Omega)$. Here, the constant C depends only on α, p, and q.

The aim of this paper is to extend Theorem 1 for more general class of V around αU. Let α_0 be the number in Theorem 1 and U be given by (2). Let $\mathcal{Q}_0$ be the projection given by $\mathcal{Q}_0 = \sum_{|n| \ge 1} \mathcal{P}_n$, where each $\mathcal{P}_n$ is the projection to the Fourier mode n in the polar coordinates for vector fields; see (16) for the precise definition of $\mathcal{P}_n$. The operator rot is defined as $\operatorname{rot} u = \partial_1 u_2 - \partial_2 u_1$, i.e., it defines the vorticity field of the velocity u. Our assumption for V is as follows.

Assumption The velocity V is of the form $V(x) = \alpha U(x) + W(x)$, where $0 < |\alpha| \le \alpha_0$ and $W \in BC^1(\Omega)^2$ satisfies

$$
\operatorname{div} W \;=\; 0 \quad \text{in } \Omega \,, \tag{W1}
$$

$$
|W(x)| \le \frac{\delta}{|x|} \,, \qquad |(\mathcal{Q}_0 W)(x)| \le \frac{\delta}{|x| \log(e + |x|)} \qquad \text{in } \Omega \,, \tag{W2}
$$

$$|\mathrm{rot}\, W(x)| \le \frac{\delta}{\big(|x|\log(e+|x|)\big)^2} \qquad \text{in } \Omega\,. \tag{W3}$$

Here, δ is a positive number, which will be assumed to be small.

Remark 1. A typical example satisfying all conditions (W1) - (W3) is

$$W(x) \;=\; \delta\frac{x}{|x|^2}\,, \tag{6}$$

which is a radial circular flow such that $\mathscr{Q}_0 W = 0$ and rot $W = 0$ for $|x| \neq 0$. Note that the velocity $\alpha U + W$ with W in (6) always defines a stationary solution to the Navier–Stokes equations in Ω with the boundary condition $u = \alpha x^{\perp} + \delta x$ on $\partial\Omega$ when $\Omega = \{x \in \mathbb{R}^2 \mid |x| > 1\}$. Recently, the global L^2 stability of W defined by (6) (i.e., $\alpha = 0$) is proved in [5] when δ is sufficiently small. The key observation in [5] is that, for the exact solution W in (6), the following Hardy-type inequality is valid:

$$|\langle v\cdot\nabla v,\, W\rangle_{L^2(\Omega)}| \le C\delta\|\nabla v\|^2_{L^2(\Omega)}\,, \qquad v \in \dot{W}^1_{0,\sigma}(\Omega)\,. \tag{7}$$

This crucial estimate is shown by Russo [10]. Then the global L^2 stability is established in [5] using the energy method. Note that, as proved in [5], inequality (7) does not hold if W is replaced by the circular flow U. Hence, the stability of $\alpha U + W$ is not covered by the results of [7] or [5].

The main result of this paper is stated as follows, where the novelty of the result is the estimates of $e^{-t\mathbf{A}_V}$ for $t \gg 1$.

Theorem 2. *Suppose that V satisfies* **Assumption** *stated above. Then there exists $\delta_0 > 0$ such that if $\delta \in [0, \delta_0]$, then*

$$\|e^{-t\mathbf{A}_V} f\|_{L^2(\Omega)} \le Ct^{-\frac{1}{q}+\frac{1}{2}}\|f\|_{L^q(\Omega)}\,, \qquad t > 0\,, \tag{8}$$

$$\|\nabla e^{-t\mathbf{A}_V} f\|_{L^2(\Omega)} \le Ct^{-\frac{1}{q}}\|f\|_{L^q(\Omega)}\,, \qquad t > 0\,. \tag{9}$$

for $1 < q < 2$ and $f \in L^2_\sigma(\Omega) \cap L^q_\sigma(\Omega)$. Here, δ_0 depends only on α, while C depends only on α and q.

Remark 2. (i) It is not clear so far whether or not the decay condition on $\mathscr{Q}_0 W$ and rot W in (W2) - (W3) can be replaced by the scale-critical ones: $|\mathscr{Q}_0 W(x)| \le \delta|x|^{-1}$ and $|\mathrm{rot}\, W(x)| \le \delta|x|^{-2}$.
(ii) The author does not know whether or not one can take $q = 2$ in (8)–(9). The difficulty appears in estimating the semigroup in large time; there is of course no difficulty in obtaining the estimates stated in Theorem 2 including the case $q = 2$ for a local time, say, $0 < t \le 1$. The reason why the strict condition $q < 2$ is required so far is that our proof is based on the estimate for the norm $\|\frac{1}{|x|}\mathrm{rot}\,\mathscr{P}_{\pm1}u(t)\|_{L^1(\Omega)}$ for $t > 0$. Note that the integrability $a \in L^2_\sigma(\Omega)$ is not enough to ensure the bound of the norm $\|\frac{1}{|x|}\mathrm{rot}\, u(t)\|_{L^1(\Omega)}$.

(iii) The condition $\alpha \neq 0$ stated in **Assumption** is essential in obtaining Theorem 2. Indeed, the nonzero α leads to a stabilizing effect in view of spatial decay of resolvent kernel. This stabilizing property enables us to treat the term coming from $\mathscr{P}_0 W$, which has a scale-critical decay as assumed in (W2), as a perturbation of the operator $\mathbf{A}_{\alpha U}$.

Once Theorem 2 is established, it is a routine work from the Banach fixed point theorem to obtain the nonlinear stability of V for the Navier–Stokes equations, whose integral form is given as

$$u(t) = e^{-t\mathbf{A}_V} a - \int_0^t e^{-(t-s)\mathbf{A}_V} \mathbb{P}(u \cdot \nabla u)\, ds\,. \qquad (\mathrm{INS_V})$$

As mentioned in (ii) of Remark 2, due to the enough regularity of W in **Assumption**, it is not difficult to show the estimates

$$\|e^{-t\mathbf{A}_V} a\|_{L^2(\Omega)} \le C t^{-\frac{1}{q}+\frac{1}{2}} \|a\|_{L^q(\Omega)}\,, \qquad 0 < t \le 1 \quad 1 < q \le 2\,, \qquad (10)$$

$$\|\nabla e^{-t\mathbf{A}_V} a\|_{L^2(\Omega)} \le C t^{-\frac{1}{q}} \|a\|_{L^q(\Omega)}\,, \qquad 0 < t \le 1\,, \quad 1 < q \le 2\,. \qquad (11)$$

Using (8), (9), (10), and (11), and from the fact that $\{e^{-t\mathbf{A}_V}\}_{t \ge 0}$ defines a C_0-analytic semigroup in $L^2_\sigma(\Omega) \cap L^q_\sigma(\Omega)$ and that $C^\infty_{0,\sigma}(\Omega)$ is dense in $L^2_\sigma(\Omega) \cap L^q_\sigma(\Omega)$, $1 < q < \infty$, we have the following theorem for the nonlinear problem. The proof of this nonlinear result will be omitted in this paper since the argument is rather standard; see, e.g., the argument in [7, Section 4].

Theorem 3. *Suppose that V satisfies* **Assumption** *with $\delta \in [0, \delta_0]$, where δ_0 is the number in Theorem 2. Then there exists $\varepsilon > 0$ such that if $a \in L^2_\sigma(\Omega) \cap L^q_\sigma(\Omega)$ and $\|a\|_{L^2(\Omega)\cap L^q(\Omega)} \le \varepsilon$ for some $1 < q < 2$, then there exists a unique solution $u \in C([0, \infty); L^2_\sigma(\Omega)) \cap C((0, \infty); W^{1,2}_0(\Omega)^2)$ to (INSv) satisfying $t^{\frac{1}{2}} \nabla u(t) \in L^\infty(0, \infty; L^2(\Omega))$ and*

$$\lim_{t \to \infty} t^{\frac{1}{q}-\frac{1}{2}+\frac{k}{2}} \|\nabla^k u(t)\|_{L^2(\Omega)} = 0\,, \qquad k = 0, 1\,. \qquad (12)$$

As a special case, Theorem 3 shows that the stationary flow $V(x) = \alpha \frac{x^\perp}{|x|^2} + \delta \frac{x}{|x|^2}$ with $0 < |\alpha| \ll 1$ and $|\delta| \ll 1$ is asymptotically stable with respect to small initial perturbations in $L^2_\sigma(\Omega) \cap L^q(\Omega)^2$, $1 < q < 2$.

This paper is organized as follows. In Section 2, we recall some basic descriptions of velocity fields in the polar coordinates. In Section 3, we recall the Biot–Savart law in Ω. Section 4 is the core of this paper, where the resolvent problem for the linearized operator $\mathbf{A}_V$ is analyzed using both the energy method and the representation formula established in [7]. In particular, Theorem 4 is the main result of Section 4. In Section 5, we prove Theorem 2 based on the resolvent estimate obtained in Section 4. Finally, some estimates for the modified Bessel functions, which are used in estimating the solution to the resolvent equations, are stated in the appendix.

2 Vector Fields in the Polar Coordinates

In this section, we recall some basic facts on the polar coordinates for vector fields. We set

$$\mathbf{e}_r = \frac{x}{|x|}\,, \qquad \mathbf{e}_\theta = \frac{x^\perp}{|x|} = \partial_\theta \mathbf{e}_r\,.$$

For $v \in L^2(\Omega)^2$, we set $v_r = v \cdot \mathbf{e}_r$ and $v_\theta = v \cdot \mathbf{e}_\theta$, and thus $v = v_r\,\mathbf{e}_r + v_\theta\,\mathbf{e}_\theta$. Let $v \in W^{1,2}(\Omega)^2$. Then we have the following fundamental identities.

$$\operatorname{div} v = \partial_1 v_1 + \partial_2 v_2 = \frac{1}{r}\partial_r(r v_r) + \frac{1}{r}\partial_\theta v_\theta\,, \tag{13}$$

$$\operatorname{rot} v = \partial_1 v_2 - \partial_2 v_1 = \frac{1}{r}\partial_r(r v_\theta) - \frac{1}{r}\partial_\theta v_r\,, \tag{14}$$

$$|\nabla v|^2 = |\partial_r v_r|^2 + |\partial_r v_\theta|^2 + \frac{1}{r^2}\big(|\partial_\theta v_r - v_\theta|^2 + |v_r + \partial_\theta v_\theta|^2\big)\,. \tag{15}$$

Next, for each $n \in \mathbb{Z}$, we denote by $\mathscr{P}_n$ the projection on the Fourier mode n with respect to the angular variable θ, that is,

$$\mathscr{P}_n v = v_{r,n} e^{in\theta}\mathbf{e}_r + v_{\theta,n} e^{in\theta}\mathbf{e}_\theta\,, \tag{16}$$

where

$$v_{r,n}(r) = \frac{1}{2\pi}\int_0^{2\pi} v_r(r\cos\theta, r\sin\theta)e^{-in\theta}\,\mathrm{d}\theta\,,$$

$$v_{\theta,n}(r) = \frac{1}{2\pi}\int_0^{2\pi} v_\theta(r\cos\theta, r\sin\theta)e^{-in\theta}\,\mathrm{d}\theta\,.$$

We set for $m \in \mathbb{N}\cup\{0\}$,

$$\mathscr{Q}_m v = \sum_{|n|=m+1}^{\infty} \mathscr{P}_n v\,. \tag{17}$$

It is easy to see that the right-hand side of (17) converges in $L^2(\Omega)^2$ and each of $\mathscr{P}_n, \mathscr{Q}_m : L^2(\Omega)^2 \to L^2(\Omega)^2$ is an orthogonal projection. Moreover, the divergence free condition as well as the no-flux condition $\mathbf{e}_r \cdot v = 0$ on the boundary $\partial\Omega$ is preserved under the action of $\mathscr{P}_n$. More precisely, we have

$$\begin{aligned} \mathscr{P}_n C_{0,\sigma}^\infty(\Omega) \subset C_{0,\sigma}^\infty(\Omega)\,, \qquad \mathscr{Q}_m C_{0,\sigma}^\infty(\Omega) \subset C_{0,\sigma}^\infty(\Omega)\,,\\ \mathscr{P}_n\mathbb{P} = \mathbb{P}\mathscr{P}_n\,, \qquad \mathscr{Q}_m\mathbb{P} = \mathbb{P}\mathscr{Q}_m\,. \end{aligned} \tag{18}$$

Therefore, for each $m \in \mathbb{N} \cup \{0\}$, we have an orthogonal decomposition of $L^2_\sigma(\Omega)$ as

$$L^2_\sigma(\Omega) = L^2_{\sigma,rad}(\Omega) \oplus \mathscr{Q}_m L^2_\sigma(\Omega) \oplus \left(\oplus_{1\le|n|\le m} \mathscr{P}_n L^2_\sigma(\Omega)\right), \tag{19}$$

where $L^2_{\sigma,rad}(\Omega) = \mathscr{P}_0 L^2_\sigma(\Omega)$ is the space of radially symmetric flows in $L^2_\sigma(\Omega)$. The Stokes operator $\mathbf{A} = -\mathbb{P}\Delta$ is then diagonalized as

$$\mathbf{A} = \mathbf{A}|_{L^2_{\sigma,rad}} \oplus \mathbf{A}|_{\mathscr{Q}_m L^2_\sigma} \oplus \left(\oplus_{1\le|n|\le m} \mathbf{A}|_{\mathscr{P}_n L^2_\sigma}\right). \tag{20}$$

Here, for an invariant subspace $X \subset L^2_\sigma(\Omega)$, the operator $\mathbf{A}|_X$ is defined as $D(\mathbf{A}|_X) = D_{L^2_\sigma}(\mathbf{A}) \cap X$ and $\mathbf{A}|_X : D(\mathbf{A}|_X) \to X, \mathbf{A}|_X u = \mathbf{A}u$ for $u \in D(\mathbf{A}|_X)$. Due to the radial symmetry of U in (2), the perturbed Stokes operator $\mathbf{A}_{\alpha U}$ is also diagonalized as in (20); see [7, Proposition 2.10].

From (15) and the definition of $\mathscr{P}_n$ it follows that for $n \in \mathbb{N} \cup \{0\}$ and for $v \in W^{1,2}(\Omega)^2$,

$$\|\nabla v\|^2_{L^2(\Omega)} = \sum_{k\in\mathbb{Z}} \|\nabla \mathscr{P}_k v\|^2_{L^2(\Omega)}, \tag{21}$$

$$\begin{aligned} |\nabla \mathscr{P}_n v|^2 =& |\partial_r v_{r,n}|^2 + \frac{1+n^2}{r^2}|v_{r,n}|^2 \\ &+ |\partial_r v_{\theta,n}|^2 + \frac{1+n^2}{r^2}|v_{\theta,n}|^2 - \frac{4n}{r^2}\Im(v_{\theta,n}\overline{v_{r,n}}). \end{aligned} \tag{22}$$

In particular, we have

$$|\nabla \mathscr{P}_n v|^2 \ge |\partial_r v_{r,n}|^2 + \frac{(|n|-1)^2}{r^2}|v_{r,n}|^2 + |\partial_r v_{\theta,n}|^2 + \frac{(|n|-1)^2}{r^2}|v_{\theta,n}|^2. \tag{23}$$

The estimates (21) and (23) imply the Hardy inequality for the part $\sum_{|n|\ne 1} \mathscr{P}_n v$, $v \in W^{1,2}(\Omega)^2$, as follows.

Lemma 1. *Let $v \in W^{1,2}(\Omega)^2$. Then*

$$\|\frac{1}{|x|}\sum_{|n|\ne1} \mathscr{P}_n v\|_{L^2(\Omega)} \le \|\nabla \sum_{|n|\ne1} \mathscr{P}_n v\|_{L^2(\Omega)}. \tag{24}$$

3 Biot–Savart Law in $\mathscr{P}_n L^2_\sigma(\Omega)$

We will analyze the operator $\mathbf{A}_V$ using the streamfunction–vorticity formulation. For a given scalar field w in Ω, the streamfunction ψ is formally defined as the solution to the Poisson equations: $-\Delta\psi = w$ in Ω, $\psi = 0$ on $\partial\Omega$. For $n \in \mathbb{Z}$ and $w \in L^2(\Omega)$, we set

$$w_n(r) = \frac{1}{2\pi}\int_0^{2\pi} w(r\cos\theta, r\sin\theta)e^{-in\theta}\,\mathrm{d}\theta\,. \tag{25}$$

In the polar coordinates, the Poisson equations for the Fourier mode n (with respect to the angular variable) is described as

$$-\psi_n^{''} - \frac{1}{r}\psi_n^{'} + \frac{n^2}{r^2}\psi_n = w_n\,, \quad r > 1\,, \qquad \psi_n(1) = 0\,. \tag{26}$$

Here, $\psi_n^{'} = \frac{\mathrm{d}\psi_n}{\mathrm{d}r}$ and $\psi_n^{''} = \frac{\mathrm{d}^2\psi_n}{\mathrm{d}r^2}$. Let $|n| \geq 1$. Then the solution $\psi_n = \psi_n[w_n]$ to the ordinary differential equation (26) decaying at spatial infinity is formally given as

$$\begin{aligned}\psi_n[w_n](r) =& \frac{1}{2|n|}\Big(- \frac{d_n[w_n]}{r^{|n|}} + \frac{1}{r^{|n|}}\int_1^r s^{1+|n|}w_n(s)\,\mathrm{d}s + r^{|n|}\int_r^\infty s^{1-|n|}w_n(s)\,\mathrm{d}s\Big)\,,\\ d_n[w_n] =& \int_1^\infty s^{1-|n|}w_n(s)\,\mathrm{d}s\,.\end{aligned} \tag{27}$$

The Biot–Savart law $V_n[w_n]$, the relation between the solenoidal velocity and the vorticity w_n, is then defined as

$$\begin{aligned} V_n[w_n] &= V_{r,n}[w_n]\,e^{in\theta}\,\mathbf{e}_r + V_{\theta,n}[w_n]\,e^{in\theta}\,\mathbf{e}_\theta\,,\\ V_{r,n}[w_n] &= \frac{in}{r}\psi_n[w_n]\,, \qquad V_{\theta,n}[w_n] = -\frac{\mathrm{d}}{\mathrm{d}r}\psi_n[w_n]\,. \end{aligned} \tag{28}$$

It is straightforward to see

$$\begin{aligned} \mathrm{div}\, V_n[w_n] = 0\,, &\qquad \mathrm{rot}\, V_n[w_n] = w_n\,e^{in\theta} \qquad \text{in } \Omega\,,\\ \mathbf{e}_r \cdot V_n[w_n] = 0 &\qquad \text{on } \partial\Omega\,. \end{aligned} \tag{29}$$

Conversely, if the vorticity field of the velocity $v \in \mathscr{P}_n L^2_\sigma(\Omega)$, $|n| \geq 1$, decays fast enough as $|x| \to \infty$, then v is given by the Biot–Savart law; e.g., see [7, Lemma 2.4]. The formula of the Biot–Savart law leads to the next lemma, which will be used later.

Lemma 2. *Let $u \in L^2_\sigma(\Omega) \cap W^{1,2}(\Omega)^2$. Let $|n| = 1$ and $\frac{1}{|x|}\mathrm{rot}\,\mathscr{P}_n u \in L^1(\Omega)$. Then*

$$\|\mathscr{P}_n u\|_{L^\infty(\Omega)} \leq C\|\frac{1}{|x|}\mathrm{rot}\,\mathscr{P}_n u\|_{L^1(\Omega)}\,. \tag{30}$$

Proof. Set $w_n = e^{-in\theta}\mathrm{rot}\,\mathscr{P}_n u$. Then we have $\mathscr{P}_n u = V_n[w_n]$ in the polar coordinates. Thus, it suffices to show for $r \geq 1$,

$$|V_{r,n}[w_n](r)| \le C\int_1^\infty |w_n(s)|\,\mathrm{d}s\,, \qquad |V_{\theta,n}[w_n](r)| \le C\int_1^\infty |w_n(s)|\,\mathrm{d}s\,.$$

These estimates are directly obtained from (28) and the definition of $\psi_n[w_n]$ in (27). The proof is complete.

Remark 3. The condition $\frac{1}{|x|}\mathrm{rot}\,\mathscr{P}_n u \in L^1(\Omega)$ for $|n| = 1$ is always satisfied when $u \in D_{L^2_\sigma \cap L^q_\sigma}(\mathbf{A})$ with some $q \in (1,2)$.

4 Resolvent Analysis

To prove Theorem 2, we consider the resolvent problem associated with (Sv): $\lambda u - \Delta u + V\cdot\nabla u + u\cdot\nabla V + \nabla p = f$, $\mathrm{div}\,u = 0$ in Ω and $u = 0$ on $\partial\Omega$. Here, $\lambda \in \mathbb{C}$ is a resolvent parameter. By taking into account the identity $u\cdot\nabla u = \frac{1}{2}\nabla|u|^2 + u^\perp \mathrm{rot}\,u$, $u^\perp = (-u_2, u_1)^\top$, with a suitable modification of the pressure p, the above resolvent problem is equivalent to

$$\begin{cases} \lambda u - \Delta u + V^\perp \mathrm{rot}\,u + u^\perp \mathrm{rot}\,V + \nabla p = f\,, & x \in \Omega\,, \\ \mathrm{div}\,u = 0\,, & x \in \Omega\,, \\ u = 0\,, & x \in \partial\Omega\,. \end{cases} \qquad (\mathrm{RS_V})$$

Set

$$\tilde{u} = \sum_{|n|\ne 1} \mathscr{P}_n u\,, \qquad \tilde{p} = \sum_{|n|\ne 1} p_n$$

where p_n is defined as in (25) for p. From $V = \alpha U + W$ in **Assumption** of Section 1 together with the invariant property $\mathscr{P}_n U^\perp \mathrm{rot}\,u = U^\perp \mathrm{rot}\,\mathscr{P}_n u$ (see [7, Lemma 2.9]) and $\mathrm{rot}\,U = 0$, the problem (RSv) is decomposed into

$$\begin{cases} \lambda\tilde{u} - \Delta\tilde{u} + \alpha U^\perp \mathrm{rot}\,\tilde{u} + \sum_{|n|\ne 1} \mathscr{P}_n\big(W^\perp \mathrm{rot}\,u + u^\perp \mathrm{rot}\,W\big) + \nabla\tilde{p} \\ \quad = \sum_{|n|\ne 1} \mathscr{P}_n f\,, \qquad x \in \Omega\,, \\ \mathrm{div}\,\tilde{u} = 0\,, \quad x \in \Omega\,, \qquad \tilde{u} = 0\,, \quad x \in \partial\Omega\,, \end{cases} \qquad (\tilde{\mathrm{RS}}_{\mathrm{V}})$$

and

$$\begin{cases} \lambda\mathscr{P}_n u - \Delta\mathscr{P}_n u + \alpha U^\perp \mathrm{rot}\,\mathscr{P}_n u + \mathscr{P}_n\big(W^\perp \mathrm{rot}\,u + u^\perp \mathrm{rot}\,W\big) + \nabla p_n \\ \quad = \mathscr{P}_n f\,, \qquad x \in \Omega\,, \\ \mathrm{div}\,\mathscr{P}_n u = 0\,, \quad x \in \Omega\,, \qquad \mathscr{P}_n u = 0\,, \quad x \in \partial\Omega\,, \end{cases} \qquad (\mathrm{nRS_V})$$

for $|n| = 1$. Roughly speaking, our strategy is: (i) to apply the energy method for $\tilde{u}$ (Section 4.1), and (ii) to apply the representation formula for $(\lambda + \mathbf{A}_{\alpha U}|_{\mathscr{P}_n L^2_\sigma})^{-1}\mathbb{P}\mathscr{P}_n$ established in [7] to estimate $\mathscr{P}_n u$ (Section 4.2).

4.1 Resolvent Estimate by Energy Method

In this subsection, we derive the estimate of $\tilde{u}$ from $(\widetilde{\mathrm{RS}}_{\mathrm{V}})$ using the standard energy method.

Lemma 3. *Let* $1 < q < 2$. *There exists* $C_0 > 0$ *such that any solution* $u \in D_{L^2_\sigma \cap L^q_\sigma}(\mathbf{A}_V)$ *to (RSv) satisfies the estimate*

$$\begin{aligned} &\Re(\lambda)\, \|\tilde{u}\|^2_{L^2(\Omega)} + (1 - |\alpha| - C_0\delta)\|\nabla\tilde{u}\|^2_{L^2(\Omega)} \\ &\le C_0\delta \sum_{|n|=1} \big(\|\mathrm{rot}\, \mathscr{P}_n u\|^2_{L^2(\Omega)} + \|\frac{1}{|x|}\mathrm{rot}\, \mathscr{P}_n u\|^2_{L^1(\Omega)}\big) + \Re\langle f, \tilde{u}\rangle_{L^2(\Omega)}\,, \end{aligned} \tag{31}$$

and

$$\begin{aligned} |\Im(\lambda)|\, \|\tilde{u}\|^2_{L^2(\Omega)} &\le (|\alpha| + C_0\delta)\|\nabla\tilde{u}\|^2_{L^2(\Omega)} \\ &\quad + C_0\delta \sum_{|n|=1} \big(\|\mathrm{rot}\, \mathscr{P}_n u\|^2_{L^2(\Omega)} + \|\frac{1}{|x|}\mathrm{rot}\, \mathscr{P}_n u\|^2_{L^1(\Omega)}\big) + |\Im\langle f, \tilde{u}\rangle_{L^2(\Omega)}|\,. \end{aligned} \tag{32}$$

Here, $\tilde{u} = \sum_{|n|\ne 1} \mathscr{P}_n u$.

Proof. We write $\|\cdot\|_{L^p}$ for $\|\cdot\|_{L^p(\Omega)}$ to simplify the notation. By taking the inner product with $\tilde{u}$ in the first equation of $(\widetilde{\mathrm{RS}}_{\mathrm{V}})$, we have

$$\lambda\|\tilde{u}\|^2_{L^2} + \|\nabla\tilde{u}\|^2_{L^2} + \alpha\langle U^\perp \mathrm{rot}\, \tilde{u}, \tilde{u}\rangle_{L^2} + \langle W^\perp \mathrm{rot}\, u + u^\perp \mathrm{rot}\, W, \tilde{u}\rangle_{L^2} = \langle f, \tilde{u}\rangle_{L^2}\,.$$

The real part of this equality gives

$$\begin{aligned} &\Re(\lambda)\|\tilde{u}\|^2_{L^2} + \|\nabla\tilde{u}\|^2_{L^2} + \alpha\Re\langle U^\perp \mathrm{rot}\, \tilde{u}, \tilde{u}\rangle_{L^2} \\ &\quad + \Re\langle W^\perp \mathrm{rot}\, u, \tilde{u}\rangle_{L^2} + \Re\langle (\sum_{|n|=1} \mathscr{P}_n u)^\perp \mathrm{rot}\, W, \tilde{u}\rangle_{L^2} = \Re\langle f, \tilde{u}\rangle_{L^2}\,, \end{aligned} \tag{33}$$

while the imaginary part yields

$$\begin{aligned} &\Im(\lambda)\|\tilde{u}\|^2_{L^2} + \alpha\Im\langle U^\perp \mathrm{rot}\, \tilde{u}, \tilde{u}\rangle_{L^2} + \Im\langle W^\perp \mathrm{rot}\, u, \tilde{u}\rangle_{L^2} \\ &\quad + \Im\langle u^\perp \mathrm{rot}\, W, \tilde{u}\rangle_{L^2} = \Im\langle f, \tilde{u}\rangle_{L^2}\,, \end{aligned} \tag{34}$$

Here, we have used $\Re\langle \tilde{u}^{\perp}\mathrm{rot}\, W, \tilde{u}\rangle_{L^2} = 0$ in the derivation of (33). From Lemma 1 for $\tilde{u}$, the term $\langle U^{\perp}\mathrm{rot}\, \tilde{u}, \tilde{u}\rangle_{L^2}$ is estimated as

$$|\langle U^{\perp}\mathrm{rot}\, \tilde{u}, \tilde{u}\rangle_{L^2}| \le \|\frac{1}{|x|}\tilde{u}\, \mathrm{rot}\, \tilde{u}\|_{L^1} \le \|\mathrm{rot}\, \tilde{u}\|_{L^2}\|\nabla\tilde{u}\|_{L^2} = \|\nabla\tilde{u}\|_{L^2}^2 . \tag{35}$$

Here, we have used $\|\nabla v\|_{L^2(\Omega)} = \|\mathrm{rot}\, v\|_{L^2(\Omega)}$ for $v \in D_{L^2_\sigma}(\mathbf{A})$. Similarly, the condition (W2) yields

$$\begin{aligned} |\langle W^{\perp}\mathrm{rot}\, u, \tilde{u}\rangle_{L^2}| &\le |\langle(\mathscr{P}_0 W)^{\perp}\mathrm{rot} u, \tilde{u}\rangle_{L^2}| + |\langle(\mathscr{Q}_0 W)^{\perp}\mathrm{rot}\, u, \tilde{u}\rangle_{L^2}| \\ &= |\langle(\mathscr{P}_0 W)^{\perp}\mathrm{rot}\, \tilde{u}, \tilde{u}\rangle_{L^2}| + |\langle(\mathscr{Q}_0 W)^{\perp}\mathrm{rot}\, u, \tilde{u}\rangle_{L^2}| \\ &\le \delta\|\mathrm{rot}\, \tilde{u}\|_{L^2}\|\nabla\tilde{u}\|_{L^2} + \delta\|\frac{\mathrm{rot}\, u}{\log(e+|x|)}\|_{L^2}\|\nabla\tilde{u}\|_{L^2} \\ &\le 2\delta\big(\|\nabla\tilde{u}\|_{L^2}^2 + \sum_{|n|=1}\|\mathrm{rot}\, \mathscr{P}_n u\|_{L^2}^2\big) . \end{aligned} \tag{36}$$

In the second line, we have used $\langle(\mathscr{P}_0 W)^{\perp}\mathrm{rot}\, \mathscr{P}_n u, \tilde{u}\rangle_{L^2} = 0$ for $|n| = 1$ in virtue of $(\mathscr{P}_0 W)^{\perp}\mathrm{rot}\, \mathscr{P}_n = \mathscr{P}_n\big((\mathscr{P}_0 W)^{\perp}\mathrm{rot}\, \mathscr{P}_n\big)$ and $\mathscr{P}_n\tilde{u} = 0$ for $|n| = 1$. Next we have from (W3), Lemma 1 for $\tilde{u}$, and Lemma 2,

$$\begin{aligned} &|\langle u^{\perp}\mathrm{rot}\, W, \tilde{u}\rangle_{L^2}| + |\langle(\sum_{|n|=1}\mathscr{P}_n u)^{\perp}\mathrm{rot}\, W, \tilde{u}\rangle_{L^2}| \\ &\le |\langle\tilde{u}^{\perp}\mathrm{rot}\, W, \tilde{u}\rangle_{L^2}| + 2|\langle(\sum_{|n|=1}\mathscr{P}_n u)^{\perp}\mathrm{rot}\, W, \tilde{u}\rangle_{L^2}| \\ &\le C\delta\|\nabla\tilde{u}\|_{L^2}^2 + 2\delta\sum_{|n|=1}\|\mathscr{P}_n u\|_{L^\infty}\|\frac{\tilde{u}}{(|x|\log(e+|x|))^2}\|_{L^1} \\ &\le C\delta\|\nabla\tilde{u}\|_{L^2}^2 + C\delta\sum_{|n|=1}\|\frac{1}{|x|}\mathrm{rot}\, \mathscr{P}_n u\|_{L^1}\|\nabla\tilde{u}\|_{L^2} \\ &\le C\delta\big(\|\nabla\tilde{u}\|_{L^2}^2 + \sum_{|n|=1}\|\frac{1}{|x|}\mathrm{rot}\, \mathscr{P}_n u\|_{L^1}^2\big) . \end{aligned} \tag{37}$$

Combining (35), (36), and (37) with (33) and (34), we obtain

$$\begin{aligned} &\Re(\lambda)\, \|\tilde{u}\|_{L^2}^2 + (1-|\alpha|-C_0\delta)\|\nabla\tilde{u}\|_{L^2}^2 \\ &\le C_0\delta\sum_{|n|=1}\big(\|\mathrm{rot}\, \mathscr{P}_n u\|_{L^2}^2 + \|\frac{1}{|x|}\mathrm{rot}\, \mathscr{P}_n u\|_{L^1}^2\big) + \Re\langle f, \tilde{u}\rangle_{L^2} , \end{aligned}$$

and

$$|\Im(\lambda)|\,\|\tilde u\|_{L^2}^2 \le (|\alpha| + C_0\delta)\|\nabla\tilde u\|_{L^2}^2 \\ + C_0\delta\sum_{|n|=1}\big(\|\mathrm{rot}\,\mathscr{P}_n u\|_{L^2}^2 + \|\frac{1}{|x|}\mathrm{rot}\,\mathscr{P}_n u\|_{L^1}^2\big) + |\Im\langle f,\tilde u\rangle_{L^2}|\,.$$

Here, C_0 is a universal constant. The proof is complete.

4.2 Estimate for Vorticity in Terms of Representation Formula

In this subsection, we study the solution $\mathscr{P}_n u$ to (nRSv) with $|n| = 1$ using the representation formula for the resolvent $(\lambda + \mathbf{A}_{\alpha U}|_{\mathscr{P}_n L^2_\sigma})^{-1}$ established in [7]. To this end, we set

$$H_n[u] = -\mathscr{P}_n\big(W^\perp \mathrm{rot}\,u + u^\perp \mathrm{rot}\,W\big)\,, \qquad f_n = \mathscr{P}_n f\,. \tag{38}$$

Lemma 4. *Let $|n| = 1$. Then we have for $u \in D_{L^2_\sigma}(\mathbf{A})$ satisfying $\frac{1}{|x|}\mathrm{rot}\,\mathscr{P}_k u \in L^1(\Omega)$, $|k| = 1$,*

$$\|H_n[u]\|_{L^1(\Omega)} \le C\delta\big(\sum_{|k|=1}\|\frac{1}{|x|}\mathrm{rot}\,\mathscr{P}_k u\|_{L^1(\Omega)} + \|\nabla\tilde u\|_{L^2(\Omega)}\big)\,. \tag{39}$$

Here $\tilde u = \sum_{|k|\neq 1}\mathscr{P}_k u$.

Proof. To show (39), we first note that

$$\begin{aligned}\mathscr{P}_{\pm1}(W^\perp \mathrm{rot}\,u) &= \mathscr{P}_{\pm1}((\mathscr{P}_0 W)^\perp \mathrm{rot}\,u) + \mathscr{P}_{\pm1}((\mathscr{Q}_0 W)^\perp \mathrm{rot}\,u)\\ &= \mathscr{P}_{\pm1}((\mathscr{P}_0 W)^\perp \mathrm{rot}\,\mathscr{P}_{\pm1}u) + \mathscr{P}_{\pm1}((\mathscr{Q}_0 W)^\perp \mathrm{rot}\,u)\,.\end{aligned} \tag{40}$$

Then, by (W2), we have

$$\begin{aligned}\|\mathscr{P}_{\pm1}(W^\perp \mathrm{rot}\,u)\|_{L^1(\Omega)} &\le \delta\big(\|\frac{1}{|x|}\mathrm{rot}\,\mathscr{P}_{\pm1}u\|_{L^1(\Omega)} + \|\frac{1}{|x|\log(e+|x|)}\mathrm{rot}\,u\|_{L^1(\Omega)}\big)\\ &\le \delta\big(2\sum_{|k|=1}\|\frac{1}{|x|}\mathrm{rot}\,\mathscr{P}_k u\|_{L^1(\Omega)} + C\|\mathrm{rot}\,\tilde u\|_{L^2(\Omega)}\big)\,.\end{aligned}$$

Next, the condition (W3) and Lemma 2 lead to

$$\begin{aligned}\|\mathscr{P}_n(u^{\perp}\mathrm{rot}\, W)\|_{L^1(\Omega)} &\le \delta \|\frac{u}{(|x|\log(e+|x|))^2}\|_{L^1(\Omega)}\\ &\le C\delta\big(\sum_{|k|=1}\|\mathscr{P}_k u\|_{L^\infty(\Omega)}+\|\nabla\tilde{u}\|_{L^2(\Omega)}\big)\\ &\le C\delta\big(\sum_{|k|=1}\|\frac{1}{|x|}\mathrm{rot}\,\mathscr{P}_k u\|_{L^1(\Omega)}+\|\nabla\tilde{u}\|_{L^2(\Omega)}\big)\,.\end{aligned}$$

Thus, (39) follows. The proof is complete.

Let $\alpha_0 > 0$ be the number in Theorem 1. By [7, Theorem 1.5 (5)], one can take a sector $\Sigma_{\pi-\varepsilon_0} = \{z \in \mathbb{C}\setminus\{0\} \mid |\arg z| < \pi - \varepsilon_0\}$ with some $\varepsilon_0 \in (0, \frac{\pi}{2})$ so that $\Sigma_{\pi-\varepsilon_0} \subset \rho(-\mathbf{A}_{\alpha U})$ for any $0 < |\alpha| \le \alpha_0$. Here, $\rho(-\mathbf{A}_{\alpha U})$ is the resolvent set of $-\mathbf{A}_{\alpha U}$ in $L^2_\sigma(\Omega)$. Then, as long as $\lambda \in \Sigma_{\pi-\varepsilon_0}$, the solution to (nRSv) is the sum of $v_n^{(j)}$, $j = 1, 2$, where

$$v_n^{(1)} = (\lambda + \mathbf{A}_{\alpha U}|_{\mathscr{P}_n L^2_\sigma})^{-1}\mathbb{P} f_n\,, \qquad v_n^{(2)} = (\lambda + \mathbf{A}_{\alpha U}|_{\mathscr{P}_n L^2_\sigma})^{-1}\mathbb{P} H_n[u]\,.$$

To state the representation formula of the resolvent $(\lambda + \mathbf{A}_{\alpha U}|_{\mathscr{P}_n L^2_\sigma})^{-1}\mathbb{P}\mathscr{P}_n$, we introduce some special functions. Let $I_\mu(z)$ be the modified Bessel function of first kind of order μ:

$$I_\mu(z) = \big(\frac{z}{2}\big)^{\mu}\sum_{m=0}^{\infty}\frac{1}{m!\,\Gamma(\mu+m+1)}\big(\frac{z}{2}\big)^{2m}\,, \qquad z \in \mathbb{C}\setminus\overline{\mathbb{R}_-}\,. \tag{41}$$

Here, $\overline{\mathbb{R}_-} = \{z \le 0\}$, and $z^\mu = e^{\mu \mathrm{Log}\, z}$ and $\mathrm{Log}\, z = \log|z| + i\arg z$ is the principal branch of the logarithm of $z = |z|e^{i\theta}$, $-\pi < \arg z = \theta < \pi$, while $\Gamma(z)$ is the Gamma function. For $\mu \notin \mathbb{Z}$, the modified Bessel function of second kind $K_\mu(z)$ of order μ is defined by the formula

$$K_\mu(z) = \frac{\pi}{2}\cdot\frac{I_{-\mu}(z) - I_\mu(z)}{\sin(\mu\pi)}\,. \tag{42}$$

It is well known that K_μ and I_μ are independent solutions to the ordinary differential equation

$$-\frac{\mathrm{d}^2\omega}{\mathrm{d}z^2} - \frac{1}{z}\frac{\mathrm{d}\omega}{\mathrm{d}z} + \big(1 + \frac{\mu^2}{z^2}\big)\omega = 0 \tag{43}$$

with the Wronskian

$$W_\mu(z) = \det\begin{pmatrix} K_\mu(z) & I_\mu(z)\\ \frac{\mathrm{d}K_\mu}{\mathrm{d}z}(z) & \frac{\mathrm{d}I_\mu}{\mathrm{d}z}(z)\end{pmatrix} = \frac{1}{z}\,. \tag{44}$$

Set

$$\mu_n(\alpha) = (n^2 + i\alpha n)^{\frac{1}{2}}, \tag{45}$$

with $\Re(\mu_n(\alpha)) > 0$. We will often write μ_n instead of $\mu_n(\alpha)$ for simplicity of notations. Then $K_{\mu_n}(\sqrt{\lambda}r)$ and $I_{\mu_n}(\sqrt{\lambda}r)$ with $\Re(\sqrt{\lambda}) > 0$ are independent solutions to the homogeneous equation of

$$-\frac{\mathrm{d}^2\omega}{\mathrm{d}r^2} - \frac{1}{r}\frac{\mathrm{d}\omega}{\mathrm{d}r} + (\lambda + \frac{n^2 + i\alpha n}{r^2})\omega = 0, \tag{46}$$

and their Wronskian is r^{-1}. As is well known, the function $K_\mu(z)$ decays exponentially as $|z| \to \infty$, $\Re(z) > 0$, while $I_\mu(z)$ grows exponentially. Next, we introduce the function $F_n(z;\alpha)$ as

$$F_n(z;\alpha) = \int_1^\infty s^{1-|n|} K_{\mu_n(\alpha)}(sz)\,\mathrm{d}s\,, \quad \Re(z) > 0\,. \tag{47}$$

By [7, Proposition 3.4] and its proof, if $\lambda \in \rho(-\mathbf{A}_{\alpha U})$, then the operator $(\lambda + \mathbf{A}_{\alpha U}|_{\mathscr{P}_n L^2_\sigma})^{-1}\mathbb{P}\mathscr{P}_n$ is expressed as

$$(\lambda + \mathbf{A}_{\alpha U}|_{\mathscr{P}_n L^2_\sigma})^{-1}\mathbb{P} f_n = -\frac{c_{n,\lambda}[f_n]}{F_n(\sqrt{\lambda};\alpha)} V_n[K_{\mu_n(\alpha)}(\sqrt{\lambda}\,\cdot)] + V_n[\Phi_{n,\lambda}[f_n]]\,. \tag{48}$$

Here, the operator $V_n[\cdot]$ is the Biot–Savart law defined by (28), while

$$\begin{aligned}\Phi_{n,\lambda}[f_n](r) = {} & -K_{\mu_n}(\sqrt{\lambda}r)\Big(\int_1^r I_{\mu_n}(\sqrt{\lambda}s)\big(\mu_n f_{\theta,n}(s) + inf_{r,n}(s)\big)\,\mathrm{d}s \\ & \qquad + \sqrt{\lambda}\int_1^r I_{\mu_n+1}(\sqrt{\lambda}s) f_{\theta,n}(s)s\,\mathrm{d}s\Big) \\ & + I_{\mu_n}(\sqrt{\lambda}r)\Big(\int_r^\infty K_{\mu_n}(\sqrt{\lambda}s)\big(\mu_n f_{\theta,n}(s) - inf_{r,n}(s)\big)\,\mathrm{d}s \\ & \qquad + \sqrt{\lambda}\int_r^\infty K_{\mu_n-1}(\sqrt{\lambda}s) f_{\theta,n}(s)s\,\mathrm{d}s\Big)\end{aligned} \tag{49}$$

and

$$c_{n,\lambda}[f_n] = \int_1^\infty s^{1-|n|}\Phi_{n,\lambda}[f_n](s)\,\mathrm{d}s\,. \tag{50}$$

Note that $f_{r,n}$ and $f_{\theta,n}$ are defined so that

$$f_n = \mathscr{P}_n f = f_{r,n}e^{in\theta}\mathbf{e}_r + f_{\theta,n}e^{in\theta}\mathbf{e}_\theta\,. \tag{51}$$

Remark 4. (i) In [7], the formula (48) is derived in the case $\mathbb{P} f_n = f_n$. But in the derivation of this formula, the divergence free condition of f_n is not needed,

and hence, the formula obtained in [7] is in fact for $(\lambda + \mathbf{A}_{\alpha U}|_{\mathscr{P}_n L^2_\sigma})^{-1}\mathbb{P}\mathscr{P}_n$ in $\mathscr{P}_n L^2(\Omega)^2$, not only for $(\lambda + \mathbf{A}_{\alpha U}|_{\mathscr{P}_n L^2_\sigma})^{-1}$ in $\mathscr{P}_n L^2_\sigma(\Omega)$.
(ii) Let $\lambda \in \Sigma_{\pi-\varepsilon}$ for some $\varepsilon \in (0, \frac{\pi}{2})$. Then for the coefficient $c_{n,\lambda}[f_n]$, we have

$$|c_{n,\lambda}[f_n]| \le \frac{C}{|\lambda|^{1-\frac{1}{q}}(1+|\lambda|^{\frac{1}{2q}})}\|f_n\|_{L^q(\Omega)}\,, \quad f_n \in L^q(\Omega)^2\,,\ 1 \le q < \infty\,. \tag{52}$$

Here, C depends only on α, n, ε, and q. For the proof of (52), see [7, Proposition 3.16]. In [7, Proposition 3.16], this estimate is stated, again, under the divergence free condition on f_n, but such additional condition is not required in deriving (52), and is needed only for the estimate in the case $q = \infty$.

Taking (49) into account, we set

$$\begin{aligned}
\Phi^{(1)}_{n,\lambda}[\phi](r) &= -K_{\mu_n}(\sqrt{\lambda}r)\int_1^r I_{\mu_n}(\sqrt{\lambda}s)\,\phi(s)\,\mathrm{d}s\,,\\
\Phi^{(2)}_{n,\lambda}[\phi](r) &= -\sqrt{\lambda}K_{\mu_n}(\sqrt{\lambda}r)\int_1^r I_{\mu_n+1}(\sqrt{\lambda}s)\,\phi(s)s\,\mathrm{d}s\,,\\
\Phi^{(3)}_{n,\lambda}[\phi](r) &= I_{\mu_n}(\sqrt{\lambda}r)\int_r^\infty K_{\mu_n}(\sqrt{\lambda}s)\,\phi(s)\,\mathrm{d}s\,,\\
\Phi^{(4)}_{n,\lambda}[\phi](r) &= \sqrt{\lambda}I_{\mu_n}(\sqrt{\lambda}r)\int_r^\infty K_{\mu_n-1}(\sqrt{\lambda}s)\,\phi(s)s\,\mathrm{d}s\,.
\end{aligned}$$

Then, by setting

$$g_n = \mu_n f_{\theta,n} + inf_{r,n}\,, \qquad j_n = \mu_n f_{\theta,n} - inf_{r,n}\,,$$

the operator $\Phi_{n,\lambda}[f_n]$ is written as

$$\Phi_{n,\lambda}[f_n] = \Phi^{(1)}_{n,\lambda}[g_n] + \Phi^{(2)}_{n,\lambda}[f_{\theta,n}] + \Phi^{(3)}_{n,\lambda}[j_n] + \Phi^{(4)}_{n,\lambda}[f_{\theta,n}]\,. \tag{53}$$

Lemma 5. *Let $|n| \ge 1$. Let $\lambda \in \Sigma_{\pi-\varepsilon}$ for some $\varepsilon \in (0, \frac{\pi}{2})$, and let $k = 1, 2, 3, 4$. Then*

$$\|\frac{1}{r}\Phi^{(k)}_{n,\lambda}[\phi]\|_{L^1} \le C|\lambda|^{-1+\frac{1}{q}}\|\phi\|_{L^q}\,, \qquad 1 \le q < 2\,, \tag{54}$$

$$\|\Phi^{(k)}_{n,\lambda}[\phi]\|_{L^2} \le C|\lambda|^{-1+\frac{1}{q}}\|\phi\|_{L^q}\,, \qquad 1 \le q \le 2\,. \tag{55}$$

where C depends only on α, n, ε, and q. Here $\|\phi\|_{L^q} = (\int_1^\infty |\phi(r)|^q r\,\mathrm{d}r)^{\frac{1}{q}}$, $1 \le q < \infty$.

Proof. We give the proof only for $\Phi^{(1)}_{n,\lambda}[\phi]$, since the proof for $\Phi^{(k)}_{n,\lambda}[\phi]$ with $k = 2, 3, 4$ is very similar. The condition $q < 2$ is needed for (54) with $k = 2, 4$. Note that $C_\varepsilon|\lambda|^{\frac{1}{2}} \le \Re(\sqrt{\lambda}) \le C'_\varepsilon|\lambda|^{\frac{1}{2}}$ for $\lambda \in \Sigma_{\pi-\varepsilon}$ with $\varepsilon \in (0, \frac{\pi}{2})$. First, we consider the case $\Re(\sqrt{\lambda}) \le 1$. By changing the order of integrals, we see

$$\int_1^\infty |\Phi^{(1)}_{n,\lambda}[\phi](r)|\,\mathrm{d}r \le \int_1^\infty \int_s^\infty |K_{\mu_n}(\sqrt{\lambda}r)|\,\mathrm{d}r\,|I_{\mu_n}(\sqrt{\lambda}s)\phi(s)|\,\mathrm{d}s$$
$$\le \int_1^{\frac{1}{\Re\sqrt{\lambda}}} \cdots\,\mathrm{d}s + \int_{\frac{1}{\Re\sqrt{\lambda}}}^\infty \cdots\,\mathrm{d}s =: I + II\,.$$

For the term I we have from Lemma 6 in the appendix,

$$I \le C\int_1^{\frac{1}{\Re(\sqrt{\lambda})}} \int_s^{\frac{1}{\Re(\sqrt{\lambda})}} |\sqrt{\lambda}r|^{-\Re(\mu_n)}\,\mathrm{d}r\,|\sqrt{\lambda}s|^{\Re(\mu_n)}|\phi(s)|\,\mathrm{d}s$$
$$+ C\int_1^{\frac{1}{\Re(\sqrt{\lambda})}} \int_{\frac{1}{\Re(\sqrt{\lambda})}}^\infty |\sqrt{\lambda}r|^{-\frac{1}{2}}e^{-r\Re(\sqrt{\lambda})}\,\mathrm{d}r\,|\sqrt{\lambda}s|^{\Re(\mu_n)}|\phi(s)|\,\mathrm{d}s$$
$$\le \frac{C}{\Re(\mu_n)-1}\int_1^{\frac{1}{\Re(\sqrt{\lambda})}} |\phi(s)|s\,\mathrm{d}s + C|\lambda|^{\frac{\Re(\mu_n)-1}{2}}\int_1^{\frac{1}{\Re(\sqrt{\lambda})}} s^{\Re(\mu_n)}|\phi(s)|\,\mathrm{d}s\,.$$

Here, we note that $\Re(\mu_n) > 1$ as long as $\alpha \ne 0$ by the definition $\mu_n = (n^2 + i\alpha n)^{1/2}$. Hence, we obtain $I \le C|\lambda|^{-1+\frac{1}{q}}\|\phi\|_{L^q}$ with C depends only on α, n, and ε. Similarly, using Lemma 6, the term II is estimated as

$$II \le C\int_{\frac{1}{\Re\sqrt{\lambda}}}^\infty |\sqrt{\lambda}s|^{-\frac{1}{2}}e^{s\Re(\sqrt{\lambda})}\int_s^\infty |\sqrt{\lambda}r|^{-\frac{1}{2}}e^{-r\Re(\sqrt{\lambda})}\,\mathrm{d}r\,|\phi(s)|\,\mathrm{d}s$$
$$\le C|\lambda|^{-1}\int_{\frac{1}{\Re\sqrt{\lambda}}}^\infty s^{-1}|\phi(s)|\,\mathrm{d}s \le C|\lambda|^{-1+\frac{1}{q}}\|\phi\|_{L^q}\,, \quad 1 \le q < \infty\,. \tag{56}$$

Collecting these above, we have shown the estimate $\|\frac{1}{r}\Phi^{(1)}_{n,\lambda}[\phi]\|_{L^1} \le C|\lambda|^{-1+\frac{1}{q}}$ $\|\phi\|_{L^q}$. The case $\Re(\sqrt{\lambda}) \ge 1$ is proved in the similar manner. Indeed, the norm $\|\frac{1}{r}\Phi^{(1)}_{n,\lambda}\|_{L^1}$ is computed as in (56), which leads to $\|\frac{1}{r}\Phi^{(1)}_{n,\lambda}\|_{L^1} \le C|\lambda|^{-1}\|\phi\|_{L^q}$. Hence, we have proved (54) for $k = 1$.

Next, let us prove (55). We recall that in the proof of [7, Lemma 3.21], it is shown that

$$\|\Phi^{(k)}_{n,\lambda}[\phi]\|_{L^q} \le C|\lambda|^{-\frac{1}{2}}\|\phi\|_{L^q}\,, \quad 1 \le q \le \infty\,, \quad \lambda \in \Sigma_{\pi-\varepsilon}\,, \quad k = 1, 2, 3, 4\,. \tag{57}$$

Hence, it suffices to show the L^2-L^1 estimate; then in virtue of the L^2-L^2 estimate in (57), the L^2-L^q estimate for $1 < q < 2$ follows from the interpolation. To show the $L^2 - L^1$ estimate, from the inequality

$$\|h\|_{L^2} \leq \|\frac{1}{r}h\|_{L^1}^{\frac{1}{2}} (\sup_{r\geq 1} r|h(r)|)^{\frac{1}{2}} ,$$

it suffices to prove $\sup_{r\geq 1} r|\Phi_{n,\lambda}^{(1)}[\phi](r)| \leq C\|\phi\|_{L^1}$. We see

$$|\Phi_{n,\lambda}^{(1)}[\phi](r)| \leq |K_{\mu_n}(\sqrt{\lambda}r)| \sup_{1\leq s\leq r} s^{-1}|I_{\mu_n}(\sqrt{\lambda}s)|\|\phi\|_{L^1} . \tag{58}$$

First, we consider the case $\Re(\sqrt{\lambda}) \leq 1$. Let $r \leq \frac{1}{\Re(\sqrt{\lambda})}$. Then Lemma 6 implies

$$|K_{\mu_n}(\sqrt{\lambda}r)| \leq C|\sqrt{\lambda}r|^{-\Re(\mu_n)} ,$$

while

$$s^{-1}|I_{\mu_n}(\sqrt{\lambda}s)| \leq C|\lambda|^{\frac{\Re(\mu_n)}{2}} s^{\Re(\mu_n)-1} \leq C|\lambda|^{\frac{\Re(\mu_n)}{2}} r^{\Re(\mu_n)-1}, \quad 1 \leq s \leq r .$$

Here, we have again used $\Re(\mu_n) > 1$ for $\alpha \neq 0$. Thus, the estimate $r|\Phi_{n,\lambda}^{(1)}[\phi](r)| \leq C\|\phi\|_{L^1}$ holds in the case $r \leq \frac{1}{\Re(\sqrt{\lambda})}$. If $r \geq \frac{1}{\Re(\sqrt{\lambda})}$ then Lemma 6 yields

$$|K_{\mu_n}(\sqrt{\lambda}r)| \leq C|\sqrt{\lambda}r|^{-\frac{1}{2}} e^{-r\Re(\sqrt{\lambda})} , \tag{59}$$

and

$$\begin{aligned}
&\sup_{1\leq s\leq r} s^{-1}|I_{\mu_n}(\sqrt{\lambda}s)| \\
&\leq \sup_{1\leq s\leq \frac{1}{\Re(\sqrt{\lambda})}} s^{-1}|I_{\mu_n}(\sqrt{\lambda}s)| + \sup_{\frac{1}{\Re(\sqrt{\lambda})}\leq s\leq r} s^{-1}|I_{\mu_n}(\sqrt{\lambda}s)| \\
&\leq C|\lambda|^{\frac{1}{2}} + C \sup_{\frac{1}{2\Re(\sqrt{\lambda})}\leq s\leq \frac{r}{2}} s^{-1}|\sqrt{\lambda}s|^{-\frac{1}{2}} e^{s\Re(\sqrt{\lambda})} + C \sup_{\frac{r}{2}\leq s\leq r} s^{-1}|\sqrt{\lambda}s|^{-\frac{1}{2}} e^{s\Re(\sqrt{\lambda})} \\
&\leq C(|\lambda|^{\frac{1}{2}} + |\lambda|^{-\frac{1}{4}} r^{-\frac{3}{2}} e^{r\Re(\sqrt{\lambda})}) .
\end{aligned}$$

Thus, we have for $r \geq \frac{1}{\Re(\sqrt{\lambda})}$,

$$|\Phi_{n,\lambda}^{(1)}[\phi](r)| \leq C\big(r^{-\frac{1}{2}}|\lambda|^{\frac{1}{4}} e^{-r\Re(\sqrt{\lambda})} + |\lambda|^{-\frac{1}{2}} r^{-2}\big)\|\phi\|_{L^1} \leq Cr^{-1}\|\phi\|_{L^1} .$$

Hence, we have shown $\sup_{r\geq 1} r|\Phi_{n,\lambda}^{(1)}[\phi](r)| \leq C\|\phi\|_{L^1}$, which implies, from (54) with $q = 1$, the $L^2 - L^1$ estimate for $k = 1$ in the case $\Re(\sqrt{\lambda}) \leq 1$. Next we consider the case $\Re(\sqrt{\lambda}) \geq 1$. The estimates (58) and (59), together with the inequality

$$\sup_{1\leq s\leq r} s^{-1}|I_{\mu_n}(\sqrt{\lambda}s)| \leq C|\lambda|^{-\frac{1}{4}} r^{-\frac{3}{2}} e^{r\Re(\sqrt{\lambda})}, \quad \Re(\sqrt{\lambda}) \geq 1 ,$$

yield

$$|\Phi^{(1)}_{n,\lambda}[\phi](r)| \le C|\lambda|^{-\frac{1}{2}} r^{-2} \|\phi\|_{L^1} .$$

In particular, the estimate $\sup_{r\ge 1} r|\Phi^{(1)}_{n,\lambda}[\phi](r)| \le C|\lambda|^{-\frac{1}{2}}\|\phi\|_{L^1}$ holds also for $\Re(\sqrt{\lambda}) \ge 1$. Collecting these, we have shown

$$\sup_{r\ge 1} r|\Phi^{(1)}_{n,\lambda}[\phi](r)| \le C\min\{1, |\lambda|^{-\frac{1}{2}}\}\|\phi\|_{L^1} , \qquad \lambda \in \Sigma_{\pi-\varepsilon} , \tag{60}$$

which finally implies (55) with $q = 1$ for $k = 1$, and thus also for $1 < q \le 2$ due to (57) and the interpolation as mentioned above. The estimates for $k = 2, 3, 4$ are proved similarly using Lemma 6, and we omit the details here. The proof is complete.

Corollary 1. *Let $\lambda \in \Sigma_{\pi-\varepsilon}$ for some $\varepsilon \in (0, \frac{\pi}{2})$. Let $|n| = 1$ and let $H_n = H_n[u]$ be as in* (38) *with $u \in D_{L^2_\sigma}(\mathbf{A})$ satisfying $\frac{1}{|x|}\mathrm{rot}\, \mathscr{P}_k u \in L^1(\Omega)$, $|k| = 1$. Then*

$$\|\frac{1}{r}\Phi_{n,\lambda}[H_n]\|_{L^1} \le C\delta\big(\sum_{|k|=1} \|\frac{1}{|x|}\mathrm{rot}\, \mathscr{P}_k u\|_{L^1(\Omega)} + \|\nabla\tilde{u}\|_{L^2(\Omega)}\big) , \tag{61}$$

$$\|\Phi_{n,\lambda}[H_n]\|_{L^2} \le C\delta\big(\sum_{|k|=1} \|\frac{1}{|x|}\mathrm{rot}\, \mathscr{P}_k u\|_{L^1(\Omega)} + \|\nabla\tilde{u}\|_{L^2(\Omega)}\big) . \tag{62}$$

Here, C depends only on α and ε.

Proof. These estimates directly follow from Lemma 4 and Lemma 5 with $q = 1$, by recalling the relation (53). The proof is complete.

Proposition 1. *Let $\lambda \in \rho(-\mathbf{A}_{\alpha U}) \cap \Sigma_{\pi-\varepsilon}$ for some $\varepsilon \in (0, \frac{\pi}{2})$. Let $|n| = 1$ and let $H_n = H_n[u]$ be as in* (38) *with $u \in D_{L^2_\sigma}(\mathbf{A})$ satisfying $\frac{1}{|x|}\mathrm{rot}\, \mathscr{P}_k u \in L^1(\Omega)$, $|k| = 1$.*
(i) *Let $\Re(\sqrt{\lambda}) \le 1$. Then*

$$\|\frac{1}{|x|}\mathrm{rot}\, (\lambda + \mathbf{A}_{\alpha U}|_{\mathscr{P}_n L^2_\sigma})^{-1}\mathbb{P}H_n\|_{L^1(\Omega)} \le C\delta\Big(\frac{|\lambda|^{-\frac{\Re(\mu_n)}{2}}}{|F_n(\sqrt{\lambda};\alpha)|} + 1\Big)\big(\sum_{|k|=1} \|\frac{1}{|x|}\mathrm{rot}\, \mathscr{P}_k u\|_{L^1(\Omega)} + \|\nabla\tilde{u}\|_{L^2(\Omega)}\big) , \tag{63}$$

$$\|\mathrm{rot}\, (\lambda + \mathbf{A}_{\alpha U}|_{\mathscr{P}_n L^2_\sigma})^{-1}\mathbb{P}H_n\|_{L^2(\Omega)} \le C\delta\Big(\frac{|\lambda|^{-\frac{\Re(\mu_n)}{2}}}{|F_n(\sqrt{\lambda};\alpha)|} + 1\Big)\big(\sum_{|k|=1} \|\frac{1}{|x|}\mathrm{rot}\, \mathscr{P}_k u\|_{L^1(\Omega)} + \|\nabla\tilde{u}\|_{L^2(\Omega)}\big) . \tag{64}$$

(ii) *Let* $\Re(\sqrt{\lambda}) \geq 1$. *Then*

$$\begin{aligned}
&\|\frac{1}{|x|}\mathrm{rot}\,(\lambda + \mathbf{A}_{\alpha U}|_{\mathscr{P}_n L^2_\sigma})^{-1}\mathbb{P}H_n\|_{L^1(\Omega)} \\
&\quad \leq C\delta\Big(\frac{|\lambda|^{-\frac{5}{4}}e^{-\Re(\sqrt{\lambda})}}{|F_n(\sqrt{\lambda};\alpha)|} + 1\Big)\Big(\sum_{|k|=1}\|\frac{1}{|x|}\mathrm{rot}\,\mathscr{P}_k u\|_{L^1(\Omega)} + \|\nabla\tilde{u}\|_{L^2(\Omega)}\Big), \qquad (65)\\
&\|\mathrm{rot}\,(\lambda + \mathbf{A}_{\alpha U}|_{\mathscr{P}_n L^2_\sigma})^{-1}\mathbb{P}H_n\|_{L^2(\Omega)} \\
&\quad \leq C\delta\Big(\frac{|\lambda|^{-1}e^{-\Re(\sqrt{\lambda})}}{|F_n(\sqrt{\lambda};\alpha)|} + 1\Big)\Big(\sum_{|k|=1}\|\frac{1}{|x|}\mathrm{rot}\,\mathscr{P}_k u\|_{L^1(\Omega)} + \|\nabla\tilde{u}\|_{L^2(\Omega)}\Big). \qquad (66)
\end{aligned}$$

Proof. As stated in the proof of [7, Theorem 3.23], we have the formula for the vorticity field in terms of the polar coordinates such as

$$\mathrm{rot}\,(\lambda + \mathbf{A}_{\alpha U}|_{\mathscr{P}_n L^2_\sigma})^{-1}\mathbb{P}H_n = -\frac{c_{n,\lambda}[H_n]}{F_n(\sqrt{\lambda};\alpha)}K_{\mu_n}(\sqrt{\lambda}r)e^{in\theta} + \Phi_{n,\lambda}[H_n](r)e^{in\theta}. \qquad (67)$$

We have already shown the estimates for $\Phi_{n,\lambda}[H_n]$, and thus, it suffices to estimate the first term. From (52), the coefficient $c_{n,\lambda}[H_n]$ is estimated as

$$|c_{n,\lambda}[H_n]| \leq \frac{C}{1+|\lambda|^{\frac{1}{2}}}\|H_n\|_{L^1(\Omega)}. \qquad (68)$$

Hence, the estimate of $c_{n,\lambda}[H_n]$ follows from Lemma 4. On the other hand, the direct calculation using Lemma 6 yields, for $\lambda \in \Sigma_{\pi-\varepsilon}$,

$$\begin{aligned}
\|\frac{1}{r}K_{\mu_n}(\sqrt{\lambda}\,\cdot)\|_{L^1} &\leq C|\lambda|^{-\frac{\Re(\mu_n)}{2}}, & \Re(\sqrt{\lambda}) \leq 1, \\
\|\frac{1}{r}K_{\mu_n}(\sqrt{\lambda}\,\cdot)\|_{L^1} &\leq C|\lambda|^{-\frac{3}{4}}e^{-\Re(\sqrt{\lambda})}, & \Re(\sqrt{\lambda}) \geq 1,
\end{aligned} \qquad (69)$$

and

$$\begin{aligned}
\|K_{\mu_n}(\sqrt{\lambda}\,\cdot)\|_{L^2} &\leq C|\lambda|^{-\frac{\Re(\mu_n)}{2}}, & \Re(\sqrt{\lambda}) \leq 1, \\
\|K_{\mu_n}(\sqrt{\lambda}\,\cdot)\|_{L^2} &\leq C|\lambda|^{-\frac{1}{2}}e^{-\Re(\sqrt{\lambda})}, & \Re(\sqrt{\lambda}) \geq 1.
\end{aligned} \qquad (70)$$

We skip the details here; see, e.g., the proof of [7, Lemma 3.2]. The estimate of the term $\frac{c_{n,\lambda}[H_n]}{F_n(\sqrt{\lambda};\alpha)}K_{\mu_n}(\sqrt{\lambda}r)$ follows from (68), (69), and (70). The proof is complete.

Let us recall that, if $|\alpha| \leq \alpha_0$, where $\alpha_0 > 0$ is the number in Theorem 1, then there exists $\varepsilon_0 \in (0, \frac{\pi}{2})$ such that $\Sigma_{\pi-\varepsilon_0} \subset \rho(-\mathbf{A}_{\alpha U})$ and $F_n(\sqrt{\lambda};\alpha) \neq 0$ for $\lambda \in \Sigma_{\pi-\varepsilon_0}$; see [7, Theorem 1.5 (4)]. Moreover, from [7, Lemma 3.29] we have already known that

$$F_n(\sqrt{\lambda};\alpha) = \beta_n \lambda^{-\frac{\mu_n}{2}}\big(1+o(1)\big)\,, \qquad \lambda \in \Sigma_{\pi-\varepsilon_0}\,,\ 0<|\lambda|\ll 1\,, \tag{71}$$

$$F_n(\sqrt{\lambda};\alpha) = \beta_n' \lambda^{-\frac{3}{4}} e^{-\sqrt{\lambda}}\big(1+O(|\lambda|^{-\frac12})\big)\,, \qquad \lambda \in \Sigma_{\pi-\varepsilon_0}\,,\ |\lambda|\gg 1\,. \tag{72}$$

Here β_n and β_n' are nonzero complex numbers independent of λ. From these observations the following resolvent estimate has been established in [7, Est. (263)]:

$$\|(\lambda + \mathbf{A}_{\alpha U}|_{\mathscr{P}_n L^2_\sigma})^{-1}\mathbb{P} f_n\|_{L^2(\Omega)} \le C|\lambda|^{-\frac32+\frac1q}\|f_n\|_{L^q(\Omega)}\,, \quad 1\le q\le 2\,, \quad \lambda\in\Sigma_{\pi-\varepsilon_0}\,, \tag{73}$$

for all $f_n \in \mathscr{P}_n L^2(\Omega)^2 \cap \mathscr{P}_n L^q(\Omega)^2$. Here C depends only on α, ε_0, and q.

Remark 5. Although Est. (73) is mentioned for $(\lambda + \mathbf{A}_{\alpha U}|_{\mathscr{P}_n L^2_\sigma})^{-1}$ in [7, Est. (263)], the proof in [7] is given for the representation in the right-hand side of (48). Thus, Est. (73) is valid also for $(\lambda + \mathbf{A}_{\alpha U}|_{\mathscr{P}_n L^2_\sigma})^{-1}\mathbb{P}\mathscr{P}_n$ (in fact, the divergence free condition $\mathbb{P} f_n = f_n$ is needed only when one shows the $L^\infty - L^\infty$ estimate for the case $|n| = 1$).

Combining (71) and (72) with Proposition 1, we obtain the following result for sufficiently small δ.

Corollary 2. *Let $0<|\alpha|\le\alpha_0$, where α_0 is the number in Theorem 1. Let $\lambda\in\Sigma_{\pi-\varepsilon_0}$, where $\Sigma_{\pi-\varepsilon_0}$ with $\varepsilon_0\in(0,\frac{\pi}{2})$ is the sector contained in the resolvent set of $-\mathbf{A}_{\alpha U}$ in $L^2_\sigma(\Omega)$ as stated above. Let $|n|=1$. Then there exists $\delta_0>0$ such that if $\delta\in[0,\delta_0]$ then any solution $u\in D_{L^2_\sigma}(\mathbf{A})$ to (RSv) satisfying the condition $\frac{1}{|x|}\mathrm{rot}\,\mathscr{P}_n u\in L^1(\Omega)$ for $|n|=1$ obeys*

$$\sum_{|n|=1}\|\frac{1}{|x|}\mathrm{rot}\,\mathscr{P}_n u\|_{L^1(\Omega)} \le C\big(\delta\|\nabla\tilde u\|_{L^2(\Omega)} + |\lambda|^{-1+\frac1q}\sum_{|n|=1}\|\mathscr{P}_n f\|_{L^q(\Omega)}\big)\,, \quad 1\le q<2\,, \tag{74}$$

$$\sum_{|n|=1}\|\mathrm{rot}\,\mathscr{P}_n u\|_{L^2(\Omega)} \le C\big(\delta\|\nabla\tilde u\|_{L^2(\Omega)} + |\lambda|^{-1+\frac1q}\sum_{|n|=1}\|\mathscr{P}_n f\|_{L^q(\Omega)}\big)\,, \quad 1\le q<2\,, \tag{75}$$

whenever $f\in L^2(\Omega)^2\cap L^q_\sigma(\Omega)$. Here, δ_0 and C depend only on α and q.

Proof. In view of (nRSv), we have the formula

$$\mathscr{P}_n u = (\lambda + \mathbf{A}_{\alpha U}|_{\mathscr{P}_n L^2_\sigma})^{-1}\mathbb{P} H_n + (\lambda + \mathbf{A}_{\alpha U}|_{\mathscr{P}_n L^2_\sigma})^{-1}\mathbb{P} f_n\,. \tag{76}$$

Since $\mathrm{rot}\,(\lambda + \mathbf{A}_{\alpha U}|_{\mathscr{P}_n L^2_\sigma})^{-1}\mathbb{P} f_n$ is written as in (67) with H_n replaced by f_n, we have

$$
\begin{aligned}
&\|\frac{1}{|x|}\mathrm{rot}\,(\lambda+\mathbf{A}_{\alpha U}|_{\mathscr{P}_n L^2_\sigma})^{-1}\mathbb{P}f_n\|_{L^1(\Omega)}\\
&\le \frac{|c_{n,\lambda}[f_n]|}{|F_n(\sqrt{\lambda};\alpha)|}\|\frac{1}{r}K_{\mu_n}(\sqrt{\lambda}\,\cdot)\|_{L^1}+\|\frac{1}{r}\Phi_{n,\lambda}[f_n]\|_{L^1}\\
&\le \frac{C}{|\lambda|^{1-\frac{1}{q}}(1+|\lambda|^{\frac{1}{2q}})}\frac{1}{|F_n(\sqrt{\lambda};\alpha)|}\|\frac{1}{r}K_{\mu_n}(\sqrt{\lambda}\,\cdot)\|_{L^1}\|f\|_{L^q(\Omega)}+\|\frac{1}{r}\Phi_{n,\lambda}[f_n]\|_{L^1}\,.
\end{aligned}
$$

Here, we have used (52) for the estimate of $c_{n,\lambda}[f_n]$. The norm $\|\frac{1}{r}K_{\mu_n}(\sqrt{\lambda}\,\cdot)\|_{L^1}$ is estimated in (69), while $F_n(\sqrt{\lambda};\alpha)$ behaves as (71) and (72) for $0<|\lambda|\ll 1$ and $|\lambda|\gg 1$, respectively. Then, since $F_n(\sqrt{\lambda};\alpha)$ is continuous on λ in $\Sigma_{\pi-\varepsilon_0}$ and $F_n(\sqrt{\lambda};\alpha)\neq 0$ in $\Sigma_{\pi-\varepsilon_0}$ if $0<|\alpha|\le\alpha_0$, we conclude that

$$
\frac{1}{|F_n(\sqrt{\lambda};\alpha)|}\|\frac{1}{r}K_{\mu_n}(\sqrt{\lambda}\,\cdot)\|_{L^1}\le C\,,\qquad \lambda\in\Sigma_{\pi-\varepsilon_0}\,. \tag{77}
$$

Here, C depends only on α and ε_0. On the other hand, Lemma 5 implies that

$$
\|\frac{1}{r}\Phi_{n,\lambda}[f_n]\|_{L^1}\le C|\lambda|^{-1+\frac{1}{q}}\|f_n\|_{L^q(\Omega)}\,,\qquad 1\le q<2\,. \tag{78}
$$

Collecting (77) and (78), we have

$$
\|\frac{1}{|x|}\mathrm{rot}\,(\lambda+\mathbf{A}_{\alpha U}|_{\mathscr{P}_n L^2_\sigma})^{-1}\mathbb{P}f_n\|_{L^1(\Omega)}\le C|\lambda|^{-1+\frac{1}{q}}\|f_n\|_{L^q(\Omega)}\,,\quad 1\le q<2\,. \tag{79}
$$

Similarly, we have

$$
\begin{aligned}
&\|\mathrm{rot}\,(\lambda+\mathbf{A}_{\alpha U}|_{\mathscr{P}_n L^2_\sigma})^{-1}\mathbb{P}f_n\|_{L^2(\Omega)}\\
&\le \frac{|c_{n,\lambda}[f_n]|}{|F_n(\sqrt{\lambda};\alpha)|}\|K_{\mu_n}(\sqrt{\lambda}\,\cdot)\|_{L^2}+\|\Phi_{n,\lambda}[f_n]\|_{L^2}\\
&\le \frac{C}{|\lambda|^{1-\frac{1}{q}}(1+|\lambda|^{\frac{1}{2q}})}\frac{1}{|F_n(\sqrt{\lambda};\alpha)|}\|K_{\mu_n}(\sqrt{\lambda}\,\cdot)\|_{L^2}\|f_n\|_{L^q(\Omega)}+C|\lambda|^{-1+\frac{1}{q}}\|f_n\|_{L^q(\Omega)}\\
&\le C|\lambda|^{-1+\frac{1}{q}}\|f_n\|_{L^q(\Omega)}\,,\quad 1\le q<2\,.
\end{aligned}
\tag{80}
$$

Note that we have used $\|\Phi_{n,\lambda}[f_n]\|_{L^2}\le C|\lambda|^{-1+\frac{1}{q}}\|f_n\|_{L^q(\Omega)}$, in virtue of Lemma 5. Next, Proposition 1, (71), and (72), combined with the fact that $F_n(\sqrt{\lambda};\alpha)$ is continuous on λ in $\Sigma_{\pi-\varepsilon_0}$ and $F_n(\sqrt{\lambda};\alpha)\neq 0$ in $\Sigma_{\pi-\varepsilon_0}$, imply that

$$
\|\frac{1}{|x|}\mathrm{rot}\,(\lambda+\mathbf{A}_{\alpha U}|_{\mathscr{P}_n L^2_\sigma})^{-1}\mathbb{P}H_n\|_{L^1(\Omega)}\le C\delta\Big(\sum_{|n|=1}\|\frac{1}{|x|}\mathrm{rot}\,\mathscr{P}_n u\|_{L^1(\Omega)}+\|\nabla\tilde{u}\|_{L^2(\Omega)}\Big)\,, \tag{81}
$$

and

$$\|\mathrm{rot}\,(\lambda+\mathbf{A}_{\alpha U}|_{\mathscr{P}_n L^2_\sigma})^{-1}\mathbb{P}H_n\|_{L^2(\Omega)} \le C\delta\Big(\sum_{|n|=1}\|\frac{1}{|x|}\mathrm{rot}\,\mathscr{P}_n u\|_{L^1(\Omega)} + \|\nabla\tilde{u}\|_{L^2(\Omega)}\Big)\,. \tag{82}$$

Hence, (76), (79), and (81) yield (74) if $\delta > 0$ is sufficiently small, while (76), (80), (82), and (74) yield (75). The proof is complete.

Recall that any $u \in D_{L^2_\sigma \cap L^q_\sigma}(\mathbf{A})$ with some $q \in (1,2)$ satisfies the condition $\frac{1}{|x|}\mathrm{rot}\,\mathscr{P}_n u \in L^1(\Omega)$ for $|n| = 1$. Then we are now in position to prove the key resolvent estimates as follows.

Theorem 4. *Let $1 < q < 2$. Let $0 < |\alpha| \le \alpha_0$, where α_0 is the number in Theorem 1. Let $\lambda \in \Sigma_{\pi-\varepsilon_0}$, where $\Sigma_{\pi-\varepsilon_0}$ with $\varepsilon_0 \in (0, \frac{\pi}{2})$ is the sector contained in the resolvent set of $-\mathbf{A}_{\alpha U}$ in $L^2_\sigma(\Omega)$ as stated above. Then there exists $\delta_0 > 0$ such that if $\delta \in [0, \delta_0]$ then λ belongs to the resolvent set of $-\mathbf{A}_V$ in $L^2_\sigma(\Omega) \cap L^q_\sigma(\Omega)$ and*

$$\|(\lambda+\mathbf{A}_V)^{-1}f\|_{L^2(\Omega)} \le C|\lambda|^{-\frac{3}{2}+\frac{1}{q}}\|f\|_{L^q(\Omega)}\,, \qquad 1<q<2\,, \tag{83}$$

$$\|\nabla(\lambda+\mathbf{A}_V)^{-1}f\|_{L^2(\Omega)} \le C|\lambda|^{-1+\frac{1}{q}}\|f\|_{L^q(\Omega)}\,, \qquad 1<q<2\,, \tag{84}$$

whenever $f \in L^2_\sigma(\Omega) \cap L^q_\sigma(\Omega)$. Here δ_0 and C depend only on α and q.

Proof. By arguing as in [7, Proposition 2.12] one can show that $\sigma(-\mathbf{A}_V) = \overline{\mathbb{R}_-} \cup \sigma_{disc}(-\mathbf{A}_V)$, where $\overline{\mathbb{R}_-} = \{z \le 0\}$ and $\sigma(-\mathbf{A}_V)$ is the set of spectrum of $-\mathbf{A}_V$ in $L^2_\sigma(\Omega) \cap L^q_\sigma(\Omega)$, while $\sigma_{disc}(-\mathbf{A}_V)$ is the set of discrete eigenvalues of $-\mathbf{A}_V$ in $L^2_\sigma(\Omega) \cap L^q_\sigma(\Omega)$ with finite algebraic multiplicities. Hence, it suffices to show the a priori estimates of the form (83) and (84) under the assumption of Theorem 4; in particular, for sufficiently small δ. First, we consider the estimate of $\tilde{u} = \sum_{|n|\neq 1}\mathscr{P}_n u$. Combining Corollary 2 with (31) and (32), we obtain

$$\begin{aligned}&\Re(\lambda)\,\|\tilde{u}\|^2_{L^2(\Omega)} + (1-|\alpha|-C_0\delta)\|\nabla\tilde{u}\|^2_{L^2(\Omega)}\\ &\le CC_0\delta\big(\delta\|\nabla\tilde{u}\|^2_{L^2(\Omega)} + |\lambda|^{-2+\frac{2}{q}}\sum_{|n|=1}\|\mathscr{P}_n f\|^2_{L^q(\Omega)}\big) + \Re\langle f,\tilde{u}\rangle_{L^2(\Omega)}\,,\end{aligned}$$

and

$$\begin{aligned}|\Im(\lambda)|\,\|\tilde{u}\|^2_{L^2(\Omega)} &\le (|\alpha|+C_0\delta)\|\nabla\tilde{u}\|^2_{L^2(\Omega)}\\ &\quad + CC_0\delta\big(\delta\|\nabla\tilde{u}\|^2_{L^2(\Omega)} + |\lambda|^{-2+\frac{2}{q}}\sum_{|n|=1}\|\mathscr{P}_n f\|^2_{L^q(\Omega)}\big) + |\Im\langle f,\tilde{u}\rangle_{L^2}|\,.\end{aligned}$$

Next, we observe from the Gagliardo–Nirenberg inequality, for $q \in (1,2)$ and $q' = q/(q-1)$,

$$\begin{aligned}
|\langle f, \tilde{u}\rangle_{L^2(\Omega)}| &\le \|f\|_{L^q(\Omega)}\|\tilde{u}\|_{L^{q'}(\Omega)} \\
&\le C\|f\|_{L^q(\Omega)}\|\tilde{u}\|_{L^2(\Omega)}^{2(1-\frac{1}{q})}\|\nabla\tilde{u}\|_{L^2(\Omega)}^{\frac{2}{q}-1} \\
&\le C_\kappa\|f\|_{L^q(\Omega)}^{\frac{2q}{3q-2}}\|\tilde{u}\|_{L^2(\Omega)}^{\frac{4(q-1)}{3q-2}} + \kappa\|\nabla\tilde{u}\|_{L^2(\Omega)}^2 .
\end{aligned} \tag{85}$$

Here, $\kappa > 0$ is a sufficiently small number. Thus, we have

$$\begin{aligned}
&\Re(\lambda)\, \|\tilde{u}\|_{L^2(\Omega)}^2 + (1 - |\alpha| - C_0'\delta - \kappa)\|\nabla\tilde{u}\|_{L^2(\Omega)}^2 \\
&\le C_0'\delta|\lambda|^{-2+\frac{2}{q}}\sum_{|n|=1}\|\mathscr{P}_n f\|_{L^q(\Omega)}^2 + C_\kappa\|f\|_{L^q(\Omega)}^{\frac{2q}{3q-2}}\|\tilde{u}\|_{L^2(\Omega)}^{\frac{4(q-1)}{3q-2}} ,
\end{aligned} \tag{86}$$

and

$$\begin{aligned}
|\Im(\lambda)|\, \|\tilde{u}\|_{L^2(\Omega)}^2 &\le (|\alpha| + C_0'\delta + \kappa)\|\nabla\tilde{u}\|_{L^2(\Omega)}^2 \\
&\quad + C_0'\delta|\lambda|^{-2+\frac{2}{q}}\sum_{|n|=1}\|\mathscr{P}_n f\|_{L^q(\Omega)}^2 + C_\kappa\|f\|_{L^q(\Omega)}^{\frac{2q}{3q-2}}\|\tilde{u}\|_{L^2(\Omega)}^{\frac{4(q-1)}{3q-2}} .
\end{aligned} \tag{87}$$

If $|\alpha| + C_0'\delta + \kappa < 1$, then (86) and (87) imply

$$\begin{aligned}
&\Big(|\Im(\lambda)| + \frac{2(|\alpha| + C_0'\delta + \kappa)}{1 - |\alpha| - C_0'\delta - \kappa}\Re(\lambda)\Big)\, \|\tilde{u}\|_{L^2(\Omega)}^2 + (|\alpha| + C_0'\delta + \kappa)\|\nabla\tilde{u}\|_{L^2(\Omega)}^2 \\
&\le C\Big(\delta|\lambda|^{-2+\frac{2}{q}}\sum_{|n|=1}\|\mathscr{P}_n f\|_{L^q(\Omega)}^2 + \|f\|_{L^q(\Omega)}^{\frac{2q}{3q-2}}\|\tilde{u}\|_{L^2(\Omega)}^{\frac{4(q-1)}{3q-2}}\Big) .
\end{aligned} \tag{88}$$

Here, C depends only on α and κ. By taking $\varepsilon_0 \subset (0, \frac{\pi}{2})$ close enough to $\frac{\pi}{2}$, we may assume that the sector $\Sigma_{\pi-\varepsilon_0}$ is strictly included in the sector

$$\Big\{z \in \mathbb{C}\setminus\{0\} \;\Big|\; |\Im(z)| > -\frac{2(|\alpha| + C_0'\delta + \kappa)}{1 - |\alpha| - C_0'\delta - \kappa}\Re(z)\Big\} ,$$

which ensures the bound

$$C|\lambda| \le |\Im(\lambda)| + \frac{2(|\alpha| + C_0'\delta + \kappa)}{1 - |\alpha| - C_0'\delta - \kappa}\Re(\lambda) \le C'|\lambda| , \qquad \lambda \in \Sigma_{\pi-\varepsilon_0} . \tag{89}$$

The estimates (88) and (89) imply (83) and (84) for the part $\sum_{|n|\neq 1}\mathscr{P}_n(\lambda + \mathbf{A}_V)^{-1}f$ when $\lambda \in \Sigma_{\pi-\varepsilon_0}$. Hence, it remains to show (83) and (84) for the part $\sum_{|n|=1}\mathscr{P}_n(\lambda + \mathbf{A}_V)^{-1}f$, which is written as (76). First, we have from Corollary 2 and (84) for the part $\sum_{|n|\neq 1}\mathscr{P}_n(\lambda + \mathbf{A}_V)^{-1}f$,

$$\sum_{|n|=1}\|\mathrm{rot}\,\mathscr{P}_n u\|_{L^2(\Omega)} \le C|\lambda|^{-1+\frac{1}{q}}\|f\|_{L^q(\Omega)} . \tag{90}$$

Thus, (84) holds for the part $\sum_{|n|=1} \mathscr{P}_n(\lambda + \mathbf{A}_V)^{-1} f$. Next, by Lemma 4 and Corollary 2 combined with (84) for the part $\sum_{|n|\neq 1} \mathscr{P}_n(\lambda + \mathbf{A}_V)^{-1} f$, we have

$$\|H_n\|_{L^1(\Omega)} \leq C\delta|\lambda|^{-1+\frac{1}{q}}\|f\|_{L^q(\Omega)} . \tag{91}$$

Then (73) with $q = 1$ and (91) yield

$$\|(\lambda + \mathbf{A}_{\alpha U}|_{\mathscr{P}_n L^2_\sigma})^{-1}\mathbb{P}H_n\|_{L^2(\Omega)} \leq C|\lambda|^{-\frac{3}{2}+\frac{1}{q}}\|f\|_{L^q(\Omega)} . \tag{92}$$

Collecting (92) and (73), we have proved (83) also for the part $\sum_{|n|=1} \mathscr{P}_n(\lambda + \mathbf{A}_V)^{-1} f$. The proof is complete.

5 Estimate for Semigroup

In this section, we establish the estimates for the semigroup $\{e^{-t\mathbf{A}_V}\}_{t\geq 0}$, and show Theorem 2. Let us recall the representation of $e^{-t\mathbf{A}_V}$ in the Dunford integral

$$e^{-t\mathbf{A}_V} = \frac{1}{2\pi i}\int_\gamma e^{t\lambda}(\lambda + \mathbf{A}_V)^{-1}\,\mathrm{d}\lambda , \qquad t > 0 . \tag{93}$$

Here, $\gamma = \gamma_b$ is the curve $\{\lambda \in \mathbb{C} \mid |\arg\lambda| = \eta ,\ |\lambda| \geq b\} \cup \{\lambda \in \mathbb{C} \mid |\arg\lambda| \leq \eta ,\ |\lambda| = b\}$ for some $\eta \in (\frac{\pi}{2}, \pi)$ and $0 < b \ll 1$, oriented counterclockwise. We take η close to $\frac{\pi}{2}$ so that γ belongs to the sector $\Sigma_{\pi-\varepsilon_0}$, and we denote by γ_0 the curve obtained by taking the limit $b \to 0$. Then for $f \in L^2_\sigma(\Omega) \cap L^q_\sigma(\Omega)$, $1 < q < 2$, we have from Theorem 4 and by taking the limit $b \to 0$,

$$\begin{aligned}\|e^{-t\mathbf{A}_V} f\|_{L^2(\Omega)} &\leq \frac{1}{2\pi}\int_{\gamma_0} \|e^{t\lambda}(\lambda + \mathbf{A}_V)^{-1} f\|_{L^2(\Omega)}|\,\mathrm{d}\lambda| \\ &\leq C\int_0^\infty r^{-\frac{3}{2}+\frac{1}{q}} e^{-(\cos\eta)tr}\,\mathrm{d}r\,\|f\|_{L^q(\Omega)} \leq Ct^{\frac{1}{2}-\frac{1}{q}}\|f\|_{L^q(\Omega)} . \end{aligned} \tag{94}$$

Similarly, we have

$$\begin{aligned}\|\nabla e^{-t\mathbf{A}_V} f\|_{L^2(\Omega)} &\leq \frac{1}{2\pi}\int_{\gamma_0} \|e^{t\lambda}\nabla(\lambda + \mathbf{A}_V)^{-1} f\|_{L^2(\Omega)}|\,\mathrm{d}\lambda| \\ &\leq C\int_0^\infty r^{-1+\frac{1}{q}} e^{-(\cos\eta)tr}\,\mathrm{d}r\,\|f\|_{L^q(\Omega)} \leq Ct^{-\frac{1}{q}}\|f\|_{L^q(\Omega)} . \end{aligned} \tag{95}$$

The proof of Theorem 2 is complete.

Appendix

Here, we state some basic estimates for the modified Bessel functions. For the proof we refer to [1]; see also [7, Lemmas A.2, A.3].

Lemma 6. (i) *Let* $\Re(\mu) > 0$. *Then for any* $M > 0$ *there is a constant* $C = C(M, \mu) > 0$ *such that if* $|z| \le M$ *then*

$$|I_\mu(z)| \le C|z|^{\Re(\mu)}, \qquad |K_\mu(z)| \le C|z|^{-\Re(\mu)}. \tag{96}$$

(ii) *Let* $\varepsilon \in (0, \frac{\pi}{2})$. *Then there is a constant* $C = C(\varepsilon, \mu) > 0$ *such that if* $z \in \Sigma_{\frac{\pi}{2}-\varepsilon}$ *and* $|z| \ge 1$ *then*

$$|I_\mu(z)| \le C|\Re(z)|^{-\frac{1}{2}} e^{\Re(z)}, \qquad |K_\mu(z)| \le C|\Re(z)|^{-\frac{1}{2}} e^{-\Re(z)}. \tag{97}$$

Moreover, for $|z| \gg 1$, *the following expansion holds.*

$$\begin{aligned} &I_\mu(z) = \frac{1}{\sqrt{2\pi z}} e^z\big(1 + l_\mu(z)\big), \qquad K_\mu(z) = \sqrt{\frac{\pi}{2z}} e^{-z}\big(1 + k_\mu(z)\big), \\ &|l_\mu(z)| + |k_\mu(z)| \le C|z|^{-1}. \end{aligned} \tag{98}$$

References

1. G.E. Andrews, R. Askey, R. Roy, *Special Functions*, vol. 71, Encyclopedia of Mathematics and Its Applications (Cambridge University Press, Cambridge, 1999)
2. W. Borchers, Zur Stabilität und Faktorisierungsmethode für die Navier-Stokes Gleichungen inkompressibler viskoser Flüssigkeiten. Habilitationsschrift, Universität Paderborn (1992)
3. W. Borchers, T. Miyakawa, On stability of exterior stationary Navier-Stokes flows. Acta Math. **174**, 311–382 (1995)
4. G.P. Galdi, Stationary Navier-Stokes problem in a two-dimensional exterior domain, in *Handbook of Differential Equations, Stationary Partial Differential Equations*, vol. I, ed. by M. Chipot, P. Quittner (North-Holland, Amsterdam, 2004), pp. 71–155
5. J. Guillod, On the asymptotic stability of steady flows with nonzero flux in two-dimensional exterior domains (2016), arXiv:1605.03864 [math.AP]
6. M. Hillairet, P. Wittwer, On the existence of solutions to the planar exterior Navier-Stokes system. J. Differ. Equ. **255**, 2996–3019 (2013)
7. Y. Maekawa, On stability of steady circular flows in a two-dimensional exterior disk. Arch. Ration. Mech. Anal. **225**, 287–374 (2017)
8. K. Pileckas, R. Russo, On the existence of vanishing at infinity symmetric solutions to the plane stationary exterior Navier-Stokes problem. Math. Ann. **352**, 643–658 (2012)
9. A. Russo, A note on the exterior two-dimensional steady-state Navier-Stokes problem. J. Math. Fluid Mech. **11**, 407–414 (2009)

10. A. Russo, On the existence of D-solutions of the steady-state Navier-Stokes equations in plane exterior domains (2011), arXiv:1101.1243
11. M. Yamazaki, Unique existence of stationary solutions to the two-dimensional Navier-Stokes equations on exterior domains, *Mathematical Analysis on the Navier-Stokes Equations and Related Topics, Past and Future - In memory of Professor Tetsuro Miyakawa*, vol. 35, Gakuto International Series in Mathematical Sciences and Applications (Gakkōtosho, Tokyo, 2011), pp. 220–241

Stochastic Effects and Time-Filtered Leapfrog Schemes for Maxwell's Equations

Alex Mahalov and Austin McDaniel

Abstract We consider Maxwell's equations where the conductivity contains fast random fluctuations in time. Using an Ornstein–Uhlenbeck process, we study the effects of correlations between the random fluctuations of two different different time scales, with one an order of magnitude smaller than the other. We show that this asymptotic regime gives rise to a limiting equation where the effects of the fluctuations in the conductivity are captured in additional terms containing deterministic and stochastic corrections. For deterministic dynamics, numerical solutions to the time dependent Maxwell's equations using a new time stepping scheme are presented. This scheme, which is based on the leapfrog method and a fourth-order time filter, significantly reduces the short oscillations generated by numerical dispersion. It uses staggering in space only, allowing explicit treatment of the electric current density terms and application of numerical smoothers. Comparisons of simulation results where Maxwell's equations are integrated in a presence of the scattering of an electromagnetic pulse by a perfectly conducting square and those obtained with the unfiltered leapfrog show that the developed method is robust and accurate.

Keywords Maxwell equations · Stochastic effects · Wave propagation
Fast numerical schemes

Dedicated to Professor Yoshikazu Giga

A. Mahalov (✉) · A. McDaniel
School of Mathematical and Statistical Sciences, Arizona State University, Tempe, AZ 85287, USA
e-mail: mahalov@asu.edu

A. McDaniel
e-mail: austin.mcdaniel@asu.edu

Y. Maekawa and S. Jimbo (eds.), *Mathematics for Nonlinear Phenomena—Analysis and Computation*, Springer Proceedings in Mathematics & Statistics 215,
DOI 10.1007/978-3-319-66764-5_7

1 Introduction

Describing and understanding wave propagation in complex media is of great interest for many applications. Because the details of the microstructure of a complex medium are often unknown, a given medium parameter is frequently decomposed into deterministic and stochastic components and modelled as the realization of a stochastic process. The stochastic characteristic parameters then define what is called a random medium. Electromagnetic waves in random media have been studied in situations in which the permittivity, and sometimes also the permeability and conductivity, in Maxwell's equations contain random fluctuations in space. The effects of small, random fluctuations in the permittivity alone have been investigated in [2]. In [6], a white-noise paraxial approximation is derived starting from Maxwell's equations in which the permittivity and permeability contain random fluctuations in every spatial direction. In contrast, a medium is called randomly layered if its defining parameters contain random fluctuations in only a single spatial variable. In [7], the reflectivity properties are studied of a randomly layered half-space in which the permittivity and conductivity are random processes. Maxwell's equations in a randomly layered media characterized by a stochastic permittivity, permeability, and conductivity have been studied in [10, 12]. In complex media such as the troposphere, tropopause, stratosphere, and ionosphere, these parameters also vary in time [11, 13, 15, 22, 23]. For example, in deep turbulence, the refractive index, and hence the permittivity, can fluctuate with a temporal frequency on the order of a few hundred hertz [24, 25]. We also mention recent research on conductivity in optically pumped materials [1, 21] as well as two-dimensional electron systems [20].

Many important applications involve the propagation of electromagnetic (EM) waves through complex media. For example, the Global Positioning System (GPS) relies on microwave signals that travel through the ionosphere, which contains turbulent structures caused by plasma instabilities that occur over seven orders of magnitude in spatial scale, from more than 10^5 m to less than 0.1 m [3, 11, 13]. The turbulent hydrodynamic mixing of ionospheric plasmas leads to density irregularities which cause fluctuations in the permittivity along the path that a signal travels. Furthermore, because of electron precipitation, ionospheric conductivities are highly variable and dynamic even during quiet periods [17]. Permittivity fluctuations are seen not only in the ionosphere but also, in particular, in the troposphere, tropopause, and lower stratosphere where atmospheric turbulence mixes layers of different temperatures and densities [15]. The root of the problem in long range ionospheric propagation scenarios is associated with strong amplitude and phase scintillation effects caused by the passage of light/radio waves/microwaves through random ionospheric media characterized by strong neutral-plasma interactions for deterministic dynamics and stochastic fluctuations.

The paper is organized as follows. In Section 2 we study Maxwell's equations in which the conductivity contains random fluctuations in time. In Sections 3 and 4 we present new time-filtered unstaggered leapfrog numerical schemes for deterministic dynamics. In our future work, we plan to combine the methods developed in this paper to study coupling of multi-scale deterministic and stochastic dynamics in three-dimensional Navier–Stokes and Maxwell's Equations for inhomogeneous media [4, 9, 13, 14, 16].

2 Maxwell's Equations with Random Conductivity

We consider electromagnetic waves propagating in a three-dimensional medium and we let $\boldsymbol{x} = (x_1, x_2, x_3) \in \mathbb{R}^3$ denote the spatial vector. Maxwell's equations for the electric field $\boldsymbol{E}(t, \boldsymbol{x}) = (E_1(t, \boldsymbol{x}), E_2(t, \boldsymbol{x}), E_3(t, \boldsymbol{x}))$ and the magnetic field $\boldsymbol{H}(t, \boldsymbol{x}) = (H_1(t, \boldsymbol{x}), H_2(t, \boldsymbol{x}), H_3(t, \boldsymbol{x}))$ are then

$$\begin{aligned}
&\nabla \times \boldsymbol{E} = -\mu \frac{\partial}{\partial t} \boldsymbol{H} \\
&\nabla \cdot \left(\varepsilon \boldsymbol{E}\right) = \rho \\
&\nabla \times \boldsymbol{H} = \sigma(t) \boldsymbol{E} + \varepsilon \frac{\partial}{\partial t} \boldsymbol{E} \\
&\nabla \cdot \left(\mu \boldsymbol{H}\right) = 0
\end{aligned} \tag{1}$$

where ε is the permittivity, μ is the permeability, $\sigma(t)$ is the random, time-dependent conductivity, and ρ is the charge density. Let ε_0 and μ_0 denote the vacuum permittivity and the permeability constant, respectively. In terms of these constants, the free-space propagation speed is $c_0 \equiv (\mu_0 \varepsilon_0)^{-1/2}$ and the impedance of free-space is $Z_0 \equiv (\mu_0/\varepsilon_0)^{1/2}$. Let l_0 be a characteristic length scale and let E_0 be a reference electric-field strength. In terms of these constants, we define the following dimensionless variables:

$$\begin{aligned}
\boldsymbol{x}' = \left(x_1', x_2', x_3'\right) \equiv l_0^{-1} \boldsymbol{x} = \left(l_0^{-1} x_1, l_0^{-1} x_2, l_0^{-1} x_3\right) \,, \qquad & t' \equiv c_0 l_0^{-1} t \,, \\
\boldsymbol{E}' \equiv E_0^{-1} \boldsymbol{E} \,, \qquad \varepsilon' \equiv \varepsilon_0^{-1} \varepsilon \,, \qquad & \rho' \equiv l_0 \varepsilon_0^{-1} E_0^{-1} \rho \,, \\
\boldsymbol{H}' \equiv Z_0 E_0^{-1} \boldsymbol{H} \,, \qquad \mu' \equiv \mu_0^{-1} \mu \,, \qquad & \sigma'(t') = \sigma'\left(c_0 l_0^{-1} t\right) \equiv Z_0 l_0 \sigma(t)
\end{aligned}$$

Expressing the system (1) in terms of these new variables gives the dimensionless Maxwell's equations

$$\nabla' \times \boldsymbol{E}' = -\mu' \frac{\partial}{\partial t'} \boldsymbol{H}' \tag{2a}$$

$$\nabla' \cdot \left(\varepsilon' \boldsymbol{E}'\right) = \rho' \tag{2b}$$

$$\nabla' \times \boldsymbol{H}' = \sigma'(t') \boldsymbol{E}' + \varepsilon' \frac{\partial}{\partial t'} \boldsymbol{E}' \tag{2c}$$

$$\nabla' \cdot \left(\mu' \boldsymbol{H}'\right) = 0 \tag{2d}$$

where

$$\nabla' \equiv l_0 \nabla = \left(\frac{\partial}{\partial x_1'}, \frac{\partial}{\partial x_2'}, \frac{\partial}{\partial x_3'} \right)$$

We drop the primes and from here on we will work with equations (2a) – (2d).

We consider a model of the conductivity that contains random fluctuations of two different time scales, with one an order of magnitude smaller than the other. In particular, let the random conductivity σ in (2) be defined by

$$\sigma(t) = \delta \beta^s \nu^s(t) + \beta^r \nu^r \left(\frac{t}{\delta^2} \right) \tag{3}$$

where ν^s and ν^r are stationary, mean-zero stochastic processes, δ is a small dimensionless parameter, and the parameters β^s and β^r determine the strengths of the random fluctuations. We will study the effects of correlations between the random fluctuations of different scales. Because ν^s and ν^r are mean-zero processes, σ fluctuates between positive and negative values. Recently it has been demonstrated that negative conductivity can occur in optically pumped materials [1, 21] as well as two-dimensional electron systems [20].

We will study the white-noise limit of the system given by (2a) and (2c) in a situation in which the conductivity fluctuations of different scales are correlated. In order to take such a limit, we make the following rescaling

$$t \to \frac{t}{\delta^2}, \quad \boldsymbol{x} \to \frac{\boldsymbol{x}}{\delta^2} \tag{4}$$

Since we will be working with stochastic differential equations (SDEs), from here on we will use SDE notation. Making the rescaling (4) in (2a) and (2c) gives

$$d\boldsymbol{H}_t = -\frac{1}{\mu} \nabla \times \boldsymbol{E}_t dt \tag{5a}$$

$$d\boldsymbol{E}_t = \frac{1}{\varepsilon} \nabla \times \boldsymbol{H}_t dt - \frac{1}{\delta} \frac{\beta^s}{\varepsilon} \nu^s_{\frac{t}{\delta^2}} \boldsymbol{E}_t dt - \frac{1}{\delta^2} \frac{\beta^r}{\varepsilon} \nu^r_{\frac{t}{\delta^4}} \boldsymbol{E}_t dt \tag{5b}$$

The white-noise limit in which the correlation times of the conductivity fluctuations go to zero corresponds to the limit of (5) as $\delta \to 0$. We note that under natural assumptions on the stationary, mean-zero process ν^r, as $\delta \to 0$,

$$\int_0^t \frac{1}{\delta^2} \nu^r \left(\frac{s}{\delta^4} \right) ds \to W(t)$$

in some probabilistic sense, where W is a Wiener process [8, 18]. Thus, since the process $\nu^r(t/\delta^4)$ fluctuates on a scale smaller than any other in the system, we approximate (5) by the system

$$
\begin{aligned}
d\boldsymbol{H}_t &= -\frac{1}{\mu}\nabla\times\boldsymbol{E}_t dt \\
d\boldsymbol{E}_t &= \frac{1}{\varepsilon}\nabla\times\boldsymbol{H}_t dt - \frac{1}{\delta}\frac{\beta^s}{\varepsilon}\nu^s_{\frac{t}{\delta^2}}\boldsymbol{E}_t dt + \frac{(\beta^r)^2}{2\varepsilon^2}\boldsymbol{E}_t dt - \frac{\beta^r}{\varepsilon}\boldsymbol{E}_t dW_t
\end{aligned} \tag{6}
$$

where W is a Wiener process and where we have included the Stratonovich correction due to the last term in (5b) that would appear in the white-noise limit.

As a model for the larger-scale fluctuations in the conductivity, we take ν^s to be a stationary Ornstein–Uhlenbeck process [5]. An Ornstein–Uhlenbeck process y is defined as the solution of the SDE

$$dy_t = -by_t dt + \Gamma dW_t \tag{7}$$

where $b > 0$. Equation (7) has a unique stationary measure and the solution of (7) with the initial condition distributed according to this stationary measure defines a stationary process called the stationary Ornstein–Uhlenbeck process. The stationary Ornstein–Uhlenbeck process y is an ergodic, Gaussian process with mean zero and autocovariance function

$$E\left[y_t y_s\right] = \frac{\Gamma^2}{2b}e^{-b|t-s|} \tag{8}$$

We let $\nu^s(t/\delta^2) = \eta(t)$ where η is the stationary Ornstein–Uhlenbeck process defined as the solution of the equation

$$d\eta_t = -\frac{b}{\delta^2}\eta_t dt + \frac{\Gamma}{\delta}dW_t \tag{9}$$

with $\eta_0 = \Theta$, where Θ is a Gaussian random variable with mean zero and variance $\Gamma^2/2b$ that is independent of the process W. We take the Wiener process W in (9) to be the same Wiener process as that in (6) in order to study the effects of correlations between the conductivity fluctuations of different scales. We note that by (8), the process $\nu^s(t/\delta^2)$ defined above has correlation time δ^2/b.

We take the Fourier transform of the system (6) in the spatial coordinates. Letting $\boldsymbol{k} = (k_1, k_2, k_3)$, we define $\hat{\boldsymbol{E}}(t,\boldsymbol{k}) = (\hat{E}_1(t,\boldsymbol{k}), \hat{E}_2(t,\boldsymbol{k}), \hat{E}_3(t,\boldsymbol{k}))$ and $\hat{\boldsymbol{H}}(t,\boldsymbol{k}) = (\hat{H}_1(t,\boldsymbol{k}), \hat{H}_2(t,\boldsymbol{k}), \hat{H}_3(t,\boldsymbol{k}))$ by

$$\hat{\boldsymbol{E}}(t,\boldsymbol{k}) = \int_{\mathbb{R}^3}\boldsymbol{E}(t,\boldsymbol{x})e^{i\boldsymbol{k}\cdot\boldsymbol{x}}d\boldsymbol{x} \tag{10}$$

and

$$\hat{\boldsymbol{H}}(t,\boldsymbol{k}) = \int_{\mathbb{R}^3}\boldsymbol{H}(t,\boldsymbol{x})e^{i\boldsymbol{k}\cdot\boldsymbol{x}}d\boldsymbol{x} \tag{11}$$

where we are using the notation $d\boldsymbol{x} = dx_1dx_2dx_3$. The transform of (6) together with (9) form the SDE system

$$\begin{aligned}
d\hat{\boldsymbol{E}}_t &= \frac{i}{\varepsilon}\boldsymbol{k}\times\hat{\boldsymbol{H}}_t dt - \frac{1}{\delta}\frac{\beta^s}{\varepsilon}\eta_t\hat{\boldsymbol{E}}_t dt + \frac{(\beta^r)^2}{2\varepsilon^2}\hat{\boldsymbol{E}}_t dt - \frac{\beta^r}{\varepsilon}\hat{\boldsymbol{E}}_t dW_t \\
d\hat{\boldsymbol{H}}_t &= -\frac{i}{\mu}\boldsymbol{k}\times\hat{\boldsymbol{E}}_t dt \\
d\eta_t &= -\frac{b}{\delta^2}\eta_t dt + \frac{\Gamma}{\delta}dW_t
\end{aligned}$$

Let $\hat{\boldsymbol{E}}^R = (Re(\hat{E}_1), Re(\hat{E}_2), Re(\hat{E}_3))$ and $\hat{\boldsymbol{E}}^I = (Im(\hat{E}_1), Im(\hat{E}_2), Im(\hat{E}_3))$ denote the real and imaginary parts of $\hat{\boldsymbol{E}}$, and define $\hat{\boldsymbol{H}}^R$ and $\hat{\boldsymbol{H}}^I$ similarly. Then $\hat{\boldsymbol{E}}^R, \hat{\boldsymbol{E}}^I, \hat{\boldsymbol{H}}^R, \hat{\boldsymbol{H}}^I$, and η satisfy the real system

$$\begin{aligned}
d\hat{\boldsymbol{E}}_t^R &= -\frac{1}{\varepsilon}\boldsymbol{k}\times\hat{\boldsymbol{H}}_t^I dt - \frac{1}{\delta}\frac{\beta^s}{\varepsilon}\eta_t\hat{\boldsymbol{E}}_t^R dt + \frac{(\beta^r)^2}{2\varepsilon^2}\hat{\boldsymbol{E}}_t^R dt - \frac{\beta^r}{\varepsilon}\hat{\boldsymbol{E}}_t^R dW_t \\
d\hat{\boldsymbol{E}}_t^I &= \frac{1}{\varepsilon}\boldsymbol{k}\times\hat{\boldsymbol{H}}_t^R dt - \frac{1}{\delta}\frac{\beta^s}{\varepsilon}\eta_t\hat{\boldsymbol{E}}_t^I dt + \frac{(\beta^r)^2}{2\varepsilon^2}\hat{\boldsymbol{E}}_t^I dt - \frac{\beta^r}{\varepsilon}\hat{\boldsymbol{E}}_t^I dW_t \\
d\hat{\boldsymbol{H}}_t^R &= \frac{1}{\mu}\boldsymbol{k}\times\hat{\boldsymbol{E}}_t^I dt \\
d\hat{\boldsymbol{H}}_t^I &= -\frac{1}{\mu}\boldsymbol{k}\times\hat{\boldsymbol{E}}_t^R dt \\
d\eta_t &= -\frac{b}{\delta^2}\eta_t dt + \frac{\Gamma}{\delta}dW_t
\end{aligned} \tag{12}$$

We study the limit as $\delta \to 0$ of the above system by using the multiscale expansion method [19], which makes use of the correspondence between SDEs and PDEs. Associated with a typical SDE is the backward Kolmogorov equation (and its adjoint, the Fokker-Planck equation), which is a PDE that describes the probability transition density. We will work with the backward Kolmogorov equation corresponding to the above SDE system. In order to write this equation compactly, we introduce some notation. Let $\nabla_{\hat{E}^R}$ denote the gradient with respect to the variables $(\hat{E}_1^R, \hat{E}_2^R, \hat{E}_3^R)$, and define $\nabla_{\hat{E}^I}$, $\nabla_{\hat{H}^R}$, and $\nabla_{\hat{H}^I}$ similarly. Also, using the usual definition of the gradient of a vector-valued function, for a scalar-valued function ϕ we let $\nabla_{\hat{E}^R}\nabla_{\hat{E}^R}\phi$ denote the Hessian matrix that is defined componentwise as

$$(\nabla_{\hat{E}^R}\nabla_{\hat{E}^R}\phi)_{ij} = \frac{\partial^2\phi}{\partial\hat{E}_i^R\partial\hat{E}_j^R}$$

Let $\nabla_{\hat{E}^I}\nabla_{\hat{E}^I}\phi$ be defined analogously and let $\nabla_{\hat{E}^I}\nabla_{\hat{E}^R}\phi$ be defined componentwise as

$$(\nabla_{\hat{E}^I}\nabla_{\hat{E}^R}\phi)_{ij} = \frac{\partial^2\phi}{\partial\hat{E}_i^R\partial\hat{E}_j^I}$$

Using this notation, the backward Kolmogorov equation corresponding to the SDE system (12) is

$$\frac{\partial \rho}{\partial t} = \left(\frac{1}{\delta^2} L_0 + \frac{1}{\delta} L_1 + L_2\right) \rho \tag{13}$$

where

$$L_0 \rho = -b\eta \frac{\partial \rho}{\partial \eta} + \frac{\Gamma^2}{2} \frac{\partial^2 \rho}{\partial \eta^2}$$

$$L_1 \rho = -\frac{\beta^s}{\varepsilon} \eta \hat{\boldsymbol{E}}^R \cdot \nabla_{\hat{E}^R} \rho - \frac{\beta^s}{\varepsilon} \eta \hat{\boldsymbol{E}}^I \cdot \nabla_{\hat{E}^I} \rho - \frac{\Gamma \beta^r}{\varepsilon} \hat{\boldsymbol{E}}^R \cdot \nabla_{\hat{E}^R} \frac{\partial \rho}{\partial \eta} - \frac{\Gamma \beta^r}{\varepsilon} \hat{\boldsymbol{E}}^I \cdot \nabla_{\hat{E}^I} \frac{\partial \rho}{\partial \eta}$$

$$\begin{aligned} L_2 \rho = &-\frac{1}{\varepsilon}\left(\boldsymbol{k} \times \hat{\boldsymbol{H}}^I\right) \cdot \nabla_{\hat{E}^R} \rho + \frac{1}{\varepsilon}\left(\boldsymbol{k} \times \hat{\boldsymbol{H}}^R\right) \cdot \nabla_{\hat{E}^I} \rho + \frac{(\beta^r)^2}{2\varepsilon^2} \hat{\boldsymbol{E}}^R \cdot \nabla_{\hat{E}^R} \rho \\ &+ \frac{(\beta^r)^2}{2\varepsilon^2} \hat{\boldsymbol{E}}^I \cdot \nabla_{\hat{E}^I} \rho + \frac{(\beta^r)^2}{2\varepsilon^2} \hat{\boldsymbol{E}}^R \cdot \left(\left(\nabla_{\hat{E}^R} \nabla_{\hat{E}^R} \rho\right) \hat{\boldsymbol{E}}^R\right) \\ &+ \frac{(\beta^r)^2}{2\varepsilon^2} \hat{\boldsymbol{E}}^I \cdot \left(\left(\nabla_{\hat{E}^I} \nabla_{\hat{E}^I} \rho\right) \hat{\boldsymbol{E}}^I\right) + \frac{(\beta^r)^2}{\varepsilon^2} \hat{\boldsymbol{E}}^R \cdot \left(\left(\nabla_{\hat{E}^I} \nabla_{\hat{E}^R} \rho\right) \hat{\boldsymbol{E}}^I\right) \\ &+ \frac{1}{\mu}\left(\boldsymbol{k} \times \hat{\boldsymbol{E}}^I\right) \cdot \nabla_{\hat{H}^R} \rho - \frac{1}{\mu}\left(\boldsymbol{k} \times \hat{\boldsymbol{E}}^R\right) \cdot \nabla_{\hat{H}^I} \rho \end{aligned}$$

We can then use homogenization methods for PDEs to obtain the limiting backward Kolmogorov equation, from which we can read off the limiting SDE. We expand ρ in powers of δ, i.e.,

$$\rho = \rho_0 + \delta \rho_1 + \delta^2 \rho_2 + \cdots \tag{14}$$

and derive the backward Kolmogorov equation for the limiting density ρ_0. First, substituting (14) in (13) and equating terms of the same order in δ gives the equations

$$O\left(\frac{1}{\delta^2}\right) : L_0 \rho_0 = 0 \tag{15}$$

$$O\left(\frac{1}{\delta}\right) : L_0 \rho_1 = -L_1 \rho_0 \tag{16}$$

$$O(1) : \frac{\partial \rho_0}{\partial t} = L_0 \rho_2 + L_1 \rho_1 + L_2 \rho_0 \tag{17}$$

Equation (15) implies that $\rho_0 = \rho_0(t, \hat{\boldsymbol{E}}^R, \hat{\boldsymbol{E}}^I, \hat{\boldsymbol{H}}^R, \hat{\boldsymbol{H}}^I)$. Next, we find ρ_1 in terms of $\rho_0(t, \hat{\boldsymbol{E}}^R, \hat{\boldsymbol{E}}^I, \hat{\boldsymbol{H}}^R, \hat{\boldsymbol{H}}^I)$ by solving equation (16), which, since ρ_0 does not depend on η, is equivalent to

$$-b\eta\frac{\partial\rho_1}{\partial\eta}+\frac{\Gamma^2}{2}\frac{\partial^2\rho_1}{\partial\eta^2}=\frac{\beta^s}{\varepsilon}\eta\hat{\boldsymbol{E}}^R\cdot\nabla_{\hat{E}^R}\rho_0+\frac{\beta^s}{\varepsilon}\eta\hat{\boldsymbol{E}}^I\cdot\nabla_{\hat{E}^I}\rho_0$$

The solution of this equation is

$$\rho_1=-\frac{\beta^s}{b\varepsilon}\eta\hat{\boldsymbol{E}}^R\cdot\nabla_{\hat{E}^R}\rho_0-\frac{\beta^s}{b\varepsilon}\eta\hat{\boldsymbol{E}}^I\cdot\nabla_{\hat{E}^I}\rho_0$$

Equation (17) implies that the function $\frac{\partial\rho_0}{\partial t}-L_1\rho_1-L_2\rho_0$ is in the range of the operator L_0. Thus, for equation (17) to have a solution, by the Fredholm alternative the above function must be orthogonal to the kernel of L_0^* (where L_0^* denotes the adjoint of L_0); that is, the following solvability condition must hold:

$$\int_{-\infty}^{\infty}\rho^\infty\left(\frac{\partial\rho_0}{\partial t}-L_1\rho_1-L_2\rho_0\right)d\eta=0 \tag{18}$$

where ρ^∞ solves

$$L_0^*\rho^\infty=b\frac{\partial}{\partial\eta}(\eta\rho^\infty)+\frac{\Gamma^2}{2}\frac{\partial^2\rho^\infty}{\partial\eta^2}=0 \tag{19}$$

The backward Kolmogorov equation for the limiting density ρ_0 comes from this solvability condition. We take ρ^∞ to be the normalized solution of (19), i.e., the solution such that

$$\int_{-\infty}^{\infty}\rho^\infty d\eta=1$$

Equation (19) is equivalent to the stationary Fokker-Planck equation corresponding to the SDE (9), so ρ^∞ is the stationary density for the Ornstein–Uhlenbeck process η that is defined as the solution of (9). The stationary distribution for η is Gaussian with mean zero and variance equal to $\Gamma^2/2b$, so that we have

$$\rho^\infty(\eta)=\frac{\sqrt{b}}{\Gamma\sqrt{\pi}}e^{-b\eta^2/\Gamma^2}$$

The solvability condition (18) then becomes

$$\begin{aligned}\frac{\partial\rho_0}{\partial t}=\int_{-\infty}^{\infty}\rho^\infty\bigg(\bigg[&-\frac{\beta^s}{\varepsilon}\eta\hat{\boldsymbol{E}}^R\cdot\nabla_{\hat{E}^R}-\frac{\beta^s}{\varepsilon}\eta\hat{\boldsymbol{E}}^I\cdot\nabla_{\hat{E}^I}-\frac{\Gamma\beta^r}{\varepsilon}\hat{\boldsymbol{E}}^R\cdot\nabla_{\hat{E}^R}\frac{\partial}{\partial\eta}\\&-\frac{\Gamma\beta^r}{\varepsilon}\hat{\boldsymbol{E}}^I\cdot\nabla_{\hat{E}^I}\frac{\partial}{\partial\eta}\bigg]\bigg(-\frac{\beta^s}{b\varepsilon}\eta\hat{\boldsymbol{E}}^R\cdot\nabla_{\hat{E}^R}\rho_0-\frac{\beta^s}{b\varepsilon}\eta\hat{\boldsymbol{E}}^I\cdot\nabla_{\hat{E}^I}\rho_0\bigg)\\&-\frac{1}{\varepsilon}\left(\boldsymbol{k}\times\hat{\boldsymbol{H}}^I\right)\cdot\nabla_{\hat{E}^R}\rho_0+\frac{1}{\varepsilon}\left(\boldsymbol{k}\times\hat{\boldsymbol{H}}^R\right)\cdot\nabla_{\hat{E}^I}\rho_0\\&+\frac{(\beta^r)^2}{2\varepsilon^2}\hat{\boldsymbol{E}}^R\cdot\nabla_{\hat{E}^R}\rho_0+\frac{(\beta^r)^2}{2\varepsilon^2}\hat{\boldsymbol{E}}^I\cdot\nabla_{\hat{E}^I}\rho_0\end{aligned}$$

$$+\frac{(\beta^r)^2}{2\varepsilon^2}\hat{\boldsymbol{E}}^R\cdot\left(\left(\nabla_{\hat{E}^R}\nabla_{\hat{E}^R}\rho_0\right)\hat{\boldsymbol{E}}^R\right)+\frac{(\beta^r)^2}{2\varepsilon^2}\hat{\boldsymbol{E}}^I\cdot\left(\left(\nabla_{\hat{E}^I}\nabla_{\hat{E}^I}\rho_0\right)\hat{\boldsymbol{E}}^I\right)$$
$$+\frac{(\beta^r)^2}{\varepsilon^2}\hat{\boldsymbol{E}}^R\cdot\left(\left(\nabla_{\hat{E}^I}\nabla_{\hat{E}^R}\rho_0\right)\hat{\boldsymbol{E}}^I\right)$$
$$+\frac{1}{\mu}\left(\boldsymbol{k}\times\hat{\boldsymbol{E}}^I\right)\cdot\nabla_{\hat{H}^R}\rho_0-\frac{1}{\mu}\left(\boldsymbol{k}\times\hat{\boldsymbol{E}}^R\right)\cdot\nabla_{\hat{H}^I}\rho_0\Big)d\eta$$

Thus, using $\int_{-\infty}^{\infty}\eta^2\rho^\infty d\eta=\Gamma^2/2b$, we have

$$\begin{aligned}\frac{\partial\rho_0}{\partial t}=&\frac{\Gamma^2(\beta^s)^2}{2b^2\varepsilon^2}\hat{\boldsymbol{E}}^R\cdot\nabla_{\hat{E}^R}\left(\hat{\boldsymbol{E}}^R\cdot\nabla_{\hat{E}^R}\rho_0\right)+\frac{\Gamma^2(\beta^s)^2}{2b^2\varepsilon^2}\hat{\boldsymbol{E}}^I\cdot\nabla_{\hat{E}^I}\left(\hat{\boldsymbol{E}}^R\cdot\nabla_{\hat{E}^R}\rho_0\right)\\&+\frac{\Gamma\beta^r\beta^s}{b\varepsilon^2}\hat{\boldsymbol{E}}^R\cdot\nabla_{\hat{E}^R}\left(\hat{\boldsymbol{E}}^R\cdot\nabla_{\hat{E}^R}\rho_0\right)+\frac{\Gamma\beta^r\beta^s}{b\varepsilon^2}\hat{\boldsymbol{E}}^I\cdot\nabla_{\hat{E}^I}\left(\hat{\boldsymbol{E}}^R\cdot\nabla_{\hat{E}^R}\rho_0\right)\\&+\frac{\Gamma^2(\beta^s)^2}{2b^2\varepsilon^2}\hat{\boldsymbol{E}}^R\cdot\nabla_{\hat{E}^R}\left(\hat{\boldsymbol{E}}^I\cdot\nabla_{\hat{E}^I}\rho_0\right)+\frac{\Gamma^2(\beta^s)^2}{2b^2\varepsilon^2}\hat{\boldsymbol{E}}^I\cdot\nabla_{\hat{E}^I}\left(\hat{\boldsymbol{E}}^I\cdot\nabla_{\hat{E}^I}\rho_0\right)\\&+\frac{\Gamma\beta^r\beta^s}{b\varepsilon^2}\hat{\boldsymbol{E}}^R\cdot\nabla_{\hat{E}^R}\left(\hat{\boldsymbol{E}}^I\cdot\nabla_{\hat{E}^I}\rho_0\right)+\frac{\Gamma\beta^r\beta^s}{b\varepsilon^2}\hat{\boldsymbol{E}}^I\cdot\nabla_{\hat{E}^I}\left(\hat{\boldsymbol{E}}^I\cdot\nabla_{\hat{E}^I}\rho_0\right)\\&-\frac{1}{\varepsilon}\left(\boldsymbol{k}\times\hat{\boldsymbol{H}}^I\right)\cdot\nabla_{\hat{E}^R}\rho_0+\frac{1}{\varepsilon}\left(\boldsymbol{k}\times\hat{\boldsymbol{H}}^R\right)\cdot\nabla_{\hat{E}^I}\rho_0+\frac{(\beta^r)^2}{2\varepsilon^2}\hat{\boldsymbol{E}}^R\cdot\nabla_{\hat{E}^R}\rho_0\\&+\frac{(\beta^r)^2}{2\varepsilon^2}\hat{\boldsymbol{E}}^I\cdot\nabla_{\hat{E}^I}\rho_0+\frac{(\beta^r)^2}{2\varepsilon^2}\hat{\boldsymbol{E}}^R\cdot\left(\left(\nabla_{\hat{E}^R}\nabla_{\hat{E}^R}\rho_0\right)\hat{\boldsymbol{E}}^R\right)\\&+\frac{(\beta^r)^2}{2\varepsilon^2}\hat{\boldsymbol{E}}^I\cdot\left(\left(\nabla_{\hat{E}^I}\nabla_{\hat{E}^I}\rho_0\right)\hat{\boldsymbol{E}}^I\right)+\frac{(\beta^r)^2}{\varepsilon^2}\hat{\boldsymbol{E}}^R\cdot\left(\left(\nabla_{\hat{E}^I}\nabla_{\hat{E}^R}\rho_0\right)\hat{\boldsymbol{E}}^I\right)\\&+\frac{1}{\mu}\left(\boldsymbol{k}\times\hat{\boldsymbol{E}}^I\right)\cdot\nabla_{\hat{H}^R}\rho_0-\frac{1}{\mu}\left(\boldsymbol{k}\times\hat{\boldsymbol{E}}^R\right)\cdot\nabla_{\hat{H}^I}\rho_0\end{aligned}\tag{20}$$

Using, for $\alpha=R,I$,

$$\hat{\boldsymbol{E}}^\alpha\cdot\nabla_{\hat{E}^\alpha}\left(\hat{\boldsymbol{E}}^\alpha\cdot\nabla_{\hat{E}^\alpha}\rho_0\right)=\hat{\boldsymbol{E}}^\alpha\cdot\nabla_{\hat{E}^\alpha}\rho_0+\hat{\boldsymbol{E}}^\alpha\cdot\left(\left(\nabla_{\hat{E}^\alpha}\nabla_{\hat{E}^\alpha}\rho_0\right)\hat{\boldsymbol{E}}^\alpha\right)$$

and

$$\hat{\boldsymbol{E}}^I\cdot\nabla_{\hat{E}^I}\left(\hat{\boldsymbol{E}}^R\cdot\nabla_{\hat{E}^R}\rho_0\right)=\hat{\boldsymbol{E}}^R\cdot\nabla_{\hat{E}^R}\left(\hat{\boldsymbol{E}}^I\cdot\nabla_{\hat{E}^I}\rho_0\right)=\hat{\boldsymbol{E}}^R\cdot\left(\left(\nabla_{\hat{E}^I}\nabla_{\hat{E}^R}\rho_0\right)\hat{\boldsymbol{E}}^I\right)$$

equation (20) becomes

$$\begin{aligned}\frac{\partial\rho_0}{\partial t}=&-\frac{1}{\varepsilon}\left(\boldsymbol{k}\times\hat{\boldsymbol{H}}^I\right)\cdot\nabla_{\hat{E}^R}\rho_0+\frac{1}{\varepsilon}\left(\boldsymbol{k}\times\hat{\boldsymbol{H}}^R\right)\cdot\nabla_{\hat{E}^I}\rho_0+\frac{\Gamma^2(\beta^s)^2}{2b^2\varepsilon^2}\hat{\boldsymbol{E}}^R\cdot\nabla_{\hat{E}^R}\rho_0\\&+\frac{\Gamma^2(\beta^s)^2}{2b^2\varepsilon^2}\hat{\boldsymbol{E}}^I\cdot\nabla_{\hat{E}^I}\rho_0+\frac{\Gamma\beta^r\beta^s}{b\varepsilon^2}\hat{\boldsymbol{E}}^R\cdot\nabla_{\hat{E}^R}\rho_0+\frac{\Gamma\beta^r\beta^s}{b\varepsilon^2}\hat{\boldsymbol{E}}^I\cdot\nabla_{\hat{E}^I}\rho_0\end{aligned}$$

$$
\begin{aligned}
&+\frac{(\beta^r)^2}{2\varepsilon^2}\hat{\boldsymbol{E}}^R\cdot\nabla_{\hat{E}^R}\rho_0+\frac{(\beta^r)^2}{2\varepsilon^2}\hat{\boldsymbol{E}}^I\cdot\nabla_{\hat{E}^I}\rho_0\\
&+\frac{1}{2}\left(\frac{\Gamma\beta^s}{b\varepsilon}+\frac{\beta^r}{\varepsilon}\right)^2\hat{\boldsymbol{E}}^R\cdot\left(\left(\nabla_{\hat{E}^R}\nabla_{\hat{E}^R}\rho_0\right)\hat{\boldsymbol{E}}^R\right)\\
&+\frac{1}{2}\left(\frac{\Gamma\beta^s}{b\varepsilon}+\frac{\beta^r}{\varepsilon}\right)^2\hat{\boldsymbol{E}}^I\cdot\left(\left(\nabla_{\hat{E}^I}\nabla_{\hat{E}^I}\rho_0\right)\hat{\boldsymbol{E}}^I\right)\\
&+\left(\frac{\Gamma\beta^s}{b\varepsilon}+\frac{\beta^r}{\varepsilon}\right)^2\hat{\boldsymbol{E}}^R\cdot\left(\left(\nabla_{\hat{E}^I}\nabla_{\hat{E}^R}\rho_0\right)\hat{\boldsymbol{E}}^I\right)\\
&+\frac{1}{\mu}\left(\boldsymbol{k}\times\hat{\boldsymbol{E}}^I\right)\cdot\nabla_{\hat{H}^R}\rho_0-\frac{1}{\mu}\left(\boldsymbol{k}\times\hat{\boldsymbol{E}}^R\right)\cdot\nabla_{\hat{H}^I}\rho_0
\end{aligned}
$$

This is the limiting backward Kolmogorov equation. The corresponding limiting SDE system is

$$
\begin{aligned}
d\hat{\boldsymbol{E}}_t^R &= -\frac{1}{\varepsilon}\boldsymbol{k}\times\hat{\boldsymbol{H}}_t^I dt+\frac{\Gamma\beta^r\beta^s}{b\varepsilon^2}\hat{\boldsymbol{E}}_t^R dt+\frac{\Gamma^2(\beta^s)^2}{2b^2\varepsilon^2}\hat{\boldsymbol{E}}_t^R dt\\
&\qquad+\frac{(\beta^r)^2}{2\varepsilon^2}\hat{\boldsymbol{E}}_t^R dt-\frac{1}{\varepsilon}\left(\frac{\Gamma\beta^s}{b}+\beta^r\right)\hat{\boldsymbol{E}}_t^R dW_t\\
d\hat{\boldsymbol{E}}_t^I &= \frac{1}{\varepsilon}\boldsymbol{k}\times\hat{\boldsymbol{H}}_t^R dt+\frac{\Gamma\beta^r\beta^s}{b\varepsilon^2}\hat{\boldsymbol{E}}_t^I dt+\frac{\Gamma^2(\beta^s)^2}{2b^2\varepsilon^2}\hat{\boldsymbol{E}}_t^I dt\\
&\qquad+\frac{(\beta^r)^2}{2\varepsilon^2}\hat{\boldsymbol{E}}_t^I dt-\frac{1}{\varepsilon}\left(\frac{\Gamma\beta^s}{b}+\beta^r\right)\hat{\boldsymbol{E}}_t^I dW_t\\
d\hat{\boldsymbol{H}}_t^R &= \frac{1}{\mu}\boldsymbol{k}\times\hat{\boldsymbol{E}}_t^I dt\\
d\hat{\boldsymbol{H}}_t^I &= -\frac{1}{\mu}\boldsymbol{k}\times\hat{\boldsymbol{E}}_t^R dt
\end{aligned}
$$

By rewriting the above system in terms of the complex state variables $\hat{\boldsymbol{E}}$ and $\hat{\boldsymbol{H}}$, we get

$$
\begin{aligned}
d\hat{\boldsymbol{E}}_t &= \frac{i}{\varepsilon}\boldsymbol{k}\times\hat{\boldsymbol{H}}_t dt+\frac{\Gamma\beta^r\beta^s}{b\varepsilon^2}\hat{\boldsymbol{E}}_t dt+\frac{\Gamma^2(\beta^s)^2}{2b^2\varepsilon^2}\hat{\boldsymbol{E}}_t dt\\
&\qquad+\frac{(\beta^r)^2}{2\varepsilon^2}\hat{\boldsymbol{E}}_t dt-\frac{1}{\varepsilon}\left(\frac{\Gamma\beta^s}{b}+\beta^r\right)\hat{\boldsymbol{E}}_t dW_t\\
d\hat{\boldsymbol{H}}_t &= -\frac{i}{\mu}\boldsymbol{k}\times\hat{\boldsymbol{E}}_t dt
\end{aligned}
$$

Taking the inverse Fourier transform (with respect to (10) and (11)) to return to the spatial coordinates gives

$$d\boldsymbol{E}_t = \frac{1}{\varepsilon}\nabla \times \boldsymbol{H}_t dt + \frac{\Gamma\beta^r\beta^s}{b\varepsilon^2}\boldsymbol{E}_t dt + \frac{\Gamma^2(\beta^s)^2}{2b^2\varepsilon^2}\boldsymbol{E}_t dt + \frac{(\beta^r)^2}{2\varepsilon^2}\boldsymbol{E}_t dt - \frac{1}{\varepsilon}\left(\frac{\Gamma\beta^s}{b} + \beta^r\right)\boldsymbol{E}_t dW_t \tag{21a}$$

$$d\boldsymbol{H}_t = -\frac{1}{\mu}\nabla \times \boldsymbol{E}_t dt \tag{21b}$$

This is the limiting system corresponding to the system (6) as $\delta \to 0$.

We note that, combined, the last three drift terms in (21a) equal

$$\frac{1}{2\varepsilon^2}\left(\frac{\Gamma\beta^s}{b} + \beta^r\right)^2 \boldsymbol{E}_t dt$$

However, we choose to write these terms separately in (21a) in order to explain where each one comes from. The last two drift terms in (21a) are the Stratonovich corrections that result from taking the correlation times of the processes $\nu^s(t/\delta^2)$ and $\nu^r(t/\delta^4)$ to zero in (5). The term

$$\frac{\Gamma\beta^r\beta^s}{b\varepsilon^2}\boldsymbol{E}_t dt$$

is a result of the correlations between the conductivity fluctuations of different scales; in particular, it comes from the fact that the equations (6) and (9) are driven by the same Wiener process.

We drop the noise term in (21) and work with the deterministic system

$$\frac{\partial}{\partial t}\varepsilon\boldsymbol{E} = \nabla \times \boldsymbol{H} - \boldsymbol{J}$$
$$\frac{\partial}{\partial t}\mu\boldsymbol{H} = -\nabla \times \boldsymbol{E}$$

In Sections 3 and 4 we present time-filtered unstaggered leapfrog schemes designed for numerical integration of Maxwell's equations.

3 Time-Filtered Unstaggered Leapfrog Schemes

3.1 Formulation for Advection and Diffusion Tests

To illustrate time-filtered unstaggered leapfrog schemes on simple examples, we consider the one-dimensional combined advection-diffusion differential equation given by:

$$\frac{\partial \psi}{\partial t} = -U\frac{\partial \psi}{\partial x} - \nu\frac{(\Delta x)^4}{\Delta t}\frac{\partial^4 \psi}{\partial x^4}, \tag{22}$$

where, the first term in the right hand side of the above equation represents advection by a simple flow U, while the second is a hyper-diffusion (which we will refer to as diffusion) term introduced as a fourth-order numerical smoother with coefficient ν. This smoother is applied to remove the dispersive errors due to the poorly resolved high wavenumber oscillations generated by centered finite difference approximations of the spatial derivative $\partial\psi/\partial x$.

This equation (22) is numerically integrated using the time stepping scheme based on the leapfrog method and a fourth-order implicit time filter (LF-MMK), and is formulated as follows:

$$\begin{aligned} \frac{\psi^{n+1} - \bar{\psi}^{n-1}}{2\Delta t} &= F(\psi^n) \\ \bar{\psi}^{n-1} &= \psi^{n-1} + \gamma_4\left(-\bar{\psi}^{n-3} + 4\bar{\psi}^{n-2} - 6\bar{\psi}^{n-1} + 4\psi^n - \psi^{n+1}\right), \end{aligned} \tag{23}$$

Here, ψ^n is the approximation to the solution at the time $n\Delta t$ and $\bar{\psi}^n$ is the solution after applying a fourth-order implicit time filter using a real constant γ that determines the strength of the filter. The term $F(\psi^n)$ represents the tendency in the right hand side of (22), which is evaluated at the time $n\Delta t$ in this scheme:

$$F(\psi^n) = \left(-U\frac{\partial \psi}{\partial x} - \nu\frac{(\Delta x)^4}{\Delta t}\frac{\partial^4 \psi}{\partial x^4}\right)^n \tag{24}$$

Although the scheme uses an implicit filter, in that both $\bar{\psi}^{n-1}$ and ψ^{n+1} in (23)-(24) are not known and depend on each other, this scheme can be formulated and implemented in a simple explicit manner.

The performance of the above time stepping scheme combined with finite difference representations for the spatial derivatives that are evaluated at $t = n\Delta t$ is tested by numerical integrations of equation (22). The first derivative uses a second and fourth-order centered difference approximations given by:

$$\left(\frac{\partial \psi}{\partial x}\right)_i^{2nd} = \frac{\psi_{i+1}^n - \psi_{i-1}^n}{2\Delta x} \tag{25}$$

$$\left(\frac{\partial \psi}{\partial x}\right)_i^{4th} = \frac{8(\psi_{i+1}^n - \psi_{i-1}^n) - (\psi_{i+2}^n - \psi_{i-2}^n)}{12\Delta x} \tag{26}$$

and the fourth derivative in the numerical smoother term is approximated by:

$$\left(\frac{\partial^4 \psi}{\partial x^4}\right)_i = \frac{-\psi_{i-2}^n + 4\psi_{i-1}^n - 6\psi_i^n + 4\psi_{i+1}^n - \psi_{i+2}^n}{(\Delta x)^4} \tag{27}$$

To show performance for advection-diffusion tests, we apply the unfiltered leapfrog (LF) and LF-MMK schemes to (22), and we compare the numerical approximations obtained from these schemes to the exact solution. In this test, we select a simple flow where U is constant. The one-dimensional domain uses 101 grid where $0 < x < 100$, and periodic boundary conditions are imposed. The initial conditions consist of a half sine pulse that has the form:

$$\psi^0(x) = \sin\left(\pi \frac{x-45}{10}\right), \quad \text{for} \quad 44 \le x \le 54$$

$$\psi^0(x) = 0, \quad \text{otherwise}$$

Equation (22) is numerically integrated within the domain using two choices of time steps Δt, flow velocity U and the grid spacing Δx such that the corresponding Courant numbers ($Cr = U\Delta t/\Delta x$) are 0.3 and 0.1 respectively. The total number of time steps is chosen such that the initial pulse is transported one revolution around the domain. Figure 1a superimposes the exact solution to Equation (22) and its numerical approximations obtained from the unfiltered leapfrog (LF) scheme using a Courant number of $Cr = 0.3$. These approximations use a second and a fourth-order accurate finite difference approximations for the first derivative in a pure advection, where the numerical smoother is removed ($\nu = 0$). The LF scheme is stable for both cases. The numerical solution approximated by using a second-order spatial central differences is very different from the exact solution. It shows generation of oscillations with high wavenumbers that are manifestation of dispersion errors due to poor finite difference representations of the derivatives. In contrast, the solution obtained by LF using fourth-order approximations for the spatial derivatives is comparatively closer to the exact solution and is more accurate than the one obtained by using a second order spatial differences. Nevertheless, the fourth order solution also produces short oscillations due to dispersion errors. However, these oscillations have smaller amplitudes and higher wavenumbers. Figure 1b shows the solutions approximated by LF using a second and a fourth-order spatial differences, and a Courant number of $Cr = 0.1$. These approximations are similar to the one shown in Fig. 1a ($Cr = 0.3$). This indicates that reducing the Courant number does not improve the numerical solutions in this case. The solutions obtained from RK3 and LF-MMK with a time filter coefficient $\gamma = 0.1$, but without any smoother ($\nu = 0$), also generate comparable errors (not shown). Thus, these errors appear to be less sensitive to the time stepping schemes used.

Traditionally, dispersion errors can be reduced by using odd-order finite difference approximations for the first derivative [26], or by applying selective numerical smoothers. However, the unfiltered leapfrog is unconditionally unstable when these methods are applied as shown in Fig. 2a, b. These figures present the numerical solutions obtained at earlier stages by LF for the advection equation where a small diffusion is included with a smoothing coefficient of $\nu = 0.005$. The diffusion terms are evaluated at $t = n\Delta t$. The numerical instability in this figure is caused by the computational modes. These modes are unconditionally unstable for diffusion even

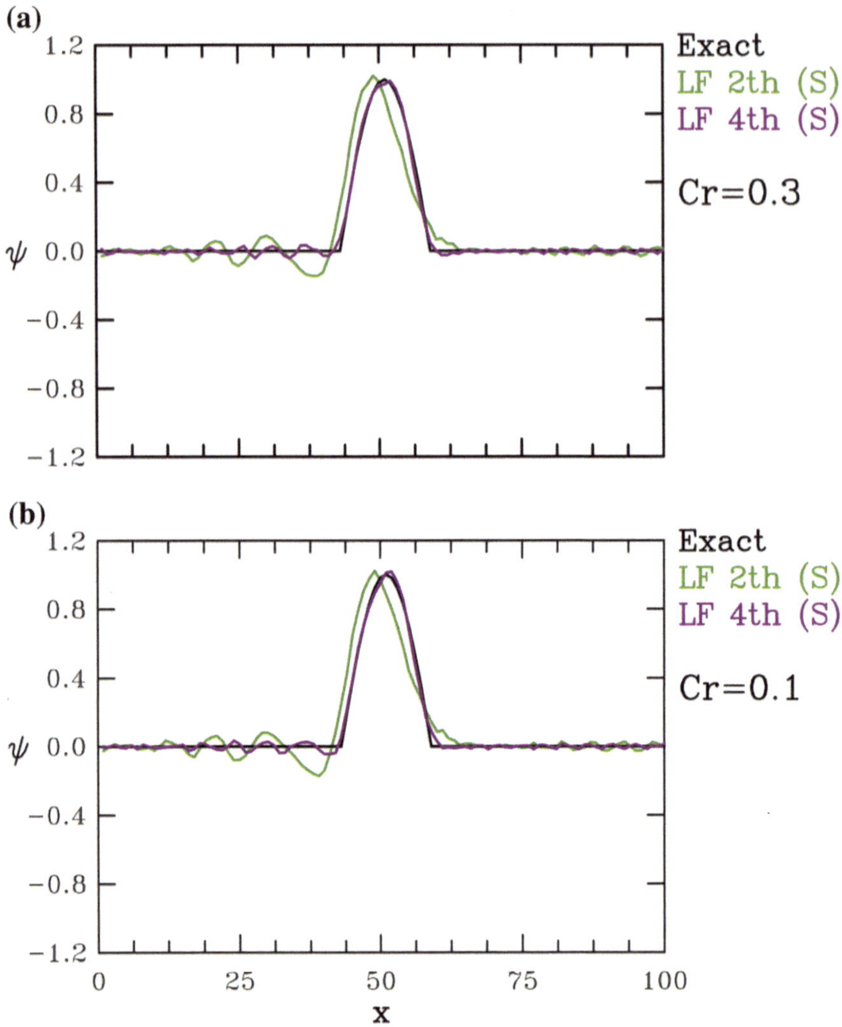

Fig. 1 Comparison of the exact solution to equation (22) with numerical approximations obtained using the unfiltered leapfrog scheme using a second and fourth-order spatial differencing and a Courant number of (a) $Cr = 0.3$, (b) $Cr = 0.1$. All the simulations do not use numerical smoothing ($\nu = 0$). (S) and (U) indicate stability and instability respectively.

in linear cases. We note that these modes are also present in LF integrations of pure advection (Fig. 1); but the physical and computational modes are both conditionally stable in this case since they do not interact with each other. However, these modes may become a serious limitation even for advection problems where nonlinearity is included.

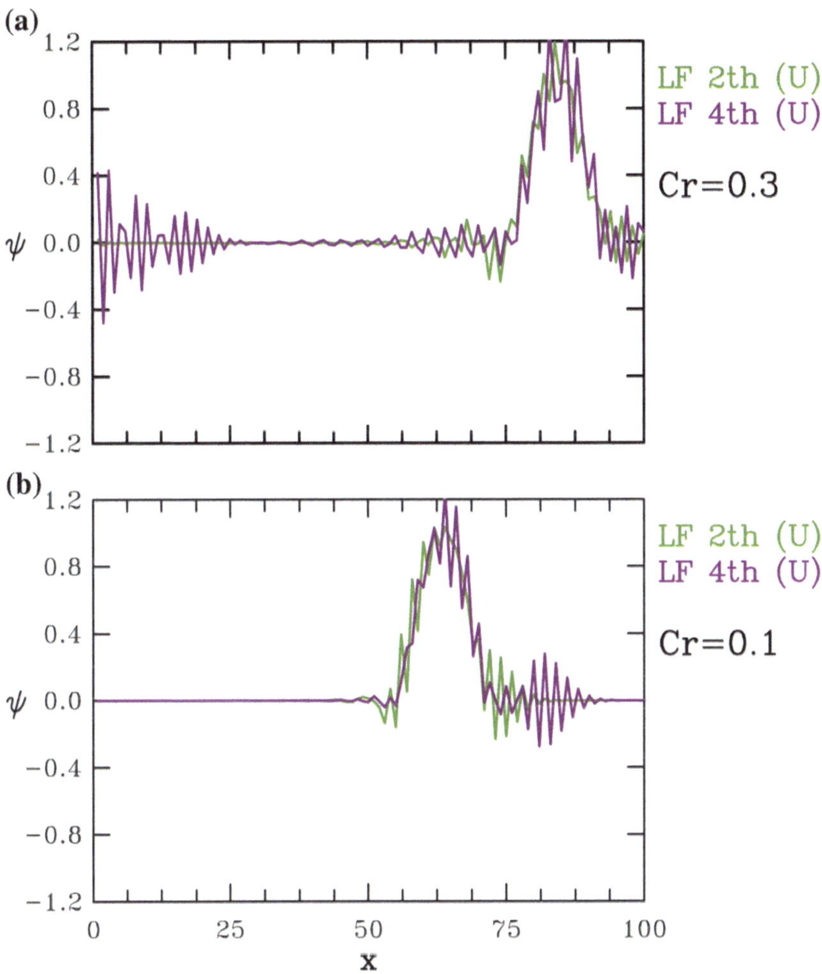

Fig. 2 Same as Fig. 1, but the numerical approximations use a numerical smoother with coefficient $\nu = 0.005$. (S) and (U) indicate stability and instability respectively.

Figure 3a, b show the numerical simulations obtained from the LF-MMK scheme applied to the combined advection-diffusion equation 22. The approximations for the spatial derivatives in these simulations are fourth-order. The LF-MMK uses a time filter and a numerical smoother coefficients of $\gamma = 0.1$ and $\nu = 0.005$ respectively. This figure shows that in contrast with LF (Fig. 2), LF-MMK is stable for both advection and diffusion contributions. The high wavenumber oscillations produced by the finite difference approximations for spatial derivatives are reduced compared to LF. The solution obtained from LF-MMK is more accurate and is superior to the one computed from the LF scheme. The solution obtained by LF-MMK is comparable with the one computed from RK3 and to the exact solution, which is also

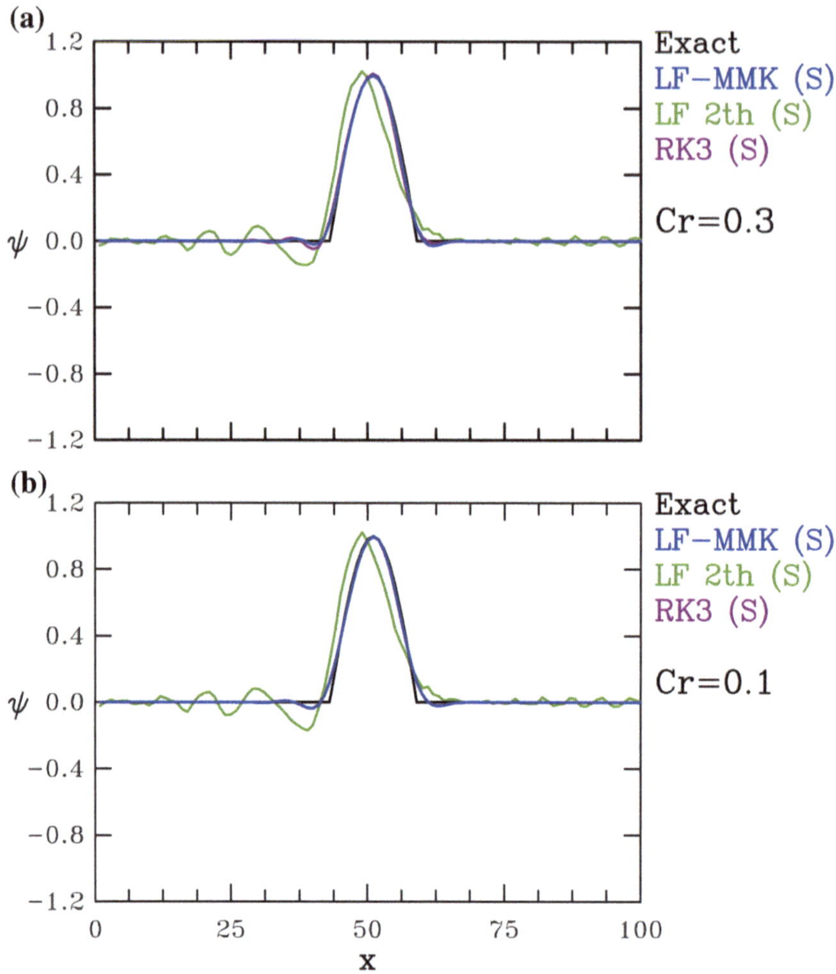

Fig. 3 Comparison of the exact solution to equation (22) with numerical approximations obtained using the LF-MMK and RK3 with a fourth-order numerical smoothing ($\nu = 0.005$), combined with a fourth-order spatial differencing and a Courant number of (a) $Cr = 0.3$, (b) $Cr = 0.1$. The numerical solution from the unfiltered leapfrog, and without smoothing is also superimposed.

superimposed in Fig. 3. The RK3 scheme is conditionally stable for both diffusion and advection equations [26]. However, it requires three evaluation functions for each time step. The LF-MMK scheme stabilizes the numerical errors due to the computational modes. The tests presented in Figs. 1, 2 and 3 demonstrate the ability of LF-MMK to stabilize the leapfrog scheme and to control the computational modes in both advection and diffusion terms.

3.2 *Von Neumann Stability Analysis*

We examine the behavior of LF-MMK combined with a fourth-order space differencing for both advection and diffusion equations. For this purpose, von Neumann stability analysis are investigated where ψ takes the form:

$$\psi_i^n = \xi^n e^{jki\Delta x} \tag{28}$$

Here, k is a real spatial wave number, ξ^n is a complex function, $j = \sqrt{-1}$ and Δx is the spatial grid spacing.

3.2.1 Advection Equation

If we apply the LF-MMK to the advection equation (Eq. 22, with $\nu = 0$), and using the amplification A defined by $\xi^{n+1} = A\xi^n$ (or $\bar{\xi}^{n+1} = A\bar{\xi}^n$), then, Eqs. (23), (24) become:

$$\frac{\bar{\xi}^n}{\xi^n} = \left(A^2 - \lambda A\right) \tag{29}$$

$$A^2\frac{\bar{\xi}^n}{\xi^n} = A^2 + \gamma_4\left(-\frac{\bar{\xi}^n}{\xi^n} + 4A\frac{\bar{\xi}^n}{\xi^n} - 6A^2\frac{\bar{\xi}^n}{\xi^n} + 4A^3 - A^4\right). \tag{30}$$

where,

$$\begin{aligned}\lambda(k\Delta x) &= j\frac{U\Delta t}{3\Delta x}\left(\sin(2k\Delta x) - 8\sin(k\Delta x)\right)\\ &= j\frac{Cr}{3}\left(\sin(2k\Delta x) - 8\sin(k\Delta x)\right)\end{aligned}$$

After eliminating $\bar{\xi}^n/\xi^n$ in these two equations, a cubic equation satisfied by the amplification factor A is derived:

$$a_3A^3 + a_2A^2 + a_1A + a_0 = 0. \tag{31}$$

where

$$\begin{aligned}a_3 &= 1 + 7\gamma\\ a_2 &= -\left(8\gamma + 6\gamma\lambda + \lambda\right)\\ a_1 &= -\left(1 - \gamma - 4\gamma\lambda\right)\\ a_0 &= \gamma\lambda\end{aligned}$$

This equation has three roots corresponding to one physical mode and two computational modes ($2\Delta t$ and $4\Delta t$ modes). Using Taylor series expansion of the amplification factor for the physical mode, one can derive the order of the amplitude errors. For the well resolved waves where $k\Delta x$ approaches zero, the magnitude of amplification factor for the physical mode is given by:

$$|A_p| = 1 - \frac{1}{2}\frac{\gamma_4}{1+3\gamma_4}(Crk\Delta x)^4.$$

The amplitude error generated by the LF-MMK scheme is of order $O[(\Delta x)^4]$.

Figure 4a, b show the magnitudes of the amplification factors corresponding to the physical and the computational modes for finite $k\Delta x$ obtained from the LF-MMK scheme using a Courant numbers of $Cr = 0.3$ and $Cr = 0.1$ respectively. The value for the filter parameter is $\gamma = 0.1$. Unlike the leapfrog scheme, LF-MMK produces two computational modes corresponding to $2\Delta t$ and $4\Delta t$ modes (Fig. 4). The physical and computational modes are stable for all values of $k\Delta x$, and for both values of the Courant numbers used. We note that as $k\Delta x$ approaches 0, the magnitude of the $2\Delta t$ computational mode is damped at the rate of $(1-\gamma)/(1+7\gamma)$. The $4\Delta t$ mode is strongly damped and vanishes as $k\Delta x$ approaches 0. The amplification factor for the physical mode is closer to 1, while those of the computational modes are less than one for all value of $k\Delta x$. This contrasts with the unfiltered leapfrog, where the $2\Delta t$ computational mode has an amplification factor with a magnitude that is equal to that of the physical mode ($|A_p| = |A_c| = 1$).

3.2.2 Diffusion Equation

Let us now examine the von Neumann stability analysis of the LF-MMK time stepping scheme combined with finite spatial differences when applied to the pure diffusion case of equation (22). It can be shown that the amplification factor A satisfies the same cubic equations as for the pure advection equation except that the function $\lambda(k\Delta x)$ is replaced by $\eta(k\Delta x)$, where

$$\eta(k\Delta x) = -8\nu\,(1 - cos(k\Delta x))^2 \tag{32}$$

Figure 5a shows the magnitudes of the amplifications factors corresponding to the physical and computational modes. Here we use values for time filter and numerical smoothing of of $\gamma = 0.1$ and $\nu = 0.005$ respectively. This figure shows that the LF-MMK scheme stabilizes and damps the amplitude of the computational modes. In contrast, the amplification factor obtained from the unfiltered leapfrog scheme for this mode is unstable (Fig. 5b). Figures 6 and 7 compare the magnitude of the amplification of the physical and computational modes obtained from LF-MMK and LF using values for the numerical smoother coefficients of $\nu = 0.01$ and $\nu = 0.0025$ respectively. They show similar behaviors as in Fig. 4 with numerical instabilities for LF and conditional stabilities for LF-MMK. Thus, increasing or reducing the

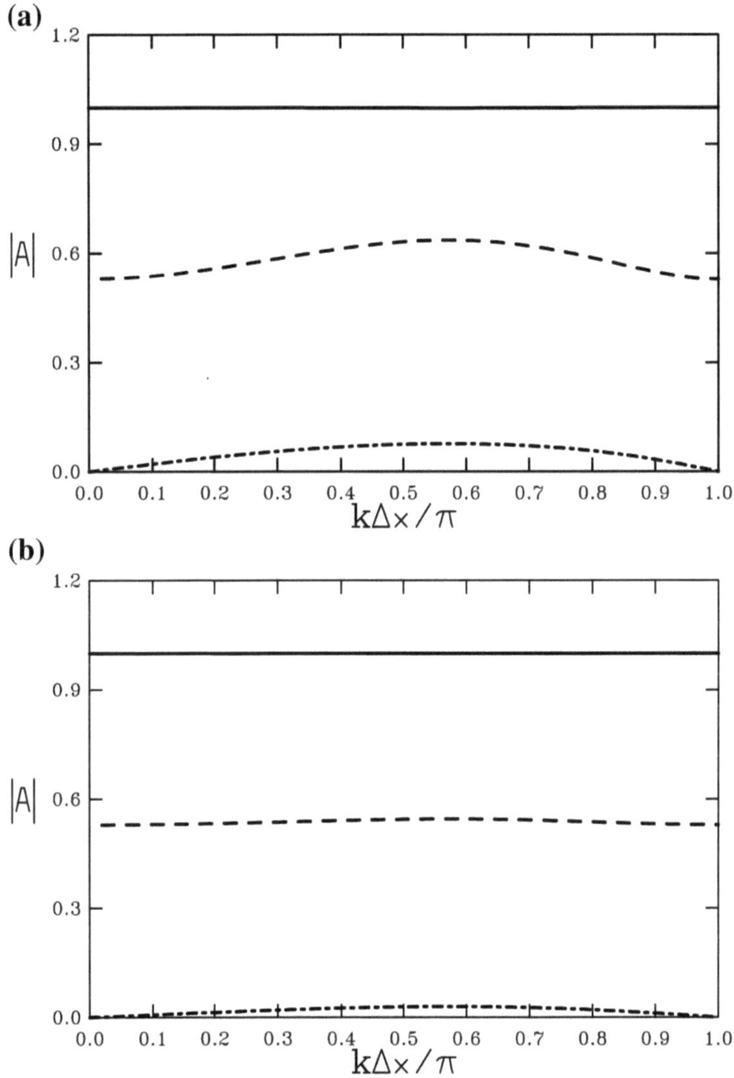

Fig. 4 Magnitudes of the amplification factors for the physical mode, the $2\Delta t$ and the $4\Delta t$ computational modes as a function of $k\Delta x$ for the LF-MMK scheme applied to the pure advection ($\nu = 0$) obtained with $\gamma = 0.1$, and (a) $Cr = 0.3$ and (b) $Cr = 0.1$.

coefficient of the smoother ν does not stabilize the computational mode in LF. This explains the numerical instability arising in the unfiltered leapfrog scheme observed in Fig. 2a, b. The leapfrog scheme cannot be used to reduce the dispersion errors due to finite spatial difference approximations. However, LF-MMK allows for control of the short oscillations generated by dispersion (Fig. 3).

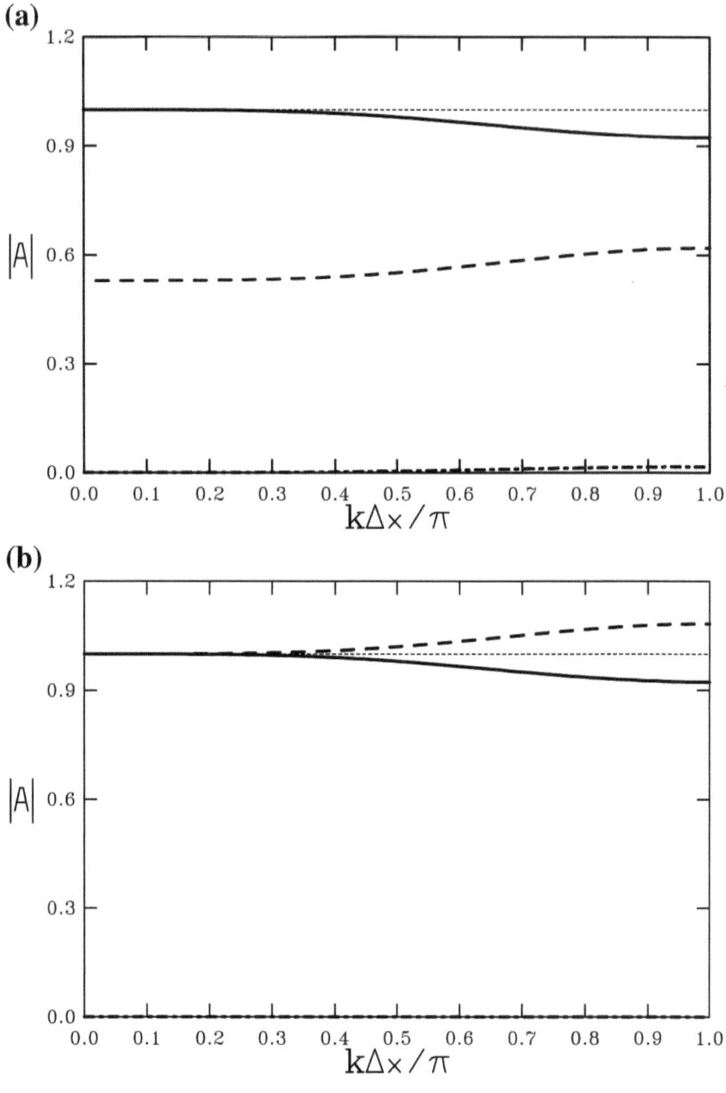

Fig. 5 Magnitudes of the amplification factors for the physical mode, the $2\Delta t$ and the $4\Delta t$ computational modes as a function of $k\Delta x$ for (a) the LF-MMK scheme applied to the pure diffusion equation ($Cr = 0$) with a diffusion coefficient of $\nu = 0.005$ and a filter parameter $\gamma = 0.1$. (b) is the same as (a) but the scheme used is the unfiltered leapfrog scheme.

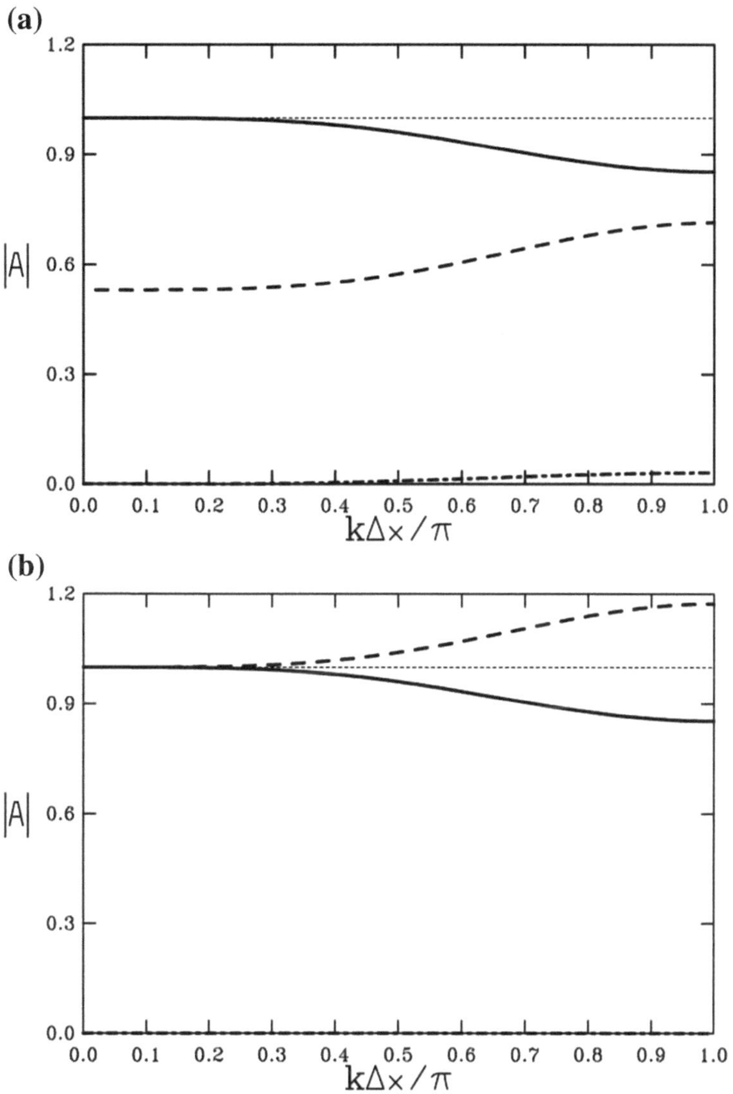

Fig. 6 Same as Fig. 5 but with $\nu = 0.01$.

4 Time-Filtered Unstaggered Leapfrog Schemes for Maxwell's Equations

Here, we present a formulation of a numerical solver for Maxwell's equations based on the leapfrog scheme and a fourth-order implicit time filter described in Section 3. Assuming an inhomogeneous material, the time dependent components of Maxwell

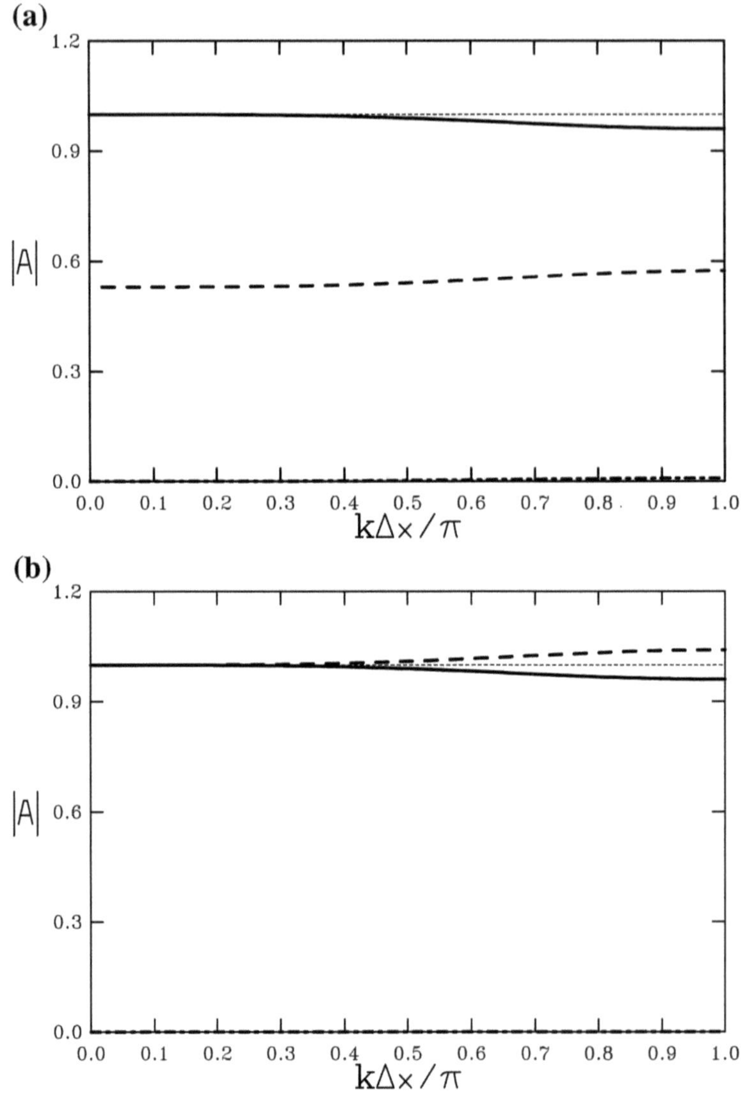

Fig. 7 Same as Fig. 5 but with $\nu = 0.0025$.

equations (Faraday's and Ampere's laws) have the form:

$$\frac{\partial \mathbf{B}}{\partial t} = -\nabla \times \mathbf{E} \tag{33}$$

$$\frac{\partial \mathbf{D}}{\partial t} = \nabla \times \mathbf{H} - \mathbf{J} \tag{34}$$

where $\nabla \cdot \mathbf{D} = \rho$ and $\nabla \cdot \mathbf{B} = 0$, together with the material laws $\mathbf{B} = \mu \mathbf{H}$, $\mathbf{J} = \sigma \mathbf{E}$, $\mathbf{D} = \varepsilon \mathbf{E}$. In these equations, $\mathbf{E}$ is the electric field, $\mathbf{D}$ is the electric flux density, $\mathbf{H}$ is the magnetic field, $\mathbf{B}$ is the magnetic flux density, $\mathbf{J}$ is the electric current density, ρ is the charge density μ is the permeability, ε is the permittivity and σ is the electric conductivity.

The spatial discretization uses a staggered three dimensional grid, where the magnetic field components (H_x, H_y, H_z) and (B_x, B_y, B_z) are defined on the cell's faces. The electric field components (E_x, E_y, E_z) and (D_x, D_y, D_z) are defined on the cell's edges. The inhomogeneous properties of the medium are characterized by a permeability (μ) and permittivity (ε) that depend on space. These are defined at the center of the grid cell. With this configuration, the numerical formulation of the new time stepping scheme using leapfrog and a fourth order implicit time filter has the form:

$$\frac{B^{n+1}_{x;i,j+1/2,k+1/2} - \bar{B}^{n-1}_{x;i,j+1/2,k+1/2}}{2\Delta t} = \frac{E^{av,n}_{y;i,j+1/2,k+1} - E^{av,n}_{y;i,j+1/2,k}}{\Delta z} - \frac{E^{av,n}_{z;i,j+1,k+1/2} - E^{av,n}_{z;i,j,k+1/2}}{\Delta y} \tag{35}$$

$$\frac{B^{n+1}_{y;i+1/2,j,k+1/2} - \bar{B}^{n-1}_{y;i+1/2,j,k+1/2}}{2\Delta t} = \frac{E^{av,n}_{z;i+1,j,k+1/2} - E^{av,n}_{z;i,j,k+1/2}}{\Delta x} - \frac{E^{av,n}_{x;i+1/2,j,k+1/2} - E^{av,n}_{x;i+1/2,j,k}}{\Delta z} \tag{36}$$

$$\frac{B^{n+1}_{z;i+1/2,j+1/2,k} - \bar{B}^{n-1}_{z;i+1/2,j+1/2,k}}{2\Delta t} = \frac{E^{av,n}_{x;i+1/2,j+1,k} - E^{av,n}_{x;i+1/2,j,k}}{\Delta y} - \frac{E^{av,n}_{y;i+1,j+1/2,k} - E^{av,n}_{y;i,j+1/2,k}}{\Delta x} \tag{37}$$

$$\frac{D^{n+1}_{x;i+1/2,j,k} - \bar{D}^{n-1}_{x;i+1/2,j,k}}{2\Delta t} = \frac{H^{av,n}_{z;i+1/2,j+1/2,k} - H^{av,n}_{z;i+1/2,j-1/2,k}}{\Delta y} - \frac{H^{av,n}_{y;i+1/2,j,k+1/2} - H^{av,n}_{y;i+1/2,j,k-1/2}}{\Delta z} - J^{n}_{x;i+1/2,j,k} \tag{38}$$

$$\frac{D^{n+1}_{y;i,j+1/2,k} - \bar{D}^{n-1}_{y;i,j+1/2,k}}{2\Delta t} = \frac{H^{av,n}_{x;i,j+1/2,k+1/2} - H^{av,n}_{x;i,j+1/2,k-1/2}}{\Delta z} - \frac{H^{av,n}_{z;i+1/2,j+1/2,k} - H^{av,n}_{z;i-1/2,j+1/2,k}}{\Delta x} - J^{n}_{y;i,j+1/2,k} \tag{39}$$

$$\frac{D^{n+1}_{z;i,j,k+1/2} - \bar{D}^{n-1}_{z;i,j,k+1/2}}{2\Delta t} = \frac{H^{av,n}_{y;i+1/2,j,k+1/2} - H^{av,n}_{y;i-1/2,j,k+1/2}}{\Delta x} - \frac{H^{av,n}_{x;i,j+1/2,k+1/2} - H^{av,n}_{x;i,j-1/2,k+1/2}}{\Delta y} - J^{n}_{z;i,j,k+1/2} \tag{40}$$

and

$$\mathbf{H}^{n+1} = \tilde{\mu}_{\mathbf{H}}^{-1}\mathbf{B}^{n+1} \tag{41}$$
$$\mathbf{E}^{n+1} = \tilde{\varepsilon}_{\mathbf{E}}^{-1}\mathbf{D}^{n+1} \tag{42}$$

The variables $\tilde{\mu}_{\mathbf{H}}$ and $\tilde{\varepsilon}_{\mathbf{E}}$, represent the averages of μ and ε at the locations of each component H_α and E_α respectively, where $\alpha = x, y, z$. This average is required because μ and ε are defined at the center of the grid cell $(i + 1/2, j + 1/2, k + 1/2)$.

The components $\bar{D}_\alpha^{n-1}$ and $\bar{B}_\alpha^{n-1}$ are not known in this numerical implementation as they depend on D_α^{n+1} and B_α^{n+1} because of the applications of the implicit filter. i.e:

$$\bar{D}_\alpha^{n-1} = D_\alpha^{n-1} + \gamma\left(-\bar{D}_\alpha^{n-3} + 4\bar{D}_\alpha^{n-2} - 6\bar{D}_\alpha^{n-1} + 4D_\alpha^{n} - D_\alpha^{n+1}\right), \tag{43}$$

however, as explained in Section 3, it is easy to formulate this scheme in an explicit manner.

The quantities $E^{av,n}$ and $H^{av,n}$ represent some suitable averages of E^n and H^n applied along the dimension where the derivative is calculated. This averaging has the effect of increasing the order of the finite difference approximations used to evaluate the spatial derivatives. For second-order approximations the averaging is not required: $E^{av,n} = E^n$ and $B^{av,n} = B^n$. For higher-order approximations, these have the forms:

$$H_{i+1/2}^{av,n} = \frac{-H_{i-1/2}^{n} + 26H_{i+1/2}^{n} - H_{i+3/2}^{n}}{24} \tag{44}$$
$$E_{i}^{av,n} = \frac{-E_{i-1}^{n} + 26E_{i}^{n} - E_{i+1}^{n}}{24} \tag{45}$$

for fourth-order, and

$$H_{i+1/2}^{av,n} = \frac{9H_{i-3/2}^{n} - 116H_{i-1/2}^{n} + 2134H_{i+1/2}^{n} - 116H_{i+3/2}^{n} + 9H_{i+5/2}^{n}}{1920} \tag{46}$$
$$E_{i}^{av,n} = \frac{9E_{i-2}^{n} - 116E_{i-1}^{n} + 2134E_{i}^{n} - 116E_{i+1}^{n} + 9E_{i+2}^{n}}{1920} \tag{47}$$

for six-order approximations. These expressions can be derived by using Taylor series expansions, and casting the finite difference approximations in flux forms. For example:

$$\begin{aligned}\left(\frac{\partial H}{\partial x}\right)_i^{4th} &= \frac{H_{i+1/2}^{av} - H_{i-1/2}^{av}}{\Delta x} \\ &= \frac{27(H_{i+1/2} - H_{i-1/2}^{n}) - (H_{i+3/2} - H_{i+3/2})}{24\Delta x}\end{aligned} \tag{48}$$

which is the staggered fourth-order approximation for the first spatial derivative.

Next we apply the numerical method above to the Yee's test case originated in [27]. In this test case, we consider the two dimensional Maxwell's equations for a transverse magnetic wave in a lossless material ($\sigma = 0$). These equations are given by

$$\begin{aligned} E_x &= E_y = 0,\, H_z = 0 \\ \varepsilon \frac{\partial E_z}{\partial t} &= \frac{\partial H_y}{\partial x} - \frac{\partial H_x}{\partial y} \\ \mu \frac{\partial H_x}{\partial t} &= -\frac{\partial E_z}{\partial y} \\ \mu \frac{\partial H_y}{\partial t} &= \frac{\partial E_z}{\partial x} \end{aligned}$$

As in Yee's test [27], we use a two dimensional staggered grid with 81 and 97 grid points in x and y dimensions respectively. A perfectly conducting square is placed inside the domain $[0, 80] \times [0, 96]$, in the area $16 < x < 48$ and $32 < y < 64$ (see Fig. 2 in [27]). The initial conditions consist of a half sine incident pulse, and are given by

$$\begin{aligned} E_z^0(x, y) &= \sin\left(\pi \frac{x-48}{8}\right) \\ H_y^0(x, y) &= \frac{1}{Z} E_z^0(x, y) \\ H_x^0(x, y) &= 0 \end{aligned}$$

where $Z = \sqrt{\mu/\varepsilon} = 376.7$.

First, we conducted two numerical simulations without including the conducting square. Figure 8 shows the numerical results from the unfiltered leapfrog scheme using a second-order centered difference approximations for the spatial derivatives. It shows short oscillations due to the dispersion of the initial pulse. The dispersion is caused by the poor representation of the finite difference system. These oscillations are also found in the numerical results calculated from the Yee scheme (see Fig. 3 in [27]). The numerical results obtained from the new scheme are also superimposed in Fig. 8. The scheme uses a time filter parameter of $\gamma = 0.1$ and fourth-order difference approximations for the spatial derivatives. A small numerical smoothing with a coefficient of $\nu = 0.005$ is introduced in the equations to remove the small oscillations due to dispersion. The numerical solution obtained by the new method is more accurate and reduces the effect of dispersion.

Figure 9 shows the numerical results obtained from both LF and LF-MMK schemes at $y = 49$, when the obstacle in included. As expected [27], the wave is reflected in both cases as it encounters the obstacle. However, the results using LF-MMK are much more accurate and less dispersive compared to those using the standard unfiltered leapfrog scheme. Figure 10 shows the value of the electric field

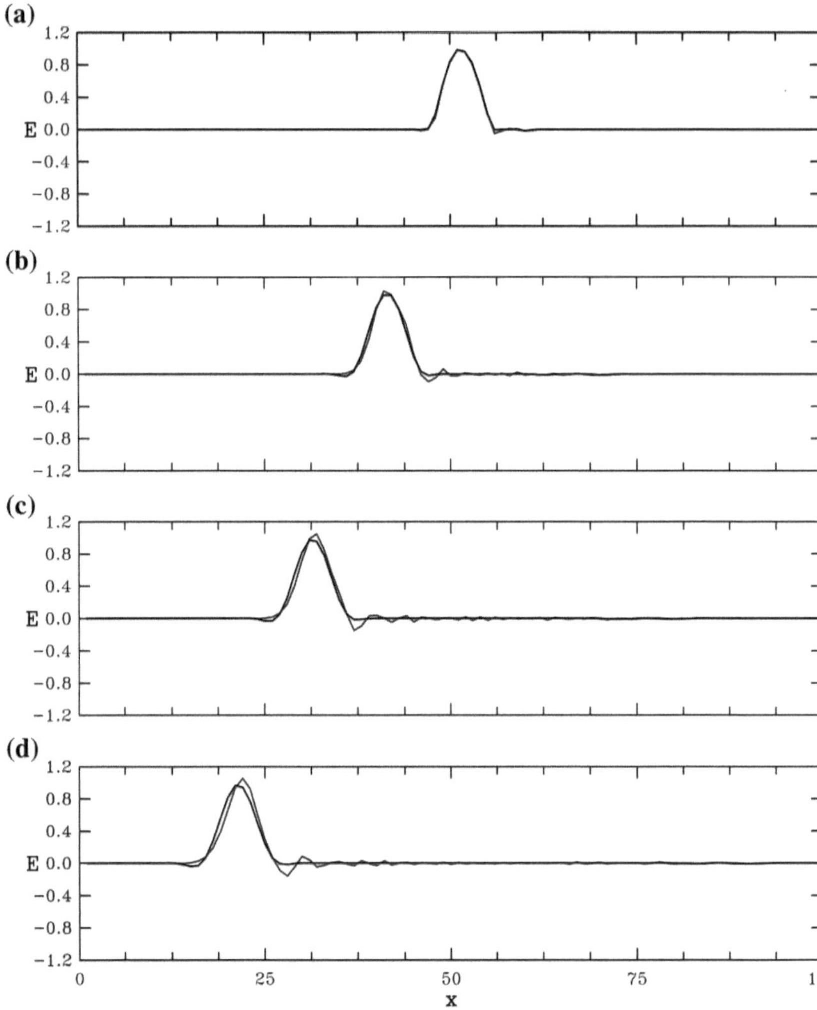

Fig. 8 Results of the computation of the electric field at different times in the absence of the obstacle using LF with a second-order and LF-MMK with a fourth-order approximations. The LF-MMK scheme uses a numerical smoother with coefficient $\nu = 0.005$.

at $y = 29$. This location is close, but below the lower boundary of the obstacle (the location of the obstacle starts at $y = 32$). In this case, both schemes were also able to resolve the partial reflection and transmission of the wave in agreement with the Yee's results (Fig. 4 in [27]).

The presented numerical method, which is based on leapfrog and a fourth-order time filter, requires only one function evaluation per time step and is simple to implement. This scheme was tested on Maxwell's equations where a new time stepping method was combined with fourth-order space centered difference approximations

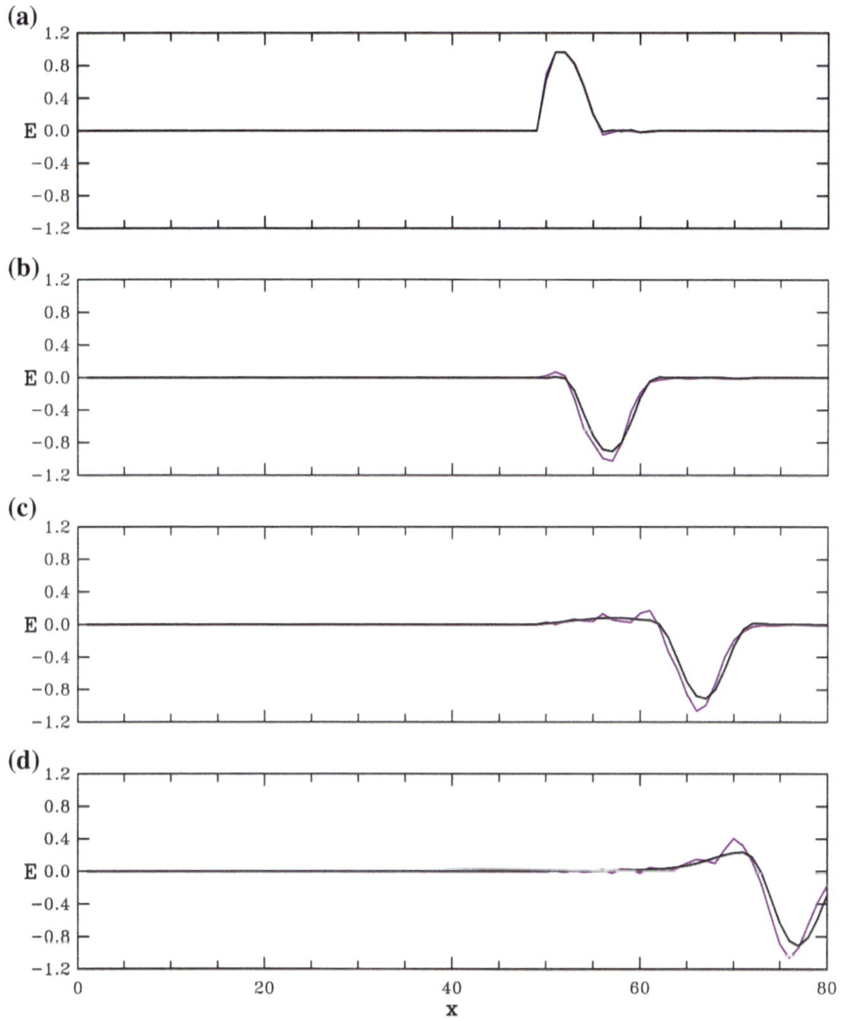

Fig. 9 Results of the computation of the electric field at different times and at $y = 50$ in the presence of the obstacle using LF with a second-order and LF-MMK with a fourth-order approximations. The LF MMK scheme uses a numerical smoother with coefficient $\nu = 0.005$.

for spatial derivatives. The von Neumann stability analysis of the new time stepping scheme combined with finite spatial differences showed superior properties compared to conventional finite-difference, time-domain schemes based on leapfrog methods. The technique significantly reduced the short oscillations generated by spurious numerical dispersion. In our future work, we will present generalizations of the numerical method based on time-filtered unstaggered leapfrog schemes presented in

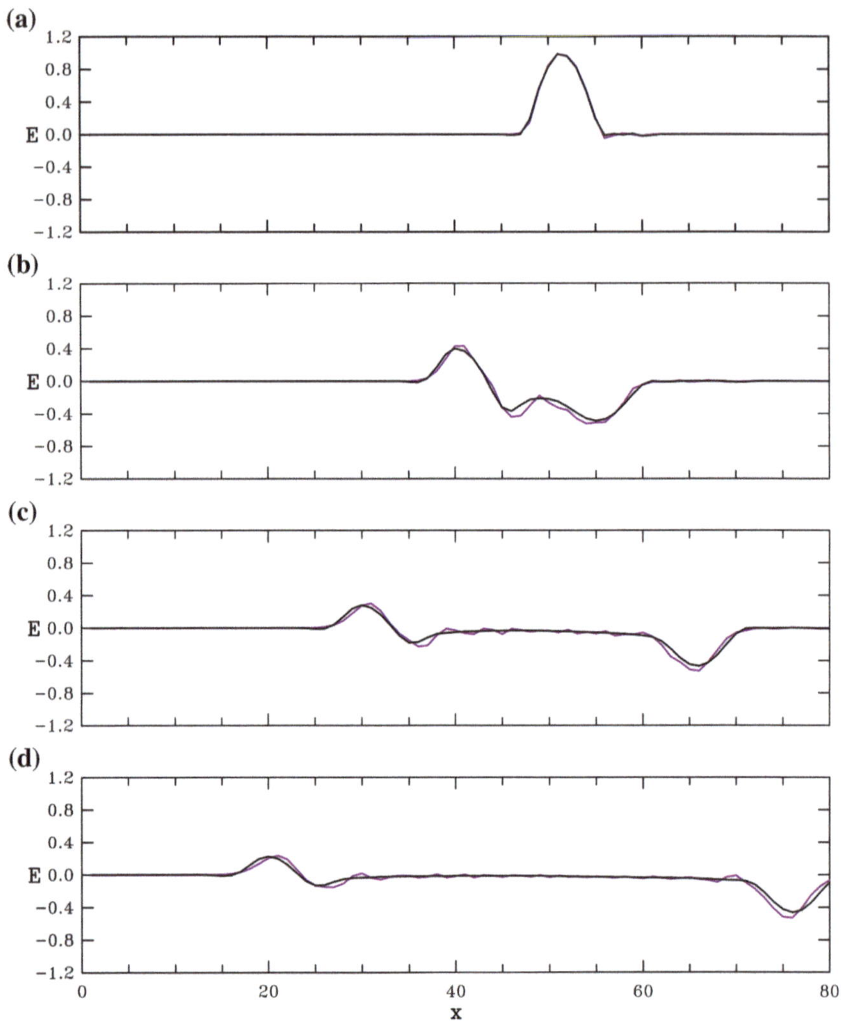

Fig. 10 Same as Fig. 9, but at $y = 29$.

this paper to solutions of stochastic three-dimensional Navier–Stokes and Maxwell's Equations for inhomogeneous media [4, 9, 16].

Acknowledgements This material is based upon work supported by the Air Force Office of Scientific Research under award number FA9550-15-1-0096.

References

1. A.S. Baturin, Y.M. Belousov, V.N. Gorelkin, V.P. Krainov, V.S. Rastunkov, Laser induced negative conductivity of diamond. Laser Phys. Lett. **3**(12), 578–583 (2006)
2. L. Borcea, J. Garnier, Polarization effects for electromagnetic wave propagation in random media. preprint (2016)
3. J. Durazo, E. Kostelich, A. Mahalov, W. Tang, Observing system experiments with an ionospheric electrodynamics model. Physica Scripta **91**(4), 044001 (2016). https://doi.org/10.1088/0031-8949/91/4/044001
4. F. Flandoli, A. Mahalov, Stochastic 3D rotating Navier-Stokes equations: averaging, convergence and regularity. Arch. Ration. Mech. Anal. **205**, 195–237 (2012)
5. C.W. Gardiner, *Handbook of Stochastic Methods for Physics* (Chemistry and the Natural Sciences. Springer, Berlin, 1985)
6. J. Garnier, K. Sølna, Paraxial coupling of electromagnetic waves in random media. Multiscale Model. Simul. **7**(4), 1928–1955 (2009)
7. J. Garnier, K. Sølna, Transmission and reflection of electromagnetic waves in randomly layeredmedia. Commun. Math. Sci. **13**(3), 707–728 (2015)
8. I.A. Ibragimov, Y.V. Linnik, *Independent and Stationary Sequences of Random Variables* (Wolters-Noordhoff, Groningen, 1971)
9. C.-Y. Jung, B. Kwon, A. Mahalov, T.B. Nguyen, Solution of Maxwell equations in media with multiple random interfaces. Int. J. Numer. Anal. Model. **11**(1), 194–213 (2014)
10. J.H. Kim, S.Y. Sohn, An asymptotic diffusion limit for electromagnetic wave reflection from a random medium. SIAM J. Appl. Math. **60**(5), 1502–1519 (2000)
11. M.C. Kelley, *The Earth's Ionosphere, Plasma Physics and Electrodynamics, International Geophysics Series*, vol. 43 (Academic Press, San Diego, 1989)
12. W. Kohler, G. Papanicolaou, M. Postel, B. White, Reflection of pulsed electromagnetic waves from a randomly stratified half-space. J. Opt. Soc. Am. A **8**(7), 1109–1125 (1991)
13. A. Mahalov, Multiscale modeling and nested simulations of three-dimensional ionospheric plasmas: Rayleigh-Taylor turbulence and nonequilibrium layer dynamics at fine scales. Physica Scripta, Phys. Scr. **89**, 098001 (22pp), Royal Swedish Academy of Sciences (2014)
14. A. Mahalov, M. Moustaoui, Time-filtered leapfrog integration of Maxwell equations using unstaggered temporal grids. J. Comput. Phys. **325**, 98–115 (2016)
15. A. Mahalov, M. Moustaoui, V. Grubišić, A numerical study of mountain waves in the upper troposphere and lower stratosphere. Atmos. Chem. Phys. **11**, 5123–5139 (2011)
16. A. Mahalov, E. Suazo, S. Suslov, Spiral laser beams in inhomogeneous media. Opt. Lett. Opt. Soc. Am. **38**(15), 2763–2769 (2013)
17. R. McGranaghan, D.J. Knipp, T. Matsuo, H. Godinez, R.J. Redmon, S.C. Solomon, S.K. Morley, Modes of high-latitude auroral conductance variability derived from DMSP energetic electron precipitation observations: empirical orthogonal function analysis. J. Geophys. Res.: Space Phys. **120**(12), 11,013-11,031 (2015)
18. G. Papanicolaou, L. Ryzhik, K. Sølna, Self-averaging from lateral diversity in the Itô Schrödinger equation. Multiscale Model. Simul. **6**(2), 468–492 (2007)
19. G.A. Pavliotis, A.M. Stuart, *Multiscale Methods: Averaging and Homogenization, Texts in Applied Mathematics*, vol. 53 (Springer, New York, 2008); Stochastic Effects and Time-Filtered Leapfrog Schemes for Maxwell's Equations 21
20. V. Ryzhii, Microwave-induced negative conductivity and zero-resistance states in twodimensional electronic systems: history and current status. Phys.-Usp. **48**(2), 191–198 (2005)
21. V. Ryzhii, M. Ryzhii, T. Otsuji, Negative dynamic conductivity of graphene with optical pumping. J. Appl. Phys. **101**, 083114 (2007)
22. W. Tang, A. Mahalov, Stochastic Lagrangian dynamics for charged flows in the E-F regions of ionosphere. Phys. Plasm. Am. Inst. Phys. **20**(3) (2013)
23. W. Tang, A. Mahalov, The response of plasma density to breaking inertial gravity wave in the lower regions of ionosphere. Phys. Plasm. Am. Inst. Phys. **21**(4) (2014)

24. R.K. Tyson, B.W. Frazier, Field Guide to Adaptive Optics (SPIE Press, 2004)
25. F. Wang, I. Toselli, O. Korotkova, Two spatial light modulator system for laboratory simulation of random beam propagation in random media. Appl. Opt. **55**(5), 1112–1117 (2016)
26. L.J. Wicker, W.C. Skamarock, Time splitting methods for elastic models using forward time schemes. Mon. Wea. Rev. **130**, 2088–2097 (2002)
27. K.S. Yee, Numerical solution of initial boundary values problems involving Maxwell's equations in isotopic media. IEEE Tran. Antennas Propagat. **14**(3), 302–307 (1966)

Weak Solutions to the Navier–Stokes Equations with Data in $\mathbb{L}(3, \infty)$

P. Maremonti

Abstract The paper concerns the existence of weak solutions to the $3d$-Navier–Stokes initial boundary value problem in exterior domains. The problem is considered with an initial data belonging to $\mathbb{L}(3, \infty)$ which is a special subspace of the Lorentz's space $L(3, \infty)$. The nature of the domain and the initial data in $L(3, \infty)$ make the result of existence not comparable with the usual Leray-Hopf theory of weak solutions. However, we are able to prove both that the weak solutions enjoy the partial regularity in the sense of Leray's structure theorem and the asymptotic limit of $|u(t)|_{3\infty}$.

Keywords Navier–Stokes equations · Weak solutions · Partial regularity

AMS Subject Classifications 35Q30 · 35D30 · 76D05

1 Introduction

For the 3-d Navier–Stokes theory a great interest is reserved to the mathematical problem of stability, in exterior domains, of steady solutions with respect to perturbations in $L(3, \infty)$, where by $L(3, \infty)$ we denote the Lorentz space. This interest arises both from physical and analytical reasons (cf. [15, 34]). However, no known result furnishes a theorem of global existence of solutions with data in $L(3, \infty)$. The question of the existence of solutions, defined for all $t > 0$ (as weak solutions) not necessarily belonging to L^2, has been recently considered in [25]. In [25] the initial data has finite Dirichlet norm, hence, *a priori*, it can be not bounded in $L^2(\Omega)$. In this note

Dedicated to Professor Y. Giga on his 60th birthday

P. Maremonti (✉)
Dipartimento di Matematica e Fisica, Seconda Università degli Studi di Napoli,
via Vivaldi 43, 81100 Caserta, Italy
e-mail: paolo.maremonti@unina2.it

Y. Maekawa and S. Jimbo (eds.), *Mathematics for Nonlinear Phenomena—Analysis and Computation*, Springer Proceedings in Mathematics & Statistics 215,
DOI 10.1007/978-3-319-66764-5_8

we give a further contribution to this question considering initial data in $\mathbb{L}(3,\infty)$.[1] Here the symbol $\mathbb{L}(3,\infty)$ denotes the completion of $\mathscr{C}_0(\Omega)$ in $L(3,\infty)(\Omega)$, where $\mathscr{C}_0(\Omega) := \{\varphi : \varphi \in C_0^\infty(\Omega) \text{ and } \nabla\cdot\varphi = 0\}$. Since $u_0 \in \mathbb{L}(3,\infty)$, in the case of an unbounded domain, *a priori*, a solution of problem (1.1) (below) does not have finite energy. So that our (weak or regular) solutions are not included in the set of so-called Leray-Hopf's weak solutions [19, 20]. Our study, in particular, is developed as premise to obtain, in a forthcoming paper jointly with F. Crispo, results of pointwise stability of the same kind of the one proved in [7]. The assumption $u_0 \in \mathbb{L}(3,\infty)$ seems the more coherent one in connection with the fact that the data is assigned as $O(|x|^{-\alpha})$, for $\alpha \in [1,3)$.

It is natural to inquire why we do not put the above questions in the framework of the L^q-theory of the Navier–Stokes equations initial boundary value problem, for which theorems of existence local in time are known (cf. [10, 14, 18]). The reason is that in exterior domains no result of global existence is known in L^q for $q > 3$. This aspect is not secondary for us in order to deduce the regularity of the weak solution in a suitable neighborhood of $t = \infty$.

In the case of the 2-*d* Navier–Stokes initial boundary value problem the equivalent question is formulated in the $L(2,\infty)$ Lorentz space. The question is studied in [18] where is proved for all $t > 0$ the existence of a unique regular solution. In the recent paper [32] is given for the Navier–Stokes Cauchy problem a result of existence of weak solution with data in $L^3(\Omega)$.

We conclude our introductive considerations saying that no connection is developed between the $L(3,\infty)$-setting and the regularity of weak solutions for which we refer e.g. to [30, 31].

In this paper we consider the Navier–Stokes initial boundary value problem

$$\begin{aligned} &u_t + u\cdot\nabla u + \nabla\pi_u = \Delta u,\ \nabla\cdot u = 0,\ \text{in } (0,T)\times\Omega,\\ &u = 0 \text{ on } (0,T)\times\partial\Omega,\ u(0,x) = u_0(x) \text{ on } \{0\}\times\Omega, \end{aligned} \tag{1.1}$$

where $\Omega \subseteq \mathbb{R}^3$ is a C^2-smooth domain and $u_0 \in \mathbb{L}(3,\infty)$. To better explain our aims, we introduce some notation and definitions. By $L^q(a,b;Y)$ we mean the space of functions $u : (a,b) \to Y$, where $(a,b) \subseteq \mathbb{R}$ and Y is a Banach space, such that $\int_a^b \|u\|_Y^q d\tau < \infty$. Analogously, by $C(a,b;Y)$ we mean the space of functions $u : (a,b) \to Y$ such that $\sup\limits_{(a,b)} \|u\|_Y < \infty$. We define $(u,g) := \int\limits_\Omega u\cdot g dx$. We set

$$\begin{aligned} X := \{&(u,\pi_u) : u \in C([0,T);\mathbb{L}(n,\infty)),\ tu_t,\ t^{\frac12}\nabla u \in L^\infty(0,T;L(n,\infty)),\ \text{and},\\ &\text{for } q\in(n,\infty],\ \mu := \tfrac{n}{2}\big(\tfrac1n - \tfrac1q\big),\ \overline{T} < T,\ t^{1+\mu}D^2u,\ t^{1+\mu}\nabla\pi \in L^\infty(0,\overline{T};L^q(\Omega))\}\,. \end{aligned}$$

[1]The results of this note were communicated in some meetings: http://www.math.sci.hokudai.ac.jp/sympo/sapporo/program150819_en.html, http://perso-math.univ-mlv.fr/users/danchin.raphael/Porq15/abstract15.pdf, https://www.maths.ox.ac.uk/events/past/2895.

Definition 1.1 A pair (u, π_u) solution to problem (1.1) that belongs to X is said regular.

Instead, for weak solution we mean:

Definition 1.2 Let $u_0 \in \mathbb{L}(3, \infty)$. A field $u : (t, x) \in (0, \infty) \times \Omega \to \mathbb{R}^3$ is said a weak solution to problem (1.1) if, for all $T > 0$,

i. $u \in L^2(0, T; W^{1,2}_{\ell oc}(\overline{\Omega})) \cap L^4(0, T; \mathbb{L}(3, \infty))$,

ii.

$$\int_0^T \big[(u,\varphi_t) - (\nabla u, \nabla\varphi) + (u \cdot \nabla\varphi, u)\big]d\tau + (u_0, \varphi(0)) = 0,$$

$$for\ all\ \varphi \in C_0^\infty([0, T) \times \Omega)\ with\ \nabla \cdot \varphi = 0,\ for\ all\ t \in (0, T),$$

iii. $(u, \nabla\phi) = 0$, *for all* $\phi \in C_0^\infty(\Omega)$ and $t \in (0, T)$,

iv. $\gamma(u) = 0$ in $H^{\frac{1}{2}}(\partial\Omega)$ a.e. in $t \in (0, T)$ and $\lim_{t\to 0}(u(t), \varphi) = (u_0, \varphi)$, for all $\varphi \in \mathscr{C}_0(\Omega)$.

In order to state the result of existence of a weak solution, we need to introduce two auxiliary problems:

$$\begin{aligned} &V_t - \Delta V = -\nabla\pi_V,\ \nabla \cdot V = 0 \text{ in } (0, T) \times \Omega,\\ &V = 0 \text{ on } (0, T) \times \partial\Omega \text{ and } V(0, x) = V_0 \text{ on } \{0\} \times \Omega, \end{aligned} \tag{1.2}$$

and

$$\begin{aligned} &v_t - \Delta v = -v \cdot \nabla v - v \cdot \nabla V - V \cdot \nabla v - V \cdot \nabla V - \nabla\pi_v,\\ &\nabla \cdot v = 0 \text{ in } (0, T) \times \Omega,\\ &v = 0 \text{ on } (0, T) \times \partial\Omega,\ \ v(0, x) = 0 \text{ on } \{0\} \times \Omega. \end{aligned} \tag{1.3}$$

Definition 1.3 A solution to problem (1.2) that belongs to X is said regular.

Definition 1.4 Let $V \in X$ in (1.3). A pair (v, π_v), such that $v : (0, \infty) \times \Omega \to \mathbb{R}^3$ and $\pi_v : (0, \infty) \times \Omega \to \mathbb{R}$, is said a suitable weak solution to problem (1.3) if, for all $T > 0$,

(i) $v \in L^2(0, T; J^{1,2}(\Omega))$ and, for some $q \in (1, \infty)$, $\pi_v \in L^q_{\ell oc}((0, T) \times \Omega)$,

$$\|v(t)\|_2^2 + 2\int_s^t \|\nabla v(\tau)\|_2^2 d\tau \le \|v(s)\|_2^2 + \int_s^t \big[(v \cdot \nabla v, V) - (V \cdot \nabla V, v)\big]d\tau < \infty,$$

$$\text{for all } t \ge s,\ \text{for } s = 0 \text{ and a.e. in } s \ge 0,$$

(ii) $\lim_{t\to 0} \|v(t)\|_2 = 0$,

(iii) for all $t, s \in (0, T)$, the pair (v, π_v) satisfies the equation:

$$\int_s^t \Big[(v, \varphi_\tau) - (\nabla v, \nabla\varphi) + ((v+V)\cdot\nabla\varphi, v+V) + (\pi_v, \nabla\cdot\varphi)\Big]d\tau$$
$$+ (v(s), \varphi(s)) = (v(t), \varphi(t)) \text{ for all } \varphi \in C_0^1([0,T)\times\Omega).$$

(iv) for all $t \geq s$, for $s = 0$ and a.e. in $s \geq 0$,

$$\int_\Omega |v(t)|^2\phi(t)dx + 2\int_s^t\int_\Omega |\nabla v|^2\phi dx d\tau \leq \int_\Omega |v(s)|^2\phi(s)dx + \int_s^t\int_\Omega |v|^2(\phi_\tau + \Delta\phi)dxd\tau$$
$$+ \int_s^t\int_\Omega \Big[(|v|^2 + V\cdot v)(v+V)\cdot\nabla\phi + 2(v+V)\cdot\nabla v\cdot V\phi + 2\pi_v v\cdot\nabla\phi\Big]dxd\tau,$$

for all nonnegative $\phi \in C_0^\infty(\mathbb{R}\times\Omega)$.

Now, we are in a position to state our chief results:

Theorem 1.1 *For all $u_0 \in \mathbb{L}(3,\infty)$ there exists a pair (u, π_u) where u is a weak solution to problem* (1.1) *and π_u is an associated pressure field, with $u := V + v$ and $\pi_u := \pi_V + \pi_v$, where (V, π_V) is the regular solution to problem* (1.2) *with initial data u_0 and (v, π_v) is a weak solution to problem* (1.3)*. In particular, $\pi_v := \pi_1 + \pi_2$ has $\nabla\pi_1 \in L^{\frac{5}{4}}((0,T)\times\Omega)$ and $\nabla\pi_2 \in L^q((0,T)\times\Omega)$, $q \in (\frac{3}{2}, \frac{5}{3})$. Moreover, there exist a $T(u_0) > 0$ such that $u \in C(0, T(u_0); \mathbb{L}(3,\infty))$, and a sequence $\{(\theta_\ell, T_\ell)\}_{\ell\in\mathbb{N}_0} \subset \mathbb{R}^+$, with $(\theta_0, T_0) \equiv (\theta_0, \infty)$, such that (u, π_u) is regular in $(\theta_\ell, T_\ell)\times\Omega$, and $\mathbb{R}_+ - \bigcup_{\ell\in\mathbb{N}_0}(\theta_\ell, T_\ell)$ has zero Lebesgue measure. Finally, we get $\lim_{t\to\infty} |u(t)|_{3\infty} = 0$.*

Corollary 1.1 *Let (u, π_u) be the weak solution of Theorem 1.1. Then (u, π_u) is a weak solution to the integral equation:*

$$\int_0^t \big[(u, \varphi_t) - (\nabla u, \nabla\varphi) + (u\cdot\nabla\varphi, u) + (\pi_u, \nabla\cdot\varphi)\big]d\tau = (u(t), \varphi(t)) - (u_0, \varphi(0)), \tag{1.4}$$

for all $\varphi \in C_0^1([0,T)\times\Omega)$, for all $t \in (0,T)$, and satisfies the generalized energy inequality:

$$\int_\Omega |u(t)|^2\phi(t)dx + 2\int_s^t\int_\Omega |\nabla u|^2\phi dx d\tau \leq \int_\Omega |u(s)|^2\phi(s)dx$$
$$+ \int_s^t\int_\Omega |u|^2(\phi_\tau + \Delta\phi)dxd\tau + \int_s^t\int_\Omega \Big[(|u|^2 + 2\pi_u)u\cdot\nabla\phi\Big]dxd\tau, \tag{1.5}$$

for all $t \geq s$, *for* $s = 0$ *and a.e. in* $s \geq 0$, *and for all nonnegative* $\phi \in C_0^\infty(\mathbb{R} \times \Omega)$.

Remark 1.1 We look for a solution to problem (1.1) of the kind $u := V + v$ with V solution to problem (1.2) and v solution to problem (1.3). Both the solutions are defined for all $t > 0$. However they are not comparable in a strict sense. Actually V and v have different integrability and regularity properties.

For the partial regularity in the sense of Caffarelli-Kohn-Nirenberg, we postpone the results related to our weak solution in a forthcoming paper jointly with F. Crispo.

In Theorem 1.1 an estimate of θ by means of $|u_0|_{n\infty}$ is not given. This is a difference with respect to the structure theorem by Leray for a weak solution. Actually, the chief difference with the L^2-setting is that our solution *a priori* does not admit a bound uniformly in time. This is due to the fact the solution exhibits in Theorem 1.1 is $u = V + v$ with v satisfies an energy inequality where the L^2-norm can grow for all $t > 0$. This also means that *a priori* the rest state can be not stable with respect a metric induced by the L^2 and $L(3, \infty)$ norms. However the rest state is attractive, thanks to the asymptotic behavior in $|u|_{3\infty}$.

2 Notation and Some Preliminary Results

We use the same symbol to denote vector or scalar functions. We set $J^q(\Omega) :=$ completion of $\mathscr{C}_0(\Omega)$ with respect to the norm of $L^q(\Omega)$, $q \in (1, \infty)$. The symbol $\mathscr{C}_0(\Omega)$ denotes the subset of $C_0^\infty(\Omega)$ whose elements are divergence free. By P_q (the index q is omitted when there is no confusion) we mean the projector from L^q into J^q. For properties and details on these spaces see for example [9]. By the symbol $L(p, q)$, $p \in [1, \infty)$, $q \in [1, \infty]$, we mean the Lorentz space [13, 35]. For $p = q = 1$, $L(1, 1) := L^1$. As well, for $p = q = \infty$, $L(\infty, \infty) := L^\infty$. For $p \in (1, \infty)$ and $q \in [1, \infty]$, we set $\mathbb{L}(p, q)$:=completion of $\mathscr{C}_0(\Omega)$ in $L(p, q)$. For $q \in [1, \infty]$ and $(r, s) \in [1, \infty) \times [1, \infty]$, by the symbols $\|\cdot\|_q$ and $|\cdot|_{rs}$ we denote the norm of the Lebesgue space and the one of the Lorentz space, respectively. Finally, we set $J^{1,q}(\Omega)$:=completion of $\mathscr{C}_0(\Omega)$ in $W^{1,q}(\Omega)$ and we denote by $\mathbb{W}(p, q)$:=completion of $C_0^\infty(\Omega)$ with respect to the norm $|\nabla \cdot|_{pq}$, provided that $p \in [1, n)$ and $q \in [1, p]$.

Lemma 2.1 *Let* $\nabla f \in L^q(\Omega)$, $q \in (1, n)$. *Then there exists constant* f_0 *and there exists a constant* c *independent of* f *such that*

$$\|f - f_0\|_{\frac{nq}{n-q}} \leq c\|\nabla f\|_q \,. \tag{2.1}$$

Proof. See [9]. □

Lemma 2.2 *Let* Ω *be an exterior domain with the cone property. Let* $m \in \mathbb{N}$ *and let* $q, r \in [1, \infty)$. *Let* $u \in L^q(\Omega)$ *and, for* $|\alpha| = m$, $D^\alpha u \in L^r(\Omega)$. *Then there exists a constant* c *independent of* u *such that*

$$\|D^\beta u\|_p \le c\|D^\alpha u\|_r^a \|u\|_q^{1-a}\,, \tag{2.2}$$

provided that for $|\beta| = j$ *the following relation holds:*

$$\frac{1}{p} = \frac{j}{n} + a\left(\frac{1}{r} - \frac{m}{n}\right) + (1-a)\frac{1}{q}\,,$$

with $a \in [\frac{j}{m}, 1]$ *either if* $p = 1$ *or if* $p > 1$ *and* $m - j - \frac{n}{p} \notin \mathbb{N} \cup \{0\}$, *while* $a \in [\frac{j}{m}, 1)$ *if* $p > 1$ *and* $m - j - \frac{n}{p} \in \mathbb{N} \cup \{0\}$.

The above lemma, proved in [6], gives an interpolation inequality of Gagliardo–Nirenberg's type. The difference with respect to the usual result is that the function u does not belong to a completion space of $C_0^\infty(\Omega)$.

Lemma 2.3 *Let* $D^2u \in L^q(\Omega)$ *and, for all bounded domain* $\Omega' \subset \Omega$ *such that* $\partial\Omega \cap \partial(\Omega - \Omega') = \emptyset$, *assume that* $u \in W^{1,q}(\Omega')$ *with zero trace on* $\partial\Omega$. *Finally, assume that* $\nabla \cdot u = 0$ *almost everywhere. Then there exists a pressure field* π_u *and a constant* c *independent of* u *such that*

$$\|D^2u\|_q + \|\nabla\pi_u\|_q + \|u\|_{W^{1,q}(\Omega')} \le c(\|P_q\Delta u\|_q + \|u\|_{L^q(\Omega')}). \tag{2.3}$$

Proof. For the proof see for example [9] or [27]. □

Lemma 2.4 *Let* $g \in \mathbb{W}(p,q)$, *with* $p \in [1,n)$ *and* $q \in [1,p]$. *Then, the following holds:*

$$|g|_{rq} \le A|\nabla g|_{pq}\,,\ r = \frac{np}{n-p},\ \textit{for all } g \in \mathbb{W}(p,q), \tag{2.4}$$

where $A := p\big(\Gamma(1+\frac{n}{2})\big)^{\frac{1}{n}}\big[\pi^{\frac{1}{2}}(n-p)\big]^{-1}$.

Proof. The above Sobolev's inequality is proved in [29]. The best constant A is produced in [3]. □

The following lemmas justify the application of Hölder's inequality in integrals involving functions of Lorentz's spaces.

Lemma 2.5 *Let be* $f \in L(p_1,q_1), g \in L(p_2,q_2)$, *with* $p_1, p_2 \in (1,\infty)$ *and* $q_1, q_2 \in [1,\infty]$.

i. If $\frac{1}{p_1} + \frac{1}{p_2} = 1 \le \frac{1}{q} = \frac{1}{q_1} + \frac{1}{q_2}$, *then*

$$|fg|_{1q} \le \int_\Omega |fg|dx \le c|f|_{p_1q_1}|g|_{p_2q_2}. \tag{2.5}$$

ii. If $\frac{1}{p_1} + \frac{1}{p_2} = \frac{1}{p}$ *and* $\frac{1}{q_1} + \frac{1}{q_2} \ge \frac{1}{q}$, *then*

$$|fg|_{pq} \le c|f|_{p_1q_1}|g|_{p_2q_2}. \tag{2.6}$$

In the inequalities c is independent of f, g, and for $p_1 = q_1$, $p_2 = q_2$, we have $c = 1$.

Proof. Estimate (2.5) is proved in [13, 29] and the best constant c is given in [16]. For estimates (2.6) see [13] section 4 p.271 or also [29]. □

Lemma 2.6 *Let $f \in L(p_1, q_1)$, $g \in L(p_2, q_2)$, $h \in L(p_3, q_3)$. If $\frac{1}{p_1} + \frac{1}{p_2} + \frac{1}{p_3} = 1$, and $\frac{1}{q_1} + \frac{1}{q_2} + \frac{1}{q_3} = 1$, then, the following inequality holds:*

$$\int_\Omega |fgh|dx \le c|f|_{p_1q_1}|g|_{p_2q_2}|h|_{p_3q_3}\,. \tag{2.7}$$

Proof. The estimate is a consequence of Lemma 2.5. Indeed, with obvious meaning of the symbols p', q', we get

$$\int_\Omega |fgh|dx \le c|fg|_{p'_3q'_3}|h|_{p_3q_3} \le c|f|_{p_1q_1}|g|_{p_2q_2}|h|_{p_3q_3}.$$

□

3 Results for the Stokes Problem

Let us consider the Stokes problem:

$$\begin{aligned} &w_t - \Delta w = -\nabla\pi_w + f,\ \nabla\cdot w = 0 \text{ in } (0,T)\times\Omega,\\ &w = 0 \text{ on } (0,T)\times\partial\Omega \text{ and } w(0,x) = w_0 \text{ on } \{0\}\times\Omega. \end{aligned} \tag{3.1}$$

We set

$$X_{pq} := \{u(t,x) : \text{a.e. in } t>0,\ u_t(t,x),\ D^\alpha u(t,x) \in L(p,q)(\Omega),\ |\alpha| \le 2\}.$$

Theorem 3.1 *Let be $f = 0$ in (3.1). For any $w_0 \in L(p,q)$, $1 < p < \infty$ and $1 \le q \le \infty$, problem (3.1) admits a unique solution $(u, \nabla\pi)$ with*

$$\begin{aligned} &u \in L^\infty(0,T; L(p,q)(\Omega)),\ \textit{with}\ \lim_{t\to 0}(u(t),\varphi) = (w_0,\varphi)\ \textit{for all}\ \varphi\in\mathscr{C}_0(\Omega),\\ &\textit{for each}\ \eta > 0\ \textit{and}\ r\in(p,\infty)\\ &(u,\nabla\pi_u)\in L^\infty(\eta,T;X_{pq})\cap C(\eta,T;J^{1,r}(\Omega))\times L^\infty(\eta,T;L(p,q)(\Omega)). \end{aligned} \tag{3.2}$$

Moreover, if $q \ne \infty$, the following estimate holds:

$$t|u_t(t)|_{pq} + t^{\mu_1}|\nabla u(t)|_{pq} + |u(t)|_{pq} \le c|w_0|_{pq}\,,\ t>0; \tag{3.3}$$

otherwise for all $q \in [1, \infty]$ *we get for* $p < r \le \infty$ *and* $1 \le s \le \infty$,

$$t^{\mu_2}|u_t(t)|_{rs} + t^{\mu_1}|\nabla u(t)|_{rs} + t^{\mu}|u(t)|_{rs} \le c|w_0|_{pq}\,,\ t > 0; \tag{3.4}$$

where $\mu = \frac{n}{2}\left(\frac{1}{p} - \frac{1}{r}\right)$, $\mu_2 = 1 + \mu$ *and* $\mu_1 = \begin{cases} \frac{1}{2} + \mu \text{ if } t \in (0, 1] \\ \frac{1}{2} + \mu \text{ if } t > 0 \text{ and } r \in [p, n], \\ \frac{n}{2p} \quad \text{ if } t > 1 \text{ and } r \ge n \end{cases}$ *and constant* c *is independent of* w_0. *Finally, if* $w_0 \in \mathbb{L}(p, \infty)$, *then, besides estimate* (3.4), *the following also holds:*

$$t|u_t(t)|_{p\infty} + t^{\mu_1}|\nabla u(t)|_{p\infty} + |u(t)|_{p\infty} \le c|w_0|_{p\infty},\ t > 0, \tag{3.5}$$

with c *independent of* w_0.

Proof. See for example [23]. □

Corollary 3.1 *Let be* $f = 0$ *in* (3.1). *Let* $u_0 \in \mathbb{L}(p, q)$ *and* $\{u_{0_k}\} \subset \mathscr{C}_0(\Omega)$ *converging to* u_0. *Then there exists a sequence* $\{(u_k, \pi_k)\}$ *of solutions to problem* (3.1) *that, for all* $(p, q) \in (1, \infty) \times [1, \infty]$, *enjoys the regularity properties* (3.2)–(3.5) *claimed in Theorem 3.1. Moreover, we get*

$$u \in C([0, T); \mathbb{L}(p, q)),\quad \lim_{t\to\infty} |u(t)|_{pq} = 0, \tag{3.6}$$

and

$$\begin{aligned} t|u_t(t) - u_{kt}(t)|_{pq} + t^{\mu_1}|\nabla u(t) - \nabla u_k(t)|_{pq} + |u(t) - u_k(t)|_{pq} \\ \le c|u_0 - u_{0k}|_{pq}\,,\ t > 0. \end{aligned} \tag{3.7}$$

For $p = q = \infty$, *we also get*

$$\|u_k(t)\|_\infty \le c\|u_{0k}\|_\infty\ t > 0, \tag{3.8}$$

with c *independent of* k. *If* $p < r \le \infty$ *and* $1 \le s \le \infty$, *we also get*

$$\begin{aligned} t^{\mu_2}|u_t(t) - u_{kt}(t)|_{rs} + t^{\mu_1}|\nabla u(t) - \nabla u_k(t)|_{rs} + t^{\mu}|u(t) - u_k(t)|_{rs} \\ \le c|u_0 - u_{0k}|_{pq}\,,\ t > 0; \end{aligned} \tag{3.9}$$

where $\mu = \frac{n}{2}\left(\frac{1}{p} - \frac{1}{r}\right)$, $\mu_2 = 1 + \mu$ *and* $\mu_1 = \begin{cases} \frac{1}{2} + \mu \text{ if } t \in (0, 1] \\ \frac{1}{2} + \mu \text{ if } t > 0 \text{ and } r \in [p, n], \\ \frac{n}{2p} \quad \text{ if } t > 1 \text{ and } r \ge n \end{cases}$ *and constant* c *is independent of* u_0 *and* k.

Proof. The proof is given in [26]. However in [26] the case $p = q = \infty$ is not considered. But estimate (3.8) is a consequence of the results proved in [1, 2] and in [24], see also [4]. □

We conclude the section with the following

Theorem 3.2 *Assume* $f \in L^q(0, T; L^q(\Omega))$ *and* $w_0 \in J^{2-\frac{2}{q},q}(\Omega)$, $q \geq 2$. *Then there exists a unique solution* (w, π_w) *to problem* (3.1) *such that*

$$\begin{aligned} &w \in C(0, T; J^q(\Omega)) \cap L^q(0, T; J^{1,q}(\Omega)),\\ &\int_0^T \big[\|w_t\|_q^q + \|D^2 w\|_q^q + \|\nabla \pi_w\|_q^q\big] d\tau \leq c(T)\Big[\int_0^T \|f\|_q^q d\tau + \|w_\circ\|^q_{2-\frac{2}{q},q}\Big], \end{aligned} \tag{3.10}$$

where, for all $\varepsilon > 0$, $c(T) := C(1 + T^{1+\varepsilon q - \frac{n}{2q}})$, *C independent of* f *and* $w_\circ$.

Proof. For the proof see e.g. [11, 28]. □

In the sequel we make use of the adjoint system of (3.1). More precisely, given a $t > 0$, on the cylinder $(0, t) \times \Omega$ we consider the system:

$$\begin{aligned} &\widehat{\varphi}_\tau + \Delta \widehat{\varphi} = -\nabla \pi_{\widehat{\varphi}}, \quad \nabla \cdot \widehat{\varphi} = 0, \text{ in } (0, t) \times \Omega,\\ &\widehat{\varphi} = 0 \text{ on } (0, t) \times \partial\Omega. \end{aligned} \tag{3.11}$$

Each solution $(\varphi(s, x), \pi_\varphi(s, x))$ of problem (3.1) given in Theorem 3.1 furnishes a solution of system (3.11) setting $\widehat{\varphi}(\tau, x) := \varphi(t - \tau, x)$, $(\tau, x) \in (0, t) \times \Omega$. Of course, by continuity in $C(0, T; \mathbb{L}(p, q))$, we get

$$\widehat{\varphi}(0, x) = \varphi(t, x) \text{ and } \widehat{\varphi}(t, x) = \varphi(0, x). \tag{3.12}$$

When we consider a solution to (3.11) of the kind $\widehat{\varphi}(\tau, x) := \varphi(t - \tau, x)$, then we always mean that $\varphi(0, x) = \varphi_0(x) \in \mathscr{C}_0(\Omega)$.

4 Existence and Some Properties of the Regular Solution to Problem (1.1)

The results of the following theorems and of the corollary are related to the existence and the regularity of solutions to problem (1.1) with initial data in $L(3, \infty)$. For these kind of results a well-known reference is [18], and one more recent is [26]. However in [18] is not considered $u_0 \in \mathbb{L}(n, \infty)$, thus estimates (4.1), (4.3) and (4.5) are not deduced. These estimates are crucial for the results of this note. For these motives Theorems 4.1, 4.2 and Corollary 4.1 are referred to [26].

Theorem 4.1 *For all* $u_0 \in \mathbb{L}(n, \infty)$ *there exist an interval* $[0, T)$ *and suitable functions* $A_q(t)$, *for* $q \in (n, \infty]$, *such that a unique solution* (u, π_u) *solves problem* (1.1), *and enjoys the properties:* $u \in C(0, T; \mathbb{L}(n, \infty))$, $t^{\frac{1}{2}} \nabla u \in L^\infty(0, T; L(n,\infty))$, *and*

$$\begin{aligned} &|u(t)|_{n\infty} \leq c|u_0|_{n\infty},\ t \in [0, T),\\ &t\ |u_t(t)|_{n\infty} \leq c|u_0|_{n\infty},\ t \in [0, T). \end{aligned} \tag{4.1}$$

Moreover, for $q \in (n, \infty]$, $\mu := \frac{n}{2}\big(\frac{1}{n} - \frac{1}{q}\big)$, $t^{1+\mu}D^2u, t^{1+\mu}\nabla\pi \in L^\infty(0, \overline{T}; L^q(\Omega))$, $\overline{T} < T \leq \infty$, *and*

$$\begin{aligned} \|u(t)\|_q &\leq A_q(t) + c|u_0|_{n\infty}t^{-\mu}, t \in [0, T),\\ t^{1+\mu}\ \|u_t(t)\|_q &\leq c|u_0|_{n\infty}, t \in [0, T). \end{aligned} \tag{4.2}$$

Theorem 4.2 *There exists an* $\eta_0 > 0$ *such that if* $u_0 \in \mathbb{L}(n, \infty)$ *with* $|u_0|_{n\infty} < \eta_0$, *then the solution* (u, π_u) *of Theorem 4.1 is defined for all* $t > 0$. *Moreover, we get*

$$\begin{aligned} |u(t)|_{n\infty} &\leq c|u_0|_{n\infty}, \text{ for all } t > 0 \text{ and } |u(t)|_{n\infty} = o(1),\\ t\ |u_t(t)|_{n\infty} &\leq c|u_0|_{n\infty}, \text{ for all } t > 0 \text{ and } t|u_t(t)|_{n\infty} = o(1), \end{aligned} \tag{4.3}$$

and

$$\begin{aligned} \|u(t)\|_q &\leq c|u_0|_{n\infty}t^{-\mu}, \text{ for all } t > 0 \text{ and } t^{-\mu}\|u(t)\|_q = o(1),\\ t^{1+\mu}\ \|u_t(t)\|_q &\leq c|u_0|_{n\infty}, \text{ for all } t > 0 \text{ and } \|t^{1+\mu}u(t)\|_q = o(1). \end{aligned} \tag{4.4}$$

For solutions of Theorem 4.1 we have

Corollary 4.1 *Let* (u, π_u) *be the solution to problem* (1.1) *obtained in Theorem 4.1. Then, for all* $\varepsilon > 0$ *there exist an interval* $(0, T(\varepsilon))$ *and a solution* u_ε, *enjoying properties* (4.1)–(4.2), *such that*

$$\begin{aligned} &|u(t) - u_\varepsilon(t)|_{n\infty} < \varepsilon\,, \text{ for all } t \in (0, T)\\ &\|u_\varepsilon(t)\|_{2n} \leq c\|u_\varepsilon(0)\|_{2n}, \text{ for all } t \in (0, T(\varepsilon)). \end{aligned} \tag{4.5}$$

To better explain the aim of Lemma 4.1 below, we point out the following arguments. In Section 5 we have to compare two solutions corresponding to the same data (say) $u_0 \in \mathbb{L}(3, \infty)$. One of these solutions belongs to the set X and is furnished by Theorem 4.1 or Theorem 4.2. The other is a weak solution. In order to perform the quoted comparison we need of a suitable "decomposition". That is for a solution u given in Theorem 4.1 we are going to prove that if

$$u_0 := V_0 + v_0, \text{ with } V_0 \in \mathbb{L}(3, \infty) \text{ and } v_0 \in J^2(\Omega), \tag{4.6}$$

then

$$u = V + v,$$

where V is solution to problem (3.1) with initial data V_0, and v verifies the following system:

$$\begin{aligned} &v_t - \Delta v = -u \cdot \nabla u - \nabla \pi_v,\ \nabla \cdot v = 0 \text{ in } (0, T) \times \Omega\\ &v = 0 \text{ on } (0, T) \times \partial\Omega,\ v(0, x) = v_0 \text{ on } \{0\} \times \Omega. \end{aligned} \tag{4.7}$$

In the 3-dimensional case we are interested in proving

Lemma 4.1 *Let u be a solution given in Theorem 4.1, with u_0 enjoying* (4.6). *Then, a solution v to* (4.7) *is such that*

$$\begin{aligned} & v \in C([0, T); J^2(\Omega)) \cap L^2(0, T; J^{1,2}(\Omega)) \\ & \textit{for all } \delta > 0,\ v_t, D^2 v, \in L^\infty(\delta, T; J^2(\Omega)). \end{aligned} \tag{4.8}$$

Proof. We start by proving that $v \in L^\infty(0, T; J^2(\Omega))$. To this end we multiply equation $(4.8)_1$ by $\widehat{\varphi}$ (for the definition of $\widehat{\varphi}$ see the end of Section 3). Integrating by parts on $(0, t) \times \Omega$, taking into (3.11) into account, we get the relation

$$(v(t), \varphi_0) = (v_0, \varphi(t)) + \int_0^t (u \cdot \nabla \widehat{\varphi}, u) d\tau =: I_1 + I_2.$$

Applying Hölder's inequality, we get

$$|I_1(t)| \leq \|v_0\|_2 \|\varphi(t)\|_2 \leq \|v_0\|_2 \|\varphi_0\|_2$$

Applying Hölder's inequality, then estimate (4.1) for u, and the semigroup properties $(3.3)_2$ of φ, we deduce ($r = 3, \sigma = 1$):

$$|I_2(t)| \leq c \int_0^t |u|^2_{3\infty} \|\nabla \widehat{\varphi}\|_{r\sigma} d\tau \leq c \|\varphi_0\|_2 \int_0^t |u|^2_{3\infty} (t-\tau)^{-\frac{3}{4}} d\tau \leq c \|\varphi_0\|_2 |u|^2_{3\infty} t^{\frac{1}{4}}.$$

The estimates for I_1 and I_2 imply that

$$\|v(t)\|_2 \leq \|v_0\|_2 + c|u|^2_{3\infty} t^{\frac{1}{4}}\ ,\ t \in (0, T)\,. \tag{4.9}$$

We remark that from the regularity of u it follows that

$$\|u \cdot \nabla u\|_2 \leq \|u\|_4 \|\nabla u\|_4 \leq c t^{-\frac{3}{4}} \|u_0\|^2_{3\infty}\,, t \in (0, T). \tag{4.10}$$

Hence we have $tu \cdot \nabla u \in L^2(0, T; L^2(\Omega))$. Multiplying equation $(4.1)_1$ by t, setting $w := tv$, we deduce that:

$$\begin{aligned} & w_t - \Delta w = -tu \cdot \nabla u + v + \nabla \pi_w,\ \nabla \cdot w = 0 \text{ in } (0, T) \times \Omega \\ & w = 0 \text{ on } (0, T) \times \partial\Omega,\ w(0, x) = 0 \text{ on } \{0\} \times \Omega. \end{aligned} \tag{4.11}$$

Estimates (4.9) and (4.10) ensure that $tu \cdot \nabla u + v \in L^2(0, T; L^2(\Omega))$. By virtue of Theorem 3.2 we deduce that the pair $(w, \pi_w) \equiv (tv, t\pi_v)$ enjoyed the properties (3.10). Hence, via estimates (3.10) on tv and tv_t, and via estimate (2.3), we deduce that, for all $\delta > 0$, $D^2 v, \nabla \pi_v \in L^2(\delta, T; L^2(\Omega))$. The above estimate in particular implies that

$$\text{for all } s > 0, \quad v \in C(s, T; J^2(\Omega)) \cap L^2(s, T; J^{1,2}(\Omega))\,. \tag{4.12}$$

Now, we complete the proof of (4.8) making $s = 0$ in (4.12). Since (4.9) ensures that $v \in J^2(\Omega)$ for all $t \in [0, T)$, taking properties of u into account, it is straightforward to deduce, via the equation $(4.7)_1$, that, for all $\phi \in J^2(\Omega)$, $(v(t), \phi))$ is a continuous function on $[0, T)$. So, via (4.9), by a classical trick we get

$$\begin{aligned} \lim_{s\to 0} \|v(s) - v_0\|_2^2 &= \lim_{s\to 0} \Big[\|v(s)\|_2^2 + \|v_0\|_2^2 - 2(v(s), v_0)) \Big] \\ &\leq 2\|v_0\|_2^2 + c \lim_{s\to 0} |u_0|_{3\infty}^4 s^{\frac{1}{2}} - 2 \lim_{s\to 0} (v(s), v_0)) = 0. \end{aligned} \tag{4.13}$$

Multiplying $(4.7)_1$ by v and integrating on $(s, t) \times \Omega$, we get the following energy relation:

$$\|v(t)\|_2^2 + 2\int_s^t \|\nabla v\|_2^2 d\tau = \|v(s)\|_2^2 - \int_s^t \big[(v \cdot \nabla v, V) - (V \cdot \nabla v, V)\big] d\tau\,. \tag{4.14}$$

Since $u_0 \in \mathbb{L}(3, \infty)$, by virtue of Corollary 3.1 we find a sequence $\{(V^k, \pi_{V^k})\}$ such that

$$|V(t) - V^k(t)|_{3,\infty} \leq c|u_0 - V^k(0)|_{3\infty}, \ \text{for all } t > 0.$$

Hence, applying Hölder's inequality, employing estimate (2.4) and the semigroup properties for V^k, we deduce $(r = 6, \sigma = 2)$

$$\begin{aligned} |(v \cdot \nabla v, V)| &\leq c|v|_{r\sigma} \|\nabla v\|_2 |V - V^k|_{3\infty} + \|v\|_2 \|\nabla v\|_2 \|V^k\|_\infty \\ &\leq c\|\nabla v\|_2^2 |u_0 - V^k(0)|_{3\infty} + \frac{1}{2}\|\nabla v\|_2^2 + \|v\|_2^2 \|V^k(0)\|_r^2 t^{-\frac{3}{r}}, \ t \in (0, T). \end{aligned}$$

Moreover, applying Hölder's inequality and the semigroup properties for V, we get

$$|(V \cdot \nabla v, V)| \leq c\|V\|_4^2 \|\nabla v\|_2 \leq \frac{1}{2}\|\nabla v\|_2^2 + |u_0|_{3\infty}^4 t^{-\frac{1}{2}}, \ t \in (0, T)\,.$$

By the above estimates, setting k sufficiently large in such a way that $c|u_0 - V^k(0)|_{3\infty} < \frac{1}{2}$, we prove that

$$\|v(t)\|_2^2 + \frac{1}{2}\int_s^t \|\nabla v\|_2^2 d\tau \leq \|v(s)\|_2^2 + \int_s^t \big[\|v\|_2^2 \tau^{-\frac{3}{r}} + |u_0|_{3\infty}^4 \tau^{-\frac{1}{2}}\big] d\tau\,, \ \text{for all } s > 0.$$

The last estimate and the limit property (4.13) ensure that the limit in $s = 0$ is well posed on the right-hand side of (4.14). We have completed the proof of the lemma. □

Corollary 4.2 *Let u be a solution given in Theorem 4.1. Then, v satisfies the energy relation* ($s \geq 0$)*:*

$$\|v(t)\|_2^2 + 2\int_s^t \|\nabla v\|_2^2 d\tau = \|v(s)\|_2^2 + \int_s^t \big[(v \cdot \nabla v, V) + (V \cdot \nabla v, V)\big] d\tau . \quad (4.15)$$

Proof. The proof trivially follows from the properties in Lemma 4.1 and employing the dominate convergence theorem for (4.14). □

5 Comparison of the Regular Solution and of One Weak

We consider a pair of solutions to (1.1), say (u, π_u) and $(\overline{u}, \pi_{\overline{u}})$, both with initial data

$$u_0 \in \mathbb{L}(3, \infty),\ u_0 := V_0 + v_0,\ \text{ with } V_0 \in \mathbb{L}(3, \infty),\ v_0 \in J^2(\Omega). \quad (5.1)$$

We assume that $\overline{u}$ enjoys the regularity claimed in Theorem 4.1 and in Corollary 4.1, and the pair $(\overline{v}, \pi_{\overline{v}})$, given in the decomposition $\overline{u} = \overline{v} + V$, enjoys the regularity claimed in Section 4. We define

$$\mathscr{W} := \big\{\varphi : \varphi \in C([0, T); J^{1,2}(\Omega)),\ \varphi_t \in L^2(0, T; L^2(\Omega)) \\ \text{and } \varphi(t, x) = 0 \text{ in a neighborhood of } T\big\}.$$

For the solution $\overline{u}$ we assume that

$$\overline{u} := w + V, \quad (5.2)$$

where

i. V is the solution to problem (3.1) with initial data V_0;
ii. $w \in L^\infty(0, T; J^2(\Omega)) \cap L^2(0, T; J^{1,2}(\Omega))$, with $w_0 := v_0 = u_0 - V_0 \in J^2(\Omega)$, and

$$\|w(t)\|_2^2 + 2\int_s^t \|\nabla w\|_2^2 d\tau \leq \|w(s)\|_2^2 + \int_s^t \big[(w \cdot \nabla w, V) + (V \cdot \nabla w, V)\big] d\tau ,$$

for $t > s$ and a.e. in $s \geq 0$.

iii.

$$\int_s^t \Big[(w,\varphi_\tau) - (\nabla w,\nabla\varphi) + (w\cdot\nabla\varphi, w) + (w\cdot\nabla\varphi, V) + (V\cdot\nabla\varphi, w) + (V\cdot\nabla\varphi, V)\Big]d\tau$$
$$+ (w(s),\varphi(s)) = (w(t),\varphi(t))\,, \quad \text{for all } \varphi(t,x) \in \mathscr{W}(\Omega_T).$$

Lemma 5.1. *Let w be the function given in items ii.–iii.. Then*

$$\lim_{t\to s^+} \|w(t) - w(s)\|_2 = 0, \ \textit{a. e. in } s > 0 \textit{ and } s = 0. \tag{5.3}$$

Proof. Applying Hölder's inequality, we get ($r = 6, \sigma = 2$)

$$|(w\cdot\nabla w, V)| \le c|w|_{r\sigma}\|\nabla w\|_2 |V|_{3\infty} \le c\|\nabla w\|_2^2 |u_0|_{3\infty}\,,$$
$$|(V\cdot\nabla w, V)| \le \|V\|_4^2 \|\nabla w\|_2 \le c|u_0|_{3\infty}^2 t^{-\frac14}\|\nabla w\|_2\,,$$

where for the term $|w|_{r\sigma}$ we have used the Sobolev inequality (2.4). So that by the integrability hypotheses on $\|\nabla w\|_2$ we easily deduce

$$\limsup_{t\to s^+} \|w(t)\|_2 \le \|w(s)\|_2\,.$$

Hence (5.3) follows via the arguments employed for (4.13). □

Theorem 5.1 *Let $\overline{u}$ and u be two solutions corresponding to the same data u_0 satisfying* (5.1). *Assume that $\overline{u}$ is the regular solution of Theorem 4.1 and u is given by* (5.2) *with w enjoying items i.–iii. Then u and $\overline{u}$ coincide on the common interval of existence.*

Proof. The proof follows the technique employed in the Leray–Sather–Serrin uniqueness theorems (see [20, 33]). In these uniqueness theorems it is assumed that a solution is smooth and the other is weak. By virtue of Theorem 4.1, for all $t > 0$, $(\overline{u}, \pi_{\overline{u}})$ is a regular solution to problem (1.1). The same regularity is enjoyed by V corresponding to u_0. Moreover, recalling that we can write $\overline{v} = \overline{u} - V$, $\overline{v}$ is a solution to problem (4.7) with the properties given in Lemma 4.1 and Corollary 4.2. We multiply equation $(4.7)_1$ by w and integrate by parts on $(s,t)\times\Omega$:

$$\int_s^t \Big[(\overline{v}_\tau, w) + (\nabla\overline{v}, \nabla w) + (u\cdot\nabla u, w)\Big]d\tau = 0. \tag{5.4}$$

Let $h(s,\tau) \ge 0$ be a smooth function such that $h(s,\tau) = 0$ for $\tau \le \frac{s}{2}$ and $h(s,\tau) = 1$ for $\tau \ge s$. For all $s > 0$, we have $hv \in \mathscr{W}(\Omega_T)$. Hence, for all $t, s > 0$, we can replace $h\overline{v}$ to the test function φ in the weak formulation for w in item iii.. After integrating by parts, we get

$$\int_s^t \Big[(w, \overline{v}_\tau) - (\nabla w, \nabla \overline{v}) - (w \cdot \nabla w, \overline{v}) - (w \cdot \nabla V, \overline{v}) - (V \cdot \nabla w, \overline{v}) - (V \cdot \nabla V, \overline{v})\Big] d\tau$$
$$+ (w(s), \overline{v}(s)) = (w(t), \overline{v}(t)).$$

Making the difference of the above identities, we get

$$2\int_s^t (\nabla w, \nabla \overline{v}) d\tau + (w(t), \overline{v}(t)) - (w(s), \overline{v}(s)) = -\int_s^t (u \cdot \nabla u, w)$$
$$-(w \cdot \nabla w, \overline{v}) - (w \cdot \nabla V, \overline{v}) - (V \cdot \nabla w, \overline{v}) - (V \cdot \nabla V, \overline{v})\Big] d\tau. \tag{5.5}$$

Summing the energy relation (4.15) related to v and the energy relation ii. related to w with equation (5.5) changed in sign, we get

$$\frac{1}{2}\|\overline{v}(t)\|_2^2 + \frac{1}{2}\|w(t)\|_2^2 + \int_s^t \Big[\|\nabla \overline{v}(\tau)\|_2^2 + \|\nabla w(\tau)\|_2^2 - 2\int_s^t (\nabla \overline{v}, \nabla w)\Big] d\tau$$
$$- \frac{1}{2}\|\overline{v}(s)\|_2^2 - \frac{1}{2}\|w(s)\|_2^2 - (\overline{v}(t), w(t)) + (\overline{v}(s), w(s))$$
$$\leq \int_s^t (u \cdot \nabla u, w) + (w \cdot \nabla w, \overline{v}) + (w \cdot \nabla V, \overline{v}) + (V \cdot \nabla w, \overline{v}) + (V \cdot \nabla V, \overline{v})$$
$$+ (w \cdot \nabla w, V) + (V \cdot \nabla w, V) + (\overline{v} \cdot \nabla \overline{v}, V) + (V \cdot \nabla \overline{v}, V)\Big] d\tau.$$

Setting $U = \overline{v} - w$, since $\overline{u} = \overline{v} + V$ the last inequality is equivalent to

$$\|U(t)\|_2^2 + 2\int_s^t \|\nabla U(\tau)\|_2^2 d\tau \leq \|U(s)\|_2^2 + 2\int_s^t \big[(U \cdot \nabla U, \overline{v}) + (U \cdot \nabla U, V)\big] d\tau$$
$$= \|U(s)\|_2^2 + 2\int_s^t (U \cdot \nabla U, \overline{u}) d\tau.$$

Applying Hölder's inequality ($r = 6$, $\sigma = 2$) and (2.4), we get

$$|(U \cdot \nabla U, \overline{u})| \leq c|U|_{r\sigma}\|\nabla U\|_2 |\overline{u}|_{3\infty} \leq c\|\nabla U\|_2^2 |\overline{u}|_{3\infty}, \text{ a.e. } \tau \in (0, T).$$

Since $\overline{u} \in C(0, T; \mathbb{L}(3, \infty))$, we can perform the limit as $s \to 0$. Recalling that $\|\overline{v}(s) - v_0\|_2 \to 0$ via (4.9), and $\|w(s) - v_0\|_2 \to 0$ via (5.3) we obtain

$$\|U(t)\|_2^2 + 2\int_0^t \|\nabla U(\tau)\|_2^2 d\tau \leq 2\int_0^t (U\cdot\nabla U, \overline{u})d\tau. \tag{5.6}$$

Employing Corollary 4.1, applying Hölder's and Sobolev's inequalities (2.4), for all $t \in (0, T(\varepsilon))$, we obtain ($r = 6, \sigma = 2$)

$$\begin{aligned}\|U(t)\|_2^2 + 2\int_0^t \|\nabla U(\tau)\|_2^2 d\tau &\leq 2c\int_0^t |U|_{r\sigma}\|\nabla u\|_2|\overline{u} - u_\varepsilon|_{3\infty} d\tau + 2\int_0^t \|U\|_3\|\nabla U\|_2\|u_\varepsilon\|_6 d\tau \\ &\leq 2(c\varepsilon + \frac{1}{4})\int_0^t \|\nabla U\|_2^2 d\tau + c\|u_\varepsilon(0)\|_6^4\int_0^t \|U\|_2^2 d\tau.\end{aligned}$$

Therefore the uniqueness follows on $(0, T(\varepsilon)]$. For $t \in [T(\varepsilon), T)$ the result is trivial. Actually, $\overline{u} \in C(T(\varepsilon), T; J^q(\Omega))$, for all $q > 3$, so that (5.6) becomes

$$\|U(t)\|_2^2 + 2\int_0^t \|\nabla U(\tau)\|_2^2 d\tau \leq 2\int_{T(\varepsilon)}^t (U\cdot\nabla U, \overline{u})d\tau \leq 2\int_{T(\varepsilon)}^t \|U\|_3\|\nabla U\|_2\|\overline{u}\|_6 d\tau,$$

for which the uniqueness follows. The theorem is completely proved. □

6 Existence of Weak Solutions to Problem (1.3)

Here, as already considered, (V, π_V) is the solution to problem (3.1) and initial data u_0 whose existence is ensured by Theorem 3.1. Instead v is the solution to problem (1.3). The properties of V make $V \cdot \nabla V \in L^2(0, T; H^{-1}(\Omega))$. We prove the existence of a weak solution to problem (1.3) as suitable limit as $\varepsilon \to 0$ of solutions v^ε to problem

$$\begin{aligned}&v_t^\varepsilon - \Delta v^\varepsilon = -J_\varepsilon(v^\varepsilon)\cdot\nabla v^\varepsilon - v^\varepsilon\cdot\nabla V - V\cdot\nabla v^\varepsilon - V\cdot\nabla V - \nabla\pi_{v^\varepsilon}, \\ &\nabla\cdot v^\varepsilon = 0 \text{ in } (0, T)\times\Omega \\ &v^\varepsilon = 0 \text{ on } (0, T)\times\partial\Omega,\ v^\varepsilon(0, x) = 0 \text{ on } \{0\}\times\Omega,\end{aligned} \tag{6.1}$$

where J_ε is the Friederich mollifier. We follow this approach, which is due to Leray [20], since for solutions to problem (1.3) we are interested in proving an energy inequality in strong form and the one of generalized kind proved in [5].

The solution to problem (6.1) can be studied by means of the so-called Faedo-Galerkin method. Since the technique is well known we limit ourselves to prove the chief estimates in order to obtain the limit solution v^ε to problem (6.1) (see subsection 6.1). To perform the proof we need some lemmas.

Lemma 6.1 *Let $v \in J^{1,2}(\Omega)$, $g \in L^2(\Omega)$, and $V \in \mathbb{L}(3, \infty) \cap L^\infty(\Omega)$ with $\nabla V \in L(3, \infty)$. Then, the following estimates hold:*

$$\begin{aligned} |(J_\varepsilon(v) \cdot \nabla v, g)| &\le c(\varepsilon)\|v\|_2^2\|\nabla v\|_2^2 + \tfrac{1}{8}\|g\|_2^2, \\ |(v \cdot \nabla V, g)| &\le c\|\nabla V\|_{3\infty}^2\|v\|_2\|P\Delta v\|_2 + \tfrac{1}{8}\|g\|_2^2, \\ |(V \cdot \nabla v, g)| &\le c\|V\|_\infty^2\|v\|_2\|P\Delta v\|_2 + \tfrac{1}{8}\|g\|_2^2, \\ \textit{for } r = 6, \sigma = 2, \; |(V \cdot \nabla V, g)| &\le c\|V\|_{r\sigma}^2\|\nabla V\|_{3\infty}^2 + \tfrac{1}{8}\|g\|_2^2. \end{aligned} \tag{6.2}$$

Proof. Applying Hölder's inequality, we get

$$\begin{aligned} |(J_\varepsilon(v) \cdot \nabla v, g)| &\le c(\varepsilon)\|v\|_2\|\nabla v\|_2\|g\|_2, \\ \text{for } r = 6, \sigma = 2, \; |(v \cdot \nabla V, g)| &\le c\|\nabla V\|_{3\infty}\|v\|_{r\sigma}\|g\|_2, \\ |(V \cdot \nabla v, g)| &\le c\|V\|_\infty\|\nabla v\|_2\|g\|_2, \\ \text{for } r = 6, \sigma = 2, \; |(V \cdot \nabla V, g)| &\le c\|V\|_{r\sigma}\|\nabla V\|_{3\infty}\|g\|_2. \end{aligned}$$

Estimates $(6.2)_{1,4}$ are deduced via Cauchy inequality. Using (2.4) for the term $|v|_{r\sigma}$, $(6.2)_{2,3}$ follow by applying the inequality $\|\nabla v\|_2 \le \|v\|_2\|P\Delta v\|_2$ and Cauchy inequality. □

Lemma 6.2 *Assume that $v \in L^\infty(0, T; J^2(\Omega)) \cap L^2(0, T; J^{1,2}(\Omega))$ with a bound $A(T) < \infty$. Let be V the solution to problem* (3.1) *with initial data $u_0 \in \mathbb{L}(3, \infty)$ given in Theorem 3.1. Then, we get*

$$\begin{aligned} &\int_0^T \|v \cdot \nabla v\|_{\frac{5}{4}}^{\frac{5}{4}} d\tau \le A^{\frac{5}{2}}(T), \quad \int_0^T \|V \cdot \nabla v\|_{\frac{5}{4}}^{\frac{5}{4}} d\tau \le c|u_0|_{3\infty}^{\frac{5}{4}} A^{\frac{5}{4}}(T) T^{\frac{5}{16}}, \\ &\int_0^T \|v \cdot \nabla V\|_{\frac{5}{4}}^{\frac{5}{4}} d\tau \le c|u_0|_{3\infty}^{\frac{5}{4}} A^{\frac{5}{4}}(T) \max\{1, T^{\frac{7}{16}}\}. \end{aligned} \tag{6.3}$$

Proof. By the Gagliardo–Nirenberg inequality (2.2) we deduce $\|v\|_{\frac{10}{3}} \le c\|\nabla v\|_2^{\frac{3}{5}}\|v\|_2^{\frac{2}{5}}$. Hence, applying Hölder's inequality, we easily get

$$\int_0^T \|v \cdot \nabla v\|_{\frac{5}{4}}^{\frac{5}{4}} d\tau \le \Big[\int_0^T \|v\|_{\frac{10}{3}}^{\frac{10}{3}} d\tau\Big]^{\frac{3}{8}} \Big[\int_0^T \|\nabla v\|_2^2 d\tau\Big]^{\frac{5}{8}} = cA^{\frac{5}{2}}(T).$$

Applying Holder's inequality, by virtue of (3.4), we get

$$\int_0^T \|V \cdot \nabla v\|_{\frac{5}{4}}^{\frac{5}{4}} d\tau \le \Big[\int_0^T \|V\|_{\frac{10}{3}}^{\frac{10}{3}} d\tau\Big]^{\frac{3}{8}} \Big[\int_0^T \|\nabla v\|_2^2 d\tau\Big]^{\frac{5}{8}} \le c\|u_0\|_{3\infty}^{\frac{5}{4}} A^{\frac{5}{4}}(t) T^{\frac{5}{16}}.$$

Finally, applying Hölder's inequality, by virtue of (3.4), we obtain

$$\int_0^T \|v\cdot\nabla V\|_{\frac{5}{4}}^{\frac{5}{4}}d\tau \le c\int_0^T \|v\|_2^{\frac{5}{4}}\|\nabla V\|_{\frac{10}{3}}^{\frac{5}{4}}d\tau \le c\|u_0\|_{3\infty}^{\frac{5}{4}}A^{\frac{5}{4}}(t)\max\{1,T^{\frac{7}{16}}\}.$$

The lemma is proved. □

Lemma 6.3 *Let $V(t,x)$ be the solution to problem* (3.1) *with initial data $u_0 \in \mathbb{L}(3,\infty)$ given in Theorem 3.1. Then, for $q\in(\frac{3}{2},\frac{5}{3})$, we get*

$$\int_0^T \|V\cdot\nabla V\|_q^q d\tau \le c|u_0|_{3\infty}^{2q}T^{\frac{5}{2}-\frac{3q}{2}}. \tag{6.4}$$

Proof. By Hölder's inequality, we easily get

$$\int_0^T \|V\cdot\nabla V\|_q^q d\tau \le \int_0^T |V|^q_{\frac{3q}{3-q},1}|\nabla V|^q_{3\infty}d\tau \le c|u_0|_{3\infty}^{2q}\int_0^T \tau^{\frac{3}{2}-\frac{3q}{2}}d\tau,$$

which proves the lemma. □

For all $R>0$, we set $\Omega_R := \Omega\cap B_R$ and $\Omega_R^c := \Omega - \Omega_R$.

Lemma 6.4 *Let $\{v^h\}$ be a sequence included in a bounded set of $L^2(0,T;W^{1,2}(\Omega)) \cap L^\infty(0,T;L^2(\Omega))$. Let $V\in C([0,T);\mathbb{L}(3,\infty))$, and assume the existence of a sequence $\{V^k\}\subset C([0,T);L^q(\Omega))$, such that, for all $q\in[3,\infty]$,*

$$\|V^k(t)\|_q \le c\|V_0^k\|_q,\ \textit{for all}\ t>0, \tag{6.5}$$

which converges to V in $C([0,T);\mathbb{L}(3,\infty))$. Then, the sequence $\{v^h\}$ admits weak limit $v\in L^2(0,T;W^{1,2}(\Omega))\cap L^\infty(0,T;L^2(\Omega))$ and the following property holds:

$$\textit{for all}\ s\ge 0,\quad \lim_h \int_s^t (J_h(v^h)\cdot\nabla v^h, V)d\tau = \int_s^t (v\cdot\nabla v, V)d\tau. \tag{6.6}$$

Proof. First we recall that in our hypotheses on $\{v^h\}$, the sequence $\{v^h\}$ admits a subsequence, again labeled by h, which weakly converges in $L^2(0,T;J^{1,2}(\Omega))$ and, weakly-star, in $L^\infty(0,T;L^2(\Omega))$ (see e.g. [36]). We denote the limit by v. By virtue of Friederich's lemma, for each ball $B_R\subset\Omega$, $\{v^h\}$ converges strongly to v in $L^2(0,T;L^2(\Omega\cap B_R))$. This result and the properties of the mollifier ensure that

$$\lim_h \int_0^T \|J_h(v^h - v)\|^2_{L^2(\Omega_R)} d\tau \le \lim_h \int_0^T \|v^h - v\|^2_{L^2(\Omega_{2R})} d\tau = 0. \tag{6.7}$$

We set $I_h := (J_h(v^h) \cdot \nabla v^h, V)$. Then, we perform the decomposition

$$\begin{aligned}
I_h &= (J_h(v^h) \cdot \nabla v^h, V - V^k) + (J_h(v^h) \cdot \nabla v^h, V^k) \\
&= B_1 + (J_h(v^h - v) \cdot \nabla v^h, V^k)_{\Omega_R} + (J_h(v^h - v) \cdot \nabla v^h, V^k)_{\Omega_R^c} + (J_h(v) \cdot \nabla v^h, V^k) \\
&= B_1 + B_2 + B_3 + ((J_h(v) - v) \cdot \nabla v^h, V^k) + (v \cdot \nabla v^h, V^k) \\
&= B_1 + B_2 + B_3 + B_4 + (v \cdot \nabla (v^h - v), V^k) + (v \cdot \nabla v, V^k) \\
&= B_1 + B_2 + B_3 + B_4 + B_5 + (v \cdot \nabla v, V^k - V) + (v \cdot \nabla v, V) \\
&= \sum_{i=1}^{6} B_i + (v \cdot \nabla v, V).
\end{aligned}$$

Since we have

$$\lim_h \int_0^T I_h(\tau) d\tau - \int_0^T (v \cdot \nabla v, V) = \lim_k \lim_R \lim_h \sum_{i=1}^{6} \int_0^T B_i(h, R, k, \tau) d\tau,$$

we are proving that

$$\lim_k \lim_R \lim_h \int_0^T B_i(h, R, k, \tau) d\tau = 0, \text{ for all } i = 1, \cdots, 6.$$

In the following we mean $r = 6$, $\sigma = 2$ and $s \geq 0$. Applying Hölder's inequality, Lemma 2.4, by the strong convergence of $\{V^k\}$ and uniform bound for $\{v_h\}$, uniformly in $R > 0$, we get

$$\begin{aligned}
|\lim_k \lim_h \int_s^t B_1(h, k) d\tau| &\le \lim_k \lim_h \int_0^t |v^h|_{r\sigma} \|\nabla v^h\|_2 |V - V^k|_{3\infty} \\
&\le \lim_k \sup_{(0,T)} |V - V^k|_{3\infty} \lim_h \int_0^t \|\nabla v^h\|_2^2 d\tau = 0.
\end{aligned}$$

Applying Hölder's inequality, via assumption $(6.5)_2$, by virtue of (6.7), we easily deduce

$$|\lim_k \lim_h \int_s^t B_2(h,k,R)d\tau| \leq \lim_k \lim_h \int_s^t \|v^h - v\|_{L^2(\Omega_{2R})}\|\nabla v^h\|_2 |V^k|_\infty d\tau$$

$$\leq c\|V_0^k\|_\infty \lim_h \int_s^t \|v^h - v\|_{L^2(\Omega_{2R})}\|\nabla v^h\|_2 = 0.$$

We set $\|V^k\|_{q,R} := \|V^k\|_{L^q(|x|>R)}$. Applying Hölder's inequality and estimate (2.2), then Cauchy's inequality, we get

$$|B_3(h,k,R)| \leq \|v^h - v\|_3 \|\nabla v^h\|_2 \|V^k\|_{6,R} \leq c\big[\|\nabla v^h\|_2^{\frac{3}{2}} + \|\nabla v\|_2^{\frac{3}{2}}\big]\|v^h - v\|_2^{\frac{1}{2}}\|V^k\|_{6,R}\,.$$

Recalling that $\|v^h\|_2^2 + \|v\|_2^2 + \int_0^T (\|\nabla v^h\|_2^2 + \|\nabla v\|_2^2)d\tau \leq M^2 < \infty$, applying Hölder's inequality on (s,t), we deduce

$$|\lim_k \lim_R \lim_h \int_s^t B_3(h,k,R)d\tau| \leq cM^2 \lim_k \lim_R \Big[\int_s^t \|V^k\|_{6,R}^4 d\tau\Big]^{\frac{1}{4}}.$$

Applying for all $k \in \mathbb{N}$ Beppo Levi's theorem to the last integral, for $R \to \infty$, finally we get

$$\lim_k \lim_R \lim_h \int_s^t B_3(h,k,R)d\tau = 0\,.$$

Since the sequence $\{J_h(v)\}$ strongly converges to v in $L^2(0,T;W^{1,2}(\Omega))$, applying Hölder's inequality and Sobolev's inequality (2.4), by virtue of assumption (6.5), we easily get

$$\lim_k \lim_R \lim_h \int_s^t |B_4(h,R,k)|d\tau \leq \lim_k \lim_h \int_s^t |J_h(v) - v|_{r\sigma}\|\nabla v^h\|_2 |V^k|_{3\infty} d\tau$$

$$\leq c \lim_k |V_0^k|_{3\infty} \lim_h \int_s^t \|\nabla v - \nabla J_h(v)\|_2 \|\nabla v^h\|_2 d\tau = 0.$$

Since $v \otimes V^k \in L^2(0, T; L^2(\Omega))$, by the weak convergence of $\{\nabla v^h\}$, we get

$$\lim_k \lim_R \lim_h \int_s^t B_5(h, k, R)d\tau = 0.$$

By the strong convergence of $\{V^k\}$ to V in $\mathbb{L}(3,\infty)$, we get

$$\lim_k \lim_R \lim_h \int_s^t B_6(k, R, h)d\tau = 0.$$

The lemma is proved. □

6.1 *The Solution to the Approximate Problem* (6.1)

We establish the existence of a solution to problem (6.1) by means of the Galerkin method in the way suggested in [13]. We limit ourselves to give the chief estimates in order to obtain a regular solution. The estimates concern a generic Galerkin solution, say $v^{j\varepsilon}$. For simplicity we suppress the labels j, ε. However, some results pointed towards the j-th Galerkin solution hold uniformly in $\varepsilon > 0$ too.

Lemma 6.5 *In* $(6.1)_1$ *assume that V is the solution to problem* (3.1) *with initial data $u_0 \in \mathbb{L}(3,\infty)$ given in Theorem 3.1. Then, uniformly in $j \in N$ and $\varepsilon > 0$ we get*

$$\|v(t)\|_2^2 + \int_0^t \|\nabla v\|_2^2 d\tau \le c e^{C(u_0)} C(u_0) + c|u_0|_{3\infty}^4 t^{\frac{1}{2}}\,,\ t > 0. \tag{6.8}$$

and

$$\|v(t)\|_2^2 + 2\int_0^t \|\nabla v\|_2^2 d\tau = 2\int_0^t (v\cdot\nabla v, V)d\tau + 2\int_0^t (V\cdot\nabla v, V)d\tau\,, \tag{6.9}$$

Proof. Multiplying the Galerkin approximate solution to $(6.1)_1$ by $v^{j\varepsilon}$, an integration by parts furnishes:

$$\|v(t)\|_2^2 + 2\int_0^t \|\nabla v(\tau)\|_2^2 d\tau = \int_0^t \big[(v\cdot\nabla v, V) + (V\cdot\nabla v, V)\big]d\tau\,.$$

We have

$$|(v\cdot\nabla v, V)| \le |(v\cdot\nabla v, V - V^k)| + |(v\cdot\nabla v, V^k)|,$$

where V^k is given in Corollary 3.1. Hence, applying Hölder's inequality and estimate (2.4), we get ($r = 6, \sigma = 2$)

$$\begin{aligned}|(v\cdot\nabla v, V)| &\le c|v|_{r\sigma}\|\nabla v\|_2|V - V^k|_{3\infty} + \|v\|_2\|\nabla v\|_2\|V^k\|_\infty\\ &\le (c|u_0 - V^k(0)|_{3\infty} + \frac{1}{4})\|\nabla v\|_2^2 + \|v\|_2^2\|V^k\|_\infty^2.\end{aligned}$$

Since we can consider $V^k(0, x) = u_0^k \in J^p(\Omega)\cap J^q(\Omega)$, with $3 \in (p, q)$, for k sufficiently large, then estimate (3.7) ensures that

$$|(v\cdot\nabla v, V)| \le \frac{1}{2}\|\nabla v\|_2^2 + \|v\|_2^2\|V^k\|_\infty^2. \tag{6.10}$$

Moreover, estimate (3.4) furnishes $\|V^k(t)\|_\infty \le c\|u_0^k\|_q t^{-\frac{1}{q}}$, hence the first term on the right of the energy relation is well posed uniformly with respect to $j \in \mathbb{N}$ and $\varepsilon > 0$. For the latter integral term we easily get

$$|(V\cdot\nabla v, V)| \le \|V\|_4^4 + \frac{1}{4}\|\nabla v\|_2^2. \tag{6.11}$$

Moreover, employing estimate (3.4) we deduce that $\|V(t)\|_4^4 \le |u_0|_{3\infty}^4 t^{-\frac{1}{2}}$, $t > 0$. Hence the latter term is also well posed for all $j \in \mathbb{N}$ and $\varepsilon > 0$. Then, via (6.10) and (6.11) we deduce the energy inequality

$$\|v(t)\|_2^2 + \frac{1}{4}\int_0^t \|\nabla v\|_2^2 d\tau \le \int_0^t (\|v\|_2^2\|V^k\|_\infty^2 + \|V\|_4^4)d\tau\,.$$

Since $V^k(0) = u_0^k \in J^p(\Omega)\cap J^q(\Omega)$, via estimate (3.4), for $r = s = \infty$, we get the following:

$$\begin{aligned}\int_0^\infty \|V^k(t)\|_\infty^2 dt &= \int_0^h \|V^k(t)\|_\infty^2 dt + \int_h^\infty \|V^k(t)\|_\infty^2 dt\\ &\le c\|u_0^k\|_q^2 h^{1-\frac{3}{q}} + \|u_0^k\|_p^2 h^{1-\frac{3}{p}} = c\|u_0^k\|_q^{q\frac{2}{3}\frac{3-p}{q-p}}\|u_0^k\|_p^{p\frac{2}{3}\frac{q-3}{q-p}} =: C(u_0),\end{aligned}$$

provided that $h := \big[\|u_0^k\|_p/\|u_0^k\|_q\big]^\alpha$, with $\alpha := \frac{2}{3}\frac{pq}{q-p}$. Recalling that $v(0) = 0$, we

$$\|v(t)\|_2^2 \le e^{C(u_0^k)}\int_0^t \|V\|_4^4 d\tau \le ce^{C(u_0)}|u_0|_{3\infty}^4 t^{\frac{1}{2}}\,,$$

and the energy estimate:

$$\|v(t)\|_2^2 + \frac{1}{4}\int_0^t \|\nabla v\|_2^2 d\tau \le c e^{C(u_0)} C(u_0) + c|u_0|_{3\infty}^4 t^{\frac{1}{2}}\,,$$

which holds for all $j \in \mathbb{N}$ and $\varepsilon > 0$. Hence uniformly in $j \in \mathbb{N}$ and $\varepsilon > 0$ we can write (6.9) and then estimate (6.8). □

The following result concerns uniquely *estimates of the Galerkin approximation sequence* $\{v^{j\varepsilon}\}$:

Lemma 6.6 *In the same hypotheses of Lemma 6.5, for all* $\varepsilon > 0$, *we get uniformly in* $j \in \mathbb{N}$ *the following estimate:*

$$t\|\nabla v(t)\|_2^2 + \int_0^t \tau\big[\|P\Delta v\|_2^2 + \|v_t\|_2^2\big]d\tau \le c\int_0^t \tau\|v\|_2^2\Big[c(\varepsilon)\|\nabla v\|_2^2 + \|\nabla V\|_{3\infty}^4 + \|V\|_\infty^4\Big]d\tau + c\int_0^t \tau\Big[|V|_{r\sigma}^2|\nabla V|_{3\infty}^2 + \|\nabla v\|_2^2\Big]d\tau. \tag{6.12}$$

Proof. In the following we set $r = 6$, $\sigma = 2$. We multiply the Galerkin approximate $(6.1)_1$ firstly by $P\Delta v$ and then by t. Integrating by parts and employing estimate (6.2), via Cauchy inequality, we get

$$t\|\nabla v(t)\|_2^2 + \frac{1}{2}\int_0^t \tau\|P\Delta v\|_2^2 d\tau \le c\int_0^t \tau\|v\|_2^2\Big[c(\varepsilon)\|\nabla v\|_2^2 + \|\nabla V\|_{3\infty}^4 + \|V\|_\infty^4\Big]d\tau + c\int_0^t \tau\Big[|V|_{r\sigma}^2|\nabla V|_{3\infty}^2 + \|\nabla v\|_2^2\Big]d\tau.$$

Analogously, multiplying (6.1) a first time by v_t and then by t, after an integration by parts and employing again (6.2), we get

$$t\|\nabla v(t)\|_2^2 + \int_0^t \tau\|v_t\|_2^2 d\tau \le \int_0^t \tau\|P\Delta v\|_2^2 d\tau + c\int_0^t \tau\|v\|_2^2\Big[c(\varepsilon)\|\nabla v\|_2^2 + \|\nabla V\|_{3\infty}^4 + \|V\|_\infty^4\Big]d\tau + c\int_0^t \tau\Big[|V|_{r\sigma}^2|\nabla V|_{3\infty}^2 + \|\nabla v\|_2^2\Big]d\tau.$$

Multiplying this last by $\frac{1}{4}$ and summing with the one obtained for $P\Delta v$, we formally deduce (6.12). So that we have to justify the integrability of the right-hand side of (6.12). Thanks to the energy relation, our analysis can be reduced to the terms with V and ∇V for which the semigroup properties (3.3) hold. Therefore, taking into account that for all the terms the singularity is in $t = 0$, and recalling the existence of the weight t and estimate (6.8) for $\|v(t)\|_2$, we can claim that the right-hand side is well posed. □

The above lemmas imply the following result of existence:

Lemma 6.7 *In* (6.1) *assume that V is the solution to problem* (3.1) *with data u_0 given in Theorem 3.1. For all $\varepsilon > 0$, there exists a regular solution v^ε to problem* (6.1) *as limit of the Galerkin approximation. In particular we get*

$$v^\varepsilon \in C([0,T); J^2(\Omega)) \cap L^2(0,T; J^{1,2}(\Omega)),$$
$$\textit{for all } \delta > 0,\ v_t^\varepsilon,\ D^2 v^\varepsilon,\ \nabla \pi_{v^\varepsilon} \in L^2(\delta, T; L^2(\Omega)).$$

6.2 *The Convergence* $\{v^\varepsilon\}$ *to a Weak Solution* v *of* (1.3)

Our result of existence of a weak solution (v, π_v) to problem (1.3) will be achieved by proving that $\{v^\varepsilon\} \subset L^2(0,T; J^{1,2}(\Omega))$ is uniformly bounded in $\varepsilon > 0$, $(v(t), \psi)$ is continuous in $t \geq 0$ for all $\psi \in J^2(\Omega)$, and the validity of the energy inequalities i.–iv. of Definition 1.4. To this end we premise the following results.

Lemma 6.8 *In* (1.3) *assume that V is the solution to problem* (3.1) *with initial data $u_0 \in \mathbb{L}(3,\infty)$ given in Theorem 3.1. Then, uniformly in $\varepsilon > 0$, we get*

$$\|v^\varepsilon(t)\|_2^2 + \int_0^t \|\nabla v^\varepsilon\|_2^2 d\tau \leq c e^{C(u_0)} C(u_0) + c|u_0|_{3\infty}^4 t^{\frac{1}{2}}\,,\ t > 0. \tag{6.13}$$

Proof. Estimate (6.13) is inequality (6.8). □

Lemma 6.9 *By the same assumptions of Lemma 6.8, uniformly in $\varepsilon > 0$, up to a function $c(t)$, we have $\pi_{v^\varepsilon} = \pi_1^\varepsilon + \pi_2^\varepsilon$ with*

$$\int_0^t \|\pi_1^\varepsilon\|_{\frac{15}{7}}^{\frac{5}{4}} d\tau \leq A_1(u_0,t) \ \textit{and} \ \int_0^t \|\pi_2^\varepsilon\|_{\frac{3q}{3-q}}^{q} d\tau \leq A_2(u_0,t)\,,\ q \in (\frac{3}{2}, 2). \tag{6.14}$$

Proof. We consider a solution $(v^\varepsilon, \pi_{v^\varepsilon}) \equiv (v_1^\varepsilon + v_2^\varepsilon, \pi_1^\varepsilon + \pi_2^\varepsilon)$, where

$$\begin{aligned}
&v^\varepsilon_{1t}-\Delta v^\varepsilon_1=-\nabla\pi^\varepsilon_1+f^\varepsilon_1,\quad \nabla\cdot v^\varepsilon_1=0,\quad \text{in } (0,T)\times\Omega\\
&v^\varepsilon_1=0 \ \text{ on } (0,T)\times\partial\Omega\quad v^\varepsilon_1(0,x)=0,\\
&v^\varepsilon_{2t}-\Delta v^\varepsilon_2=-\nabla\pi^\varepsilon_2+f^\varepsilon_2,\quad \nabla\cdot v^\varepsilon_1=0,\quad \text{in } (0,T)\times\Omega\\
&v^\varepsilon_2=0 \ \text{ on } (0,T)\times\partial\Omega\quad v^\varepsilon_2(0,x)=0,
\end{aligned}$$

with

$$f^\varepsilon_1:=-J_\varepsilon(v^\varepsilon)\cdot\nabla v^\varepsilon-V\cdot\nabla v^\varepsilon-v^\varepsilon\cdot\nabla V\,,\quad \text{and}\quad f^\varepsilon_2:=-V\cdot\nabla V\,.$$

By virtue of Lemmas 6.2 and 6.3, and employing estimates (3.5) for V and (6.9) for v, for $q\in(\frac32,\frac53)$, we get

$$\int_0^t\|f^\varepsilon_1\|^{\frac54}_{\frac54}d\tau\le A_1(u_0,t) \text{ and } \int_0^t\|f^\varepsilon_2\|^q_q d\tau\le A_2(u_0,t)\,,\ \text{uniformly in } \varepsilon>0.$$

So that, by virtue of Theorem 3.2 we deduce the existence of $(v^\varepsilon_1,\pi^\varepsilon_1)$ and $(v^\varepsilon_2,\pi^\varepsilon_2)$, and via estimate (3.10), employing (2.1), we get (6.14) respectively for π^ε_1 and π^ε_2. By uniqueness, up to a function $c(t)$, we get $\pi_{v^\varepsilon}=\pi^\varepsilon_1+\pi^\varepsilon_2$. The lemma is completely proved. □

Lemma 6.10 *By the same assumptions of Lemma 6.8, for all $t>0$, we get*

$$\lim_{R\to\infty}\|v^\varepsilon(t)\|_{L^2(\Omega^c_R)}=0\,,\ \textit{uniform in } \varepsilon>0\,. \tag{6.15}$$

Proof. Let h_R be a smooth function with range $[0,1]$ such that

$$h_R=0 \text{ for } |x|\le R,\quad h_R=1 \text{ for } |x|\ge 2R,\quad \text{and}\quad |\nabla h_R|\le cR^{-1}.$$

We multiply (6.1) by $h^2_R v^\varepsilon$. Integrating by parts on $(0,t)\times\Omega$ we obtain (we delete ε for simplicity)

$$\begin{aligned}
\tfrac12\|v(t)h_R\|_2^2+\int_0^t\|\nabla v h_R\|_2^2d\tau&=\int_0^t(v^2,\Delta h^2_R)d\tau+\int_0^t(\big[J_\varepsilon(v)+V\big]\cdot\nabla h^2_R,|v|^2)d\tau\\
&+\int_0^t((v+V)\cdot\nabla V,vh^2_R)d\tau+\int_0^t(\pi_{v^\varepsilon}v,\nabla h^2_R)d\tau=\sum_{i=1}^4 I_i(t,R,\varepsilon)\,.
\end{aligned}\tag{6.16}$$

By virtue of (6.8), we deduce that

$$I_1(t,R,\varepsilon)\le c(u_0,t)R^{-2},\ \text{for all } \varepsilon>0\,.$$

Applying Hölder's inequality and estimate (2.2), by virtue of the energy relation (6.8), and the semigroup properties of V, we get

$$I_2(t, R, \varepsilon) \leq cR^{-1} \int_0^t \big[\|v\|_3^3 + \|V\|_\infty \|v\|_2^2\big] d\tau \leq c(u_0, t)R^{-1}, \text{ for all } \varepsilon > 0.$$

Integrating by parts, we get

$$I_3 = -\int_0^t [((v \cdot \nabla v, Vh_R^2) + (V \cdot \nabla v, Vh_R^2) + ((v + V) \cdot v, V \cdot \nabla h_R^2)] d\tau = \int_0^t \sum_{i=1}^3 I_3^i d\tau.$$

By virtue of Corollary 3.1, applying Holder's inequality and estimate (2.4), for $r = 6$, $s = 1$, uniformly in $\varepsilon > 0$, we deduce

$$\begin{aligned}|I_3^1| &\leq |vh_R|_{rs} \|\nabla vh_R\|_2 |V - V^k|_{3\infty} + \|v\|_6 \|\nabla v\|_2 \|h_R V^k\|_3 \\ &\leq \frac{1}{2}\|\nabla vh_R\|_2^2 + cR^{-1}\|v\|_2^2 \|V - V^k\|_{3\infty} + c\|\nabla v\|_2^2 \|h_R V^k\|_3,\end{aligned}$$

where we have considered k sufficiently large such that $|V - V^k|_{3\infty} < \frac{1}{2c}$ for all $t > 0$. Applying Cauchy's inequality, for I_3^2 we obtain

$$|I_3^2| \leq \frac{1}{4}\|\nabla vh_R\|_2^2 + \|Vh_R\|_4^4, \text{ for all } \varepsilon > 0.$$

Applying Hölder's inequality, we get

$$|I_3^3| \leq cR^{-1}\big[\|v\|_2^2 \|V_\infty + \|V\|_4^2 \|v\|_2\big], \text{ for all } \varepsilon > 0.$$

Remarking that, for all $t > 0$, $\|V^k h_R\|_3$ and $\|Vh_R\|_4$ are monotone functions in R with

$$\lim_R \|V^k h_R\|_3 = \lim_R \|Vh_R\|_4 = 0,$$

by Beppo Levi's theorem, we get

$$\Big|\lim_R \int_0^t I_3(\tau)d\tau\Big| \leq \lim_R \int_0^t \big[|I_3^1(\tau)| + |I_3^2(\tau)| + |I_3^3(\tau)|\big] d\tau = 0, \text{ for all } \varepsilon > 0.$$

Finally, thanks to Lemma 6.9, applying Hölder's inequality, we get

$$|I_4| \le \int_0^t |((\pi_1+\pi_2)v, \nabla h_R^2)|d\tau$$

$$\le cR^{-\frac{9}{10}} \int_0^t \|\pi_1\|_{\frac{15}{7}} \|v\|_2 d\tau + cR^{\frac{3}{2}-\frac{3}{q}} \int_0^t \|\pi_2\|_{\frac{3q}{3-q}} \|v\|_2 d\tau\,, \text{ for all } \varepsilon > 0.$$

The above estimates for I_i, $i = 1, \cdots, 4$ and (6.16) furnish (6.15). □

Lemma 6.11 *Let* $V \in X$ *in* (1.3). *Then there exists a weak solution in the sense of Definition 1.4 and the following bound holds*

$$\|v(t)\|_2^2 + \int_0^t \|\nabla v\|_2^2 d\tau \le ce^{C(u_0)}C(u_0) + c|u_0|_{3\infty}^4 t^{\frac{1}{2}}\,,\ t > 0. \tag{6.17}$$

Moreover, we can associate a pressure field $\pi_v := \pi_1 + \pi_2$ *with*

$$\int_0^t \|\pi_1\|_{\frac{15}{7}}^{\frac{5}{4}} d\tau \le A_1(u_0,t) \ \ and \ \int_0^t \|\pi_2\|_{\frac{3q}{3-q}}^q d\tau \le A_2(u_0,t)\,,\ q \in (\frac{3}{2}, 2), \tag{6.18}$$

where $A_i(u_0,t), i = 1, 2,$ *are given in* (6.14)

Proof. We can limit all the considerations to the existence, for $\varepsilon \to 0$, of (v, π_v) as limit of $(v^\varepsilon, \pi_{v^\varepsilon})$. The existence of the weak limit π_v is a consequence of estimates (6.14), in particular we get (6.18). As far as v is concerned, we limit ourselves to the convergence of the energy inequalities i. and iv. in strong form, and limit property ii. related to Definition 1.4. Actually, energy inequality i. is crucial for the convergence of $\{v^\varepsilon\}$ to v, in order to regard v as solution to the integral equation iii. of Definition 1.3. By virtue of Lemma 6.8 we have that, for all $T > 0$, $\{v^\varepsilon\}$ is bounded in $L^2(0,T; J^{1,2}(\Omega))$. Hence there exists a weak limit v in $L^2(0,T; J^{1,2}(\Omega))$ and it attains the bound (6.17). In order to prove i. of Definition 1.4, we consider the limit of the energy relation

$$\|v^\varepsilon(t)\|_2^2 + 2\int_s^t \|\nabla v^\varepsilon\|_2^2 d\tau = \|v^\varepsilon(s)\|_2^2$$

$$+ 2\int_s^t (v^\varepsilon \cdot \nabla v^\varepsilon, V)d\tau + 2\int_s^t (V \cdot \nabla v^\varepsilon, V)d\tau \ t \ge s, \text{ a.a. } s \ge 0. \tag{6.19}$$

Of course, we have

$$\|v(t)\|_2^2 + 2\int_s^t \|\nabla v\|_2^2 d\tau \le \lim_{\varepsilon\to 0}{}' \Big[\|v^\varepsilon(t)\|_2^2 + 2\int_s^t \|\nabla v^\varepsilon\|_2^2 d\tau\Big].$$

Employing Lemma 6.10, we get

$$\|v^\varepsilon(s)\|_2^2 = \|v^\varepsilon(s)\|^2_{L^2(B_R)} + \|v^\varepsilon(s)\|^2_{L^2(\Omega - B_R)} \le \|v^\varepsilon(s)\|^2_{L^2(B_R)} + o(s, R)\,.$$

Then, by Friederich's lemma for $\{v^\varepsilon\} \subset L^2(0, T; J^{1,2}(\Omega))$, we get

$$\lim_{\varepsilon\to 0} \|v^\varepsilon(s)\|_2^2 \le \|v(s)\|_{L^2(B_R)} + o(s, R), \text{ a.e. in } s > 0 \text{ and for } s = 0.$$

Hence, in the limit as $R \to \infty$, we deduce

$$\lim_{\varepsilon\to 0} \|v^\varepsilon(s)\|_2^2 \le \|v(s)\|_2^2, \text{ a.e. in } s > 0 \text{ and for } s = 0.$$

By virtue of Lemma 6.4, for all $s \ge 0$, we get

$$\lim_{\varepsilon\to 0} \int_s^t (v^\varepsilon \cdot \nabla v^\varepsilon, V) d\tau = \int_s^t (v \cdot \nabla v, V) d\tau\,.$$

Finally, since $V \otimes V \in L^2((0, T) \times \Omega)$, then by weak convergence we get

$$\lim_{\varepsilon\to 0} \int_s^t (V \cdot \nabla v^\varepsilon, V) d\tau = \int_s^t (V \cdot \nabla v, V) d\tau\,.$$

Taking into account the above limit properties, we deduce the energy inequality in strong form:

$$\|v(t)\|_2^2 + 2\int_s^t \|\nabla v\|_2^2 d\tau \le \|v(s)\|_2^2 + 2\int_s^t (v \cdot \nabla v, V) d\tau + 2\int_s^t (V \cdot \nabla v, V) d\tau$$
$$t \ge s, \text{ a.e. in } s > 0 \text{ and for } s = 0. \qquad (6.20)$$

As it is known, estimate (6.20) is enough to prove that v satisfies item iii. of Definition 1.4. On the other hand, integral equation iii., for all $T > 0$ and $\psi \in J^2(\Omega)$, gives $(v(s), \psi) \in C(0, T)$. Hence, by a classical trick via the continuity of $(v(t), \psi)$ and (6.20) we deduce item ii.. The last step is the proof of the generalized energy inequality (item iv.) of Definition 1.4. Letting $\varepsilon \to 0$, the claim is achieved proving the convergence of the terms in the following relation:

$$\int_\Omega |v^\varepsilon(t)|^2\phi(t) + 2\int_s^t\int_\Omega |\nabla v^\varepsilon|^2\phi = \int_\Omega |v^\varepsilon(s)|^2\phi(s)dx + \int_s^t\int_\Omega |v^\varepsilon|^2(\phi_\tau + \Delta\phi)dxd\tau$$
$$+\int_s^t\int_\Omega \Big[(|v^\varepsilon|^2 + V\cdot v^\varepsilon)(v^\varepsilon + V)\cdot\nabla\phi + 2(v^\varepsilon + V)\cdot\nabla v^\varepsilon\cdot V\phi + 2\pi_{v^\varepsilon}v^\varepsilon\cdot\nabla\phi\Big]dxd\tau,$$

for all $t \geq s$, for $s = 0$ and a.e. in $s \geq 0$, and for all nonnegative $\phi \in C_0^\infty(\mathbb{R}\times\Omega)$. In the case of $V = 0$ the convergence is developed in some papers (cf. e.g. [5, 8, 21]). Hence we can limit ourselves to prove the convergence of the terms including V:

$$\int_s^t\int_\Omega \Big[(|v^\varepsilon|^2 + V\cdot v^\varepsilon)V\cdot\nabla\phi + V\cdot v^\varepsilon v^\varepsilon\cdot\nabla\phi\Big]dxd\tau + \int_s^t\int_\Omega V\cdot\nabla v^\varepsilon\cdot V\phi)dxd\tau\,.$$

Since $\{v^\varepsilon\}$ strongly converges to v in $L^2(0,T;L^2(\Omega_R))$, $\{v^h\}\subset L^\infty(0,T;L^2(\Omega)$, and $v\in L^\infty(0,T;L^2(\Omega))$, via estimate (2.2) we easily get that, for all $r\in(1,\infty)$ strongly converges in $L^r(0,T:L^2(\Omega))$ and, for all $s\in[2,6)$, in $L^{\frac{4s}{3(s-2)}}(0,T;L^s(\Omega))$. Hence, applying Höder's inequality and properties (3.4) for V, we get

$$\lim_{\varepsilon\to0}|\int_s^t\int_\Omega (|v^\varepsilon|^2 - |v|^2)V\cdot\nabla\phi dxd\tau| = \lim_{\varepsilon\to0}|\int_s^t\int_\Omega\Big[|v^\varepsilon - v|^2 + 2(v^\varepsilon - v)\cdot v\Big]V\cdot\nabla\phi dxd\tau|$$
$$\lim_{\varepsilon\to0}\int_s^t\Big[\|v^\varepsilon - v\|_2^2\|V\|_\infty + 2\|v^\varepsilon - v\|_2\|v\|_2\|V\|_\infty d\tau\Big] = 0.$$

Moreover, since $\{v^\varepsilon\}$ weakly converges to v in $L^2(0,T;L^2(\Omega))$ and, by virtue of (3.4), we have $V\otimes V\in L^2(0,T;L^2(\Omega))$, we immediately deduce

$$\int_s^t\int_\Omega\Big[V\cdot(v^\varepsilon - v)V\cdot\nabla\phi dxd\tau = 0.$$

By analogous arguments one can discuss the remaining integral term. We consider achieved the proof. □

7 Proof of Theorem 1.1 and Corollary 1.1

Proof of Theorem 1.1
(1) Existence. We are looking for $u := V + v$ and $\pi_u := \pi_V + \pi_v$, where (V,π_V), with initial data $u_0\in\mathbb{L}(3,\infty)$, is the solution established in Theorem 3.1 and (v,π_v)

is the weak solution to problem (1.3) established in Lemma 6.11. Therefore we can satisfy all the items of Definition 1.2. Moreover, V and v enjoy the integrability properties stated in Theorem 1.1.

Now we establish the partial regularity of u in the sense of the structure theorem.

(2) Initial condition.

First of all we prove that

$$\lim_{t\to 0}|u(t)-u_0|_{3\infty}=0\,. \tag{7.1}$$

Actually, for all $u_0\in\mathbb{L}(3,\infty)$, we can consider two solutions: one is a weak solution whose existence is achieved in the above item (1), another is the local regular solution whose existence is proved in Theorem 4.1. On the other hand, by virtue of the uniqueness proved in Section 5, these solutions coincide on $(0,T(u_0))$, which is the interval of existence of the regular solution. Hence our weak solution assumes the initial data by continuity in the $\mathbb{L}(3,\infty)$-norm and in this way (7.1) is proved. □

(3) Regularity

Now, we prove the structure theorem analogous to the one proved by Leray. Since $u=V+v$, then, for all $T>0$, we get in particular $u\in L^4(0,T;\mathbb{L}(3,\infty))$. Indeed, we only have to verify that $v\in L^4(0,T;\mathbb{L}(3,\infty))$. But this is true since $v\in L^2(0,T;J^{1,2}(\Omega))$. As a consequence we get that almost everywhere in $t>0$, say for $t\in\Theta$ and $(0,\infty)-\Theta$ with zero measure, $u(t,x)\in\mathbb{L}(3,\infty)$. Now, we prove that for all $s\in\Theta$ there exists an interval of regularity for (u,π_u) of the kind $(s,s+T)$. Since for $s\in\Theta$ we have $u(s,x)\in\mathbb{L}(3,\infty)$, by virtue of Theorem 4.1 we prove the existence on some interval $(s,s+T(u(s)))$ of a regular solution, say, $(\overline{u},\pi_{\overline{u}})$ to problem (1.1). Our result is achieved if we are able to prove the coincidence between (u,π_u) and $(\overline{u},\pi_{\overline{u}})$ on the interval $(s,s+T(u(s)))$. Since $\overline{u}(s,x)=u(s,x)$, then we get $\overline{u}(s,x)-V(s,x)=u(s,x)-V(s,x)=v(s,x)\in J^2(\Omega)$. We set $\overline{v}(t,x):=\overline{u}(t,x)-V(t,x)$. By virtue of Lemma 4.1, keeping in mind that $\overline{v}(s,x)\equiv v(s,x)\in J^2(\Omega)$, then we deduce that $\overline{v}$ enjoys properties (4.8). As a consequence of Corollary 4.2 we obtain for $\overline{v}$

$$\|\overline{v}(t)\|_2^2+2\int_s^t\|\nabla\overline{v}\|_2^2d\tau=\|\overline{v}(s)\|_2^2-\int_s^t\big[(\overline{v}\cdot\nabla\overline{v},V)-(V\cdot\nabla\overline{v},V)\big]d\tau\,. \tag{7.2}$$

Hence we are in a position to apply the uniqueness theorem of Section 5. Therefore we have proved the coincidence of $(\overline{u},\pi_{\overline{u}})$ with (u,π_u). This gives the first part of the structure theorem. For the second part, that is the existence of the interval (T_0,∞) of regularity, we prove that there exists an instant $T_0>0$ such that $|u(T_0)|_{3\infty}$ is sufficiently small so that we can apply Theorem 4.2 for the result with small data. Since $u=V+v$, we prove the smallness of V and v separately and in different ways. Since $u_0\in\mathbb{L}(3,\infty)$ we can apply the result of Corollary 3.1, so that (3.6) implies

$$\lim_{t\to\infty}|V(t)|_{3\infty}=0\,.$$

Hence we get

$$\text{for all } \varepsilon>0, \text{ there exists a } \overline{T}_0>0 \text{ such that } |V(t)|_{3\infty}<\varepsilon,\ \forall t\geq \overline{T}_0. \tag{7.3}$$

Now we consider $v(t,x)$. We start from the energy relation (6.20) written for $t>0$ and $s=0$:

$$\tfrac{1}{2}\|v(t)\|_2^2+\int_0^t\|\nabla v\|_2^2d\tau\leq\int_0^t(v\cdot\nabla v,V)d\tau+\int_0^t(V\cdot\nabla v,V)d\tau=:I_1+I_2. \tag{7.4}$$

We introduce the approximation $V^k(t,x)$ for $V(t,x)$. Applying Hölder's inequality, we obtain (for $r=6,\sigma=2$)

$$|I_1|\leq c\int_0^t|v|_{r\sigma}\|\nabla v\|_2|V-V^k|_{3\infty}d\tau+c\int_0^h|v|_{r\sigma}\|\nabla v\|_2|V^k|_{3\infty}+c\int_h^t\|v\|_6\|\nabla v\|_2\|V^k\|_3d\tau.$$

Hence, applying the Sobolev inequality (2.4) and the semigroup properties (3.4) of V^k, we get

$$|I_1|\leq\int_0^t\|\nabla v\|_2^2|V-V^k|_{3\infty}d\tau+ch^{-\frac{3}{2}\left(\frac{1}{r}-\frac{1}{3}\right)}\|u_{k0}\|_r\int_0^t\|\nabla v\|_2^2d\tau+c|u_{0_k}|_{3\infty}\int_0^h\|v\|_2^2d\tau.$$

We make k sufficiently large in such a way that the difference $|V-V_k|_{3\infty}$ is sufficiently small, and subsequently h sufficiently large such that, via the semigroup properties (3.4) of V_k, $c\|u_{0_k}\|_r h^{-\frac{3}{2}\left(\frac{1}{r}-\frac{1}{3}\right)}$ is small. Thus we get

$$|I_1|\leq\frac{1}{4}\int_0^t\|\nabla v\|_2^2d\tau+c|u_{0_k}|_{3\infty}\int_0^h\|\nabla v\|_2^2d\tau.$$

We consider I_2. Introducing again the sequence $\{V^k\}$, applying Hölder's inequality and the semigroup properties (3.4) of V^k, we obtain

$$\begin{aligned}|I_2|&\leq\int_0^t\|V-V^k\|_4\|\nabla v\|_2\|V\|_4d\tau+\int_0^t\|V^k\|_4\|\nabla v\|_2\|V\|_4d\tau\\&\leq\frac{1}{4}\int_0^t\|\nabla v\|_2^2d\tau+c|u_0|_{3\infty}^2|u_0-u_{0_k}|_{3\infty}^2t^{\frac{1}{2}}+c|u_0|_{3\infty}^2\|u_{0_k}\|_{\frac{12}{5}}^2t^{\frac{1}{4}}.\end{aligned}$$

Therefore, from (7.4) via the above estimates of I_1 and I_2 we obtain the inequality:

$$\|v(t)\|_2^2+\int_0^t\|\nabla v\|_2^2d\tau\le c|u_0|_{3\infty}\Big[\int_0^h\|\nabla v\|_2^2d\tau+|u_0-u_{0_k}|_{3\infty}t^{\frac12}+\|u_{0_k}\|_{\frac{12}{5}}t^{\frac14}\Big],\ t>h.$$

Hence, for all $\overline{T}\ge\overline{T}_0\ge h$, we get

$$\|v(\overline{T})\|_2^2\le c|u_0|_{3\infty}\Big[\int_0^h\|\nabla v\|_2^2d\tau+|u_0|_{3\infty}|u_0-u_{0_k}|^2_{3\infty}(\overline{T})^{\frac12}+|u_0|_{3\infty}\|u_{0_k}\|^2_{\frac{12}{5}}(\overline{T})^{\frac14}\Big].$$

Moreover, we have the existence of $\overline{\overline{T}}\in[\overline{T},2\overline{T}]$ such that

$$\overline{T}\|\nabla v(\overline{\overline{T}})\|_2^2\le c|u_0|_{3\infty}\Big[\int_0^h\|\nabla v\|_2^2d\tau+|u_0|_{3\infty}|u_0-u_{0_k}|^2_{3\infty}(2\overline{T})^{\frac12}+|u_0|_{3\infty}\|u_{0_k}\|^2_{\frac{12}{5}}(2\overline{T})^{\frac14}\Big].$$

Recalling that $\|v\|_3\le c\|\nabla v\|_2^{\frac12}\|v\|_2^{\frac12}$, for a suitable $\overline{T}$ we find a T_0 such that

$$|v(T_0)|_{3\infty}\le\|v(T_0)\|_3<\varepsilon.$$

Since $T_0\ge\overline{T}\ge\overline{T}_0$, the last estimate and estimate (7.3) furnish the existence of a T_0 such that

$$|u(T_0)|_{3\infty}\le|V(T_0)|_{3\infty}+|v(T_0)|_{3\infty}<2\varepsilon.$$

As a consequence of Theorem 4.1, we deduce the existence, for all $t>T_0$, of a regular solution corresponding to $u(T_0)$, and the uniqueness theorem of Section 5 ensures that this regular solution coincides with the weak solution. The structure theorem is completely proved. □

Proof of Corollary 1.1

Of course we can limit our considerations to the estimate of the generalized energy inequality (1.5). We follow a technique already employed in [7] that in some sense it goes back to the one introduced by Leray in [20]. Taking into account that $u:=V+v$ and $\pi_u:=\pi_V+\pi_v$, we start by considering the generalized energy inequality for solution (V,π_V) to the Stokes problem (3.1):

$$\int_\Omega|V(t)|^2\phi(t)+2\int_s^t\int_\Omega|\nabla V|^2\phi\le\int_\Omega|V(s)|^2\phi(s)dx$$
$$+\int_s^t\int_\Omega\Big[|V|^2(\phi_\tau+\Delta\phi)+2\pi_V V\cdot\nabla\phi\Big]dxd\tau.\tag{7.5}$$

Now we multiply equation $(3.1)_1$ by $v\phi$, where v is the weak solution of our decomposition. Integrating by parts on $(s, t) \times \Omega$, we get

$$\int_s^t (V_\tau, v\phi) + (\nabla V, \nabla[v\phi]) = \int_s^t \int_\Omega (\pi_V, v \cdot \nabla\phi) dx d\tau. \tag{7.6}$$

Analogously, setting $\varphi := V\phi$ in the integral equation iii. of v, we get

$$\int_s^t \Big[(v, V_\tau\phi + V\phi_\tau) - (\nabla v, \nabla[V + \phi]) + ((v + V)\cdot\nabla[V\phi], v + V) + (\pi_v, V \cdot \nabla\phi)\Big] d\tau$$
$$+ (v(s), V(s)\phi(s)) = (v(t), V\phi(t)) \,. \tag{7.7}$$

Multiplying by 2 the difference between (7.7) and (7.6), we obtain

$$\int_s^t \Big[(v, V\phi_\tau) - 2(\nabla v, \nabla[V + \phi]) + ((v + V)\cdot\nabla[V\phi], v+V) + (\pi_v + \pi_V, V \cdot \nabla\phi)\Big] d\tau$$
$$+ (v(s), V(s)\phi(s)) = (v(t), V\phi(t)) \,. \tag{7.8}$$

We sum (7.5) and iv. related to v and then we subtract twice (7.8). Recalling that $u = V + v$, after an easy handing we arrive at (1.5). The proof of the corollary is completed. □

Acknowledgements - This research was partly supported by GNFM-INdAM, and by MIUR via the PRIN 2012 *"Nonlinear Hyperbolic Partial Differential Equations, Dispersive and Transport Equations: Theoretical and Applicative Aspects"*.

References

1. K. Abe, Y. Giga, The L^∞-Stokes semigroup in exterior domains. J. Evol. Eq. **14**, 1–28 (2014)
2. K. Abe, Y. Giga, Analyticity of the Stokes semigroup in spaces of bounded functions. Acta Math. **211**, 1–46 (2013)
3. A. Alvino, Sulla diseguaglianza di Sobolev in spazi di Lorentz. Boll. Un. Mat. Ital. A **14**, 148–156 (1977)
4. M. Bolkart, M. Hieber, Pointwise upper bounds for the solution of the Stokes equation on $L^\infty_\sigma(\Omega)$ and applications. J. Funct. Anal. **268**, 1678–1710 (2015)
5. L. Caffarelli, R. Kohn, L. Nirenberg, Partial regularity of suitable weak solutions of the Navier-Stokes equations. Commun. Pure Appl. Math. **35**, 771–831 (1982)
6. F. Crispo, P. Maremonti, An interpolation inequality in exterior domains. Rend. Sem. Mat. Univ. Padova **112**, 11–39 (2004)
7. F. Crispo, P. Maremonti, On the spatial asymptotic decay of a suitable weak solution to the Navier-Stokes Cauchy problem. Nonlinearity **29**, 1355–1383 (2016). https://doi.org/10.1088/0951-7715/29/4/1355
8. R. Farwig, H. Kozono, H. Sohr, An L^q-approach to Stokes and Navier-Stokes equations in general domains. Acta Math. **195**, 21–53 (2005)

9. G.P. Galdi, *An introduction to the Mathematical theory of the Navier-Stokes Equations* (Springer, Berlin, 1994)
10. Y. Giga, T. Miyakawa, Solutions in L^r of the Navier-Stokes initial value problem. Arch. Ration. Mech. Anal. **89**, 267–281 (1985)
11. Y. Giga, H. Sohr, Abstract L^p estimates for the Cauchy problem with applications to the Navier-Stokes equations in exterior domains. J. Funct. Anal. **102**, 72–94 (1991)
12. J. Heywood, The Navier-Stokes equations: on the existence, regularity and decay of solutions. Indiana Univ. Math. J. **29**, 639–681 (1980)
13. R.A. Hunt, On $L(p, q)$ spaces. L'enseignement Mathematique **12**, 249–276 (1966)
14. T. Kato, Strong L^p-solutions of the Navier-Stokes equation in R^m, with applications to weak solutions. Math. Z. **187**, 471–480 (1984)
15. H. Koba, *On $L^{3,\infty}$-Stability of the Navier-Stokes System in Exterior Domains*, arXiv:1504.0814301
16. V. Kolyada, J. Soria, Hölder type inequalities in Lorentz spaces. Annali di Matematica **189**, 523–538 (2010)
17. H. Kozono, Global L^n -solution and its decay property for the Navier-Stokes equations in half-space R^n_+. J. Diff. Eq. **79**, 79–88 (1989)
18. H. Kozono, M. Yamazaki, Local and global unique solvability of the Navier-Stokes exterior problem with Cauchy data in the space $L^{n,\infty}$. Houston J. Math. **21**, 755–799 (1995)
19. O.A. Ladyzhenskaya, *The Mathematical Theory of Viscous Incompressible Viscous Fluid* (NewYork, Gordono Breach, 1969)
20. J. Leray, Sur le mouvement d'un liquide visqueux emplissant l'espace. Acta Math. **63**, 193–248 (1934)
21. P. Maremonti, Partial regularity of a generalized solution to the Navier-Stokes equations in exterior domain. Commun. Math. Phys. **110**, 75–87 (1987)
22. P. Maremonti, A remark on the Stokes problem with initial data in L^1. J. Math. Fluid Mech **13**, 469–480 (2011)
23. P. Maremonti, A remark on the Stokes problem in Lorentz spaces. Discret. Contin. Dyn. Syst. Ser. S **6**, 1323–1342 (2013)
24. P. Maremonti, On the Stokes problem in exterior domains: the maximum modulus theorem. Discret. Contin. Dyn. Syst. **34**, 2135–2171 (2014)
25. P. Maremonti, On weak D-solutions to the non-stationary Navier-Stokes equations in a three-dimensional exterior domain. Ann. Univ. Ferrara Sez. VII Sci. Mat. **60**, 209–223 (2014)
26. P. Maremonti, Regular solutions to the Navier-Stokes equation with an initial data in $L(n, \infty)$, Ricerche di Matematica **66**, 65–97 (2017). https://doi.org/10.1007/s11587-016-0287-7
27. P. Maremonti, V.A. Solonnikov, An estimate for the solutions of Stokes equations in exterior domains, Zap. Nauch. Sem. LOMI, 180, 105–120, trasl. J. Math. Sci. **68**(1994), 229–239 (1990)
28. P. Maremonti, V.A. Solonnikov, On nonstationary Stokes problem in exterior domains. Ann. Sc. Norm. Sup. Pisa **24**, 395–449 (1997)
29. R. O'Neil, Convolution operators and L(p, q) spaces. Duke Math. J. **30**, 129–142 (1963)
30. G. Seregin, *Lecture Notes on Regularity Theory for the Navier-Stokes Equations* (World Scientific Publishing Co Pte. Ltd., Hackensack, NJ, 2015)
31. G. Seregin, *Selected Topics of Local Regularity Theory for Navier-Stokes Equations, Topics in Mathematical Fluid Mechanics*, Lecture Notes in Math., 2073, (Springer, Heidelberg, 2013), pp. 239–313
32. G. Seregin, V. Šverák, *On global weak solutions to the Cauchy problem for the Navier-Stokes equations with large L_3-initial data*, Nonlinear Analisys, online 2 March 2016
33. J. Serrin, *The Initial Value Problem for the Navier-Stokes Equations*, ed. by R.E. Langer (University of Wisconsin press, 1963), pp. 69–98
34. Y. Shibata, On an exterior initial boundary value problem for Navier-Stokes equations. Quart. Appl. Math. **57**, 117–155 (1999)
35. E.M. Stein, G. Weiss, *Introduction to Fourier Analysis on Euclidean Spaces* (Princeton University Press, Princeton, 1990)
36. R. Temam, *Navier-Stokes Equations, Theory and Numerical Analysis*, 3rd rev. edn. (North-Holland, Amsterdam, 1984)

Energy Solutions to One-Dimensional Singular Parabolic Problems with BV Data are Viscosity Solutions

Atsushi Nakayasu and Piotr Rybka

Abstract We study one-dimensional very singular parabolic equations with periodic boundary conditions and initial data in BV, which is the energy space. We show existence of solutions in this energy space and then we prove that they are viscosity solutions in the sense of Giga–Giga.

Keywords Very singular parabolic equation · Finite energy solution
Viscosity solution · Functions of bounded variation

1 Introduction

We study a Cauchy problem for one-dimensional singular parabolic equations of the form

$$\partial_t u = (\mathcal{L}(u_x))_x, \quad (x, t) \in Q_T := \mathbb{T} \times (0, T), \tag{1}$$

$$u(x, 0) = u_0(x), \quad x \in \mathbb{T}, \tag{2}$$

where $\mathcal{L} : \mathbb{R} \to \mathbb{R}$ is monotone increasing and bounded. Such equations arise in models of crystal growth or image analysis. It is reasonable to assume that $u_0 \in BV$. For the sake of simplicity, we consider periodic boundary conditions. Here, $\mathbb{T}$ is the one-dimensional flat torus $\mathbb{T} = \mathbb{R}/\mathbb{Z}$.

Since $\mathcal{L}$ is merely monotone increasing, it may have jumped. Typical examples are

---

A. Nakayasu (✉)
Graduate School of Mathematical Science, University of Tokyo, Komaba 3-8-1, Meguro-ku, Tokyo 153-8914, Japan
e-mail: ankys@ms.u-tokyo.ac.jp

P. Rybka
Institute of Applied Mathematics and Mechanics, Warsaw University, ul. Banacha 2, 07-097 Warsaw, Poland
e-mail: rybka@mimuw.edu.pl

Y. Maekawa and S. Jimbo (eds.), *Mathematics for Nonlinear Phenomena—Analysis and Computation*, Springer Proceedings in Mathematics & Statistics 215,
DOI 10.1007/978-3-319-66764-5_9

$$\partial_t u = (\operatorname{sgn}(u_x))_x, \tag{3}$$

and

$$\partial_t u = (\operatorname{sgn}(u_x - 1) + \operatorname{sgn}(u_x + 1))_x. \tag{4}$$

Equation (3) is a prototype for the class, we have in mind, (1). We also point out that equation (3) can be formally written as $\partial_t u = 2\delta(u_x)u_{xx}$, where δ is the Dirac delta function.

Since problem (1) is one-dimensional, then for a given increasing $\mathcal{L}$, we can always find convex W, which we will call the energy density such that

$$\mathcal{L}(p) = \frac{dW}{dp}(p), \qquad \text{for a.e. } p \in \mathbb{R}.$$

Thus, we can write (1) as an L^2-gradient flow

$$\partial_t u = -\nabla E[u] \tag{5}$$

with the corresponding energy, defined by

$$E[u] = \begin{cases} \int_{\mathbb{T}} W(Du) & \text{if } u \in BV(\mathbb{T}), \\ +\infty & \text{if } u \in L^2(\mathbb{T}) \setminus BV(\mathbb{T}). \end{cases} \tag{6}$$

In order to avoid unnecessary technicalities at this stage, we will restrict our attention to piecewise linear W's which are coercive, i.e. $W(p) \to \infty$ when $|p| \to \infty$. We shall see (cf. (9)) that indeed $E[u]$ is well-defined. We will look for solutions for which $t \mapsto E[u(t)]$ is bounded over $[0, T)$, we will call them *finite energy solutions*. We notice that (6) is a functional defined on measures. There is substantial literature on this topic, including [7, 27] and others. Our W is very special, but the general case of W requires additional considerations, see [2, 7, 14, 27].

In case of (3), we have $W(p) = |p|$, this is why it is called the *total variation flow*. Equation (3) has been studied quite extensively, [15, 19, 24, 25, 29, 31, 34]. These authors did not allow jumps in the initial datum u_0. For the sake of completeness, we should mention the growing literature on the higher-dimensional versions of (3). Let us mention only [3–6, 9, 21].

On the other hand, (4) was studied in [28] with initial condition $u_0 \in BV$. There, the authors showed that if $u_0 \in BV$, then the unique solution u, which was constructed there, belongs to $L^\infty(0, T; BV)$, see [28, Theorem 2.1]. Moreover, this kind of regularity is optimal, because there are discontinuous solutions.

In case of (4), we have $W(p) = |p + 1| + |p - 1|$ then, we define $E[u]$ for $u \in BV$ in a natural way, introduced in [28], which can be easily extended to convex piecewise linear W. However, for the sake of simplifying the exposition, we will use a piecewise linear structure of W, generalizing the method used in [28]. A difficulty here is the composition of nonlinear function $\mathcal{L}$ with a measure Du, when $u \in BV$.

We also showed in [28] that finite energy solutions are viscosity solutions in the sense of [23]. However, the proof of this fact presented in [28] was direct and involved. Here, we would like to generalize this fact. However, in order to develop intuition we will restrict our attention to the case of convex, piecewise linear and coercive W. This will be a necessary preparation before tackling the general case of W with linear growth. The direct method of [28] is not likely to work.

We will show that problem (1)–(2) is well posed for $u_0 \in BV$, which is the energy space. We will exploit for this purpose the gradient flow structure of (1). We will construct solutions by means of regularizing the initial conditions. Further, we use existence result of [31]. In order to pass to the limit with the regularizing parameter we employ Minty's trick. This will be done in Theorem 1 below. However, our main result is Theorem 2 stating that finite energy solutions to (1) are also viscosity solutions in the sense of [23]. With this result, we obtain new tools to study properties of solutions to (1). Incidentally, this theorem shows that for regular data, i.e. when not only u_0 is in BV but also $\frac{du_0}{dx} \in BV$, then the notion of finite energy solutions, the almost classical solutions introduced in [30] and the viscosity solutions to (1) coincide.

The rational behind our present work is the usefulness of viscosity solutions. Originally, the viscosity solution theory was introduced in early 1980s by Crandall and Lions, [12] as a weak solution of first-order Hamilton–Jacobi equation and generalized to some regular second-order equations around 1990; see [11]. A big success of this theory was the study of the mean curvature flow equation independently by Evans-Spruck [13] and Chen–Giga–Goto [10]. The viscosity solution for the very singular diffusion equation is first introduced by Giga–Giga [16–18]. The authors establish a theorem on a unique existence of the viscosity solution of a class of very singular diffusion equations including (1) with continuous initial data u_0. The present paper provides more of examples existence theorems, this time with discontinuous initial conditions.

The plan of the paper is as follows: in Section 2, we show a lower semicontinuity of functional E, then existence of solutions in the energy space by the regularization method. Section 3 is devoted to introducing the necessary notions pertinent to the viscosity solutions. In Section 4, we construct monotone sequences of continuous functions approximating the initial conditions. This guarantees the existence of continuous solutions to (1)–(2). Then, passing to the limit with the regularizing parameter yields the desired result.

2 Finite Energy Solutions

We could use the abstract semigroup theory, see [8], exploiting the gradient flow structure, (5), to prove existence and uniqueness of solutions to (1)–(2) with data in L^2 or BV. For this purpose, it suffices to show that E defined by (6) is convex, proper and lower semicontinuous with respect to the L^2 topology. However, in this paper, not only the fact that u_0 belongs to BV is important but also how u_0 is approximated by smooth functions. This is why we will use the method of regularization to construct finite energy solutions. By 'finite energy solutions' we will understand semigroup solutions with data from the vector space, where E is finite. We restrict our attention to W, which is piecewise linear, convex and coercive.

2.1 Functional E

Understanding E is crucial for the present paper. We consider only W given by

$$W(p) = \sum_{j=1}^{N^+} \alpha_j^+ (p - \beta_j^+)^+ + \sum_{j=1}^{N^-} \alpha_j^- (p - \beta_j^-)^-, \tag{7}$$

where $N^\pm$ are positive integers, all $\alpha_j^\pm$ are positive and $\beta_j^\pm$ are real such that

$$\beta_{N^-}^- < \ldots < \beta_2^- < \beta_1^- \leq \beta_1^+ < \beta_2^+ < \ldots < \beta_{N^+}^+.$$

We adopt the usual definition, $x^+ = \max\{0, x\}$ and $x^- = (-x)^+ \geq 0$ for a real number x, so that $x = x^+ - x^-$.

First, we will explain that these are all functions of interest to us. We notice that all functions given by (7) satisfy $0 = \min\{W(p) : \ p \in \mathbb{R}\} = W(p)$, for $p \in [\beta_1^-, \beta_1^+]$.

Proposition 1. *Let us suppose that $W : \mathbb{R} \to \mathbb{R}$ is piecewise linear, convex and there is $\tilde{M} > 0$ such that $\frac{dW}{dp}(p) \equiv C_1$ for $p < -\tilde{M}$ and $\frac{dW}{dp}(p) \equiv C_2$ for $p > \tilde{M}$. If $\min\{W(p) : \ p \in \mathbb{R}\} = 0$, then, W has the form (7).*

Proof. We will first find $\beta_i^\pm$'s. We notice that coercivity and convexity of W imply that

$$a_0 := \min\{p : \ W(p) = 0\}$$

is well defined. Since by assumption W is linear for large arguments, then by definition, there are real a_i, $i = -M, \ldots, N$, (we also set $a_{N+1} = +\infty$) such that

$$\begin{aligned} W(p) &= s_i(x - a_i) + W(a_i), && \text{for } x \in [a_i, a_{i+1}), i = -M+1, \ldots, N, \\ W(p) &= s_{-M}(x - a_{-M+1}) + W(a_{-M+1}) && \text{for } x \in (-\infty, a_{-M+1}). \end{aligned}$$

If $s_0 = 0$, then we set

$$\begin{aligned} \beta_i^+ &= a_i, \quad i = 1, \ldots, N, \ \text{ and } N^+ := N, \\ \beta_i^- &= a_{1-i}, \quad i = 1, \ldots, M, \ \text{ and } N^- := M. \end{aligned}$$

Otherwise, we define

$$\begin{aligned} \beta_i^+ &= a_{i-1}, \quad i = 1, \ldots, N+1, \ \text{ and } N^+ := N+1, \\ \beta_i^- &= a_{1-i}, \quad i = 1, \ldots, M, \ \text{ and } N^- := M. \end{aligned}$$

Monotonicity of the derivative of W implies monotonicity of sequence s_i, $i = -M, \dots, N$. The definition of $\alpha_j^\pm$'s depends on s_0. If $s_0 = 0$, then we set $\alpha_1^+ = s_1$. Since $s_{i+1} > s_i$, then there is positive α_{i+1}^+ such that

$$s_{i+1} = \alpha_{i+1}^+ + s_i = \sum_{k=1}^{i} \alpha_k^+, \qquad i = 1, \dots, N^+ - 1.$$

We set $\alpha_1 = -s_{-1} > 0$. Similarly, we define α_i^-, $i = 2, \dots, N^-$, i.e.

$$s_{-i-1} + \alpha_{i+1}^- = s_{-i}, \qquad i = 1, \dots, N^- - 1.$$

On the other hand, if $s_0 > 0$, then we set $\alpha_1^+ = s_0$ and we proceed as in the case $s_0 = 0$, when defining the remaining $\alpha_j^\pm$'s. □

Formula (7) and positivity of N^+ and N^- imply the existence of positive c_0, c_1, c_2 such that

$$c_1|p| - c_2 \le W(p) \le c_0(|p| + 1). \tag{8}$$

The advantage of (7) is clearly seen when it comes to defining $\int_{\mathbb{T}} W(Du)$, because it refers to well-defined operations on measures. Indeed, if μ is a (finite) signed measure, then $\mu^\pm = \frac{1}{2}(|\mu| \pm \mu)$, where $|\mu|$ is the variation of μ. Thus, for $u \in BV(\mathbb{T})$, the following formula is correct, despite the lack of homogeneity of W,

$$\int_{\mathbb{T}} W(Du) = \int_{\mathbb{T}} \left(\sum_{i=1}^{N^+} \alpha_i^+ (Du + \beta_i^+)^+ + \sum_{j=1}^{N^-} \alpha_j^- (Du + \beta_i^-)^- \right). \tag{9}$$

We notice that obviously E is convex. It is also lower semicontinuous.

Proposition 2. *E is lower semicontinuous with respect to the convergence in the L^2 topology.*

Proof. Let us suppose that $u_n \to u$ in L^2. Since for any $v \in BV(\mathbb{T})$ and $a \in \mathbb{R}$, we have

$$\int_{\mathbb{T}} (Dv + a) = a|\mathbb{T}|,$$

then in order to prove our claim, it is sufficient to check that

$$\varliminf_{n\to+\infty} \int_{\mathbb{T}} |Du_n + a| \ge \int_{\mathbb{T}} |Du + a|.$$

We may assume $\int_{\mathbb{T}} |Du_n| \le M < \infty$, $n \in \mathbb{N}$, for otherwise there is nothing to prove. This bound and the lower semicontinuity of the BV norm imply

$$\varliminf_{n\to+\infty}\int_{\mathbb{T}}|Du_n| \geq \int_{\mathbb{T}}|Du|,$$

as a result, $u \in BV$.

We recall that if $u \in BV(\mathbb{T})$, then

$$Du = \frac{du}{dx} \llcorner \mathcal{L}^1 + D^c u + \sum_{i\in J} a_i \delta_{x_i},$$

where $\frac{du}{dx} \llcorner \mathcal{L}^1$ is absolutely continuous with respect to $\mathcal{L}^1$ part of Du, $D^c u$ is the continuous part singular with respect to $\mathcal{L}^1$ and $\sum_{i\in J} a_i \delta_{x_i}$ is the jump part. Since the number of jump discontinuity points of u and u_n, $n \in \mathbb{N}$, is at most countable, then there exists $x_0 \in \mathbb{T}$, a common continuity point of all u_n, $n \in \mathbb{N}$ and u. That is,

$$|Du_n|(\{x_0\}) = 0 = |Du|(\{x_0\}) \qquad \text{for all } n \in \mathbb{N}. \tag{10}$$

Once we identify $\mathbb{T}$ with $[0, 1)$, then for the sake of simplicity of notation we may assume that $x_0 = 0$. We notice that

$$Du_n + a = D(u_n + ax) - a\delta_0 \quad \text{on } [0, 1) = \mathbb{T}.$$

Furthermore, due to (10), we have

$$|D(u_n + ax) - a\delta_0| = |D(u_n + ax)| \llcorner \mathbb{T} \setminus \{0\}.$$

We notice that $(u_n + ax)\chi_{\mathbb{T}\setminus\{0\}}$ is a bounded sequence in $BV(\mathbb{T})$, hence the lower semicontinuity of the norm implies that

$$\varliminf_{n\to+\infty}\int_{\mathbb{T}}|D(u_n + ax)| \geq \int_{\mathbb{T}}|D(u + ax)|. \qquad \square$$

Remark 1. Density W is not homogeneous, but since W is piecewise linear, one could easily see that E coincides with its relaxation, cf. [2, 14, 27].

2.2 Existence of Finite Energy Solutions by the Regularization Method

In this section our goal is to examine the existence of solutions to (1)–(2), when the nonlinearity $\mathcal{L}$ is the derivative of a convex function W, given by (7). We studied problem (1)–(2) in [31], for rather general nonlinearities $\mathcal{L}$ and Dirichlet boundary conditions, while assuming that $u_0 \in L^1$ and $\frac{du_0}{dx} \in BV$. We stress that results of [31] are valid also in the case of periodic boundary conditions. In [28] we considered general data $u_0 \in BV$, while focusing on special $\mathcal{L}$,

$$\mathcal{L}(p) = \text{sgn}(p+1) + \text{sgn}(p-1).$$

Here, we will use [31] to prove the existence of finite energy solutions.

Of course, one could study W with a different growth rate at infinity, but we will not do this here. We will consider only W which is convex and piecewise linear, given by (7). In principle, one of N^+ and N^- could be zero, but it is easy to reduce it to the case, when both numbers are positive. We will not address this issue here.

We introduce the definition of solutions to (1)–(2) with periodic boundary conditions. It is based on the semigroup approach, see [3, 8].

Definition 1. We shall say that function $u \in L^2(0, T; L^2(\mathbb{T}))$ is a *finite energy* solution to (1)–(2) if $u \in L^\infty(0, T; BV(\mathbb{T}))$, $u_t \in L^2(0, T; L^2(\mathbb{T}))$ and $u(\cdot, 0) = u_0$. In addition, there is $\Omega \in L^2(0, T; W^{1,2}(\mathbb{T}))$ such that

$$-\Omega_x(\cdot, t) \in \partial E(u(\cdot, t)), \tag{11}$$

where $\partial E(u)$ is a subdifferential of E, i.e.

$$\partial E(u) = \{h \in L^2(\Omega) : \ E(v) - E(u) \geq (v - u, h), \quad \text{for all } v \in L^2(\Omega)\}.$$

Moreover,

$$\langle u_t, \varphi \rangle = -\int_{\mathbb{T}} \Omega \varphi_x \, dx \tag{12}$$

for all test functions $\varphi \in C^\infty(\mathbb{T})$ and for almost every $t > 0$.

In fact, this is the definition of the notion of solutions used in the theory of nonlinear semigroups, see [5, 8]. We refer the interested reader to these books for more information.

We notice that the time regularity, postulated in Definition 1 implies that solutions to (1) are in $C([0, T]; L^2(\mathbb{T}))$. Hence, we can impose initial conditions (2). At this point we stress that Ω may be regarded as a selection of the composition of multivalued operators $\mathcal{L} \circ u_x$.

We notice that we consider here less regular solutions than those in [31]. In particular, we cannot use the same notion of solution as there. Indeed, we required in [31] that $\Omega(x, t) \in \partial W(u_x(x, t))$, but this does not make sense when the distributional derivative, Du, is a measure. In this case (11) is appropriate.

The general semigroup theory yields existence and uniqueness of solutions. However, our key point is approximation of BV data by smooth functions. We want to control this process, because it is important for the development of viscosity solutions. For this purpose, we show the existence result.

Theorem 1. *Let us suppose that $u_0 \in BV(\mathbb{T})$, then there exists a finite energy solution to (1)–(2). Moreover, u is a solution in the sense of semigroup theory, hence it is a unique solution.*

Proof. We want to use results established in [31], where

$$u_0 \in L^1(\mathbb{T}) \quad \text{and} \quad \frac{du_0}{dx} \in BV(\mathbb{T}). \tag{13}$$

For this purpose we take an approximating sequence, u_0^ϵ, converging to u_0 in L^2 and satisfying (13). We may assume that $\|u_0^\epsilon\|_{BV} \le \|u_0\|_{BV} + 1$. By [31, Theorem 1] there exist u^ϵ, unique solutions to (1)–(2) with initial condition u_0^ϵ. According to this result, there is $\Omega^\epsilon \in L^2(0,T;H^1(\Omega))$ such that

$$\int_{\mathbb{T}} [u_t^\epsilon \varphi + \Omega^\epsilon \varphi_x]\, dx = 0 \qquad \text{for } a.e.\ t > 0, \tag{14}$$

for all $\phi \in C^\infty(\mathbb{T})$. Moreover, Ω^ϵ satisfies

$$\int_{\mathbb{T}} [W(u_x^\epsilon + h_x) - W(u_x^\epsilon)]\, dx \ge \int_{\mathbb{T}} \Omega^\epsilon h_x\, dx,$$

for all $h \in C^\infty(\mathbb{T})$, hence (11) holds.

During the course of proof of [31, Theorem 1] we showed the following bounds depending only on $\|u_0^\epsilon\|_{BV}$, i.e. on $\|u_0\|_{BV}$,

$$\int_0^T \int_{\mathbb{T}} |u_t^\epsilon|^2\, dxdt, \quad \int_{\mathbb{T}} W(u_x^\epsilon)\, dx, \quad \int_0^T \int_{\mathbb{T}} |\Omega_x^\epsilon|^2\, dxdt \le M(\|u_0\|_{BV}). \tag{15}$$

Estimates (15) yield existence of convergent subsequences,

$$\begin{array}{l} u_t^\epsilon \rightharpoonup u_t \ \text{ in } L^2(Q_T), \\ \Omega^\epsilon \rightharpoonup \Omega \text{ in } L^2(0,T;H^1(\Omega)). \end{array} \tag{16}$$

Since $|\Omega^\epsilon(x,t)| \le K$ we additionally obtain that

$$\Omega^\epsilon \overset{*}{\rightharpoonup} \Omega \qquad \text{in } L^\infty(Q_T).$$

We also notice that (15) yields continuity of u^ϵ, i.e. $u^\epsilon \in C([0,T];L^2(\mathbb{T}))$. Moreover, the bound $\|u_t^\epsilon\|_{L^2(0,T;L^2(\mathbb{T}))} \le M$ given by (15) implies that the family u^ϵ is equicontinuous. If we combine it with the common estimate the initial data, we see that

$$\sup_{[0,T]} \|u^\epsilon\|_{L^2(\mathbb{T})} \le M$$

independently of ϵ. Inequalities (15) provide a common bound on $E(u^\epsilon)$, hence due to (8), we have a bound on the BV norm of $u^\epsilon(\cdot,t)$,

$$\int_{\mathbb{T}} |Du^\epsilon(\cdot,t)| \le C\left(1 + \int_{\mathbb{T}} W(u_x^\epsilon(\cdot,t))\, dx\right). \tag{17}$$

Since $BV(\mathbb{T})$ is compactly embedded in $L^2(\mathbb{T})$, then by Aubin Lemma (see [33, Theorem 1]), we can select a subsequence $\{u^\epsilon\}$ (not relabeled) such that

$$u^\epsilon \to u \qquad \text{in } C([0,T]; L^2(\mathbb{T})), \tag{18}$$

in particular $u(\cdot, 0) = u_0$, because $u_0^\epsilon \to u$ in $L^2(\mathbb{T})$. This fact combined with (17) and the lower semicontinuity of the BV norm imply that

$$u \in L^\infty(0, T; BV(\Omega)).$$

From (16) and (14) we infer that

$$\int_0^T \int_{\mathbb{T}} [u_t\varphi + \Omega\varphi_x]\, dxdt = 0 \tag{19}$$

for all $\phi \in C^\infty(Q_T)$.

We have to check that Ω satisfies (11). For this purpose we will use Minty's trick. If v is any element of $D(\partial E(v))$ and $-\Omega(v)_x \in \partial E(v)$, then (11) and the monotonicity of ∂E give us,

$$\int_{\mathbb{T}} (-\Omega_x^\epsilon + \Omega_x(v))(u_x^\epsilon - v_x)\, dx \geq 0.$$

We may multiply this inequality by a nonnegative function $\psi \in C_0^\infty(0, T)$ and integrate over $(0, T)$,

$$\int_0^T \int_{\mathbb{T}} \psi(-\Omega_x^\epsilon + \Omega(v))(u_x^\epsilon - v_x)\, dxdt \geq 0. \tag{20}$$

We notice that inequality (19) requires only that φ and $\varphi_x \in L^2(Q_T)$. Smooth functions are dense in this space, so we may use $\varphi = \psi(u - v)$ in (19).

When we want to pass to the limit in (20), we see that the following term is the most difficult,

$$\int_0^T \int_{\mathbb{T}} \Omega^\epsilon (u_x^\epsilon - v_x)\, dxdt.$$

Since u^ϵ satisfies (14), then after having substituted $\varphi = u^\epsilon - v$ we obtain,

$$\int_0^T \int_{\mathbb{T}} [u_t^\epsilon \psi(u^\epsilon - v) + \Omega(v)\psi(u_x^\epsilon - v_x)\, dxdt \geq 0. \tag{21}$$

We notice that due to (18), u^ϵ converges strongly in $L^2(Q_T)$, hence

$$\lim_{\epsilon \to 0} \int_0^T \int_{\mathbb{T}} u_t^\epsilon \psi(u^\epsilon - v)\, dxdt = \int_0^T \int_{\mathbb{T}} u_t \psi(u - v)\, dxdt.$$

We calculate the limit of the second term in (21),

$$\lim_{\epsilon\to 0}\int_0^T\int_{\mathbb{T}} \Omega(v)\psi(u^\epsilon_x - v_x)\,dxdt = -\lim_{\epsilon\to 0}\int_0^T\int_{\mathbb{T}} \Omega(v)_x\psi(u^\epsilon - v)\,dxdt$$
$$= -\int_0^T\int_{\mathbb{T}} \Omega(v)_x\psi(u-v)\,dxdt = \int_0^T\int_{\mathbb{T}} \Omega(v)(Du - Dv).$$

The last equality is an application of the following well-known result,

Lemma 1 ([26]). *If $w \in H^1(a,b)$ and $\varphi \in BV(a,b)$, then*

$$\int_a^b w_x\varphi dx = -\int_a^b wD\varphi + \gamma(w\varphi)(b) - \gamma(w\varphi)(a).$$

Here, $\gamma u(c)$ denotes the trace of $u \in BV(a,b)$ at $c \in \{a,b\}$.

In the present case, we identify $\mathbb{T}$ with the interval $[0,1)$. Due to the periodicity of the ingredients, the integral over $[0,1)$ equals the integral over $[\delta, 1+\delta)$ and we can select such δ that $x=\delta$, $x = 1+\delta$ are points of continuity of $u-v$, as considered over $\mathbb{R}$.

Combining these pieces of information we reach the following conclusion,

$$0 \le -\int_0^T\int_{\mathbb{T}} \psi(\Omega_x - \Omega(v)_x)(u-v)\,dxdt.$$

Moreover, since $\psi \ge 0$ is arbitrary, then we reach,

$$0 \le \int_{\mathbb{T}} (\Omega_x - \Omega(v)_x)(u-v)\,dx.$$

Now, the maximal monotonicity of ∂E implies that $-\Omega_x \in \partial E(u)$.

Thus, we we showed that u is a solution in the sense of semigroups, see [8], hence uniqueness follows. □

2.3 Regularity of Finite Energy Solutions

We recall the following fact, which will be very useful for us.

Proposition 3 ([31]). *Let us suppose that $u_0 \in BV$ as well as $\frac{du_0}{dx} \in BV$. Then there is $\alpha > 0$ such that $u \in C^{2\alpha,\alpha}$.*

We refer the reader for the proof to [31, formula (2.19)].

2.4 Stability of Finite Energy Solutions

We show a very useful result for further studies.

Proposition 4. *Let us suppose that (u^n, Ω^n) is a sequence of energy solutions to (1)–(2), which is bounded in L^∞, u_n converges to u almost everywhere and $\Omega^n \rightharpoonup \Omega$ in $L^2(0, T; H^1(\mathbb{T}))$. Then, u and Ω form an energy solution to (1)–(2) with initial data u_0.*

Proof. Since u_t^n and Ω^n satisfy (19), then we deduce from the weak convergence of Ω^n that u_t^n also converges weakly in $L^2(Q_T)$ and $\|u_t^n\|_{L^2(Q_T)}$ are commonly bounded. By Aubin Lemma, [33, Theorem 1], this implies convergence of u^n to u in $C([0, T]; L^2(\mathbb{T}))$. Finally, we deduce that $-\Omega_x \in \partial E(u)$ by following the argument of Theorem 1. □

3 Viscosity Solutions

In this section, we review the viscosity solution theory to the singular diffusion equation of the generalized form

$$\partial_t u + F((W'(u_x))_x) = 0 \quad \text{in } U_T := (0, T) \times U. \tag{22}$$

Here, U is an open set in $\mathbb{T}$ and F is a function such that

(F1) (Continuity) $F \in C(\mathbb{R})$,
(F2) (Ellipticity) F is non-increasing.

We refer the readers to the papers [20, 23] for the details.

Let $\bar{\mathbb{R}}$ denote extended real numbers $\mathbb{R} \cup \{\pm\infty\}$, and for an $\bar{\mathbb{R}}$-valued function u on U_T define its upper (resp. lower) semicontinuous envelope u^* (resp. u_*) by

$$u^*(x,t) := \overline{\lim_{(y,s)\to(x,t)}} u(y,s) \qquad (\text{resp. } u_*(x,t) := \underline{\lim_{(y,s)\to(x,t)}} u(y,s)).$$

3.1 The Notion of Viscosity Solutions to (1)

We recall a notion of faceted functions. Let P denote the set of the jump points of the derivative of W. A function $f \in C^1(U)$ is *faceted* at a point $\hat{x} \in U$ with *slope* $p \in \mathbb{R}$ (or *p-faceted* at $\hat{x}$) if there exists a closed nontrivial finite interval $I = [c_l, c_r] \subset U$ containing $\hat{x}$ (i.e. $c_l, c_r \in U$ satisfy $c_l < c_r$ and $c_l \le \hat{x} \le c_r$) such that

$$f'(x) = p \quad \text{for all } x \in I,$$
$$f'(x) \neq p \quad \text{for all } x \in J \setminus I$$

for a neighborhood $J = (b_l, b_r) \subset U$ of I. The closed interval I is called a *faceted region* of f containing $\hat{x}$ and it is denoted by $R(f, \hat{x})$. Let $C^2_P(U)$ denote the set of all $f \in C^2(U)$ such that f is p-faceted at $\hat{x}$ whenever $p = f'(\hat{x}) \in P$. We also define the *left transition number* $\chi_l = \chi_l(f, \hat{x})$ and the *right transition number* $\chi_r = \chi_r(f, \hat{x})$ for a p-faceted function f at $\hat{x}$ by

$$\chi_l = \begin{cases} +1 & \text{if } f' < p \text{ on } (b_l, c_l), \\ -1 & \text{if } f' > p \text{ on } (b_l, c_l), \end{cases} \qquad \chi_r = \begin{cases} +1 & \text{if } f' > p \text{ on } (c_r, b_r), \\ -1 & \text{if } f' < p \text{ on } (c_r, b_r). \end{cases}$$

Now, for $\Delta > 0, I \subset U, \xi_l, \xi_r \in [-1, 1]$ and $Z \in C^{1,1}(I)$, we consider the obstacle problem to minimize

$$J[\xi] := \begin{cases} \int_I |\xi'(x)|^2 dx & \text{if } \xi \in H^1(I), \\ \infty & \text{if } \xi \in L^2(I) \setminus H^1(I) \end{cases}$$

over

$$K = \{\xi \in H^1(I) \mid \xi - Z \in [-\frac{\Delta}{2}, \frac{\Delta}{2}] \text{ on } I, \xi(c_i) = Z(c_i) + (-1)^{k(i)}\chi_i \frac{\Delta}{2}\},$$

where $i = l, r$ and $k(l) = 1, k(r) = 0$.

We note that it is easy to see that if the function Z is affine, the minimizer is affine as well. This is the case considered in this paper but, in general, even if $Z \in C^{1,1}(I)$, then it is known that the unique minimizer belongs to $Z \in C^{1,1}(I)$; see [23]. Also, we note that the derivative of minimizer $\bar{\xi}$ with C^1 smoothness is invariant with respect to shifts of Z by an additive constant. In view of these remarks, we write

$$\Lambda^{Z'}_{\chi_l \chi_r}(x; I, \Delta) = \bar{\xi}'(x) \quad \text{for } x \in I.$$

For $f \in C^2_P(U)$ and $\hat{x} \in U$, we define the *nonlocal curvature* $\Lambda_W(f)(\hat{x})$ as below. On the one hand, if $f'(\hat{x}) \notin P$, we set

$$\Lambda_W(f)(\hat{x}) = W''(f'(\hat{x}))f''(\hat{x})$$

as expected. On the other hand, if $p := f'(\hat{x}) \in P$, i.e. f is p-faceted at $\hat{x}$, then we set

$$\Lambda_W(f)(\hat{x}) = \Lambda^{\sigma}_{\chi_l \chi_r}(\hat{x}; I, \Delta)$$

where Δ is the jump in the derivative of W at p, i.e. $\Delta = \frac{d^+W}{dq}(p) - \frac{d^-W}{dq}(p)$, $I = R(f, \hat{x})$, $\chi_l = \chi_l(f, \hat{x})$, $\chi_r = \chi_r(f, \hat{x})$, $\sigma(x) = 0$.

Let us denote, by $A_P(U_T)$, the set of all *admissible* functions on U_T, i.e. functions φ of the form $\varphi(x, t) = f(x) + g(t)$ where $f \in C^2_P(U)$ and $g \in C^1(0, T)$.

Definition 2 (Viscosity solutions). Let u be an $\bar{\mathbb{R}}$-valued function on U_T. We say that u is a *viscosity subsolution* (resp. *supersolution*) of (22) if whenever $\varphi \in A_P(U_T)$ and $(\hat{x}, \hat{t}) \in U_T$ satisfy

$$\max_{U_T}(u^* - \varphi) = (u^* - \varphi)(\hat{x}, \hat{t}) = 0, \qquad (\text{resp. } \min_{U_T}(u_* - \varphi) = (u_* - \varphi)(\hat{x}, \hat{t}) = 0)$$

then the inequality

$$\varphi_t(\hat{x}, \hat{t}) + F(\Lambda_W(\varphi(\cdot, \hat{t}))(\hat{x})) \leq 0, \qquad (\text{resp. } \varphi_t(\hat{x}, \hat{t}) + F(\Lambda_W(\varphi(\cdot, \hat{t}))(\hat{x})) \geq 0)$$

holds. We say that u is a *viscosity solution* if u is both a viscosity subsolution and a viscosity supersolution.

We close this subsection with an explicit example of a viscosity solution starting from general initial data u_0, not necessarily continuous.

Example 1. Let us consider the total variation flow (3) with the initial datum

$$u_0(x) = \begin{cases} 1 & \text{if } x \in (0, a), \\ 0 & \text{if } x \in (a, 1), \end{cases}$$

where $a \in (0, 1)$ is a given constant. One will realize that this problem (3), (2) has a solution of the form

$$u(x,t) = \begin{cases} 1 - \frac{2}{a}t & \text{if } t \leq a(1-a)/2, x \in (0, a), \\ \frac{2}{1-a}t & \text{if } t \leq a(1-a)/2, x \in (a, 1), \\ a & \text{if } t \geq a(1-a)/2. \end{cases}$$

In fact it is not difficult to check that u is a viscosity solution allowing a discontinuity.

3.2 *Comparison and Stability of Viscosity Solutions*

We recall the comparison and stability results applicable in the present work.

Proposition 5. (Comparison, [23, Theorem 7]) *We assume (F1) and (F2) and we take $U = \mathbb{T}$. Let u and v respectively be a viscosity sub- and supersolution of* (22) *such that $u^* < +\infty$ and $v_* > -\infty$. If $u^*|_{t=0} \leq v_*|_{t=0}$ on $\mathbb{T}$, then $u^* \leq v_*$ in Q_T.*

Proposition 6. (Stability under extremum, [20, Theorems 4.1 and 5.4]) *Assume (F1) and (F2). Let S be a family of viscosity subsolutions (resp. supersolutions) of* (22)*. Then,*

$$u(x,t) := \sup_{v \in S} v(x,t) \qquad (\textit{resp. } u := \inf_{v \in S} v(x,t))$$

is a real-valued viscosity subsolution (resp. supersolution) of (22).

In summary, one is able to construct a unique solution of (22) with continuous initial data.

3.3 Related Work

Pioneer work for the very singular diffusion equations including (22) is given by Giga–Giga in [16, 17], which study the equations with spatially homogeneous external force. The authors introduced a notion of viscosity solutions, and established comparison, existence and stability of solutions in the sense of [16].

The definition of solutions in [16] looks different from one in the present paper. However, these two definitions are equivalent as far as one considers the equation without a spatially inhomogeneous external force. Note that the value of Λ_W is determined explicitly when the external force is independent of the spatial variable. On the other hand, when the external force term depends on the spatial variable, one will encounter an obstacle problem. By considering the obstacle problem carefully, a comparison principle is given in [23], while a general existence result based on Perron method is given in [20].

As a further development of the viscosity solutions theory for singular diffusion equations, we point out that higher-dimensional total variation flows of a non-divergence form are studied in [22].

4 Finite Energy Solutions are Viscosity Solutions

We prove here our main result.

Theorem 2. *If* $u_0 \in BV(\mathbb{T})$*, then the corresponding finite energy solution to (1)–(2), constructed in Theorem 1, is also a viscosity solution in the sense of Definition 2.*

The proof is divided into a number of steps.

4.1 Continuous Initial Data

We begin our analysis with the case when the initial data u_0 is continuous.

Proposition 7. *If $u_0 \in BV(\mathbb{T})$ and $\frac{du_0}{dx} \in BV(\mathbb{T})$ so that $u_0 \in C(\mathbb{T})$, then the corresponding finite energy solution to (1)–(2) is also a viscosity solution in the sense of Definition 2.*

Proof. In order to prove this claim we recall the approximation argument introduced in [31]. Namely, the authors of [31] considered the following equation, where $\delta > 0$,

$$\begin{aligned} u_t^\delta &= \left(W_p^\delta(u_x^\delta)\right)_x + \delta u_{xx}^\delta \quad (x,t) \in Q_T, \\ u_0^\delta(x,0) &= u_0(x) \qquad\qquad x \in \mathbb{T}, \end{aligned}$$

where $W^\delta = W * \eta_\delta$ and η_δ is the standard mollifier kernel. Note that the smooth solution u^δ is a viscosity solution. Since $W^\delta + \delta\frac{p^2}{2} \to W$ locally uniformly, in view of the definition, we can invoke the convergence result established by Giga–Giga [17, Theorem 1.1] to see that u^δ converges to a function $\tilde{u} \in C(\mathbb{T} \times [0,T])$ locally uniformly and it is a viscosity solution of (1). Now, we have seen that u^δ also converges to the finite energy solution u in $C^{\alpha,\alpha/2}(\mathbb{T} \times [0,T])$ for a positive α in Proposition 3. Therefore, two limits u and $\tilde{u}$ are equal so that u is a viscosity solution.

4.2 Preparation of the Initial Conditions

We prove here that we can approximate the BV data by continuous functions in an appropriate way.

Proposition 8. *Let us suppose that $u_0 \in BV(\mathbb{T})$. Then,*
(a) there exists an increasing sequence v_0^k, $k \in \mathbb{N}$ of continuous functions and $\frac{dv_0^k}{dx} \in BV(\mathbb{T})$, such that $v_0^k \le u_0$ a.e. and

$$\lim_{k\to 0} v_0^k(x) - u_0(x) \quad \textit{for a.e. } x \in \mathbb{T};$$

(b) there exists a decreasing sequence w_0^k, $k \in \mathbb{N}$ of continuous functions and $\frac{dw_0^k}{dx} \in BV(\mathbb{T})$, such that $w_0^k \ge u_0$ a.e. and

$$\lim_{k\to 0} w_0^k(x) = u_0(x) \quad \textit{for a.e. } x \in \mathbb{T}.$$

The proof will be in a few steps. The first step is the Lemma below.

Lemma 2. *Let us suppose that $u : [a,b) \to \mathbb{R}$ is bounded and increasing, then there exists an increasing sequence of continuous functions $\{u_k\}_{k=1}^\infty$ such that $u_k(x) \le u(x)$, $\frac{du_k}{dx} \in BV(\mathbb{T})$, and*

$$\lim_{k\to\infty} u_k(x) = u(x) \quad \textit{for a.e. } x \in \mathbb{T}.$$

Proof. We extend u to $\mathbb{R}$ by the following formula,

$$\tilde{u}(x) = \begin{cases} u(b^-), & \text{for } x \geq b, \\ u(x), & \text{for } x \in (a, b), \\ u(a), & \text{for } x \leq a, \end{cases}$$

where $u(b^-)$ stands for the left side limit of u at b.

Let us denote by ρ a standard mollifying kernel, whose support is $[-1, 1]$. We set

$$u_k(x) = (\tilde{u} * \rho_{1/k})(x - \frac{1}{k}).$$

Obviously, u_k are continuous. We have to check its properties. We notice, that supp $\rho \subset [-1, 1]$ and monotonicity of $\tilde{u}$ imply

$$u_k(x) = \int_{\mathbb{R}} \tilde{u}(x - \frac{1}{k} - y)\rho(y)\, dy \leq \int_{\mathbb{R}} u(x)\rho(y)\, dy = u(x) \qquad x \in (a, b).$$

By the same argument, we have

$$u_{k+1}(x) \geq u_k(x).$$

We know that $u * \rho_{1/k}$ converges to u in L^1, hence there exists a subsequence, still denoted by u_k converging to u almost everywhere. Since the shift operator, $v(\cdot) \mapsto v(\cdot - h)$ is continuous in L^1, we conclude that $u_k(x)$ converges to $u(x)$ for almost every $x \in [a, b)$. □

Lemma 3. *Let us suppose that $f : [a, b] \to \mathbb{R}$ is such that f restricted to $[a, b)$ is increasing and $f(b) \in [f(a), f(b^-))$. Then, there exists an increasing sequence $\{f_k\}_{k=1}^{\infty}$ converging almost everywhere to f.*

Proof. We can take $u = f|_{[a,b)}$. Lemma 2 yields an increasing sequence u_k converging a.e. to u. We define it by the formula below,

$$f_k(x) = \begin{cases} u_k(x) & x \in [a, b - 1/k), \\ \ell_k(x - b) + f(b^-) & x \in [b - 1/k, b], \end{cases}$$

where

$$\ell_k = k(f(b^-) - u_k(b - 1/k)).$$

It is easy to see that f_k has the desired properties. □

Lemma 4. *Let us suppose that $f : [a, b] \to \mathbb{R}$ is such that f restricted to $[a, b)$ is decreasing and $f(b) \in [f(b^-), f(a))$. Then, there exists a decreasing sequence $\{f_k\}_{k=1}^{\infty}$ converging almost everywhere to f.*

Proof. This result follows from Lemma 3 applied to $-f$. □

Now, we are ready to prove Proposition 8. It is sufficient to establish (a) since (b) will follow from (a) applied to $-u_0$.

If we are given $u_0 \in BV(\mathbb{T})$, then we may choose its 'good representative' see [1, Theorem 3.28], which will be bounded, and we define $f : [0, 1] \to \mathbb{R}$ by the following formula,

$$f(x) = \begin{cases} u_0(x) & x \in [0, 1) \\ u_0(0) & x = 1. \end{cases}$$

Of course, f has a finite variation over $(0, 1)$, hence $f = f^+ - f^-$, where $f^\pm$ are increasing. By Lemmas 3 and 4, there exist $v_k^+, v_k^-, w_k^+, w_k^-$ such that

$$v_k^+ \le f^+ \le w_k^+, \qquad v_k^- \le f^- \le w_k^-.$$

We set $v_k = v_k^+ - w_k^-$ and $w_k = w_k^+ - v_k^-$, these sequences monotonically converge to u_0 almost everywhere. Obviously, by the definition, we have

$$v_k \le f^+ - f^- \le w_k.$$

□

4.3 BV Initial Data

Let us consider v^ϵ a unique continuous finite energy solution corresponding to v_0^ϵ, given by Proposition 8. Respectively, let w^ϵ be a unique continuous finite energy solution corresponding to w_0^ϵ, given by Proposition 8. We set

$$v = \sup v^\epsilon, \qquad w = \sup w^\epsilon.$$

Lemma 5. *If v and w are defined above, then v and w are finite energy solutions.*

Proof. Due to monotonicity of families v^ϵ and w^ϵ, we see that v and w are in fact pointwise limits, so that we can use Proposition 4 implying that v and w are finite energy solutions. Hence, our claim follows. □

Lemma 6. *If v and w are defined above, then v is a viscosity subsolution and w is a viscosity supersolution of* (1), *respectively.*

Proof. This lemma follows immediately from Proposition 6. □

Proof (Proof of Theorem 2*).* We showed the following inequality for viscosity subsolution v and viscosity supersolution w,

$$v \le w.$$

At the same time v and w are finite energy solutions to (1) with the same initial condition u_0. Uniqueness of finite energy solutions implies $v = w$, hence finite energy solutions to (1)–(2) are viscosity solutions to this equation. □

Remark 2. We constructed in Theorem 1 solutions to (1)–(2) by the regularization method for data in BV. Then, we showed in Theorem 2 that these solutions are also viscosity solutions. In [31, Theorem 1] by the same method, we constructed solutions to (1)–(2) assuming more regular data, i.e. u_0 and $\frac{du_0}{dx} \in BV$ and we showed that they are *almost classical*, [31, Theorem 3], see also [32]. Our Theorem 2 implies that all these types of solutions coincide, at least for more regular initial conditions. Finally, we point out that they are in fact solutions in the sense of the nonlinear semigroup theory, i.e. in a weak sense.

Remark 3. We discovered an inconsistency in the definition of a weak solution in [28] and a gap in the proof of the existence result there. The problem is that the definition and the proof of [28, Theorem] are correct only when Du is a function not a measure. This problem is solved in the proof of Theorem 1. Despite this gap, the remaining analysis in [28] is correct.

Acknowledgements Both authors thank the grant agencies for partial support during the preparation of this paper and the universities, which hosted them. AN was supported by a Grant-in-Aid for JSPS Fellows No. 25-7077 as well as by MNiSW through 2853/7.PR/2013/2 grant. PR was in part supported by the EC IRSES program "Flux". Both authors were in part supported by NCN through grant 2011/01/B/ST1/01197. A part of the research was performed during AN visits to the University of Warsaw supported by the Warsaw Center of Mathematics and Computer Science through the program 'Guests'. Some work was done during a PR visit to the University of Tokyo.

Both authors thank the referee for his/her constructive comments, which help to improve the text.

References

1. L. Ambrosio, N. Fusco, D. Pallara, *Functions of Bounded Variation and Free Discontinuity Problems* (The Clarendon Press, Oxford University Press, New York, 2000)
2. L. Ambrosio, D. Pallara, Integral representations of relaxed functionals on $BV(\mathbb{R}^n, \mathbb{R}^k)$ and polyhedral approximation. Indiana Univ. Math. J. **42**, 295–321 (1993)
3. F. Andreu, V. Caselles, J.M. Mazón, S. Moll, The Dirichlet problem associated to the relativistic heat equation. Mathematische Annalen **347**, 135–199 (2010)
4. F. Andreu, C. Ballester, V. Caselles, J.M. Mazón, The Dirichlet problem for the total variation flow. J. Funct. Anal. **180**, 347–403 (2001)
5. F. Andreu-Vaillo, V. Caselles, J.M. Mazón, *Parabolic quasilinear equations minimizing linear growth functionals* (Birkhäuser, Basel, 2004)
6. G. Bellettini, V. Caselles, M. Novaga, The total variation flow in $\mathbb{R}^N$. J. Differ. Equ. **184**, 475–525 (2002)
7. G. Bouchitté, M. Valadier, Integral representation of convex functionals on a space of measures. J. Funct. Anal. **80**, 398–420 (1988)
8. Brézis, H.: Opérateurs maximaux monotones et semi-groupes de contractions dans les espaces de Hilbert. North-Holland Publishing Co., Amsterdam-London; American Elsevier Publishing Co., Inc., New York (1973)

9. A. Briani, A. Chambolle, M. Novaga, G. Orlandi, On the gradient flow of a one-homogeneous functional. Conflu. Math. **3**, 617–635 (2011)
10. Y.G. Chen, Y. Giga, S. Goto, Uniqueness and existence of viscosity solutions of generalized mean curvature flow equations. J. Differ. Geom. **33**, 749–786 (1991)
11. M.G. Crandall, H. Ishii, P.-L. Lions, User's guide to viscosity solutions of second order partial differential equations. Bull. Am. Math. Soc. **27**, 1–67 (1992)
12. M.G. Crandall, P.-L. Lions, Viscosity solutions of Hamilton-Jacobi equations. Trans. Am. Math. Soc. **277**, 1–42 (1983)
13. L.C. Evans, J. Spruck, Motion of level sets by mean curvature. I. J. Differ. Geom. **33**, 635–681 (1991)
14. I. Fonseca, P. Rybka, Relaxation of multiple integrals in the space $BV(\Omega,^p)$. Proc. R. Soc. Edinburgh Sect. A **121**, 321–348 (1992)
15. Fukui, T., Giga, Y.: Motion of a graph by nonsmooth weighted curvature. In: Lakshmikantham, V. (ed.) World Congress of Nonlinear Analysts '92, Vol. I–IV (Tampa, FL, 1992), pp. 47-56. de Gruyter, Berlin (1996)
16. M.-H. Giga, Y. Giga, Evolving graphs by singular weighted curvature. Arch. Ration. Mech. Anal. **141**, 117–198 (1998)
17. M.-H. Giga, Y. Giga, Stability for evolving graphs by nonlocal weighted curvature. Commun. Partial Differ. Equ. **24**, 109–184 (1999)
18. M.-H. Giga, Y. Giga, Generalized motion by nonlocal curvature in the plane. Arch. Ration. Mech. Anal. **159**, 295–333 (2001)
19. M.-H. Giga, Y. Giga, R. Kobayashi, Very singular diffusion equations, in *Taniguchi Conference on Mathematics Nara'98*, ed. by M. Maruyama, T. Sunada (Math. Soc. Japan, Tokyo, 2001), pp. 93–125
20. M.-H. Giga, Y. Giga, A. Nakayasu, On general existence results for one-dimensional singular diffusion equations with spatially inhomogeneous driving force, in *Geometric Partial Differential Equations*, ed. by A. Chambolle, M. Novaga, E. Valdinoci. Ed. Norm., Pisa (2013), pp. 145–170
21. M.-H. Giga, Y. Giga, N. Požár, Anisotropic total variation flow of non-divergence type on a higher dimensional torus. Adv. Math. Sci. Appl. **23**, 235–266 (2013)
22. M.-H. Giga, Y. Giga, N. Požár, Periodic total variation flow of non-divergence type in $\mathbb{R}^n$. J. Math. Pures Appl. **9**(102), 203–233 (2014)
23. M.-H. Giga, Y. Giga, P. Rybka, A comparison principle for singular diffusion equations with spatially inhomogeneous driving force for graphs. Arch. Ration. Mech. Anal. **211**, 419–453 (2014)
24. Y. Giga, P. Górka, P. Rybka, Bent rectangles as viscosity solutions over a circle. Nonlinear Anal. **125**, 518–549 (2015)
25. K. Kielak, P.B. Mucha, P. Rybka, Almost classical solutions to the total variation flow. J. Evol. Eqs **13**, 21–49 (2013)
26. S. Łojasiewicz, *An Introduction to the Theory of Real Functions* (Wiley Ltd, Chichester, 1988)
27. J.-P. Mandallena, Quasiconvexification of geometric integrals. Ann. Mat. Pura Appl. **4**(184), 473–493 (2005)
28. M. Matusik, P. Rybka, Oscillating facets. Port. Math. **73**, 1–40 (2016)
29. P.B. Mucha, Stagnation, creation, breaking. Adv. Math. Sci. Appl. **24**, 223–235 (2014)
30. P.B. Mucha, P. Rybka, A caricature of a singular curvature flow in the plane. Nonlinearity **21**, 2281–2316 (2008)
31. P.B. Mucha, P. Rybka, Well-posedness of sudden directional diffusion equations. Math. Methods Appl. Sci. **36**, 2359–2370 (2013)
32. P.B. Mucha, P. Rybka, Almost classical solutions of static Stefan type problems involving crystalline curvature, in *Nonlocal and Abstract Parabolic Equations and their Applications*, ed. by P.B. Mucha, M. Niezgódka, P. Rybka (Polish Acad. Sci. Inst. Math, Warsaw, 2009), pp. 223–234
33. J. Simon, Compact sets in the space $L^p(0, T; B)$. Ann. Mat. Pura Appl. **4**(146), 65–96 (1987)
34. H. Spohn, Surface dynamics below the roughening transition. J. Physique **I**(3), 68–81 (1993)

Local Well-Posedness for the Cauchy Problem to Nonlinear Heat Equations of Fujita Type in Nearly Critical Besov Space

Takayoshi Ogawa and Yuuki Yamane

Abstract We show the local well-posedness of the Cauchy problem to a nonlinear heat equation of Fujita type in lower space dimensions. It is well known that the nonnegative solution corresponding to the Fujita critical exponent $p = 1 + \frac{2}{n}$ does not exist in the critical scaling invariant space $L^1(\mathbb{R}^n)$. We show if the initial data is in a modified Besov spaces, then the corresponding mild solution to the equation with the Fujita critical exponent $p = 1 + \frac{2}{n}$ exists and the problem is locally well-posed in the same space of the initial data. Besides we also show the problem is ill-posed in the scaling invariant Besov and inhomogeneous Besov spaces. This is known in L^1 space and extension of the result known in the Lebesgue spaces.

Keywords Nonlinear heat equation · Fujita critical exponent
Local well-posedness · Modified Besov space

AMS Subject Classification Primary 35K15 · 35K55 · 35Q60 Secondary 78A35

1 Introduction

1.1 Solvability Result for the Critical Spaces

We are concerned with the local well-posedness issue of the nonlinear heat equation of Fujita type in nearly critical function spaces. F. Weissler [17, 18] showed that the Cauchy problem for the semilinear heat equation of the Fujita type is locally and globally well-posed in the Lebesgue spaces:

T. Ogawa (✉)
Mathematical Institute, Tohoku University, Sendai 980-8578, Japan
e-mail: ogawa@math.tohoku.ac.jp

Y. Yamane
Hakuou Sinior Highschool, Kurikoma, Miyagi 989-5502, Japan
e-mail: yuuki_yamane@hakuou.myswan.ne.jp

Y. Maekawa and S. Jimbo (eds.), *Mathematics for Nonlinear Phenomena—Analysis and Computation*, Springer Proceedings in Mathematics & Statistics 215,
DOI 10.1007/978-3-319-66764-5_10

$$\begin{cases} \partial_t u - \Delta u = u^p, & t > 0, x \in \mathbb{R}^n, \\ u(0, x) = u_0(x) \geq 0, & x \in \mathbb{R}^n. \end{cases} \tag{1}$$

The critical problem to (1) originally goes back to the celebrated result to the incompressible Navier-Stokes equation by H. Fujita and T. Kato [4] (see also Kato-Fujita [9]) and generalization to (1) was realized by Y. Giga [5].

The global existence and well-posedness issue for the Cauchy problem (1) are considered by H. Fujita's pioneering work [3] and his result revealed that there appears the critical exponent $p = 1 + \frac{2}{n}$ which separates the large time behavior of the solutions. Namely, if the power of the nonlinear term p satisfies $p < 1 + \frac{2}{n}$, then any positive solution to (1) blows up in a finite time even the size of the initial data is small (H. Fujita [3], see also [6, 10, 16]). In general if $p > 1 + \frac{2}{n}$, then the local existence and life span of the solution can be determined by the size of norm of the initial data.

The nonlinear heat equation has a scaling invariant property: For a solution u to (1), and a scaling parameter $\lambda > 0$

$$u_\lambda(t, x) = \lambda^{\frac{2}{p-1}} u(\lambda^2 t, \lambda x), \quad \lambda > 0 \tag{2}$$

also satisfies the equation itself and the invariant space under the scaling (2) is given by

$$\|u_\lambda\|_{L^\theta(\mathbb{R}_+; L^r(\mathbb{R}^n))} = \|u\|_{L^\theta(\mathbb{R}_+; L^r(\mathbb{R}^n))}$$

with the scaling balance

$$\frac{2}{\theta} + \frac{n}{r} = \frac{2}{p-1}.$$

Such a scaling invariance suggests that even the local well-posedness is also related to the criticality of the power of the nonlinearity (1).

Definition. Let the spacial dimension $n \geq 1$ and the power of the nonlinear term $p > 1$ and set $q_c \geq 1$

$$\frac{n}{q_c} = \frac{2}{p-1}. \tag{3}$$

We call a pair of the exponent (θ, r) is L^{q_c}-admissible (of Serrin type) if it holds

$$\frac{2}{\theta} + \frac{n}{r} = \frac{n}{q_c},$$

where $p = 1 + \frac{2}{n}$ and $q_c \neq 1$ (Fig. 1).
We show the relation of the L^q-admissible (θ, r) for lower dimensions by the following figures (Figs. 2 and 3):

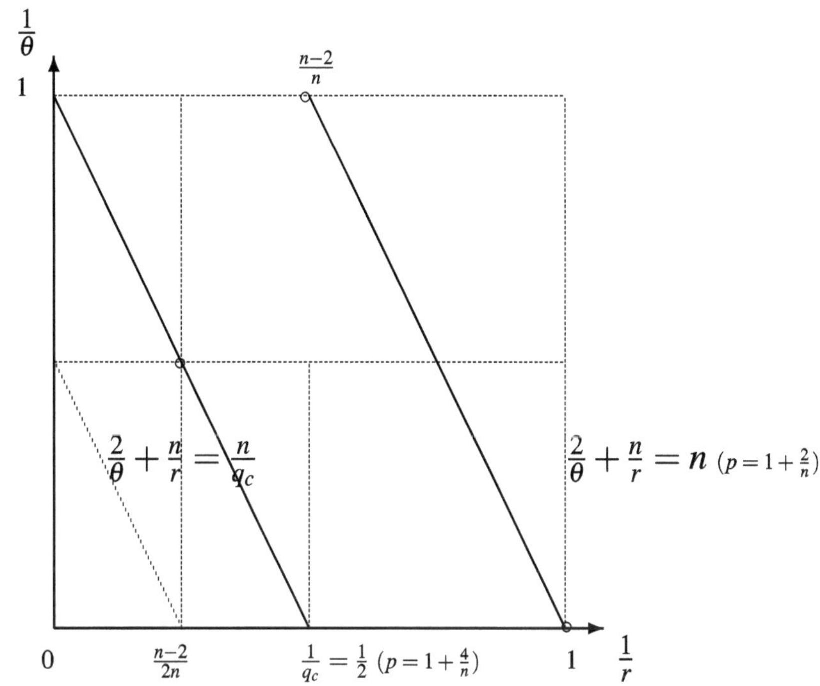

Fig. 1 L^q-admissible (θ, r)

Definition. Let $n \geq 1$, $1 \leq p \leq \infty$. Then we call u a *mild solution to* (1) if the corresponding integral equation

$$u(t) = e^{t\Delta}u_0 + \int_0^t e^{(t-s)\Delta}u^p(s)ds \tag{4}$$

holds in $C([0, T); L^p(\mathbb{R}^n))$, where $e^{t\Delta}$ denotes the heat semigroup given by the convolution of the heat kernel in $L^p(\mathbb{R}^n)$.

Definition. For $1 \leq \theta, r \leq \infty$ and for $T > 0$, with $I = (0, T)$, let space-time estimate L^θ

$$L^\infty_\theta(I; L^r) \equiv \left\{ f; I \times \mathbb{R}^n \to \mathbb{R}; \sup_{t \in I} t^{\frac{1}{\theta}} \| f(t) \|_{L^r} < \infty \right\}$$

with the norm by

$$\| f \|_{L^\infty_\theta(I;L^r)} \equiv \sup_{t \in I} t^{\frac{1}{\theta}} \| f(t) \|_{L^r}.$$

The critical well-posedness due to Weissler can be summarized as follows:

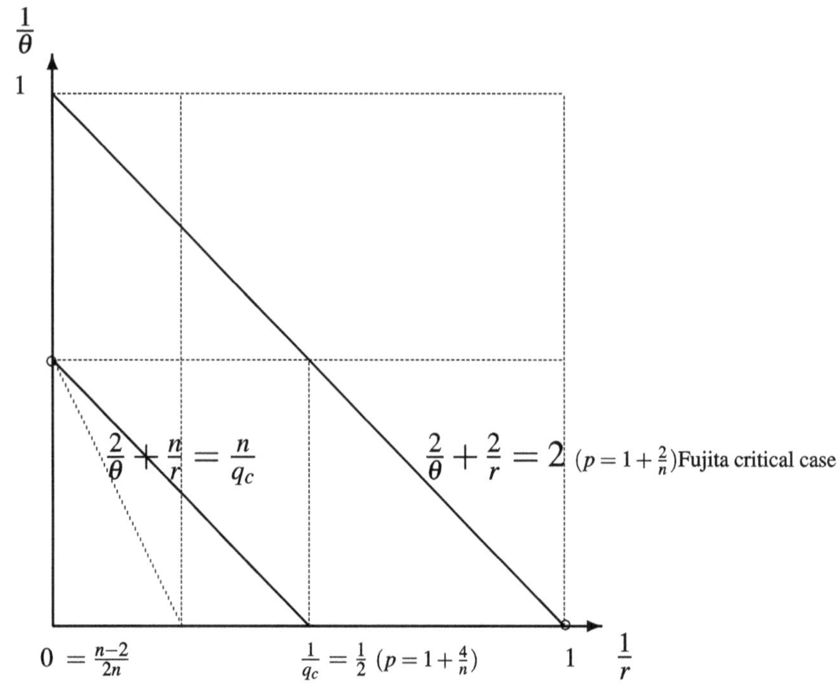

Fig. 2 L^q-admissible (θ, r) in $n = 2$.

Proposition 1.1 **(Weissler** [17]) *Let* $n \geq 3$ *and* $q < r$. *Assume that*

$$\begin{cases} q_c < 1 & \textit{for } q \geq 1, \\ q_c = 1 & \textit{for } q > 1, \\ q_c > 1 & \textit{for } q \geq q_c, \end{cases}$$

where q_c *is defined by* (3). *Let* (θ, r) *be an* L^{q_c}*-admissible with* $r > q$. *Then for any initial data* $u_0 \in L^q(\mathbb{R}^n)$, *there exists* $T = T(u_0) > 0$ *and an* L^q*-mild solution* u *to* (1) *in* $C(I; L^q(\mathbb{R}^n)) \cap L^\infty_\theta(I; L^r(\mathbb{R}^n))$ *and satisfies the following conditions:*

(1) The solution is continuous on the data in $C(I; L^q(\mathbb{R}^n)) \cap L^\infty_\theta(I; L^r)$.
(2) $u \in C^1(I; L^q) \cap C(I; W^{2,q}(\mathbb{R}^n))$ *and the solution is a strong solution to* (1) *in* L^q.
(3) For the maximal existence time $T_* \leq \infty$, *if* $T_* < \infty$ *then*

$$\varlimsup_{t \to T_*} \|u(t)\|_q = \infty.$$

The result in Proposition 1.1 does not include the limiting case $r = q_c = 1$ if the nonlinear exponent is the Fujita critical case $p = 1 + \frac{2}{n}$. In fact, H. Brezis and T. Cazenave [1] showed that any nonnegative solution does not exist in a setting

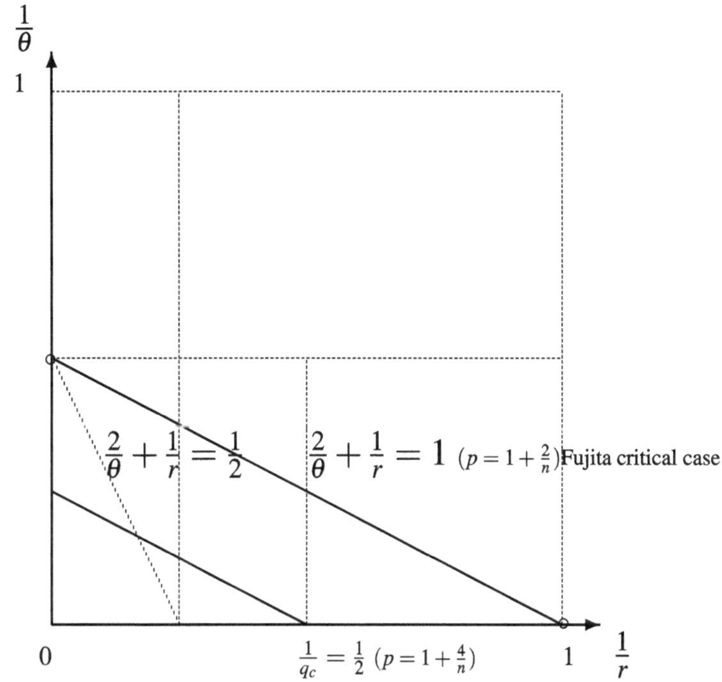

Fig. 3 L^q-admissible (θ, r) in $n = 1$.

for the initial boundary value problem in a bounded domain with the 0-Dirichlet boundary condition. For the initial value problem, one can easily obtain that for any parameter $M > 0$, there exists a initial data u_0 in L^1 with $\|u_0\|_1 = 1$ such that

$$\left\| \int_0^t e^{(t-s)\Delta} u^{1+\frac{2}{n}}(s)ds \right\|_{L^\infty(I;L^1)} \geq M.$$

Indeed,

$$\begin{aligned} L &= \sup_{t>0} \int_{\mathbb{R}^n} \left| \int_0^t \int_{\mathbb{R}^n} \frac{1}{(4\pi(t-s))^{\frac{n}{2}}} e^{-\frac{|x-y|^2}{4(t-s)}} u(s,y)^{1+\frac{2}{n}} dyds \right| dx \\ &\quad \text{(letting } t-s=r\text{)} \\ &= \sup_{t>0} \int_0^t \int_{\mathbb{R}^n} \left(\int_{\mathbb{R}^n} \frac{1}{(4\pi tr)^{\frac{n}{2}}} e^{-\frac{|x-y|^2}{4tr}} dx \right) u(t-r,y)^{1+\frac{2}{n}} dydr \\ &= \sup_{t>0} \int_0^t \int_{\mathbb{R}^n} u(t-r,y)^{1+\frac{2}{n}} dydr. \end{aligned} \tag{5}$$

If we assume the non-negativity of the solution to (1), then by a comparison theorem,

$$u(t) = e^{t\Delta}u_0 + \int_0^t e^{(t-s)\Delta}u^p(s)ds \geq e^{t\Delta}u_0. \tag{6}$$

Hence for any $\mu > 0$, we choose $u_0(x) = G_\mu(x)$, it follows $e^{t\Delta}G_s(x) = G_t(x) * G_s(x) = \mathscr{F}^{-1}[e^{-t|\xi|^2}e^{-s|\xi|^2}] = G_{t+s}(x)$ by the semigroup property of the heat kernel and the solution for the critical semilinear heat equation can be estimated from (5) and (6):

$$\begin{aligned}
L &\geq \sup_{t>0}\int_0^t\int_{\mathbb{R}^n}\left(\int_{\mathbb{R}^n} G_{t-r}(y-z)G_\mu(z)dz\right)^{1+\frac{2}{n}} dydr\\
&= \sup_{t>0}\int_0^t\int_{\mathbb{R}^n}\left(\frac{1}{(4\pi(t-r+\mu))^{\frac{n}{2}}}e^{-\frac{|y|^2}{4(t-r+\mu)}}\right)^{1+\frac{2}{n}} dydr\\
&\geq \int_0^t \frac{1}{4\pi(t-r+\mu)}\int_{\mathbb{R}^n}\frac{1}{(4\pi(t-r+\mu))^{\frac{n}{2}}}e^{-(1+\frac{2}{n})\frac{|y|^2}{4(t-r+\mu)}}dydr \qquad (7)\\
&= \left(1+\frac{2}{n}\right)^{-\frac{n}{2}}\int_0^t \frac{1}{4\pi(t-r+\mu)}\int_{\mathbb{R}^n} G_{(t-r+\mu)}(y)dydr\\
&= \left(1+\frac{2}{n}\right)^{-\frac{n}{2}}\int_0^t \frac{1}{4\pi(t-r+\mu)}dr = \frac{1}{4\pi}\left(1+\frac{2}{n}\right)^{-\frac{n}{2}}\log\left(1+\frac{t}{\mu}\right).
\end{aligned}$$

The estimate (7) shows that for $t > 0$, there exists an initial data $u_0 \in L^1(\mathbb{R}^n)$ with $\|u_0\|_1 = \|G_\mu\|_1 \equiv 1$ and the norm of the corresponding solution $\|u\|_{L^\infty(I;L^1(\mathbb{R}^n))}$ can be arbitrary large when $\mu \to 0$. This observation shows that the local existence time of the solution at the critical case does not determined only by the size of norm of the data.

On the other hand, if we consider the problem under the setting where the scaling invariance fails such as $L^q(\mathbb{R}^n)$ $(q > 1)$, we may find a solution with $\theta = \frac{2q'}{n}$, for instance.

Y. Giga [5] generalized the existence and uniqueness result for the nonlinear heat equation in the space-time Bochner spaces

$$L^\theta(0, T; L^q(\mathbb{R}^n)), \qquad \frac{2}{\theta} + \frac{n}{p} = \frac{n}{2}, \quad 1 \leq \theta < \infty.$$

This result can be considered as the real interpolation result of Weissler [18]. However, this remarkable extension yields that the analogous L^q-well-posedness holds for the Cauchy problem for the incompressible Navier-Stokes equations and he obtained the local well-posedness even in the end-point critical case $r = q_c = 1$. Namely, the problem is well-posed even for the critical case $n = 2$ and $q = 1$ which is corresponding to the Fujita critical case $p = 1 + \frac{2}{n}$ with $q = 1$. This result reveals the striking difference between nonlinear structure of the nonlinear heat equation and the one to the incompressible Navier-Stokes equation (or its vortex form) in the local existence theory. Nowadays, this result can be extended to some other type of nonlinear evolution equation such as the drift-diffusion system [8].

A finer classification of the well-posedness for (1) is found in the result by R. Laister, J.C. Robinson, M. Sierżęga and A. Vidal-López [13] in the Lebesgue spaces and H. Kozono and M. Yamazaki [11] in the Morrey-Besov (Kozono-Yamazaki) spaces.

For the Fujita type problem, one may consider the extension result for the local well-posedness for (1) in slightly smaller space than $L^1(\mathbb{R}^n)$. However, this is not the case even for smaller class such as $B^0_{1,1}(\mathbb{R}^n)$, where $B^s_{q,\sigma}(\mathbb{R}^n)$ is the inhomogeneous Besov space defined by a norm

$$\|f\|_{B^s_{q,\sigma}} \equiv \left(\|\psi * f\|_q^\sigma + \sum_{j\geq 0} 2^{sj\sigma} \|\phi_j * f\|_q^\sigma \right)^{1/\sigma}$$

for $1 \leq q, \sigma \leq \infty$ and $s \in \mathbb{R}$, where $\{\phi_j\}_{j\geq 0}$ is the Littlewood-Paley dyadic decomposition of unity and $\psi \equiv I - \sum_{j\geq 0} \phi_j$. We note that the inhomogeneous Besov space

$$B^0_{1,1}(\mathbb{R}^n) \subsetneq L^1(\mathbb{R}^n)$$

and it is also an invariant space under the scaling transformation (2). The heat kernel $G_\mu(x)$ belongs to all the scaling critical Besov spaces as $G_\mu \in B^0_{1,1}(\mathbb{R}^n) \subset B^{-n+\frac{n}{q}}_{q,1}(\mathbb{R}^n)$ $(1 < q \leq \infty)$ and the observation in (7) is also valid in the Besov space $B^0_{1,1}(\mathbb{R}^n)$ by the embedding (1.1), the norm of the initial data $\|G_\mu\|_{B^0_{1,1}}$ is depending on μ as

$$\|G_\mu\|_{B^0_{1,1}} = O(\log\langle\mu^{-\frac{1}{2}}\rangle) \tag{8}$$

when $\mu \to 0$ (see Appendix below) and this does not directly give the ill-posedness on this problem in the critical Besov spaces. Cazenave and Weissler [2] consider the Cauchy problem for the nonlinear heat equations (1) in the homogeneous Besov space represented by the heat semigroup and show the global small solution in $\dot{B}^{-2\beta}_{q,\infty}(\mathbb{R}^n)$ with $2\beta = \frac{2}{p-1} - \frac{n}{q}$ for $p > 1 + \frac{2}{n}$ and $q > q_c$ (this is partially shown in the earlier work [18]). However, the Fujita critical case $p = 1 + \frac{2}{n}$ is excluded because of the non-existence results for the positive solutions. T. Iwabuchi and the first author showed in [7] that the two-dimensional Fujita critical problem (1) is ill-posed in the scaling critical Sobolev space $H^{-1}(\mathbb{R}^2)$ in the sense that the solution is strongly discontinuous at $t \to 0$. The argument there is also valid for the special case of the scaling invariant Besov space such as $B^{-1}_{2,1}(\mathbb{R}^2)$. This result suggests that the scaling critical Besov space is not enough regular to ensure the local well-posedness to the problem (1). In this paper, we introduce a generalized Besov space that gives the almost scaling invariant but a slightly narrower condition on the high frequency of the function and establish the time local well-posedness for the Cauchy problem (1). We should like to note that our argument does not depend on the method of the comparison principle, and hence one can observe that the similar result also holds for the system of the nonlinear heat equation with the Fujita critical exponents.

On the other hand, we should like to notice that the ill-posedness of the problem is considered for the case $n = 2$ and $p = 1 + \frac{2}{n}$ in the Sobolev space with the negative regularity index $H^{-1}(\mathbb{R}^2)$ in [7]. The argument there also suggests that the problem is ill-posed even in the narrower Besov space $B^{-1}_{2,1}(\mathbb{R}^2)$ and the strong discontinuity at the initial time can be shown below. Indeed, the argument in (7) can be verified in the scaling critical Besov spaces. We show this in the appendix below.

If we restrict the data in the homogeneous Besov space that maintains the scaling invariant, then the solution to (1) may exist. This is corresponding to the case of the solution that is not necessarily nonnegative restriction.

2 Local Well-Posedness for the Critical Case

We consider the local well-posedness for the critical exponent $p = 1 + \frac{2}{n}$ of (1) in the similar scaling property as in L^1. The existence of the nonnegative solution for the problem (1) in the other space than L^1.

Definition (Modified Besov space). Let $1 \le q, \sigma \le \infty$, $s, \tilde{s} \in \mathbb{R}$. The inhomogeneous Besov space $B^{s,\log\tilde{s}}_{q,\sigma}$ is defined by

$$\|f\|_{B^{s,\log\tilde{s}}_{q,\sigma}} \equiv \left(\|\psi * f\|_q^\sigma + \sum_{j\ge 0} \left(j^{\tilde{s}} 2^{sj} \|\phi_j * f\|_q \right)^\sigma \right)^{1/\sigma} < \infty$$

where $\{\phi_j\}_{j\in\mathbb{Z}}$ is the Littlewood-Paley dyadic decomposition. Analogously, the homogeneous Besov space $\dot{B}^{s,\log\tilde{s}}_{q,\sigma}$ is given by the condition

$$\|f\|_{\dot{B}^{s,\log\tilde{s}}_{q,\sigma}} \equiv \left(\sum_{j\in\mathbb{Z}} \left(j_+^{\tilde{s}} 2^{sj} \|\phi_j * f\|_q \right)^\sigma \right)^{1/\sigma} < \infty,$$

where $j_+ = \max(j, 1)$.

If we choose the regularity exponent as $s = -n + \frac{n}{q}$, then the critical space to the Cauchy problem (1) coincides $B^{-n+\frac{n}{r}}_{r,1}(\mathbb{R}^n)$ and then $B^{-n+\frac{n}{r},\log\tilde{s}}_{r,1}(\mathbb{R}^n)$ is slightly narrower space than $B^{-n+\frac{n}{r}}_{r,1}(\mathbb{R}^n)$ for the high-frequency part of the function.

Theorem 2.1 *Let $p = 1 + \frac{2}{n}$, $1 \le r \le 2$ and $\tilde{s} > 1$.*

(1) Assume that $n = 1, 2$. For any initial data $u_0 \in B^{-n+\frac{n}{r},\log\tilde{s}}_{r,1}(\mathbb{R}^n)$, there exists a time $T > 0$ and a unique mild solution u in (4) to the problem (1) in $C([0,T); B^{-n+\frac{n}{r},\log\tilde{s}}_{r,1}(\mathbb{R}^n)) \cap L^{\frac{2}{n}}((0,T); B^{\frac{n}{r},\log\tilde{s}}_{r,1}(\mathbb{R}^n))$ satisfies the following conditions:

(1-1) $u \in C^1((0,T); B_{r,1}^{-n+\frac{n}{r},\log\tilde{s}}) \cap C((0,T); B_{r,1}^{\frac{n}{r},\log\tilde{s}})$,
(1-2) *the solution map* $u_0 \to u$ *is Lipschitz continuous.*

(2) Let $N \in \mathbb{N}$ *and* $n = 2$. *Then there exists a sequence of time* $T_N > 0$ *and sequence of initial data* $u_{0,N} \in B_{2,1}^{-\frac{n}{2}}(\mathbb{R}^n)$ *such that the corresponding sequence of the (mild) solution* (4) *to* (1), u_N *is in* $C([0,T); L^2(\mathbb{R}^n))$ *and satisfies the following:*

(2-1) $u_{0,N} \to 0$ *in* $B_{2,1}^{-n+\frac{n}{2}}(\mathbb{R}^n)$, $T_N \to 0$ $(N \to \infty)$.
(2-2)

$$\lim_{N\to\infty} \|u_N(T_N)\|_{B_{2,1}^{-n+\frac{n}{2}}} = \infty.$$

Namely, the equation is not well-posed in the class $B_{2,1}^{-n+\frac{n}{2}}(\mathbb{R}^n)$.

It is known that for the case $n = 2$, $r = 2$, the Cauchy problem (1) for the class $C([0,T) : H^{-1}(\mathbb{R}^2))$ is ill-posed for a data $H^{-1}(\mathbb{R}^2)$, namely the continuous dependence of the initial data to the solution does not hold. The latter half of the above Theorem states that the ill-posedness to the continuity of the initial data is strongly destroyed:

$$B_{2,1}^{-1,\log\tilde{s}}(\mathbb{R}^2) \subsetneq B_{2,1}^{-1}(\mathbb{R}^2) \subsetneq H^{-1}(\mathbb{R}^2) = B_{2,2}^{-1}(\mathbb{R}^2).$$

Remark: It is shown that the uniqueness of a mild solution is given by the following regularity criterion $L^{\frac{2}{n}}((0,T); B_{q,1}^{\frac{n}{q}}(\mathbb{R}^n))$ with the relation. This condition has the same scale as the invariant scaling is invariant as

$$\frac{2}{\theta} + \frac{n}{r} = n = \frac{2}{p-1}$$

and letting with $r = q, \theta = \frac{2}{n}$ and $s = \frac{n}{q}$,

$$\frac{2}{2/n} + \frac{n}{q} - \frac{n}{q} = n.$$

For the case $L^\infty(\mathbb{R}_+; B_{r,1}^s)$,

$$\frac{2}{\theta} + \frac{n}{r} - s = n$$

and letting $s = -n + \frac{n}{r}, \theta = \infty$

$$\frac{2}{\infty} + \frac{n}{r} + n - \frac{n}{r} = n$$

or letting $s = \frac{n}{r}, \theta = \frac{2}{n}, r = q$,

$$\frac{2}{\frac{2}{n}} + \frac{n}{q} - \frac{n}{r} = n.$$

Remark: The higher dimensional cases $n \geq 3$ are open so far.

We should notice that Weissler [18] proved the remarkable existence and uniqueness result to (1) for the data $u_0 \in L^1(\mathbb{R}^n) \cap B^{-n}_{\infty,\frac{2}{n}}(\mathbb{R}^n)$ for $n = 1, 2$. However, the consistency and the continuous dependence on the initial data in the space seems to be not known. The method in [18] is depending on the comparison principle and seems to be not applicable to the system of the nonlinear heat equation while ours is applicable.

Since the function space is not the scaling critical, the global existence of the small data solution is not directly obtained. Indeed, the existence of the global solution should be true however our method cannot be applied for obtaining the global solution.

3 Preliminaries

We first illustrate the product formula in the modified Besov spaces.

Lemma 3.1 *Let $1 \leq p, q \leq \infty$, $\frac{1}{p} + \frac{1}{q} \geq 1$ and $\tilde{s} > 1$.*

(1) If $-\frac{n}{q} \leq s < 0$, then there exists a constant $C > 0$ such that for all $f \in \dot{B}^{\frac{n}{q},\log\tilde{s}}_{q,\infty}(\mathbb{R}^n) \cap \dot{B}^0_{\infty,1}(\mathbb{R}^n)$ and $g \in \dot{B}^{s,\log\tilde{s}}_{p,1}(\mathbb{R}^n)$,

$$\|fg\|_{\dot{B}^{s,\log\tilde{s}}_{p,1}} \leq C\big(\|f\|_{\dot{B}^{\frac{n}{q},\log\tilde{s}}_{q,\infty}} + \|f\|_{\dot{B}^0_{\infty,1}}\big)\|g\|_{\dot{B}^{s,\log\tilde{s}}_{p,1}}. \tag{9}$$

In particular the following estimate holds:

$$\|fg\|_{\dot{B}^{s,\log\tilde{s}}_{p,1}} \leq C\|f\|_{\dot{B}^{\frac{n}{q},\log\tilde{s}}_{q,\infty}}\|g\|_{\dot{B}^{s,\log\tilde{s}}_{p,1}}. \tag{10}$$

(2) If $s = 0$, then there exists a constant $C > 0$ such that the following estimate holds:

$$\|fg\|_{\dot{B}^{0,\log\tilde{s}}_{p,1}} \leq C\|f\|_{\dot{B}^{\frac{n}{q},\log\tilde{s}}_{q,1}}\|g\|_{\dot{B}^{0,\log\tilde{s}}_{p,1}}.$$

(3) There exists a constant $C > 0$ such that for any $f \in B^{\frac{n}{q}}_{q,1}(\mathbb{R}^n)$ and $g \in B^{s,\log\tilde{s}}_{p,1}(\mathbb{R}^n)$,

$$\|fg\|_{B^{s,\log\tilde{s}}_{p,1}} \leq C\|f\|_{B^{\frac{n}{q},\log\tilde{s}}_{q,\infty}}\|g\|_{B^{s,\log\tilde{s}}_{p,1}}. \tag{11}$$

Remark. For $\tilde{s} > 1$, one can exchange the regularity weight by sharper one such as $j\big(\log(1+j)\big)^{\tilde{s}}$ or $j\log(1+j)\big(\log(1+\log(1+j))\big)^{\tilde{s}}$ and the product rule may hold similar way. In the end, we expect that the weight condition can be relaxed into the following Dini type condition: Let the weight function by $w(j)$, then

$$\sum_{j\ge 1}\frac{1}{w(j)}<\infty.$$

Proof of Lemma 3.1. (1) The estimate (10) is a direct consequence from (9) and the embedding $\dot{B}^{\frac{n}{q}}_{q,1}\subset \dot{B}^{\frac{n}{q}}_{q,\infty}\cap \dot{B}^{0}_{\infty,1}$ where $1\le q<\infty$.
To see (9), we employ the Bony para-product formula. For any $f\ g\in\mathscr{S}(\mathbb{R}^n)$,

$$\begin{aligned}
f(x)\ g(x)=&\sum_{k\in\mathbb{Z}}(\phi_k * f(x))(P_k g(x))+\sum_{k\in\mathbb{Z}}(P_k f(x))(\phi_k * g(x))\\
&+\sum_{k\in\mathbb{Z}}\sum_{|\ell-k|\le 2}(\phi_k * f(x))(\phi_\ell * g(x)) \qquad (12)\\
\equiv& h_1+h_2+h_3,
\end{aligned}$$

where $P_k g=\sum_{\ell\le k-3}\phi_\ell * g=\psi_{2^{-(k-3)}} * g\ \lambda>0$ and $\psi_\lambda=\lambda^{-n}\psi(x/\lambda)$. We notice that the following relation of the support plays a basic role in the decomposition (12).

$$\operatorname{supp}\mathscr{F}((\phi_k * f)(P_k g))\subset\{\xi\in\mathbb{R}^n;\ 2^{k-2}\le|\xi|\le 2^{k+2}\}.$$

Applying the Minkowski inequality and $1/p_1+1/p_2=1/p$, by setting $j_+=\max(j,1)$, we see that

$$\begin{aligned}
\|h_1\|_{\dot{B}^{s,\log\tilde{s}}_{p,1}}=&\sum_{j\in\mathbb{Z}}j_+^{\tilde{s}}2^{sj}\Big\|\sum_{|k-j|\le 2}\phi_j*((\phi_j*f)(P_k g))\Big\|_p\\
\le&\sum_{j\in\mathbb{Z}}j_+^{\tilde{s}}2^{sj}\sum_{|\ell|\le 2}\Big\|\phi_j*((\phi_{j+\ell}*f)(P_{j+\ell}g))\Big\|_p\\
\le& C\sum_{j\in\mathbb{Z}}\sum_{|\ell|\le 2}(j+\ell)_+^{\tilde{s}}2^{s(j+\ell)}\|\phi_j\|_1\|\phi_{j+\ell}*f\|_{p_1}\|P_{j+\ell}g\|_{p_2}\\
&(\text{letting } j+\ell=k)\\
\le& C\sum_{|\ell|\le 2}\sum_{k\in\mathbb{Z}}\|\phi_k*f\|_{p_1}\sup_{k\in\mathbb{Z}}k_+^{\tilde{s}}2^{sk}\|P_k g\|_{p_2} \qquad (13)\\
\le& C\sum_{k\in\mathbb{Z}}\|\phi_k*f\|_{p_1}\sup_{k\in\mathbb{Z}}\sum_{m\le k-3}\left(\frac{k_+}{m_+}\right)^{\tilde{s}}m_+^{\tilde{s}}2^{sm}2^{s(k-m)}\|\phi_m*g\|_{p_2}\\
&(\text{if } s<0 \text{ then } m\le k-3 \text{ and } (k_+/m_+)^{\tilde{s}}2^{s(k-m)}\le C_{r,s} \text{ gives})\\
\le& C\sum_{k\in\mathbb{Z}}\|\phi_k*f\|_{p_1}\sup_{k\in\mathbb{Z}}\sum_{m\le k-3}m_+{}^{\tilde{s}}2^{sm}\|\phi_m*g\|_{p_2}\\
\le& C\|f\|_{\dot{B}^0_{p_1,1}}\|g\|_{\dot{B}^{s,\log\tilde{s}}_{p_2,1}}.
\end{aligned}$$

We have used $s < 0$ in the estimate in (13). When $s = 0$, it follows that

$$
\begin{aligned}
\|h_1\|_{\dot{B}^{s,\log \tilde{s}}_{p,1}} =&\sum_{j\in\mathbb{Z}} j_+^{\tilde{s}} 2^{sj} \Big\| \sum_{|k-j|\le 2} \phi_j * ((\phi_j * f)(P_k g))\Big\|_p \\
\le& C\sum_{j\in\mathbb{Z}}\sum_{|\ell|\le 2} (j+\ell)_+^{\tilde{s}} 2^{s(j+\ell)} \|\phi_j\|_1 \|\phi_{j+\ell} * f\|_{p_1} \|P_{j+\ell} g\|_{p_2} \\
&\text{(letting } j+\ell = k) \\
\le& C\sum_{|\ell|\le 2}\sum_{k\in\mathbb{Z}} k_+^{\tilde{s}} \|\phi_k * f\|_{p_1} \sup_{k\in\mathbb{Z}} 2^{sk} \|P_k g\|_{p_2} \\
\le& C\sum_{k\in\mathbb{Z}} k_+^{\tilde{s}} \|\phi_k * f\|_{p_1} \sup_{k\in\mathbb{Z}} \sum_{m\le k-3} 2^{sm} 2^{s(k-m)} \|\phi_m * g\|_{p_2} \\
&\text{(when } s = 0,\ 2^{sm}2^{s(k-m)} = 1 \text{ and)} \\
\le& C\sum_{k\in\mathbb{Z}} k_+^{\tilde{s}} \|\phi_k * f\|_{p_1} \sup_{k\in\mathbb{Z}} \sum_{m\le k-3} m_+^{\tilde{s}} \|\phi_m * g\|_{p_2} \\
\le& C\|f\|_{\dot{B}^{s,\log \tilde{s}}_{p_1,1}} \|g\|_{\dot{B}^{s,\log \tilde{s}}_{p_2,1}}.
\end{aligned}
\tag{14}
$$

The second term of the right-hand side in (12) can be estimated using $1/q_1 + 1/q_2 = 1/p$ that

$$
\begin{aligned}
\|h_2\|_{\dot{B}^{s,\log \tilde{s}}_{p,1}} =&\sum_{j\in\mathbb{Z}} j_+^{\tilde{s}} 2^{sj} \Big\| \sum_{|k-j|\le 2} \phi_j * ((P_k f)(\phi_j * g))\Big\|_p \\
\le& \sum_{j\in\mathbb{Z}} j_+^{\tilde{s}} 2^{sj} \sum_{|\ell|\le 2} \Big\| \phi_j * ((P_{j+\ell} f)(\phi_{j+\ell} * g))\Big\|_p \\
\le& C\sum_{j\in\mathbb{Z}}\sum_{|\ell|\le 2} (j+\ell)_+^{\tilde{s}} 2^{s(j+\ell)} \|\phi_j\|_1 \|P_{j+\ell} f\|_{q_1} \|\phi_{j+\ell} * g\|_{q_2} \\
\le& C\|f\|_{q_1} \sum_{k\in\mathbb{Z}} k_+^{\tilde{s}} 2^{sk} \|\phi_k * g\|_{q_2} \\
\le& C\|f\|_{q_1} \|g\|_{\dot{B}^{s,\log \tilde{s}}_{q_2,1}}.
\end{aligned}
\tag{15}
$$

Here the exponents (p_1, p_2) and (q_1, q_2) are subject to

$$\frac{1}{p_1} + \frac{1}{p_2} = \frac{1}{q_1} + \frac{1}{q_2} = \frac{1}{p}$$

but chosen independently. Finally, for the third term of the right hand side of (12), we note that

$$\operatorname{supp} \mathscr{F}(\phi_k * f \cdot \phi_\ell * g) \subset \{\xi \in \mathbb{R}^n;\ |\xi| \le 2^{\max\{k,\ell\}+2}\}$$

and

$$\phi_j * ((\phi_k * f)(\phi_\ell * g)) = 0, \quad \text{where} \quad \max\{k, \ell\} \le j - 3.$$

Then by setting $\tilde{\phi}_k = \phi_{k-1} + \phi_k + \phi_{k+1}$ and $k = j + k'$, it follows $k \ge j - 2$ and $k' \ge -2$ that

$$\begin{aligned}\|h_3\|_{\dot{B}^{s,\log\tilde{s}}_{p,1}} =& \sum_{j\in\mathbb{Z}} j_+^{\tilde{s}} 2^{sj} \Big\| \sum_{j-2\le\max\{k,\ell\}} \sum_{|k-\ell|\le 2} \phi_j * ((\phi_k * f)(\phi_\ell * g)) \Big\|_p \\ \le& \sum_{j\in\mathbb{Z}} j_+^{\tilde{s}} 2^{sj} \sum_{k'\ge -2} \|\phi_j\|_a \|\phi_{j+k'} * f\|_{q_1} \|\widetilde{\phi}_{j+k'} * g\|_{q_2},\end{aligned}$$

where q_1 and q_2 satisfy

$$\frac{1}{p} + 1 = \frac{1}{a} + \frac{1}{q_1} + \frac{1}{q_2}.$$

Choosing $a = q'$ and $\|\phi_j\|_a = 2^{nj(1-\frac{1}{q'})}\|\phi\|_{q'}$,

$$\begin{aligned}\|h_3\|_{\dot{B}^{s,\log\tilde{s}}_{p,1}} \le& C \sum_{j\in\mathbb{Z}} j_+^{\tilde{s}} 2^{sj} \sum_{k'\ge -2} 2^{\frac{nj}{q}} \|\phi_{j+k'} * f\|_{q_1} \|\widetilde{\phi}_{j+k'} * g\|_{q_2} \\ \le& C \sum_{j\in\mathbb{Z}} j_+^{\tilde{s}} 2^{sj} \sum_{k'\ge -2} 2^{\frac{nj}{q}} 2^{-\frac{nk'}{q}} 2^{\frac{nk'}{q}} \|\phi_{j+k'} * f\|_{q_1} \|\widetilde{\phi}_{j+k'} * g\|_{q_2} \\ \le& C \sum_{j\in\mathbb{Z}} j_+^{\tilde{s}} \sum_{k'\ge -2} 2^{s(j+k')} 2^{\frac{n}{q}(j+k')} 2^{-(s+\frac{n}{q})k'} \|\phi_{j+k'} * f\|_{q_1} \|\widetilde{\phi}_{j+k'} * g\|_{q_2} \\ \le& C \sum_{j\in\mathbb{Z}} \Bigg(\sum_{k'\ge -2} \left(\frac{j_+}{(j+k')_+}\right)^{\tilde{s}} \left(\frac{(j_+ + k')_+}{k'_+}\right)^{\tilde{s}} 2^{\frac{n}{q}(j+k')} \|\phi_{j+k'} * f\|_{q_1} \\ & \times (j+k')_+^{\tilde{s}} 2^{s(j+k')} 2^{-(s+\frac{n}{q})k'} \|\widetilde{\phi}_{j+k'} * g\|_{q_2} \Bigg) \\ \le& C \sup_{j\in\mathbb{Z}} \Big(\sup_{k'\ge -2} (j_+ + k')_+^{\tilde{s}} 2^{\frac{n}{q}(j+k')} \|\phi_{j+k'} * f\|_{q_1} \Big) \\ & \times \sum_{j\in\mathbb{Z}} \sum_{k'>-2} \left(\frac{(j+k')_+}{k'_+}\right)^{\tilde{s}} 2^{-(s+\frac{n}{q})k'} 2^{s(j+k')} \|\widetilde{\phi}_{j+k'} * g\|_{q_2} \\ \le& C \sup_{j\in\mathbb{Z}} \Big(\sup_{k'\ge -2} (j_+ + k')_+^{\tilde{s}} 2^{\frac{n}{q}(j+k')} \|\phi_{j+k'} * f\|_{q_1} \Big) \\ & \times \sum_{j\in\mathbb{Z}} \Bigg(\sum_{k'>-2} {k'_+}^{-\tilde{s}} 2^{-(s+\frac{n}{q})k'} \Bigg) (j+k')_+^{\tilde{s}} 2^{s(j+k')} \|\widetilde{\phi}_{j+k'} * g\|_{q_2} \\ & \text{(setting } j + k' = j') \\ \le& C \sup_{j'\in\mathbb{Z}} (j_+ + k')_+^{\tilde{s}} 2^{\frac{n}{q}(j+k')} \|\phi_{j+k'} * f\|_{q_1}\end{aligned}$$

$$\times \sum_{j'\in\mathbb{Z}} \left(\sum_{k'>-2} {k'_+}^{-\tilde{s}} 2^{-(s+\frac{n}{q})k'} \right) {j'}_{+}^{\tilde{s}} 2^{sj'} \|\widetilde{\phi}_{j'} * g\|_{q_2}. \tag{16}$$

Hence if $s + \frac{n}{q} > 0$ then

$$\sum_{k'>-2} {k'_+}^{-\tilde{s}} 2^{-(s+\frac{n}{q})k'} \leq \sum_{k'>-2} 2^{-(s+\frac{n}{q})k'}$$

converges. Furthermore if $s + \frac{n}{q} = 0$ then $2^{-(s+\frac{n}{q})k'} = 1$, but since $\tilde{s} > 1$,

$$\sum_{k'>-2} {k'_+}^{-\tilde{s}} 2^{-(s+\frac{n}{q})k'} \leq C + \sum_{k'>0} {k'_+}^{-\tilde{s}}$$

converges. Hence

$$\begin{aligned}
\|h_3\|_{\dot{B}^{s,r}_{p,1}} \leq& C \sup_{j\in\mathbb{Z}} \Big(\sup_{k'\geq -2} (j_+ + k')_+^{\tilde{s}} 2^{\frac{n}{q}(j+k')} \|\phi_{j+k'} * f\|_{q_1} \Big) \\
&\times \sum_{j\in\mathbb{Z}} {j'_+}^{\tilde{s}} 2^{sj'} \|\widetilde{\phi}_{j'} * g\|_{q_2} \\
\leq& C \Big\{ \sup_{j<-k'} \big(\sup_{k'\geq -2} (j_+ + k')_+^{\tilde{s}} 2^{\frac{n}{q}(j+k')} \|\phi_{j+k'} * f\|_{q_1} \big) \\
&+ \sup_{-k'\leq j\leq 0} \big(\sup_{k'\geq -2} (j_+ + k')_+^{\tilde{s}} 2^{\frac{n}{q}(j+k')} \|\phi_{j+k'} * f\|_{q_1} \big) \\
&+ \sup_{0<j} \big(\sup_{k'\geq -2} (j_+ + k')_+^{\tilde{s}} 2^{\frac{n}{q}(j+k')} \|\phi_{j+k'} * f\|_{q_1} \big) \Big\} \\
&\times \sum_{j\in\mathbb{Z}} {j'_+}^{\tilde{s}} 2^{sj'} \|\widetilde{\phi}_{j'} * g\|_{q_2} \qquad (17) \\
\leq& C \Big\{ \sup_{j<-k'} \big(\sup_{k'\geq -2} (j_+ + k')_+^{\tilde{s}} 2^{\frac{n}{q}(j+k')} \|\phi_{j+k'} * f\|_{q_1} \big) \\
&+ \sup_{-k'\leq j\leq 0} \big(\sup_{k'\geq -2} (j_+ + k')_+^{\tilde{s}} 2^{\frac{n}{q}(j+k')} \|\phi_{j+k'} * f\|_{q_1} \big) \\
&+ \big(\sup_{k'\geq -2} {k'_+}^{\tilde{s}} 2^{\frac{n}{q}(k')} \|\phi_{k'} * f\|_{q_1} \big) \Big\} \\
&\times \sum_{j\in\mathbb{Z}} {j'_+}^{\tilde{s}} 2^{sj'} \|\widetilde{\phi}_{j'} * g\|_{q_2}
\end{aligned}$$

(here the term of f includes the gap for convergence if $j < 0$)

$$\leq C \|f\|_{\dot{B}^{\frac{n}{q},\log\tilde{s}}_{q_1,\infty}} \|g\|_{\dot{B}^{s,\log\tilde{s}}_{q_2,1}}.$$

Choosing $q_2 = p$ and $q_1 = q$ in (17),

$$\|h_3\|_{\dot{B}^{s,\log\bar{s}}_{p,1}} \le C\|f\|_{\dot{B}^{\frac{n}{q},\log\bar{s}}_{q_1,\infty}}\|g\|_{\dot{B}^{s,\log\bar{s}}_{p,1}}, \tag{18}$$

where we used $-\frac{n}{q} \le s \le 0$. Since

$$\frac{1}{p} + \frac{1}{q} \le 1,$$

we need to assume that $p \ge 1$. Letting $p_1 = \infty$, $p_2 = 1$ in (13) and $q_1 = \infty, q_2 = p$ in (15), the estimates (18) and (12) imply

$$\begin{aligned}
\|fg\|_{\dot{B}^{s,\log\bar{s}}_{p,1}} \le& \|h_1\|_{\dot{B}^{s,\log\bar{s}}_{p,1}} + \|h_2\|_{\dot{B}^{s,\log\bar{s}}_{p,1}} + \|h_3\|_{\dot{B}^{s,\log\bar{s}}_{p,1}} \\
\le& C\|f\|_{\dot{B}^{0,\log\bar{s}}_{p,1}}\|g\|_{\dot{B}^{s,\log\bar{s}}_{p,1}} + C\|f\|_\infty\|g\|_{\dot{B}^{s,\log\bar{s}}_{p,1}} + C\|f\|_{\dot{B}^{\frac{n}{q},\log\bar{s}}_{q,\infty}}\|g\|_{\dot{B}^{s,\log\bar{s}}_{p,1}} \\
\le& C(\|f\|_{\dot{B}^0_{\infty,1}} + \|f\|_{\dot{B}^{\frac{n}{q},\log\bar{s}}_{q,\infty}})\|g\|_{\dot{B}^{s,\log\bar{s}}_{p,1}}.
\end{aligned}$$

(3) To see the inhomogeneous estimate (11), we follow the analogous estimate in the homogeneous case. For $f \in B^{\frac{n}{q}}_{q,1}$ $g \in B^{s,\log\bar{s}}_{p,1}$,

$$\begin{aligned}
f(x)\ g(x) =& \sum_{k\in\mathbb{N}}(\phi_k * f(x))(P_k g(x)) + \sum_{k\in\mathbb{N}}(P_k f(x))(\phi_k * g(x)) \\
&+ \sum_{k\in\mathbb{N}}\sum_{|\ell-k|\le 2}(\phi_k * f(x))(\phi_\ell * g(x)) \\
\equiv& h_1 + h_2 + h_3,
\end{aligned} \tag{19}$$

where we set $P_k g = \sum_{\ell\le k-3}\phi_\ell * g = \psi_{2^{-(k-3)}} * g$ $\lambda > 0$ and $\psi_\lambda = \lambda^{-n}\psi(x/\lambda)$. Sincc

$$\operatorname{supp}\mathscr{F}((\phi_k * f)(P_k g)) \subset \{\xi \in \mathbb{R}^n; 2^{k-2} \le |\xi| \le 2^{k+2}\},$$

we see that the first term of the right-hand side (19) can be estimated by the Minkowski inequality and $1/p_1 + 1/p_2 = 1/p$ that

$$\begin{aligned}
\|h_1\|_{B^{s,\log\bar{s}}_{p,1}} =& \sum_{j\in\mathbb{N}} j^{\bar{s}}2^{sj}\Big\|\sum_{|k-j|\le 2}\phi_j * ((\phi_k * f)(P_k g))\Big\|_p \\
\le& \sum_{j\in\mathbb{N}} j^{\bar{s}}2^{sj}\sum_{|\ell|\le 2}\Big\|\phi_j * ((\phi_{j+\ell} * f)(P_{j+\ell} g))\Big\|_p \\
\le& \sum_{j\in\mathbb{N}}\sum_{|\ell|\le 2} j^{\bar{s}}2^{sj}\|\phi_j\|_1\|\phi_{j+\ell} * f\|_{p_1}\|P_{j+\ell} g\|_{p_2} \\
\le& C\sum_{j\in\mathbb{N}}\sum_{|\ell|\le 2}\|\phi_{j+\ell} * f\|_{p_1}\sup_{j\ge 0} j^{\bar{s}}2^{sj}\Big\|\sum_{k\le j+\ell-3}\phi_k * g\Big\|_{p_2}
\end{aligned} \tag{20}$$

$$
\begin{aligned}
&\le C\sum_{j\in\mathbb{N}}\|\phi_{j+\ell}*f\|_{p_1}\sup_{j\ge 0}\sum_{k\le j}\left(\frac{j}{k}\right)^{\tilde{s}}2^{s(j-k)}k^{\tilde{s}}2^{sk}\|\phi_k*g\|_{p_2}\\
&\qquad(\text{if } s<0 \text{ then } \left(\frac{j}{k}\right)^{\tilde{s}}2^{s(j-k)}\le C)\\
&\le C\|f\|_{B^0_{p_1,1}}\|g\|_{B^{s,\log\tilde{s}}_{p_2,1}}.
\end{aligned}
$$

When $s=0$, we put an extra weight function on f and the estimate can be done similar way to (14). For the second term, we have from (19), that for $1/q_1+1/q_2=1/p$

$$
\begin{aligned}
\|h_2\|_{B^{s,\log\tilde{s}}_{p,1}} &= \sum_{j\in\mathbb{N}} j^{\tilde{s}}2^{sj}\Big\|\sum_{|k-j|\le 2}\phi_j*((P_kf)(\phi_k*g))\Big\|_p\\
&\le \sum_{j\in\mathbb{N}} j^{\tilde{s}}2^{sj}\sum_{|\ell|\le 2}\Big\|\phi_j*((\phi_{j+\ell}*g)(P_{j+\ell}f))\Big\|_p\\
&\le \sum_{j\in\mathbb{N}} j^{\tilde{s}}\sum_{|\ell|\le 2}2^{sj}\|\phi_j\|_1\|\phi_{j+\ell}*g\|_{q_2}\|P_{j+\ell}f\|_{q_1} \qquad (21)\\
&\le C2^{-2s}\|f\|_{q_1}\sum_{j\in\mathbb{N}}\sum_{|\ell|\le 2}(j+\ell)^{\tilde{s}}2^{s(j+\ell)}\|\phi_{j+\ell}*g\|_{q_2}\\
&\le C\|f\|_{q_1}\|g\|_{B^{s,\log\tilde{s}}_{q_2,1}}.
\end{aligned}
$$

We note that the pair of the exponents (p_1, p_2) and (q_1, q_2) are independently chosen. For the third term of (12), we first remark that

$$
\operatorname{supp}\mathscr{F}(\phi_k*f\cdot\phi_\ell*g)\subset\{\xi\in\mathbb{R}^n;\,|\xi|\le 2^{\max\{k,\ell\}+2}\}
$$

and

$$
\phi_j*((\phi_k*f)(\phi_\ell*g))=0,\quad\text{where}\quad \max\{k,\ell\}\le j-3.
$$

Then letting $\tilde{\phi}_k=\phi_{k-1}+\phi_k+\phi_{k+1}$ and changing the suffix $k=j+k'$, we see for $k'\ge -2$ that

$$
\begin{aligned}
\|h_3\|_{B^{s,\log\tilde{s}}_{p,1}} &= \sum_{j\in\mathbb{N}} j^{\tilde{s}}2^{sj}\Big\|\sum_{j-2\le\max\{k,\ell\}}\sum_{|k-\ell|\le 2}\phi_j*((\phi_k*f)(\phi_\ell*g))\Big\|_p\\
&\le \sum_{j\in\mathbb{N}} j^{\tilde{s}}2^{sj}\sum_{k'\ge -2}\|\phi_j\|_r\|\phi_{j+k'}*f\|_{q_1}\|\tilde{\phi}_{j+k'}*g\|_{q_2},
\end{aligned}
$$

where q_1 and q_2 satisfy

$$
\frac{1}{p}+1=\frac{1}{r}+\frac{1}{q_1}+\frac{1}{q_2}. \qquad (22)
$$

Since $\|\phi_j\|_r = 2^{nj(1-\frac{1}{r})}\|\phi\|_r$ we choose $r' = q$ and

$$\begin{aligned}
\|h_3\|_{B^s_{p,1}} \le & C\sum_{j\in\mathbb{N}} j^{\tilde{s}}2^{sj}\sum_{k'\ge -2} 2^{\frac{nj}{r'}}\|\phi_{j+k'}*f\|_{q_1}\|\widetilde{\phi}_{j+k'}*g\|_{q_2} \\
\le & C\sum_{j\in\mathbb{N}} j^{\tilde{s}}2^{sj}\sum_{k'\ge -2} 2^{\frac{nj}{r'}}2^{-\frac{nk'}{q}}2^{\frac{nk'}{q}}\|\phi_{j+k'}*f\|_{q_1}\|\widetilde{\phi}_{j+k'}*g\|_{q_2} \\
\le & C\sup_{j\in\mathbb{N}}\sup_{k'>-2}(j+k')^{\tilde{s}}2^{\frac{n}{q}(j+k')}\|\phi_{j+k'}*f\|_{q_1} \\
& \times\sum_{j\in\mathbb{N}}\sum_{k'\ge -2} 2^{-(\frac{n}{q}+s)k'}k'^{-\tilde{s}}\left(\frac{k'}{j+k'}\right)^{\tilde{s}}\left(\frac{j}{j+k'}\right)^{\tilde{s}}(j+k')^{\tilde{s}}2^{s(j+k')}\|\widetilde{\phi}_{j+k'}*g\|_{q_2} \\
\le & C\sup_{j,k}(j+k')^{\tilde{s}}2^{\frac{n}{q}(j+k')}\|\phi_{j+k'}*f\|_{q_1} \\
& \qquad\times\sum_{j'\in\mathbb{N}+k'}\left(\sum_{k'\ge -2}2^{-(\frac{n}{q}+s)k'}k'^{-\tilde{s}}\right)j'^{\tilde{s}}2^{sj'}\|\widetilde{\phi}_{j'}*g\|_{q_2} \\
\le & C\|f\|_{B^{\frac{n}{q},\log\tilde{s}}_{q_1,\infty}}\|g\|_{B^{s,\log\tilde{s}}_{q_2,1}}.
\end{aligned}$$

Here we need the restriction $-\frac{n}{q}\le s\le 0$. From (22), it holds

$$\frac{1}{r'} = 1-\frac{1}{r} = \frac{1}{q_1}+\frac{1}{q_2}-\frac{1}{p}.$$

Then choosing $q_2 = p$ and $q_1 = q$, $r' = q$ and

$$\|h_3\|_{B^{s,\log\tilde{s}}_{p,1}} \le C\|f\|_{B^{\frac{n}{q},\log\tilde{s}}_{q,\infty}}\|g\|_{B^{s,\log\tilde{s}}_{p,1}}. \tag{23}$$

Here it is required that $q<\infty$. Now we let $p_1=\infty$, $p_2 = 1$ in (20), $q_1=\infty$, $q_2 = p$ in (21) and by (23) and (19),

$$\begin{aligned}
\|fg\|_{B^{s,\log\tilde{s}}_{p,1}} \le & \|h_1\|_{B^{s,\log\tilde{s}}_{p,1}} + \|h_2\|_{B^{s,\log\tilde{s}}_{p,1}} + \|h_3\|_{B^{s,\log\tilde{s}}_{p,1}} \\
\le & C\|f\|_{B^0_{\infty,1}}\|g\|_{B^{s,\log\tilde{s}}_{p,1}} + C\|f\|_\infty\|g\|_{B^{s,\log\tilde{s}}_{p,1}} + C\|f\|_{B^{\frac{n}{q},\log\tilde{s}}_{q,\infty}}\|g\|_{B^{s,\log\tilde{s}}_{p,1}} \\
\le & C(\|f\|_{B^0_{\infty,1}} + \|f\|_{B^{\frac{n}{q},\log\tilde{s}}_{q,\infty}})\|g\|_{B^{s,\log\tilde{s}}_{p,1}}.
\end{aligned}$$

This concludes the required estimate holds. □

4 Maximal Regularity for the Heat Equations

It is known that the Cauchy problem of the heat equation satisfies maximal regularity estimate.

$$\begin{cases} \partial_t u - \Delta u = f, & t > 0,\ x \in \mathbb{R}^n, \\ u(0,x) = u_0(x), & x \in \mathbb{R}^n. \end{cases} \tag{24}$$

The following form of the estimate is a generalization of end-point maximal regularity to (24) shown in [14, chapter 13].

Theorem 4.1 (Generalized end-point maximal regularity) *Let $1 \le \nu \le \rho \le \infty$, $1 \le p \le \infty$ and $I = (0,T) \subset \mathbb{R}_+$ be an open interval. For the external function $f \in L^\nu(I; \dot{B}^{2/\nu}_{p,\nu}(\mathbb{R}^n))$ and the initial data $u_0 \in \dot{B}^{2-2/\rho}_{p,\rho}(\mathbb{R}^n)$, let u be a solution of the Cauchy problem of the heat equation:*

$$\begin{cases} \partial_t u - \Delta u = f, & t > 0,\ x \in \mathbb{R}^n, \\ u(0,x) = u_0(x), & x \in \mathbb{R}^n. \end{cases} \tag{25}$$

(1) If $f \equiv 0$ then for any $1 \le \sigma \le \infty$, it holds

$$\|\partial_t u\|_{L^\rho(I;\dot{B}^0_{p,\sigma})} + \|\nabla^2 u\|_{L^\rho(I;\dot{B}^0_{p,\sigma})} \le C\|u_0\|_{\dot{B}^{2-\frac{2}{\rho}}_{p,\rho}},$$

where $\nabla^2 = \partial_{x_i}\partial_{x_j}$ and the constant $C > 0$ is independent of T.

(2) If $u_0 = 0$ then for any $\nu \le \sigma \le \rho$, it holds

$$\big\|\partial_t u\big\|_{L^\rho(I;\dot{B}^0_{p,\sigma})} + \big\|\nabla^2 u\big\|_{L^\rho(I;\dot{B}^0_{p,\sigma})} \le C\|f\|_{L^\nu(I;\dot{B}^{\frac{2}{\nu}-\frac{2}{\rho}}_{p,\sigma})},$$

where C is independent of T.

For the inhomogeneous Besov spaces, the following estimate also holds:

Theorem 4.2 *Let $1 \le p \le \infty$, $T < \infty$ and set $I = (0,T)$. There exists a constant $C_0 > 0$ and $C_T > 0$ such that for any initial data $u_0 \in B^0_{p,1}(\mathbb{R}^n)$,*

$$C_0 \sum_{j \ge k} \|\phi_j * u_0\|_p \le \int_0^T \|\Delta e^{s\Delta_0} u_0\|_p ds \le C_T \|u_0\|_{B^0_{p,1}}.$$

Here the constant C_0 is only depending on n but not on T and $C_T = O(\log T)$ with $k = [-\frac{\log T}{2\log 2}]$. In particular, maximal L^1 regularity for the heat equation over the interval $I = (0,T)$ holds. Namely for the same constant $C_T > 0$ appeared in the above,

$$\big\|\partial_t u\big\|_{L^1(I;B^0_{p,1})} + \big\|\Delta u\big\|_{L^1(I;B^0_{p,1})} \le C_T\Big(\|u_0\|_{B^0_{p,1}} + \|f\|_{L^1(I;B^0_{p,1})}\Big),$$

where $C_T = O(\log T)$ $(T \to \infty)$.

The proof of Theorem 4.1 and Theorem 4.2 can be seen in [14, Chapter 13] and [15].

From Theorems 4.1 and 4.2, we immediately obtain the following corollary for the modified Besov spaces:

Corollary 4.3 (End-point maximal regularity in the modified Besov spaces) *Let $1 \le \nu \le \rho \le \infty$, $1 \le p \le \infty$, $\tilde{s} \in \mathbb{R}$ and $I = (0,T) \subset \mathbb{R}_+$ be an open time interval. For the external forth $f \in L^\nu(I; \dot{B}^{2/\nu,\log\tilde{s}}_{p,\nu}(\mathbb{R}^n))$ and the initial data $u_0 \in \dot{B}^{2-2/\rho,\log\tilde{s}}_{p,\rho}(\mathbb{R}^n)$, assume that u is a solution of the Cauchy problem of the heat equations* (25).

(1) If $f \equiv 0$ then for any $1 \le \sigma \le \infty$, there exists a constant $C > 0$ such that

$$\|\partial_t u\|_{L^\rho(I;\dot{B}^{0,\log\tilde{s}}_{p,\sigma})} + \|\nabla^2 u\|_{L^\rho(I;\dot{B}^{0,\log\tilde{s}}_{p,\sigma})} \le C\|u_0\|_{\dot{B}^{2-\frac{2}{\rho},\log\tilde{s}}_{p,\rho}}$$

holds, where $\nabla^2 = \partial_{x_i}\partial_{x_j}$ and $C > 0$ is independent of T.

(2) If $u_0 = 0$ then for any $\nu \le \sigma \le \rho$,

$$\left\|\partial_t u\right\|_{L^\rho(I;\dot{B}^{0,\log\tilde{s}}_{p,\sigma})} + \left\|\nabla^2 u\right\|_{L^\rho(I;\dot{B}^{0,\log\tilde{s}}_{p,\sigma})} \le C\|f\|_{L^\nu(I;\dot{B}^{\frac{2}{\nu}-\frac{2}{\rho},\log\tilde{s}}_{p,\sigma})},$$

where C is independent of T.

(3) Let $f \in L^\nu(I; B^{2/\nu,\log\tilde{s}}_{p,\nu}(\mathbb{R}^n))$ and $u_0 \in B^{2-2/\rho,\log\tilde{s}}_{p,\rho}(\mathbb{R}^n)$ and assume that u be a solution to the Cauchy problem (25)*. Then L^1 maximal regularity holds for the problem* (25) *on $I = (0,T)$. Namely, there exists a constant $C_T > 0$ such that*

$$\left\|\partial_t u\right\|_{L^1(I;B^{0,\log\tilde{s}}_{p,1})} + \left\|\Delta u\right\|_{L^1(I;B^{0,\log\tilde{s}}_{p,1})} \le C_T\Big(\|u_0\|_{B^{0,\log\tilde{s}}_{p,1}} + \|f\|_{L^1(I;B^{0,\log\tilde{s}}_{p,1})}\Big),$$

where $C_T = O(\log T)$ $(T \to \infty)$.

Maximal regularity is a useful tool to treat the solvability problem of the quasi-linear parabolic equations. Theorem 4.1 is a useful form for applying the semilinear problem. In our application, we need to extend the result to the inhomogeneous case of modified Besov spaces.

5 The Critical Well-Posedness

The inhomogeneous Besov space $B^0_{1,1}(\mathbb{R}^n)$ is the invariant scaling space as L^1 for the high-frequency side and it includes a positive function. However, the product rule does not hold for general and it seems to be difficult to construct a local solution in such a space. Hence, we consider the local well-posedness of the Cauchy problem in the modified inhomogeneous Besov spaces.

In view of the invariant scaling (2), the homogeneous Besov space $\dot{B}^0_{1,1}(\mathbb{R}^n)$ is the invariant spaces. However, this space does not narrow the positive solution since the mean value of the solution is 0. For any $1 \le q < \infty$,

$$\dot{B}^0_{1,1}(\mathbb{R}^n) \subsetneq \dot{B}^{-n+\frac{n}{r}}_{r,1}(\mathbb{R}^n).$$

We then consider the rather stringent case than the result in Theorem 2.1, and show the local well-posedness to the problem (1) when $1 \le r \le 2$ in the modified inhomogeneous Besov spaces $B^{-n+\frac{n}{r},\log\tilde{s}}_{r,1}(\mathbb{R}^n)$. We show the following result:

Theorem 5.1 *Let $n = 1, 2$, $p = 1 + \frac{2}{n}$, $1 \le r \le 2$ and $\tilde{s} > 1$. For any initial data $u_0 \in B^{-n+\frac{n}{r},\log\tilde{s}}_{r,1}(\mathbb{R}^n)$, there exists a local time $T > 0$ and a unique mild solution u to (1) in $C([0,T); B^{-n+\frac{n}{r},\log\tilde{s}}_{r,1}(\mathbb{R}^n)) \cap L^{\frac{2}{n}}((0,T); B^{\frac{n}{r},\log\tilde{s}}_{r,1}(\mathbb{R}^n))$ and satisfies the following conditions:*

(1) $u \in C((0,T); B^{-n+\frac{n}{r},\log\tilde{s}}_{r,1}(\mathbb{R}^n)) \cap C^1((0,T); B^{-n+\frac{n}{r},\log\tilde{s}}_{r,1}(\mathbb{R}^n))$.
(2) For the maximal existence time $T_* > 0$, if $T_* < \infty$ then

$$\overline{\lim_{t\to T_*}} \|u(t)\|_{B^{-n+\frac{n}{r},\log\tilde{s}}_{r,1}} = \infty.$$

Proof of Theorem 2.1. We show that the map

$$\Phi[u] \equiv e^{t\Delta}u_0 + \int_0^t e^{(t-s)\Delta}u(s)^2 ds$$

is contraction in the closed subset in the modified Besov space. We assume that $1 \le r \le 2$ and for a constant defined by $M = 2C_0\|u_0\|_{B^{-n+\frac{n}{r},\log\tilde{s}}_{r,1}} > 0$ and a time constant $T > 0$ and $I = (0,T)$, show that

$$X^r_T = \left\{ f \in C\big([0,T); B^{-n+\frac{n}{r},\log\tilde{s}}_{r,1}(\mathbb{R}^n)\big) \cap L^{\frac{2}{n}}\left(I; B^{\frac{n}{r},\log\tilde{s}}_{r,1}(\mathbb{R}^n)\right), \quad |||f|||_{X^r_T} \le M \right\},$$

where $I \equiv (0,T)$ and

$$|||f|||_{X^r_T} \equiv \|f\|_{L^\infty(I;B^{-n+\frac{n}{r},\log\tilde{s}}_{r,1})} + \|f\|_{L^{\frac{2}{n}}(I;B^{\frac{n}{r},\log\tilde{s}}_{r,1})}.$$

and claim that the map Φ is a construction on X^r_T. We introduce a metric on X^r_T by

$$d(f,g) \equiv |||f - g|||_{X^r_T},$$

then it is straightforward to check that the set X^r_T is a complete metric space.

To apply Lemma 3.1, we set $s = -n + \frac{n}{r}$ and $1 \le q \le \frac{n}{|s|}$, then by the restriction $-\frac{n}{q} \le s \le 0$,

$$-\frac{n}{q} \le -n + \frac{n}{r} \le 0 \iff r' = \left(1 - \frac{1}{r}\right)^{-1} \le q \iff q = r'$$

This stems from the calculus in the Besov space in Lemma 3.1 and unavoidable. It follows then that

$$\begin{aligned}\|u^{1+\frac{2}{n}}(s)\|_{B_{r,1}^{-n+\frac{n}{r},\log\bar{s}}} &\le \|u(s)\|_{B_{q,1}^{\frac{n}{q},\log\bar{s}}} \|u(s)^{\frac{2}{n}}\|_{B_{r,1}^{-n+\frac{n}{r},\log\bar{s}}} \\ &\le \|u(s)\|^{\frac{2}{n}}_{B_{q,1}^{\frac{n}{q},\log\bar{s}}} \|u(s)\|_{B_{r,1}^{-n+\frac{n}{r},\log\bar{s}}}.\end{aligned}$$

Then applying maximal regularity to treat the term $\|u(s)\|^{\frac{2}{n}}_{B_{q,1}^{\frac{n}{q},\log\bar{s}}}$ and controlling by $\|u(s)\|_{B_{r,1}^{-n+\frac{n}{r},\log\bar{s}}}$, we request that $r \le q$. Under the restriction $r \le q \le r'$, we require that $1 \le r \le 2$.

If $L^\theta(\mathbb{R}_+; B^s_{r,1})$ then

$$\frac{2}{\theta} + \frac{n}{r} - s = n.$$

By substituting $s = \frac{n}{r}$, $\theta = \frac{2}{n}$,

$$\frac{2}{2/n} + \frac{n}{q} - \frac{n}{r} = n$$

If the space dimension is either $n = 1, 2$, then the power of the nonlinear term can be an integer such as 3, 2,

$$\begin{aligned}\|\Phi[u]\|_{L^\infty(I;B_{r,1}^{-n+\frac{n}{r},\log\bar{s}})} &\le \|e^{t\Delta}u_0\|_{L^\infty(I;B_{r,1}^{-n+\frac{n}{r},\log\bar{s}})} + \left\|\int_0^t e^{(t-s)\Delta}u^p(s)ds\right\|_{L^\infty(I;B_{r,1}^{-n+\frac{n}{r},\log\bar{s}})} \\ &\le C_0\|u_0\|_{B_{r,1}^{-n+\frac{n}{r},\log\bar{s}}} + C\int_I \|u^p(s)\|_{B_{r,1}^{-n+\frac{n}{r},\log\bar{s}}} ds \\ &\le C_0\|u_0\|_{B_{r,1}^{-n+\frac{n}{r},\log\bar{s}}} + C\int_I \|u(s)\|_{B_{q,1}^{\frac{n}{q},\log\bar{s}}} \|u(s)^{\frac{2}{n}}\|_{B_{r,1}^{-n+\frac{n}{r},\log\bar{s}}} ds \\ &\le C_0\|u_0\|_{B_{r,1}^{-n+\frac{n}{r},\log\bar{s}}} + C\int_I \|u(s)\|^{\frac{2}{n}}_{B_{q,1}^{\frac{n}{q},\log\bar{s}}} \|u(s)\|_{B_{r,1}^{-n+\frac{n}{r},\log\bar{s}}} ds \\ &\le C_0\|u_0\|_{B_{r,1}^{-n+\frac{n}{r},\log\bar{s}}} + C\|u\|^{\frac{2}{n}}_{L^{\frac{2}{n}}(I;B_{q,1}^{\frac{n}{q},\log\bar{s}})} \|u\|_{L^\infty(I;B_{r,1}^{-n+\frac{n}{r},\log\bar{s}})}.\end{aligned} \tag{26}$$

While for the auxiliary norm to assure the uniqueness of the solution, if $n = 2$,

$$\begin{aligned}
\|\Phi[u]\|_{L^{\frac{2}{n}}(I;B_{q,1}^{\frac{n}{q},\log\tilde{s}})} \leq & \|e^{t\Delta}u_0\|_{L^{\frac{2}{n}}(I;B_{q,1}^{\frac{n}{q},\log\tilde{s}})} + \left\| \int_0^t e^{(t-s)\Delta} u^{1+\frac{2}{n}}(s)ds \right\|_{L^{\frac{2}{n}}(I;B_{q,1}^{\frac{n}{q},\log\tilde{s}})} \\
= & \|e^{t\Delta}u_0\|_{L^1(I;B_{q,1}^{\frac{2}{q},\log\tilde{s}})} + \left\| \int_0^t e^{(t-s)\Delta} u^2(s)ds \right\|_{L^1(I;B_{q,1}^{\frac{2}{q},\log\tilde{s}})} \\
\leq & C_0\|u_0\|_{B_{q,1}^{-2+\frac{2}{q},\log\tilde{s}}} + C\left\|u^2(s)\right\|_{L^1(I;B_{q,1}^{-2+\frac{2}{q},\log\tilde{s}})} \\
\leq & C_0\|u_0\|_{B_{r,1}^{-2+\frac{2}{r},\log\tilde{s}}} + C\int_{\mathbb{R}_+} \|u(s)\|_{B_{q,1}^{\frac{2}{q},\log\tilde{s}}} \|u(s)\|_{B_{q,1}^{-2+\frac{2}{q},\log\tilde{s}}} ds \\
\leq & C_0\|u_0\|_{B_{r,1}^{-2+\frac{2}{r},\log\tilde{s}}} + C\|u\|_{L^1(I;B_{r,1}^{\frac{2}{r},\log\tilde{s}})} \|u\|_{L^\infty(I;B_{r,1}^{-2+\frac{2}{r},\log\tilde{s}})}.
\end{aligned} \tag{27}$$

Analogously for the case $n = 1$, we choose that $q = r$ (similarly, we need $r \leq 2$),

$$\begin{aligned}
\|\Phi[u]\|_{L^{\frac{2}{n}}(I;B_{r,1}^{\frac{n}{r},\log\tilde{s}})} \leq & \|e^{t\Delta}u_0\|_{L^2(I;B_{r,1}^{\frac{1}{r},\log\tilde{s}})} + \left\| \int_0^t e^{(t-s)\Delta} u^{1+\frac{2}{n}}(s)ds \right\|_{L^2(I;\dot{B}_{r,1}^{\frac{1}{r},\log\tilde{s}})} \\
= & \|e^{t\Delta}u_0\|_{L^2(I;B_{r,1}^{\frac{1}{r},\log\tilde{s}})} + \left\| \int_0^t e^{(t-s)\Delta} u^3(s)ds \right\|_{L^2(I;B_{r,1}^{\frac{1}{r},\log\tilde{s}})} \\
\leq & C_0\|u_0\|_{B_{r,1}^{-1+\frac{1}{r},\log\tilde{s}}} + C\left\|u^3(s)\right\|_{L^1(I;B_{r,1}^{-1+\frac{1}{r},\log\tilde{s}})} \\
\leq & C_0\|u_0\|_{B_{1,1}^{-n+\frac{n}{r},\log\tilde{s}}} + C\int_{\mathbb{R}_+} \|u(s)\|^2_{B_{r,1}^{\frac{1}{r},\log\tilde{s}}} \|u(s)\|_{B_{r,1}^{-1+\frac{1}{r},\log\tilde{s}}} ds \\
\leq & C_0\|u_0\|_{B_{r,1}^{-n+\frac{n}{r},\log\tilde{s}}} + C\|u\|^2_{L^2(I;B_{r,1}^{\frac{1}{r},\log\tilde{s}})} \|u\|_{L^\infty(I;B_{r,1}^{-1+\frac{1}{r},\log\tilde{s}})}.
\end{aligned} \tag{28}$$

From (26), (27)–(28) by choosing $T > 0$ small enough, (namely $|I| < \varepsilon$),

$$\|u\|_{L^{\frac{2}{n}}(I;B_{r,1}^{\frac{n}{r},\log\tilde{s}})} \equiv m < \frac{1}{2C}$$

and it follows that

$$\begin{aligned}
|||\Phi[u]|||_{X_T^r} \leq & C_0\|u_0\|_{\dot{B}_{r,1}^{-n+\frac{n}{r},\log\tilde{s}}} + Cm|||u|||_{X_T^r} \\
\leq & \frac{1}{2}M + \frac{1}{2}M = M.
\end{aligned}$$

Hence the map Φ is onto $X_{r,T}$. Furthermore

$$\begin{aligned}
&\|\Phi[u_1]-\Phi[u_2]\|_{L^\infty(I;B_{r,1}^{-n+\frac{n}{r},\log\bar{s}})}\\
&\le\left\|\int_0^t e^{(t-s)\Delta}\max(u_1(s)^{\frac{2}{n}},u_2(s)^{\frac{2}{n}})(u_1-u_2)(s)ds\right\|_{L^\infty(I;B_{r,1}^{-n+\frac{n}{r},\log\bar{s}})}\\
&\le C\int_I\left(\|u_1(s)\|^{\frac{2}{n}}_{B_{r,1}^{\frac{n}{r},\log\bar{s}}}+\|u_2(s)\|^{\frac{2}{n}}_{B_{r,1}^{\frac{n}{r},\log\bar{s}}}\right)\|(u_1-u_2)(s)\|_{B_{r,1}^{-n+\frac{n}{r},\log\bar{s}}}ds\\
&\le 2Cm^{\frac{2}{n}}\|u_1-u_2\|_{L^\infty(I;B_{r,1}^{-n+\frac{n}{r},\log\bar{s}})}.
\end{aligned}\tag{29}$$

Similarly

$$\begin{aligned}
&\|\Phi[u_1]-\Phi[u_2]\|_{L^{\frac{2}{n}}(I;B_{r,1}^{\frac{n}{r},\log\bar{s}})}\\
&\qquad\le\left\|\int_0^t e^{(t-s)\Delta}(u_1^p-u_2^p)(s)ds\right\|_{L^{\frac{2}{n}}(I;B_{r,1}^{\frac{n}{r},\log\bar{s}})}\\
&\qquad\le C\int_{\mathbb{R}_+}\|(u_1+u_2)(u_1-u_2)(s)\|_{\dot{B}_{r,1}^{-n+\frac{n}{r},\log\bar{s}}}ds\\
&\qquad\le C\int_I\left(\|u_1(s)\|^{\frac{2}{n}}_{B_{r,1}^{\frac{n}{r},\log\bar{s}}}+\|u_2(s)\|^{\frac{2}{n}}_{B_{r,1}^{\frac{n}{r},\log\bar{s}}}\right)\|(u_1-u_2)(s)\|_{B_{r,1}^{-n+\frac{n}{r},\log\bar{s}}}ds\\
&\qquad\le 2Cm^{\frac{2}{n}}\|u_1-u_2\|_{L^\infty(I;B_{r,1}^{-n+\frac{n}{r},\log\bar{s}})}.
\end{aligned}\tag{30}$$

From (29) and (30),

$$|||\Phi[u_1]-\Phi[u_2]|||_{X_T^r}\le 2Cm^{\frac{2}{n}}|||u_1-u_2|||_{X_T^r}$$

and setting $m^{\frac{2}{n}}\le\frac{1}{4C}$, $\Phi[\cdot]$ is contraction from X_T^r to X_T^r. Hence there exists a unique fixed point for the map Φ in X_T^r and it is the mild solution to (1). This proves the local well-posedness.

Concerning the ill-posedness, the result follows from the analogous argument for the nonlinear Schrödinger equation for $n=2$ (cf. [7]). The example of initial data $u_{0,N}\in B_{2,1}^{-1}(\mathbb{R}^n)$ given there also works for the case of nonlinear heat equations with the quadratic nonlinear term u^2. Besides the basic ill-posed space in [7] is $H^{-1}(\mathbb{R}^2)$ which is slightly larger than $B_{2,1}^{-1}(\mathbb{R}^2)$ and it holds that for any $f\in B_{2,1}^{-1}(\mathbb{R}^2)$

$$\|f\|_{H^{-1}(\mathbb{R}^2)}\le C\|f\|_{B_{2,1}^{-1}(\mathbb{R}^2)}.$$

This enables us to control the error terms in the similar way as in [7]. Meanwhile, the room to choose the free parameter R, one can arrange that the initial data appeared in [7] remains bounded in $B_{2,1}^{-1}(\mathbb{R}^2)$. Hence, the analogous strong discontinuity of the solution can be established. □

6 Appendix

For a positive solution with the initial data G_μ, we have seen the ill-posedness of the problem in (7). First, we verify that $G_\mu \in B^0_{1,1}(\mathbb{R}^n)$. Applying the logarithmic interpolation inequality

$$\|f\|_{B^0_{p,1}} \le C\left(1 + \|f\|_{B^0_{p,\infty}} \log\left(e + \|f\|_{B^1_{p,\infty}}\right)\right)$$

for $f \in B^0_{p,1}(\mathbb{R}^n) \cap B^1_{p,\infty}(\mathbb{R}^n)$ (cf. [12]), one can observe by $W^{k,p}(\mathbb{R}^n) \subset B^k_{p,\infty}(\mathbb{R}^n)$ that

$$\begin{aligned} \|G_\mu\|_{B^0_{1,1}} \le& C\left(1 + \|G_\mu\|_{B^0_{1,\infty}} \log\left(e + \|G_\mu\|_{B^1_{1,\infty}}\right)\right) \\ \le& C\left(1 + \|G_\mu\|_{L^1} \log\left(1 + e + \|\nabla G_\mu\|_{L^1}\right)\right) \\ \le& C \log(e + \mu^{-\frac{1}{2}}). \end{aligned} \tag{31}$$

To see the lower bound, since

$$\begin{aligned} c_n\|\phi_j * G_\mu\|_1 \ge& \sup_{\xi \in \mathbb{R}^n} c_n \left|\int_{\mathbb{R}^n} e^{-ix\cdot\xi}\phi_j * G_\mu dx\right| \\ =& \sup_{\xi \in \mathbb{R}^n} c_n \left|\widehat{\phi_j}(\xi)\widehat{G_\mu}(\xi)\right| \\ =& c_n e^{-4\mu 2^{2j}}, \end{aligned}$$

we see that

$$\begin{aligned} \|G_\mu\|_{B^0_{1,1}} =& \sum_{j\ge 0} \|\phi_j * G_\mu\|_1 \ge C \sum_{j \ge 0} c e^{-c\mu 2^{2j}} \\ \ge& C\int_1^\infty e^{-c\mu|\xi|^2}\frac{d\xi}{\xi} = C\int_{\sqrt{\mu}}^\infty e^{-c|\tilde{\xi}|^2}\frac{d\tilde{\xi}}{\tilde{\xi}} = O(\log\langle\mu^{-\frac{1}{2}}\rangle) \end{aligned} \tag{32}$$

as $\mu \to 0$. The estimates (32) and (31) implies the estimate (8).

Acknowledgements The authors thank Professor Kazuhiro Ishige, Professor Tsukasa Iwabuchi, and Dr. Ryuichi Sato for their stimulation discussion on the local well-posedness. The work of Takayoshi Ogawa is partially supported by JSPS grant-in-aid for Scientific Research (S) #25220702.

References

1. H. Brezis, T. Cazenave, A nonlinear heat equation with singular initial data. J. Anal. Math. **68**, 277–304 (1996)
2. T. Cazenave, F. Weissler, Asymptotically self-similar global solutions of the nonlinear Schrödinger and heat equations. Math. Z. **228**, 83–120 (1998)
3. H. Fujita, On the blowing-up of solutions of the Cauchy problem for $u_t = \Delta u + u^{1+\alpha}$. J. Fac. Sci. Univ. Tokyo Ser. I **13**, 109–124 (1966)
4. H. Fujita, T. Kato, On Navier-Stokes initial value problem 1. Arch. Ration. Mech. Anal. **46**, 269–315 (1964)
5. Y. Giga, Solutions for semilinear parabolic equations in L^p and regularity of weak solutions of the Navier-Stokes system. J. Differ. Equ. **61**, 186–212 (1986)
6. K. Hayakawa, On nonexistence of global solutions of some semilinear parabolic equations. Proc. Jpn. Acad. **49**, 503–505 (1973)
7. T. Iwabuchi, T. Ogawa, Ill-posedness issue for nonlinear Schrödinger equation with the quadratic nonlinearity for low dimensions. Trans. Am. Math. Soc. **367**, 2613–2630 (2015)
8. T. Iwabuchi, T. Ogawa, Ill-posedness issue for the drift diffusion system in the homogeneous Besov spaces. Osaka J. Math. **53**, 919–939 (2016)
9. T. Kato, H. Fujita, On the non-stationary Navier-Stokes system. Rend. Sem. Math. Univ. Padova **32**, 243–260 (1962)
10. K. Kobayashi, T. Sirao, H. Tanaka, On the growing up problem for semilinear heat equations. J. Math. Soc. Jpn. **29**, 407–424 (1977)
11. H. Kozono, M. Yamazaki, Semilinear heat equations and Navier-Stokes equation with distribution in new function spaces as initial data. Commun. Partial Differ. Equ. **19**, 959–1014 (1994)
12. H. Kozono, T. Ogawa, Y. Taniuchi, The critical Sobolev inequalities in Besov spaces and regularity criterion to some semi-linear evolution equations. Math. Z. **242**, 251–278 (2002)
13. R. Laister, J.C. Robinson, M. Sierżęga, A. Vidal-López, A complete characterization of local existence for semilinear heat equations in Lebesgue spaces. Math. Ann. (2015), published on line
14. T. Ogawa, Nonlinear evolutionary partial differential equations, preprint. Tohoku University (2017)
15. T. Ogawa, S. Shimizu, Maximal L^1 regularity for the Cauchy problem of a parabolic equation with variable coefficients. Math. Ann. **365**(1), 661–705 (2016)
16. S. Sugitani, On nonexistence of global solutions for some nonlinear integral equations. Osaka J. Math. **12**, 45–51 (1975)
17. F. Weissler, Local existence and non-existence for semilinear parabolic equations in L^p. Indiana Univ. Math. J. **29**, 79–102 (1980)
18. F. Weissler, Existence and non-existence of global solutions for a semilinear heat equation. Isr. J. Math. **38**, 29–40 (1981)

Spatial Lipschitz Continuity of Viscosity Solution to Level Set Equation for Evolving Spirals by Eikonal-Curvature Flow

Takeshi Ohtsuka

Abstract Level set equation for evolving spirals by an eikonal-curvature flow is considered. In this paper, we prove the spatial Lipschitz continuity of viscosity solutions to the level set equation for the evolving spirals provided that a suitable approximation of initial data exists. It is established with Bernstein's method for regularized equation approximating the level set equation, and limiting procedure of solutions tending the approximating parameter to zero.

Keywords Viscosity solution · Lipschitz continuity · Bernstein's method
Eikonal-curvature ow · Level set method for spirals

1 Introduction

A level set method for evolution of spirals Γ_t in the plane by eikonal-curvature flow

$$V = C - \kappa \quad \text{on } \Gamma_t \tag{1}$$

is introduced by [11, 12]. In this paper, we study on spatial Lipschitz continuity of solutions to the level set equation for evolution of Γ_t.

We now recall the level set method due to [11, 12]. Let $\Omega \subset \mathbb{R}^2$ be a bounded domain with smooth boundary. Assume that there exist $N(\geq 1)$ centers of spirals in Ω, which are denoted by $a_1, a_2, \ldots, a_N \in \Omega$. In this method, we regard an open neighborhood U_j of a_j as the center of spirals, and set $W = \Omega \setminus \bigcup_{j=1}^N \overline{U}_j$. Assume that ∂U_j is smooth, $\overline{U}_j \subset \Omega$ and $\overline{U}_i \cap \overline{U}_j = \emptyset$ if $i \neq j$ for $i, j = 1, 2, \ldots, N$, which implies that ∂W is smooth. We consider that Γ_t evolves on $\overline{W}$, and assume that the end points of Γ_t always stay on ∂W. In this evolution, we also assume the orthogonality condition,

---

Dedicated to Professor Yoshikazu Giga on the occasion of his 60th birthday

---

T. Ohtsuka (✉)
Division of Pure and Applied Science, Faculty of Science and Technology, Gunma University, Aramaki-machi 4-2 Maebashi, Gunma 371-8510, Japan
e-mail: tohtsuka@gunma-u.ac.jp

Y. Maekawa and S. Jimbo (eds.), *Mathematics for Nonlinear Phenomena—Analysis and Computation*, Springer Proceedings in Mathematics & Statistics 215,
DOI 10.1007/978-3-319-66764-5_11

$$\Gamma_t \perp \partial W. \tag{2}$$

We now introduce a sheet structure function, which is due to Kobayashi [9],

$$\theta(x) = \sum_{j=1}^{N} m_j \arg(x - a_j)$$

with $m_j \in \mathbb{Z} \setminus \{0\}$, where m_j denotes the signed number of spirals come out from a_j; see [12] for details of the definition of the signed number of spirals. We set

$$\Gamma_t = \{x \in \overline{W};\ u(t,x) - \theta(x) \equiv 0 \mod 2\pi\mathbb{Z}\}$$

with an auxiliary function $u(t,x)$ to describe spiral curves Γ_t. Then, the equations (1) and (2) are represented as

$$u_t - |\nabla(u-\theta)| \left\{ \mathrm{div} \frac{\nabla(u-\theta)}{|\nabla(u-\theta)|} + C \right\} = 0 \text{ in } (0,T) \times W, \tag{3}$$

$$\langle \boldsymbol{\nu}, \nabla(u-\theta) \rangle = 0 \text{ on } (0,T) \times \partial W, \tag{4}$$

where $\boldsymbol{\nu}$ is the outer unit normal vector field of ∂W, and $\langle \cdot, \cdot \rangle$ denotes the usual inner product in $\mathbb{R}^2$. Note that θ should be considered as a multiple valued function to describe spiral curves completely. However, (3) or (4) has only the derivatives of θ, which are defined as a single valued function. Thus, (3)–(4) is well-defined, and a lot of mathematical analysis of (3)–(4) has been done. The author [11] established the comparison principle and the existence of viscosity solutions globally-in-time for continuous initial data. Goto, Nakagawa and the author [7] established the uniqueness of Γ_t in the level set sense with respect to Γ_0, and existence of continuous initial data for suitable class of Γ_0. Tsai, Giga and the author [12] give numerical results on behavior of Γ_t, and a practical way to construct an initial data for given Γ_0.

When we describe the motion of Γ_t by the level set method, it is important for the numerical accuracy on location of Γ_t that the solution u to (3)–(4) is Lipschitz continuous with a small Lipschitz constant with respect to the inverse of the numerical grid size. Of course, the uniqueness of Γ_t for Γ_0 by [7] admits re-initialization of u to smoother one when its gradient becomes large. However, it is natural to reduce such a re-initialization. By the above motivation, we study on spatial Lipschitz continuity of solutions to (3)–(4).

It is a powerful option for this purpose to introduce a parabolic equation approximating (3) with the boundary condition (4). In this paper, we prove a gradient bound of a solution to the approximating problem, and prove a kind of Lipschitz bound of a solution to (3)–(4) provided that a suitable approximation of initial data exists. The basic strategy to prove a gradient bound is the maximum principle method due to Bernstein [1, 2]. See [14] or [10] for the idea deriving the gradient bounds with the maximum principle. This gradient bound also implies the spatial equi-continuity of the solutions provided that the gradient of the initial data is bounded. Then, we also show the equi-continuity on time variable, and uniform bound of solutions for the convergence results. This idea is come from [5]. However, we need the bound of $\nabla^2 u_0$ in [5] when we prove the equi-continuity on time variable. We often assume only the continuity of initial data for the existence and uniqueness of viscosity solution to (3)–(4); see [11]. Therefore, we prove the equi-continuity on time variable without a C^2 bound of an initial data. Its key idea is the construction of viscosity super- and sub-solution whose initial data is an each time data of the solution by the Perron's method for viscosity solution due to [8] or [3].

This paper is organized as follows. We study on the approximating problem in §2.1–§2.4, and prove the main result in §2.5. We give some concluding remarks in §3.

2 Approximating Problem

We now introduce an approximating problem for (3)–(4) of the form

$$u_t^\varepsilon - \sqrt{\varepsilon^2 + |\nabla(u^\varepsilon - \theta)|^2}\left\{\mathrm{div}\frac{\nabla(u^\varepsilon - \theta)}{\sqrt{\varepsilon^2 + |\nabla(u^\varepsilon - \theta)|^2}} + C\right\} = 0 \quad \text{in } (0, T) \times W, \tag{5}$$

$$\langle \boldsymbol{\nu}, \nabla(u^\varepsilon - \theta)\rangle = 0 \quad \text{on } (0, T) \times \partial W, \tag{6}$$

$$u^\varepsilon(0, \cdot) = g \quad \text{on } \overline{W} \tag{7}$$

with an approximating small parameter $\varepsilon \in (0, 1)$. The goal of this paper is to prove the following theorem.

Theorem 2.1. *Let u be a viscosity solution to (3)–(4) on $(0, T) \times \overline{W}$ with $u(0, \cdot) = u_0 \in C(\overline{W})$. Assume that there exists $L > 0$ such that*

$$|u_0(x) - u_0(y)| \le L|x - y| \quad \text{for } x, y \in \overline{W}. \tag{8}$$

Assume that there exists $g_\varepsilon \in C^1(\overline{W})$ satisfying

(A1) $\lim_{\varepsilon\to 0} \|g_\varepsilon - u_0\|_\infty = 0$,
(A2) $\limsup_{\varepsilon\to 0} \|\nabla g_\varepsilon\|_\infty \le L$,
(A3) *(5)–(7) with $g = g_\varepsilon$ has a unique solution $u^\varepsilon \in C^1([0, T] \times \overline{W}) \cap C^3((0, T) \times \overline{W})$,*

where $\|f\|_\infty := \sup_{x\in\overline{W}} |f(x)|$ and $\|\nabla f\|_\infty = \sup_{x\in\overline{W}} |\nabla f(x)|$ for a function $f \in C^1(\overline{W})$. Then, there exists a constant $c > 0$ which depends only on W such that

$$|u(t, x) - \theta(x) - (u(t, y) - \theta(y))| \le e^{ct}(L + \|\nabla\theta\|_\infty)|x - y| \tag{9}$$

for $t \in (0, T)$ *and* $x, y \in \overline{W}$ *provided that* $\lambda x + (1-\lambda)y \in \overline{W}$ *for* $\lambda \in [0, 1]$, *where* $\theta(x)$ *in* (9) *is a branch of* θ *so that* θ *is continuous at least on* $\{\lambda x + (1-\lambda)y;\ \lambda \in [0, 1]\}$.

Note that the estimate (9) is independent of the choice of branch of θ.

For this purpose, we consider (5)–(7) with $g \in C^1(\overline{W})$ assuming that

(I1) (5)–(7) has a unique solution $u^\varepsilon \in C^1([0, T] \times \overline{W}) \cap C^3((0, T) \times \overline{W})$,
(I2) there exists $R > 0$ such that $\|g\|_\infty + \|\nabla g\|_\infty \le R$.

The goal of this section is to prove a gradient bound, the equi-continuity and uniform bound of solutions u^ε to (5)–(7) with respect to $\varepsilon \in (0, 1)$ and the initial data satisfying (I1)–(I2). We first give some preliminaries in §2.1. The gradient bound will be obtained in §2.2, which also implies the spatial equi-continuity. We next show the equi-continuity of $t \mapsto u^\varepsilon(t, x)$ provided that (I1)–(I2) hold by constructing sub- and super-solution.

Remark 2.2. It is easy to find $g_\varepsilon \in C^1(\overline{W})$ satisfying (A1) and (A2) from u_0 satisfying (8). In fact, we first extend u_0 to $\tilde{u}_0 \colon \mathbb{R}^2 \to \mathbb{R}$ as

$$\tilde{u}_0(x) = \inf_{\xi \in \overline{W}} (u_0(\xi) + L|x - \xi|).$$

Note that $\tilde{u}_0 = u_0$ on $\overline{W}$, and $\tilde{u}_0$ satisfies (8) with the same Lipschitz constant $L > 0$. We next mollify $\tilde{u}_0$; set $\tilde{u}_\varepsilon = \varphi_\varepsilon * \tilde{u}_0$ with $\varphi_\varepsilon \in C^\infty(\mathbb{R}^2)$ satisfying $\varphi_\varepsilon \ge 0$, $\int_{\mathbb{R}^2} \varphi_\varepsilon = 1$ and $\mathrm{supp}\varphi_\varepsilon \subset \{x \in \mathbb{R}^2;\ |x| < \varepsilon\}$. Then, $\tilde{u}_\varepsilon$ satisfies (A1) and (A2) since $\tilde{u}_\varepsilon$ satisfies (8) for $x, y \in \mathbb{R}^2$.

The crucial difficulty finding g_ε is the condition (A3), in particular guaranteeing the regularity of solution. At least g_ε should satisfy (6) as the compatibility condition to establish the regularity. One of the ideas to obtain g_ε satisfying (6) is to solve

$$\begin{aligned} w_t - \Delta w = 0 &\quad \text{in } (0, T) \times W, \\ \langle \nu, \nabla(w - \theta) \rangle = 0 &\quad \text{on } (0, T) \times \partial W, \\ w(0, \cdot) = \tilde{u}_0 &\quad \text{on } \overline{W}. \end{aligned}$$

Then, $g_\varepsilon = w(\varepsilon, \cdot)$ may be candidate of g_ε satisfying (A1)–(A3).

2.1 Preliminaries

We first prepare the distance function $d(x)$ from ∂W, which has a role controlling the boundary condition when we construct some sub- or super-solutions estimating u^ε or $\nabla(u^\varepsilon - \theta)$.

Define

$$d(x) = \begin{cases} \mathrm{dist}(x, \partial W) \text{ if } x \in \overline{W}, \\ -\mathrm{dist}(x, \partial W) \text{ otherwise,} \end{cases}$$

where $\mathrm{dist}(x, D) = \inf_{y \in D} |x - y|$ for $D \subset \mathbb{R}^2$. Then, since ∂W is smooth, there exists $\delta_0 > 0$ such that $d(x)$ is smooth on ∂W^{δ_0}, where $\partial W^{\delta} := \{x \in \mathbb{R}^2;\ \mathrm{dist}(x, \partial W) < \delta\}$ for $\delta > 0$. We now fix $\delta \in (0, \delta_0)$, and define

$$\phi(x) = \zeta\left(\frac{d(x)}{\delta}\right), \tag{10}$$

$$\zeta(\sigma) = \begin{cases} \dfrac{(\sigma - 1)^4 - 1}{4} & \text{if } \sigma < 1, \\ -\dfrac{1}{4} & \text{otherwise.} \end{cases}$$

Then, we obtain the following properties.

Lemma 2.3. *It holds that* $\phi \in C^2(\overline{W})$, *and*

$$-\frac{1}{4} \le \phi \le 0 \quad \text{on } \overline{W}, \tag{11}$$

$$|\nabla \phi| \le \frac{1}{\delta} \quad \text{on } \overline{W}, \tag{12}$$

$$\langle \boldsymbol{\nu}, \nabla \phi \rangle = \frac{1}{\delta} \quad \text{on } \partial W. \tag{13}$$

Proof. We observe that $\phi \in C^\infty(\partial W^{\delta_0})$, and $\phi \equiv -1/4$ on $\overline{W} \setminus \partial W^{\delta}$, which implies that $\phi \in C^2(\overline{W} \setminus \partial W^{\delta})$, by definition of ϕ. Hence, we obtain $\phi \in C^2(\overline{W})$. We also obtain (11) by the definition of ϕ.

The properties (12) and (13) are obtained from straightforward calculation of $\nabla \phi$. In fact, we have

$$\nabla \phi = \begin{cases} \zeta'\left(\dfrac{d(x)}{\delta}\right)\dfrac{\nabla d(x)}{\delta} & \text{if } x \in \partial W^{\delta}, \\ 0 & \text{if } x \in \overline{W} \setminus \partial W^{\delta}. \end{cases}$$

Then, (12) or (13) follows from $|\nabla d| \le 1$, or $\nabla d = -\boldsymbol{\nu}$ on ∂W and $\zeta'(0) = -1$, respectively. □

2.2 Gradient Bound

We here prove the following a priori bound of $\nabla(u^{\varepsilon} - \theta)$, which is not only for the existence of classical solution to (5)–(7) but also the equi-continuity on $x \mapsto u^{\varepsilon}(t, x)$ or Lipschitz continuity of solutions to (3)–(4).

Theorem 2.4. *Let* $u^{\varepsilon} \in C^1([0, \infty) \times \overline{W}) \cap C^3((0, \infty) \times \overline{W})$ *be a solution to* (5)–(7) *with initial data* $g \in C^1(\overline{W})$ *satisfying (I1)–(I2). Then, there exists* $c > 0$ *which depends only on* W *such that*

$$\|\nabla(u^{\varepsilon}(t,\cdot)-\theta)\|_{\infty} \leq e^{ct}\|\nabla(g-\theta)\|_{\infty} \quad \textit{for } t \geq 0. \tag{14}$$

Proof. We give an estimate of $\chi = |\nabla(u^{\varepsilon}-\theta)|^2$ with the maximum principle. In this proof, we shall use a formal notation $h(t,x) = u^{\varepsilon}(t,x) - \theta(x)$ for simplicity. Note that only derivatives of h are well-defined since θ is a multiple valued function. The equation (5) with h is represented as

$$\begin{aligned} h_t &= \Delta h - \frac{\langle \nabla^2 h \nabla h, \nabla h\rangle}{\varepsilon^2 + |\nabla h|^2} + C\sqrt{\varepsilon^2 + |\nabla h|^2} \\ &= \Delta h - \frac{\langle \nabla \chi, \nabla h\rangle}{2(\varepsilon^2+\chi)} + C\sqrt{\varepsilon^2+\chi}. \end{aligned} \tag{15}$$

We now divide the proof into 4 steps.

Step 1. We first consider $\nabla\chi$ on ∂W. From (6) we observe that

$$\frac{\nabla(u^{\varepsilon}-\theta)}{|\nabla(u^{\varepsilon}-\theta)|} = \frac{\nabla h}{|\nabla h|} = \boldsymbol{\tau} \quad \text{on } \partial W,$$

where $\boldsymbol{\tau}$ is the unit tangential vector field of ∂W. Since $x \mapsto \langle \boldsymbol{\nu}, \nabla h\rangle$ is defined only on ∂W, we now regard this function as the function $s \mapsto \langle \boldsymbol{\nu}, \nabla h\rangle$ with the arclength parameter s of ∂W. Then, we observe that

$$\begin{aligned} \frac{\mathrm{d}}{\mathrm{d}s}\langle \boldsymbol{\nu}, \nabla h\rangle = \langle \boldsymbol{\nu}_s, \nabla h\rangle + \langle \boldsymbol{\nu}, \nabla^2 h \boldsymbol{\tau}\rangle &= -\kappa_{\partial W}\langle \boldsymbol{\tau}, \nabla h\rangle + \left\langle \boldsymbol{\nu}, \frac{\nabla^2 h \nabla h}{|\nabla h|}\right\rangle \\ &= -\kappa_{\partial W}|\nabla h| + \frac{1}{2|\nabla h|}\langle \boldsymbol{\nu}, \nabla\chi\rangle, \end{aligned}$$

where $\kappa_{\partial W}$ is the curvature of ∂W in the direction of $\boldsymbol{\nu}$. If W were convex, then $\kappa_{\partial W}$ would be nonpositive. However, in particular, $\kappa_{\partial W}$ may be positive somewhere on ∂U_j. Since (6) is represented as $\langle \boldsymbol{\nu}, \nabla h\rangle = 0$, we obtain

$$0 = -\kappa_{\partial W}|\nabla h| + \frac{1}{2|\nabla h|}\langle \boldsymbol{\nu}, \nabla\chi\rangle,$$

which implies

$$\langle \boldsymbol{\nu}, \nabla\chi\rangle = 2\chi\kappa_{\partial W}. \tag{16}$$

Since ∂W is smooth, then $\kappa_{\partial W}$ is bounded. Then, there exists $M_1 > 0$ such that

$$\langle \boldsymbol{\nu}, \nabla\chi\rangle \leq 2M_1\chi \quad \text{on } (0,T)\times\partial W. \tag{17}$$

Step 2. We next demonstrate that χ is a subsolution to a homogeneous linear parabolic equation without 0-th order term in $(0,T)\times W$. From (15) we have

$$
\begin{aligned}
\chi_t &= 2\langle \nabla h_t, \nabla h\rangle \\
&= 2\left\langle \nabla\left(\Delta h - \frac{\langle\nabla\chi, \nabla h\rangle}{2(\varepsilon^2+\chi)} + C\sqrt{\varepsilon^2+\chi}\right), \nabla h\right\rangle \\
&= 2\langle \nabla(\Delta h), \nabla h\rangle - \frac{\langle \nabla^2\chi\nabla h, \nabla h\rangle}{\varepsilon^2+\chi} \\
&\qquad - \frac{\langle \nabla^2 h\nabla\chi, \nabla h\rangle}{\varepsilon^2+\chi} + \frac{\langle\nabla\chi, \nabla h\rangle^2}{(\varepsilon^2+\chi)^2} + C\frac{\langle\nabla\chi, \nabla h\rangle}{\sqrt{\varepsilon^2+\chi}},
\end{aligned}
$$

since $\nabla\langle\nabla\chi, \nabla h\rangle = \nabla^2\chi\nabla h + \nabla^2 h\nabla\chi$. We now calculate $\Delta\chi$ and obtain

$$
\begin{aligned}
\Delta\chi = 2\sum_{k=1}^{2}\frac{\partial}{\partial x_k}\left(\sum_{j=1}^{2}\frac{\partial h}{\partial x_j}\frac{\partial^2 h}{\partial x_k\partial x_j}\right) &= 2\sum_{k=1}^{2}\sum_{j=1}^{2}\left[\left(\frac{\partial^2 h}{\partial x_k\partial x_j}\right)^2 + \frac{\partial h}{\partial x_j}\frac{\partial^3 h}{\partial x_j\partial x_k^2}\right] \\
&= 2(\mathrm{tr}\{(\nabla^2 h)^2\} + \langle\nabla(\Delta h), \nabla h\rangle),
\end{aligned}
$$

which implies

$$
2\langle\nabla(\Delta h), \nabla h\rangle = \Delta\chi - 2\mathrm{tr}\{(\nabla^2 h)^2\}.
$$

Moreover, we observe that

$$
\begin{aligned}
-\frac{\langle\nabla^2 h\nabla\chi, \nabla h\rangle}{\varepsilon^2+\chi} + \frac{\langle\nabla\chi, \nabla h\rangle^2}{(\varepsilon^2+\chi)^2} &\le -\frac{\langle\nabla\chi, \nabla^2 h\nabla h\rangle}{\varepsilon^2+\chi} + \frac{|\nabla\chi|^2|\nabla h|^2}{(\varepsilon^2+\chi)^2} \\
&= -\frac{|\nabla\chi|^2}{2(\varepsilon^2+\chi)} + \frac{\chi|\nabla\chi|^2}{(\varepsilon^2+\chi)^2} \\
&= \frac{|\nabla\chi|^2}{2(\varepsilon^2+\chi)} - \frac{\varepsilon^2|\nabla\chi|^2}{(\varepsilon^2+\chi)^2} \le \frac{|\nabla\chi|^2}{2(\varepsilon^2+\chi)}.
\end{aligned}
$$

Since $|\nabla\chi|^2 = 4|\nabla^2 h\nabla h|^2 = 4\langle(\nabla^2 h)^2\nabla h, \nabla h\rangle \le 4|\nabla h|^2\mathrm{tr}\{(\nabla^2 h)^2\}$, we obtain

$$
\begin{aligned}
&2\langle\nabla(\Delta h), \nabla h\rangle - \frac{\langle\nabla^2 h\nabla\chi, \nabla h\rangle}{\varepsilon^2+\chi} + \frac{\langle\nabla\chi, \nabla h\rangle^2}{(\varepsilon^2+\chi)^2} \\
&\le \Delta\chi - 2\chi\mathrm{tr}\{(\nabla^2 h)^2\} + \frac{2\chi\,\mathrm{tr}\{(\nabla^2 h)^2\}}{\varepsilon^2+\chi} \\
&= \Delta\chi - \frac{2\varepsilon^2}{\varepsilon^2+\chi}\mathrm{tr}\{(\nabla^2 h)^2\} \le \Delta\chi
\end{aligned}
$$

since $\mathrm{tr}\{(\nabla^2 h)^2\} \ge 0$. Consequently, we observe that χ satisfies

$$
\chi_t \le \Delta\chi - \frac{\langle\nabla^2\chi\nabla h, \nabla h\rangle}{\varepsilon^2+\chi} + \frac{C\langle\nabla\chi, \nabla h\rangle}{\sqrt{\varepsilon^2+\chi}} \quad \text{in } (0, T)\times W. \tag{18}
$$

Step 3. Define $\tilde{\chi}(t,x) = e^{-\phi(x)}\chi(t,x)$, where ϕ is defined as (10). We demonstrate that there exists $c > 0$ and smooth vector-valued function $H : \overline{W} \to \mathbb{R}^2$ such that

$$\tilde{\chi}_t \leq \Delta\tilde{\chi} - \frac{\langle \nabla^2\tilde{\chi}\nabla h, \nabla h\rangle}{\varepsilon^2 + \chi} + \langle \nabla\tilde{\chi}, H\rangle + 2c\tilde{\chi} \quad \text{in } (0,T)\times W, \tag{19}$$

$$\langle \boldsymbol{\nu}, \nabla\tilde{\chi}\rangle \leq 0 \quad \text{on } (0,T)\times \partial W \tag{20}$$

for some sufficiently small $\delta > 0$.

To determine $\delta > 0$ we first consider (20). From (17) and (13) we have

$$\langle \boldsymbol{\nu}, \nabla\tilde{\chi}\rangle = -e^{-\phi}\chi\langle \boldsymbol{\nu}, \nabla\phi\rangle + e^{-\phi}\langle \boldsymbol{\nu}, \nabla\chi\rangle \leq -\frac{\tilde{\chi}}{\delta} + 2e^{-\phi}\chi M_1 = -\frac{\tilde{\chi}}{\delta} + 2M_1\tilde{\chi}$$

on $(0,T)\times\partial W$. Thus, we obtain (20) provided that $\delta < 1/(2M_1)$, and then we now fix δ so.

We next verify (19). From (18) we have

$$\tilde{\chi}_t = e^{-\phi}\chi_t \leq e^{-\phi}\Delta\chi - \frac{e^{-\phi}\langle \nabla^2\chi\nabla h, \nabla h\rangle}{\varepsilon^2 + \chi} + \frac{Ce^{-\phi}\langle \nabla\chi, \nabla h\rangle}{\sqrt{\varepsilon^2 + \chi}}$$

in $(0,T)\times W$. We now calculate $\nabla\tilde{\chi}$, $\nabla^2\tilde{\chi}$ and obtain

$$\begin{aligned}
\nabla\tilde{\chi} &= -e^{-\phi}\chi\nabla\phi + e^{-\phi}\nabla\chi = -\tilde{\chi}\nabla\phi + e^{-\phi}\nabla\chi, \\
\nabla^2\tilde{\chi} &= \tilde{\chi}(\nabla\phi\otimes\nabla\phi - \nabla^2\phi) - e^{-\phi}(\nabla\chi\otimes\nabla\phi + \nabla\phi\otimes\nabla\chi) + e^{-\phi}\nabla^2\chi,
\end{aligned}$$

where $p\otimes q = (p_i q_j)_{1\leq i,j\leq 2}$ for $p = {}^t(p_1, p_2)$, $q = {}^t(q_1, q_2) \in \mathbb{R}^2$. Then, we first obtain

$$\begin{aligned}
e^{-\phi}\langle \nabla\chi, \nabla h\rangle &= \langle \nabla\tilde{\chi}, \nabla h\rangle + \tilde{\chi}\langle \nabla\phi, \nabla h\rangle, \\
e^{-\phi}\Delta\chi &= \Delta\tilde{\chi} - \mathrm{tr}[\tilde{\chi}(\nabla\phi\otimes\nabla\phi - \nabla^2\phi) - e^{-\phi}(\nabla\chi\otimes\nabla\phi + \nabla\phi\otimes\nabla\chi)] \\
&= \Delta\tilde{\chi} - \tilde{\chi}(|\nabla\phi|^2 - \Delta\phi) + 2e^{-\phi}\langle \nabla\chi, \nabla\phi\rangle \\
&= \Delta\tilde{\chi} + 2\langle \nabla\tilde{\chi}, \nabla\phi\rangle + \tilde{\chi}(|\nabla\phi|^2 + \Delta\phi),
\end{aligned}$$

here we have used $e^{-\phi}\langle \nabla\chi, \nabla\phi\rangle = \langle \nabla\tilde{\chi}, \nabla\phi\rangle + \tilde{\chi}|\nabla\phi|^2$ from the first formula of the above. We next calculate $e^{-\phi}\langle \nabla^2\chi\nabla h, \nabla h\rangle$. Note that $(x\otimes y)z = \langle y, z\rangle x$ for $x, y, z \in \mathbb{R}^2$. Then, we obtain

$$\begin{aligned}
e^{-\phi}\langle \nabla^2\chi\nabla h, \nabla h\rangle =& \langle \nabla^2\tilde{\chi}\nabla h, \nabla h\rangle - \tilde{\chi}(\langle \nabla\phi, \nabla h\rangle^2 - \langle \nabla^2\phi\nabla h, \nabla h\rangle) \\
&+ 2e^{-\phi}\langle \nabla\chi, \nabla h\rangle\langle \nabla\phi, \nabla h\rangle \\
=& \langle \nabla^2\tilde{\chi}\nabla h, \nabla h\rangle - \tilde{\chi}(\langle \nabla\phi, \nabla h\rangle^2 - \langle \nabla^2\phi\nabla h, \nabla h\rangle) \\
&+ 2(\langle \nabla\tilde{\chi}, \nabla h\rangle + \tilde{\chi}\langle \nabla\phi, \nabla h\rangle)\langle \nabla\phi, \nabla h\rangle
\end{aligned}$$

$$
\begin{aligned}
&=\langle \nabla^2 \tilde{\chi} \nabla h, \nabla h\rangle + 2\langle \nabla\tilde{\chi}, \nabla h\rangle\langle\nabla\phi, \nabla h\rangle \\
&\quad + \tilde{\chi}(\langle\nabla\phi, \nabla h\rangle^2 + \langle\nabla^2\phi\nabla h, \nabla h\rangle).
\end{aligned}
$$

By combining above we obtain

$$
\tilde{\chi}_t \le \Delta\tilde{\chi} - \frac{\langle\nabla^2\tilde{\chi}\nabla h, \nabla h\rangle}{\varepsilon^2+\chi} + \langle\nabla\tilde{\chi}, H\rangle + H_0\tilde{\chi}
$$

with

$$
\begin{aligned}
H &:= 2\nabla\phi - 2\frac{\langle\nabla\phi, \nabla h\rangle}{\varepsilon^2+\chi}\nabla h + \frac{C}{\sqrt{\varepsilon^2+\chi}}\nabla h, \\
H_0 &:= |\nabla\phi|^2 + \Delta\phi - \frac{\langle\nabla\phi, \nabla h\rangle^2 + \langle\nabla^2\phi\nabla h, \nabla h\rangle}{\varepsilon^2+\chi} + \frac{C\langle\nabla\phi, \nabla h\rangle}{\sqrt{\varepsilon^2+\chi}}.
\end{aligned}
$$

Note that

$$
|\nabla\phi|^2 - \frac{\langle\nabla\phi, \nabla h\rangle^2}{\varepsilon^2+\chi} \le |\nabla\phi|^2 \le \frac{1}{\delta^2}
$$

by (12). Moreover, we observe that

$$
\left|\Delta\phi - \frac{\langle\nabla^2\phi\nabla h, \nabla h\rangle}{\varepsilon^2+\chi}\right| = \left|\operatorname{tr}\left\{\left(I - \frac{\nabla h\otimes\nabla h}{\varepsilon^2+|\nabla h|^2}\right)\nabla^2\phi\right\}\right| \le \|\nabla^2\phi\| \tag{21}
$$

where

$$
\|\nabla^2\phi\| = \sup_{\overline{W}}\max_{|\xi|=1}\langle\nabla^2\phi\xi, \xi\rangle.
$$

In fact, note that $\operatorname{tr}(XY) \ge 0$ for symmetric matrices $X \ge O$ and $Y \ge O$. Then, we obtain

$$
\left|\operatorname{tr}\left\{\left(I - \frac{\nabla h\otimes\nabla h}{\varepsilon^2+|\nabla h|^2}\right)\nabla^2\phi\right\}\right| \le \|\nabla^2\phi\|\operatorname{tr}\left(I - \frac{\nabla h\otimes\nabla h}{\varepsilon^2+|\nabla h|^2}\right) = \|\nabla^2\phi\|
$$

since $-\|\nabla^2\phi\|I \le \nabla^2\phi \le \|\nabla^2\phi\|I$ and $I - (\varepsilon^2+|p|^2)^{-1}(p\otimes p) \ge O$ for $p \in \mathbb{R}^2$. Remark that $\|\nabla^2\phi\|$ depends only on δ, and independent of ε. We also have

$$
\left|\frac{C\langle\nabla\phi, \nabla h\rangle}{\sqrt{\varepsilon^2+\chi}}\right| \le C|\nabla\phi| \le \frac{C}{\delta}.
$$

Note that $\tilde{\chi} \ge 0$, and then we obtain

$$
H_0\tilde{\chi} \le 2c\tilde{\chi} \quad \text{with } 2c = \frac{1}{\delta^2} + \|\nabla^2\phi\| + \frac{C}{\delta}. \tag{22}
$$

Hence, we obtain (19).

Step 4. We are now in the position to prove Theorem 2.4. Set $\bar{\chi} = e^{-2ct}\tilde{\chi} = e^{-2ct-\phi}\chi$, then $\bar{\chi}$ satisfies

$$\bar{\chi}_t \leq \Delta\bar{\chi} - \frac{\langle \nabla^2\bar{\chi}\nabla h, \nabla h\rangle}{\varepsilon^2 + \chi} + \langle \nabla\bar{\chi}, H\rangle \quad \text{in } (0,T)\times W,$$
$$\langle \boldsymbol{\nu}, \nabla\bar{\chi}\rangle \leq 0 \quad \text{on } (0,T)\times \partial W.$$

Then, we obtain $\bar{\chi}(t,x) \leq \bar{\chi}(0,x)$ for $(t,x)\in(0,T)\times\overline{W}$ by the maximum principle as in [13], which implies that

$$e^{-2ct-\phi(x)}|\nabla(u^\varepsilon - \theta)|^2 \leq e^{-\phi(x)}|\nabla(g-\theta)|^2.$$

Hence, we obtain (14). □

Remark 2.5. 1. Theorem 2.4 implies the equi-continuity on $x \mapsto u^\varepsilon(t,x)$ for $\varepsilon \in (0,1)$ and the initial data satisfying (I1)–(I2).
2. The coefficient e^{ct} in (14), which is not appeared in [5], is yielded by the estimate of the boundary condition of χ or $\tilde{\chi}$, and $c = O(1/\delta^2)$ as $\delta \to 0$ by (22). However, if Ω is convex, then $\langle \boldsymbol{\nu}, \nabla\chi\rangle \leq 0$ on $\partial\Omega$ by (16). Thus, this growing coefficient is caused by the concavity of ∂W, in particular removing U_j from Ω. The gradient bound will be refined if we obtain a better estimate for boundary condition on ∂U_j provided that Ω is convex.

2.3 Uniform Bound

It is easy to derive a uniform bound of u^ε with respect to $\varepsilon > 0$ and initial data satisfying (I1)–(I2). We first prepare the following lemma.

Lemma 2.6. *Assume that* $\delta \in (0,\delta_0)$ *satisfies*

$$\frac{1}{\delta} \geq \sup_{\partial W}\langle \boldsymbol{\nu}, \nabla\theta\rangle. \tag{23}$$

Let us define

$$b^\pm(t,x) = \pm(\Lambda t + \phi(x) + M)$$

with constants $\Lambda > 0$ *and* $M \in \mathbb{R}$. *Then there exists* $\Lambda_1 > 0$ *depending only on* δ *such that* b^+ *(resp.* b^-*) is a super- (resp. sub-) solution to* (5)–(6) *provided that* $\Lambda \geq \Lambda_1$.

Proof. We demonstrate only that b^+ is a supersolution to (5)–(6) since the proof on b^- is parallel. Note that $b^\pm \in C^2([0,\infty)\times\overline{W})$.

The assumption (23) and (13) yield that

$$\langle \boldsymbol{\nu}, \nabla(\phi - \theta)\rangle = \frac{1}{\delta} - \langle \boldsymbol{\nu}, \nabla\theta\rangle \geq 0 \quad \text{on } (0,T) \times \partial W.$$

It remains to verify

$$b_t^+ - \sqrt{\varepsilon^2 + |\nabla(b^+ - \theta)|^2}\left\{ \operatorname{div} \frac{\nabla(b^+ - \theta)}{\sqrt{\varepsilon^2 + |\nabla(b^+ - \theta)|^2}} + C \right\} \geq 0 \quad \text{in } (0,T) \times W. \tag{24}$$

We observe that

$$\begin{aligned}
&\left| \sqrt{\varepsilon^2 + |\nabla(b^+ - \theta)|^2}\left(\operatorname{div} \frac{\nabla(b^+ - \theta)}{\sqrt{\varepsilon^2 + |\nabla(b^+ - \theta)|^2}} + C \right) \right| \\
&\leq \left| \operatorname{tr}\left\{ \left(I - \frac{\nabla(\phi - \theta) \otimes \nabla(\phi - \theta)}{\varepsilon^2 + |\nabla(\phi - \theta)|^2} \right) \nabla^2(\phi - \theta) \right\} \right| + C\sqrt{1 + \|\nabla(\phi - \theta)\|_\infty^2} \\
&\leq \left\| \nabla^2(\phi - \theta) \right\| + C\sqrt{1 + \|\nabla(\phi - \theta)\|_\infty^2} =: \Lambda_1
\end{aligned}$$

by the similar estimate as in (21). Thus, we obtain (24) provided that $\Lambda \geq \Lambda_1$. □

Note that $\Lambda_1 > 0$ depends only on W. By the comparison principle we obtain the uniform bound of u^ε with respect to $\varepsilon \in (0,1)$ and initial data satisfying (I1)–(I2).

Theorem 2.7. *Assume that* (I1)–(I2) *hold. Let u^ε be a viscosity solution to* (5)–(7). *Then, there exists $R_1 > 0$, which is independent of $\varepsilon > 0$ and depends only on T and R in* (I2), *such that $\|u^\varepsilon\|_\infty \leq R_1$ on $[0,T] \times \overline{W}$.*

Proof. Fix $\delta \in (0, \delta_0)$ satisfying (23). Let

$$\bar{b}^\pm(t,x) = \pm\left(\Lambda_1 t + \phi(x) + \frac{1}{4} + \|g\|_\infty \right).$$

Then, $\bar{b}^-$ and $\bar{b}^+$ are viscosity sub- and super-solution to (5)–(6) on $(0,T) \times \overline{W}$ by Lemma 2.6, respectively. Moreover, we have

$$\bar{b}^-(0,x) \leq u^\varepsilon(0,x) = g(x) \leq \bar{b}^+(0,x) \quad \text{for } x \in \overline{W}$$

by (11). Then, by the comparison principle we obtain

$$\bar{b}^-(t,x) \leq u^\varepsilon(t,x) \leq \bar{b}^+(t,x) \quad \text{for } (t,x) \in [0,T] \times \overline{W}.$$

This implies that

$$\|u^\varepsilon\|_\infty \le \Lambda_1 T + \frac{1}{4} + R =: R_1. \square$$

2.4 Equi-continuity on the Time Variable

In this subsection, we prove the following theorem.

Theorem 2.8. *Let* $u^\varepsilon \in C^\infty([0,\infty) \times \overline{W})$ *be a solution to* (5)–(7). *Then, the function* $t \mapsto u^\varepsilon(t,x)$ *is uniformly equi-continuous on* $[0,\infty) \times \overline{W}$ *with respect to* $\varepsilon \in (0,1)$ *and the initial data* $g \in C^1(\overline{W})$ *satisfying (I1)–(I2).*

To prove Theorem 2.8 we construct a sub- and super-solution to (5)–(7) like as the Perron's method due to [8].

We now prepare some properties for the construction of viscosity sub- and super-solution to (5)–(7). Note that there exists $C_0 > 0$ satisfying

$$\langle \boldsymbol{\nu}(x), x - y \rangle \ge -C_0|x-y|^2 \quad \text{for } x \in \partial W,\ y \in \overline{W}, \tag{25}$$

since ∂W is smooth.

Fix a positive constant ρ such that $\rho < \rho_0 := (1/2)\min\{\mathrm{dist}(a_j, \overline{W});\ j = 1, \ldots, N\}$. We now demonstrate that, for every $\mu > 0$, there exists a positive constant $M_\mu = M_\mu(\mu, \|\nabla(g-\theta)\|_\infty)$ such that

$$|g(x) - g(y) - (\theta(x) - \theta(y))| \le \mu + M_\mu|x-y|^2 \quad \text{for } y \in \overline{W},\ x \in \overline{B_{\rho,\overline{W}}(y)}, \tag{26}$$

where $B_{\rho,\overline{W}}(y) := B_\rho(y) \cap \overline{W}$ and $B_\rho(y) = \{x \in \mathbb{R}^2;\ |x-y| < \rho\}$. We first extend a function g to $\tilde{g}\colon \widetilde{W} := \{x \in \mathbb{R}^2;\ \mathrm{dist}(x, W) < \rho_0\} \to \mathbb{R}$ such that $\tilde{g} \in C^1_c(\widetilde{W})$, $\tilde{g} = g$ on W,

$$\|\nabla(\tilde{g} - \theta)\|_{\infty,\widetilde{W}} \le C_W \|\nabla(g-\theta)\|_{\infty,W}, \tag{27}$$

where C_W is a constant depending only on W and $\|\nabla f\|_{\infty,D} = \sup_{x\in D}|\nabla f(x)|$ for $f\colon \mathbb{R}^2 \supset D \to \mathbb{R}$. We refer [6, Lemma 6.37] for the extension and give a few remarks here. To obtain (27) we attempt to extend $g - \theta$ into $\widetilde{W}$. However, it does not work well since $g - \theta$ is multivalued on $\overline{W}$. To overcome this difficulty, we first introduce finite coverings for ∂W, which is denoted by $B_j = B_{\rho_0}(z_j)$ with $z_j \in \partial W$ for $j = 1, 2, \ldots, K$. We also introduce a continuous branch of θ on B_j, which is denoted by θ_j. Then, we extend $g - \theta_j \in C^1(\overline{B_j \cap W})$ to a function $\tilde{w}_j \in C^1_c(B_j)$ satisfying $\tilde{w}_j = g - \theta_j$ on $\overline{B_j \cap W}$ and $\|\nabla \tilde{w}_j\|_{\infty,B_j} \le C_W \|\nabla(g-\theta)\|_{\infty,B_j\cap\overline{W}}$ by the method as in [6, Lemma 6.37]. Finally, we combine $\tilde{w}_j$ on B_j and g on W by using the partition of unity. Let $B_0 \subset\subset W$ be an open set such that $\cup_{j=0}^K B_j \supset \overline{W}$, and η_j for $j = 0, 1, \ldots, K$ be a locally finite partition of unity subordinate to the covering $\{B_j\}$, i.e., η_j satisfies $\eta_j \in C^\infty_c(B_j)$, $\eta_j \ge 0$ and $\sum \eta_j = 1$ in W. Then, the function

$$\tilde{g} = \eta_0 g + \sum_{j=1}^{K} \eta_j(\tilde{w}_j + \theta_j)$$

is our desired extention of g.

We now prove (26). For $y \in \overline{W}$ and $x \in \overline{B_{\rho,\overline{W}}(y)}$, we observe that

$$\begin{aligned}|g(x) - g(y) - (\theta(x) - \theta(y))| &= |\tilde{g}(x) - \tilde{g}(y) - (\theta(x) - \theta(y))| \\ &\leq \|\nabla(\tilde{g} - \theta)\|_{\infty,\widetilde{W}}|x - y| \\ &\leq C_W\|\nabla(g - \theta)\|_{\infty,W}|x - y|\end{aligned}$$

by Taylor expansion of $\tilde{g} - \theta$ in $\overline{B_\rho(y)} \subset \widetilde{W}$; note that $\overline{B_\rho(y)}$ is convex and θ is defined as a smooth function on $\overline{B_\rho(y)}$. Hence, we obtain

$$|g(x) - g(y) - (\theta(x) - \theta(y))| \leq \mu + \frac{C_W^2\|\nabla(g - \theta)\|_{\infty,W}^2}{4\mu}|x - y|^2$$

by Young's inequality.

We now define $\psi_{\mu,y}^{\pm}\colon [0, T] \times B_{\rho,\overline{W}}(y) \to \mathbb{R}$ as

$$\psi_{\mu,y}^{\pm}(t, x) = \pm(\Lambda t + Me^{\phi(x)}|x - y|^2 + \mu) + g(y) + \theta(x) - \theta(y)$$

for $y \in \overline{W}$ and $\varepsilon > 0$ with constants $\Lambda > 0$ and $M > 0$ satisfying

$$\begin{aligned}&Me^{-\frac{1}{4}} \geq M_\mu, \\ &\psi_{\mu,y}(t, x) > R_1 \quad \text{for } (t, x) \in [0, T] \times (\partial B_\rho(y) \cap \overline{W}),\end{aligned}$$

where R_1 is as in Theorem 2.7. Then, we first obtain the following Lemma.

Lemma 2.9. *Assume that $\delta \in (0, 2C_0)$. Then, there exists a constant $\Lambda_2 > 0$, which depends only on μ, δ and $\|\nabla(g - \theta)\|_\infty$ and is independent of $\varepsilon > 0$, such that $\psi_{\mu,y}^+$ (resp. $\psi_{\mu,y}^-$) is a viscosity super- (resp. sub-) solution to* (5)–(6) *in $(0, T) \times B_{\rho,\overline{W}}(y)$ provided that $\Lambda \geq \Lambda_2$.*

Proof. We here verify only the result on $\psi_{\mu,y}^+$ since the proof on $\psi_{\mu,y}^-$ is parallel. We shall also omit the subscriptions μ and y of $\psi_{\mu,y}^+$, and just write ψ^+ for the simplicity.

We may assume that $Me^\phi \geq Me^{-\frac{1}{4}} \geq 1$. Note that $\psi^+ \in C^2((0, T) \times B_{\rho,\overline{W}}(y))$. By the straightforward calculation we have

$$\begin{aligned}\psi_t^+ &= \Lambda, \\ \nabla(\psi^+ - \theta) &= Me^\phi(|x - y|^2\nabla\phi + 2(x - y)), \\ \nabla^2(\psi^+ - \theta) &= Me^\phi\left[|x - y|^2\nabla^2\phi + |x - y|^2\nabla\phi \otimes \nabla\phi\right. \\ &\qquad \left. +2\{(x - y) \otimes \nabla\phi + \nabla\phi \otimes (x - y)\} + 2I\right]. \end{aligned} \tag{28}$$

If $x = y \in W$, then $\nabla(\psi^+ - \theta) = 0$ and $\nabla^2(\psi^+ - \theta) = 2I$, which implies

$$\psi_t^+ + F_\varepsilon(\nabla(\psi^+ - \theta), \nabla^2(\psi^+ - \theta)) = \Lambda - 4 - C\varepsilon \geq 0$$

provided that $\Lambda \geq 4 + C\varepsilon$, where

$$F_\varepsilon(p, X) = -\mathrm{tr}\left\{\left(I - \frac{p \otimes p}{\varepsilon^2 + |p|^2}\right) X\right\} - C\sqrt{\varepsilon^2 + |p|^2}.$$

If $x = y \in \partial W$, then

$$\langle \boldsymbol{\nu}, \nabla(\psi^+ - \theta)\rangle = 0.$$

Assume that $x \neq y$. Set $q = |x - y|\nabla\phi + 2|x - y|^{-1}(x - y)$, and calculate $q \otimes q$ to obtain

$$\begin{aligned} q \otimes q =& |x - y|^2 \nabla\phi \otimes \nabla\phi + 2(\nabla\phi \otimes (x - y) + (x - y) \otimes \nabla\phi) \\ &+ 4\frac{(x - y) \otimes (x - y)}{|x - y|^2}. \end{aligned}$$

Then, $\nabla(\psi^+ - \theta)$ and $\nabla^2(\psi^+ - \theta)$ are simplified respectively as

$$\begin{aligned} \nabla(\psi^+ - \theta) &= Me^\phi q, \\ \nabla^2(\psi^+ - \theta) &= Me^\phi \left(q \otimes q - 4\frac{(x - y) \otimes (x - y)}{|x - y|^2} + |x - y|^2 \nabla^2\phi + 2I\right). \end{aligned}$$

If $x \in W$, then we obtain

$$\begin{aligned} &\psi_t^+ + F_\varepsilon(\nabla(\psi^+ - \theta), \nabla^2(\psi^+ - \theta)) \\ &= \Lambda + Me^\phi \left\{F_{\varepsilon,0}(Me^\phi q, q \otimes q) - F_{\varepsilon,0}(Me^\phi q, 4|x - y|^{-2}(x - y) \otimes (x - y))\right. \\ &\qquad \left. + F_{\varepsilon,0}(Me^\phi q, |x - y|^2 \nabla^2\phi) + F_{\varepsilon,0}(Me^\phi q, 2I)\right\} - C\sqrt{\varepsilon^2 + M^2 e^{2\phi}|q|^2}, \end{aligned} \tag{29}$$

where

$$F_{\varepsilon,0}(p, X) = -\mathrm{tr}\left\{\left(I - \frac{p \otimes p}{\varepsilon^2 + |p|^2}\right) X\right\} \quad \text{for } p \in \mathbb{R},\ X \in \mathbb{S}^2.$$

Here we have used $F_{\varepsilon,0}(p, X + Y) = F_{\varepsilon,0}(p, X) + F_{\varepsilon,0}(p, Y)$. We first calculate that

$$
\begin{aligned}
F_{\varepsilon,0}(\lambda p_1, \mu p_2 \otimes p_2) &= -\mu \operatorname{tr}\left[p_2 \otimes p_2 - \frac{\lambda^2 (p_1 \otimes p_1)(p_2 \otimes p_2)}{\varepsilon^2 + \lambda^2 |p_1|^2} \right] \\
&= -\mu \left(|p_2|^2 - \frac{\lambda^2 \langle p_1, p_2 \rangle^2}{\varepsilon^2 + \lambda^2 |p_1|^2} \right)
\end{aligned}
$$

for $\lambda, \mu \in \mathbb{R}$ and $p_1, p_2 \in \mathbb{R}^2$, and then we obtain

$$
Me^{\phi} F_{\varepsilon,0}(Me^{\phi} q, q \otimes q) = -Me^{\phi}|q|^2 \frac{\varepsilon^2}{\varepsilon^2 + M^2 e^{2\phi}|q|^2} \geq -\frac{\varepsilon^2}{Me^{\phi}} \geq -Me^{\phi}.
$$

with the assumption $Me^{\phi} \geq 1 > \varepsilon$. On the other hand, by the Schwarz inequality we also obtain

$$
\begin{aligned}
&- Me^{\phi} F_{\varepsilon,0}(Me^{\phi} q, 4|x-y|^{-2}(x-y) \otimes (x-y)) \\
&= \frac{4Me^{\phi}}{|x-y|^2} \left(|x-y|^2 - \frac{M^2 e^{2\phi} \langle q, x-y \rangle^2}{\varepsilon^2 + M^2 e^{2\phi}|q|^2} \right) \\
&\geq 4Me^{\phi} \left(1 - \frac{M^2 e^{2\phi}|q|^2}{\varepsilon^2 + M^2 e^{2\phi}|q|^2} \right) = \frac{4Me^{\phi}\varepsilon^2}{\varepsilon^2 + M^2 e^{2\phi}|q|^2} \geq 0.
\end{aligned}
$$

On the third and fourth term of (29) we observe that

$$
\begin{aligned}
Me^{\phi} F_{\varepsilon,0}(Me^{\phi} q, |x-y|^2 \nabla^2 \phi) &\geq -Me^{\phi} \|\nabla^2 \phi\| |x-y|^2, \\
Me^{\phi} F_{\varepsilon,0}(Me^{\phi} q, 2I) &= -2Me^{\phi} \left(2 - \frac{M^2 e^{2\phi}|q|^2}{\varepsilon^2 + M^2 e^{2\phi}|q|^2} \right) \geq -4Me^{\phi},
\end{aligned}
$$

here we have used the similar estimate as (21) for the third term. Consequently, we obtain

$$
\begin{aligned}
&\psi_t^+ + F_\varepsilon(\nabla(\psi^+ - \theta), \nabla^2(\psi^+ - \theta)) \\
&\geq \Lambda - Me^{\phi}(\|\nabla^2 \phi\| |x-y|^2 + 5) - C\sqrt{1 + \|Me^{\phi} q\|_\infty^2}.
\end{aligned}
$$

By (11) we now set

$$
\Lambda_2 = M(\rho^2 \|\nabla^2 \phi\| + 5) + C\sqrt{1 + M^2 \|q\|_\infty^2}.
$$

Then, we obtain

$$
\psi_t^+ + F_\varepsilon(\nabla(\psi^+ - \theta), \nabla^2(\psi^+ - \theta)) \geq 0 \quad \text{in } (0, T) \times W
$$

provided that $\Lambda \geq \Lambda_2$.

If $x \in \partial W \cap B_\rho(y)$, then we obtain

$$\begin{aligned}\langle \boldsymbol{\nu}, \nabla(\psi^+ - \theta)\rangle &= Me^{\phi}(|x-y|^2\langle \boldsymbol{\nu}, \nabla\phi\rangle + 2\langle \boldsymbol{\nu}, x-y\rangle) \\ &\geq Me^{\phi}|x-y|^2\left(\frac{1}{\delta} - 2C_0\right) \geq 0\end{aligned}$$

by combining (28), (13), (25), and the assumption $\delta < 2C_0$. Hence, we obtain the result of Lemma 2.9. □

Remark 2.10. Remark that Λ_2 is determined with $M = O(\|\nabla(g-\theta)\|_\infty^2/\mu)$ (as $\|\nabla(g-\theta)\|_\infty \to \infty$ or $\mu \to 0$), ρ, $\|\nabla^2\phi\| = O(1/\delta^2)$ and $\|q\|_\infty = O(1/\delta)$ (as $\delta \to 0$). Note that $\delta \in (0, 2C_0)$ and $\rho \in (0, \rho_0)$ will be fixed in the proof of Theorem 2.8. Therefore, the important dependency of Λ_2 is $\Lambda_2 = O(\|\nabla(g-\theta)\|_\infty^2/\mu)$ as $\|\nabla(g-\theta)\|_\infty \to \infty$ or $\mu \to 0$.

By the stability of viscosity solutions, we can easily extend $\psi^\pm$ to the function on whole domain.

Corollary 2.11. *Assume that* $\delta \in (0, 2C_0)$, *and define*

$$\tilde{\psi}^+_{\mu,y}(t,x) = \begin{cases} \min\{\bar{b}^+(t,x), \psi^+_{\mu,y}(t,x)\} & \text{if } (t,x) \in [0,T] \times B_{\rho,\overline{W}}(y), \\ \bar{b}^+(t,x) & \text{if } (t,x) \in [0,T] \times (\overline{W} \setminus B_\rho(y)), \end{cases}$$

$$\tilde{\psi}^-_{\mu,y}(t,x) = \begin{cases} \max\{\bar{b}^-(t,x), \psi^-_{\mu,y}(t,x)\} & \text{if } (t,x) \in [0,T] \times B_{\rho,\overline{W}}(y), \\ \bar{b}^-(t,x) & \text{if } (t,x) \in [0,T] \times (\overline{W} \setminus B_\rho(y)) \end{cases}$$

with $\Lambda = \Lambda_2$. *Then,* $\tilde{\psi}^+_{\mu,y}$ *and* $\tilde{\psi}^-_{\mu,y}$ *are viscosity super- and sub-solution to* (5)–(6) *on* $(0,T) \times \overline{W}$.

Proof. We also prove only on $\tilde{\psi}^+_{\mu,y}$ since the proof on $\tilde{\psi}^-_{\mu,y}$ is parallel.

By definition of $\psi^+_{\mu,y}$ there exists an open set $D \supset \overline{W} \setminus B_\rho(y)$ such that $\tilde{\psi}^+_{\mu,y} \equiv \bar{b}^+(t,x)$ in $(0,T) \times (D \cap \overline{W})$, which implies that $\tilde{\psi}^+_{\mu,y}$ is a viscosity supersolution to (5)–(6) in $(0,T) \times (D \cap \overline{W})$. On the other hand, $\tilde{\psi}^+_{\mu,y}$ is also a viscosity supersolution to (5)–(6) in $(0,T) \times B_{\rho,\overline{W}}(y)$ since it is defined by a minimum of supersolutions over there. Hence, we obtain Corollary 2.11. □

By using $\tilde{\psi}^\pm_{\mu,y}$, we obtain a viscosity super- and sub-solution to (5)–(7).

Lemma 2.12. *Assume* $\delta \in (0, 2C_0)$ *and define*

$$\bar{\psi}^+(t,x) = \inf_{\mu>0, y\in\overline{W}} \tilde{\psi}^+_{\mu,y}(t,x), \quad \bar{\psi}^-(t,x) = \sup_{\mu>0, y\in\overline{W}} \tilde{\psi}^-_{\mu,y}(t,x)$$

with $\Lambda = \Lambda_2$. *Then,* $\bar{\psi}^+$ *or* $\bar{\psi}^-$ *is a viscosity super- and sub-solution to* (5)–(6), *respectively. Moreover, the following properties hold;*

(i) $\bar{\psi}^\pm(0,x) = g(x)$,
(ii) $\lim_{t\to 0}\bar{\psi}^\pm(t,x) = g(x)$, *and the convergence is uniform for* $x \in \overline{W}$.

Proof. The proof of super- and sub-solution part on $\bar{\psi}^{\pm}$ and (i) follows from the usual stability result and Perron's method in a theory of viscosity solutions. We now verify (ii). We here verify only $\lim_{t\to 0}\bar{\psi}^{+}(t,x) = g(x)$ since the proof on $\bar{\psi}^{-}$ is parallel.

Note that $g(x) \le \bar{\psi}^{+}(t,x)$ for every $(t,x) \in (0,T)\times\overline{W}$. In fact, for $\mu > 0$ and $y \in \overline{W}$ satisfying $x \in B_\rho(y)$ we have

$$g(x) < \Lambda_2 t + g(x) < \Lambda_2 t + M|x-y|^2 + \mu + g(y) + \theta(x) - \theta(y) = \psi^{+}_{\mu,y}(t,x)$$

by (26). Then, we obtain $g(x) \le \bar{\psi}^{+}(t,x)$ by taking minimum with the above and $b^{+}(t,x)$, and taking infimum of that with $\mu > 0$ and $y \in \overline{W}$.

Let $\mu > 0$, and consider

$$\psi^{+}_{\mu/2,x}(t,x) = \tilde{\Lambda}_2 t + \frac{\mu}{2} + g(x),$$

where $\tilde{\Lambda}_2$ is the constant chosen in the proof of Lemma 2.9 with the parameter $\mu/2$ instead of μ. This yields that

$$g(x) \le \bar{\psi}^{+}(t,x) \le \psi^{+}_{\mu/2,x}(t,x) = \tilde{\Lambda}_2 t + \frac{\mu}{2} + g(x).$$

Thus, for every $\mu > 0$ we now choose $t < \mu/(2\tilde{\Lambda}_2)$, and then

$$|\bar{\psi}^{+}(t,x) - g(x)| < \mu.$$

Hence, we obtain (ii). □

We now recall Remark 2.10, and we also have a priori estimate on $\nabla(u^\varepsilon - \theta)$ with $\|\nabla(g-\theta)\|_\infty$ as in Theorem 2.4. By combining the above we obtain Theorem 2.8.

Proof of Theorem 2.8. Let us choose $\mu > 0$ arbitrary, $t_1, t_2 \in (0,T)$ with $t_1 < t_2$, and $x \in \overline{W}$. We now fix $\delta \in (0, 2C_0)$, and then construct $\bar{\psi}^{\pm}(t,x)$ as in Lemma 2.12 with $g(x) = u^\varepsilon(t_1,x)$. Then, the comparison principle implies that

$$\bar{\psi}^{-}(t_2-t_1,x) - u^\varepsilon(t_1,x) \le u^\varepsilon(t_2,x) - u^\varepsilon(t_1,x) \le \bar{\psi}^{-}(t_2-t_1,x) - u^\varepsilon(t_1,x).$$

Note that $\tilde{\Lambda}_2$ in the proof of Lemma 2.12 depends only on $\|\nabla(u^\varepsilon-\theta)\|_\infty \le e^{cT}\|\nabla(g-\theta)\|_\infty$ by Theorem 2.4 and $\mu > 0$. In particular, we can choose a global constant $\tilde{\Lambda}_2$ for functions $u^\varepsilon(t,\cdot)$ with $t \in [0,T]$, $\varepsilon > 0$ and initial data g satisfying (I1)–(I2). Then, if $|t_1 - t_2| < \mu/(2\tilde{\Lambda}_2)$, we obtain

$$|u^\varepsilon(t_2,x) - u^\varepsilon(t_1,x)| < \mu.$$

Note that $\mu/(2\tilde{\Lambda}_2)$ is independent of ε and $x \in \overline{W}$. Hence, we obtain the equi-continuity of $t \mapsto u^\varepsilon(t,x)$ for $\varepsilon > 0$ and initial data g satisfying (I1)–(I2), and the continuity is uniform for $x \in \overline{W}$. □

2.5 Proof of Theorem 2.1

We are in the position to prove Theorem 2.1.

Let u^ε be a solution to (5)–(6) with $u^\varepsilon(0,\cdot) = g_\varepsilon$ satisfying (A1)–(A3), which implies (I1)–(I2) for $\varepsilon \in (0,1)$. Then, we obtain the uniform bound and equi-continuity of u^ε on $[0,T] \times \overline{W}$ by Theorems 2.7, 2.4 and 2.8 for $\varepsilon \in (0,1)$. This yields a subsequence of u^ε, which is still denoted by u^ε, satisfying $\lim_{\varepsilon\to 0} u^\varepsilon = u \in C([0,T] \times \overline{W})$ uniformly in $[0,T] \times \overline{W}$. According to the stability of viscosity solutions on limiting procedure, we also observe that u is a viscosity solution to (3)–(4) with $u(0,\cdot) = u_0$. See [4] or [3] for the stability.

We now derive (9). In fact,

$$
\begin{aligned}
&|u(t,x) - \theta(x) - (u(t,y) - \theta(y))| \\
&\le |u(t,x) - u^\varepsilon(t,x)| + |u^\varepsilon(t,x) - \theta(x) - (u^\varepsilon(t,y) - \theta(y))| \\
&+ |u^\varepsilon(t,y) - u(t,y)| \\
&\le 2\|u^\varepsilon - u\|_\infty + e^{ct}(\|\nabla(g_\varepsilon - \theta)\|_\infty)|x-y| \\
&\le 2\|u^\varepsilon - u\|_\infty + e^{ct}(\|\nabla g_\varepsilon\|_\infty + \|\nabla\theta\|_\infty)|x-y|
\end{aligned}
$$

by Theorem 2.4. Hence, we obtain (9) by tending $\varepsilon \to 0$. □

3 Concluding Remarks

1. Note that $\|\nabla\theta\|_\infty = O(1/\min_j \operatorname{dist}(a_j, \partial W))$ as $\min_j \operatorname{dist}(a_j, \partial W) \to 0$. In [12] we impose that every a_j is on the numerical lattice, and remove only a grid point of a_j from the lattice as U_j. This means that $\|\nabla\theta\|_\infty = O(1/\Delta x)$ as $\Delta x \to 0$, where Δx is the spatially grid size of the lattice. Unfortunately, our estimate is still too rough for our goal stated in §1.
2. We can prove the convergence of u^ε to u without choosing a converging subsequence of u^ε by the uniqueness of solution to (3)–(4), as long as the approximating equation implies the uniform bound and equi-continuity of u^ε. This fact is a background theory considering regularized equation of (3)–(4) for numerical simulations. See [12] for a details of numerical simulations for (3)–(4).

Acknowledgements The author is grateful to Professor Yoshikazu Giga, Katsuyuki Ishii, and Shigeru Sakaguchi for valuable comments. The work of the author is partly supported by JSPS Grant Kiban(C) 26400158. The author is also grateful to the anonymous referee for careful reading as well as valuable comments.

References

1. S. Bernstein, Sur la généralisation du problème de Dirichlet II. Math. Ann. **69**(1), 82–136 (1910)
2. S. Bernstein, Sur les équations du calcul des variations. Ann. Sci. École Norm. Sup. **3**(29), 431–485 (1912)
3. Y.G. Chen, Y. Giga, S. Goto, Uniqueness and existence of viscosity solutions of generalized mean curvature flow equations. J. Differ. Geom. **33**(3), 749–786 (1991)
4. M.G. Crandall, H. Ishii, and P.-L. Lions, User's guide to viscosity solutions of second order partial differential equations. Bull. Am. Math. Soc. (N.S.) **27**(1), 1–67 (1992)
5. Y. Giga, M. Ohnuma, M.-H. Sato, On the strong maximum principle and the large time behavior of generalized mean curvature flow with the Neumann boundary condition. J. Differ. Equs. **154**(1), 107–131 (1999)
6. D. Gilbarg, N.S. Trudinger, Elliptic partial differential equations of second order, in *Classics in Mathematics* (Springer, Berlin, 2001) (Reprint of the 1998 edition)
7. S. Goto, M. Nakagawa, T. Ohtsuka, Uniqueness and existence of generalized motion for spiral crystal growth. Indiana Univ. Math. J. **57**(5), 2571–2599 (2008)
8. H. Ishii, Perron's method for Hamilton-Jacobi equations. Duke Math. J. **55**(2), 369–384 (1987)
9. R. Kobayashi, A brief introduction to phase field method. AIP Conf. Proc. **1270**, 282–291 (2010)
10. G.M. Lieberman, in *Second order Parabolic Differential Equations* (World Scientific Publishing Co., Inc, River Edge, 1996)
11. T. Ohtsuka, A level set method for spiral crystal growth. Adv. Math. Sci. Appl. **13**(1), 225–248 (2003)
12. T. Ohtsuka, Y.-H.R. Tsai, Y. Giga, A level set approach reflecting sheet structure with single auxiliary function for evolving spirals on crystal surfaces. J. Sci. Comput. **62**(3), 831–874 (2015)
13. M.H. Protter, H.F. Weinberger, *Maximum Principles in Differential Equations* (Prentice-Hall Inc., Englewood Cliffs, 1967)
14. R.P. Sperb, in *Maximum Principles and their Applications* Mathematics in Science and Engineering, vol. 157 (Academic Press, Inc., New York, 1981) [Harcourt Brace Jovanovich, Publishers]

A Hyperbolic Obstacle Problem with an Adhesion Force

Seiro Omata

Abstract We will treat free boundary problems of a wave type in this section. Examples of the physical phenomena that we have in mind are a motion of a soap film attached to a water surface or a droplet motion on a planner surface. The surface acts as an obstacle and there may exist adhesion forces when the film or the droplet detach from the obstacle. We consider the case with a positive contact angle in an equilibrium state. We also calculate the moving contact angle according to a dynamical action functional.

Keywords Free boundary problems · Obstacle problems · Droplet motion

1 Removing an Adhesive Tape from a Solid Plane

1.1 Tape Peeling Energy

In this paper, we consider the problem of removing an adhesive tape from a solid plane. We shall show that this problem can be modeled by a hyperbolic type obstacle problem with an adhesive force.

This approach leads to a dynamical contact angle condition, i.e., a dynamical Young's law. In the stationary case, the contact angle is determined by the difference of the surface tension of the tape between the parts attached to the surface and detached from the surface. Therefore, it can be generalized to many problems including the bubble motion with a junction or a droplet motion on the plane.

The simplest setting is as follows: (1) Stretch an adhesive tape uniformly (with a constant tension force S), (2) attach it to a solid plane, and finally (3) lift the tape up

Dedicated to Professor Yoshikazu Giga's 60th birthday.

S. Omata (✉)
Faculty of Mathematics and Physics, Institute of Science and Engineering, Kanazawa University
Kakuma, Kanazawa, Ishikawa 920-1192, Japan
e-mail: omata@kenroku.kanazawa-u.ac.jp

Y. Maekawa and S. Jimbo (eds.), *Mathematics for Nonlinear Phenomena—Analysis and Computation*, Springer Proceedings in Mathematics & Statistics 215,
DOI 10.1007/978-3-319-66764-5_12

Fig. 1 Tape peeling model

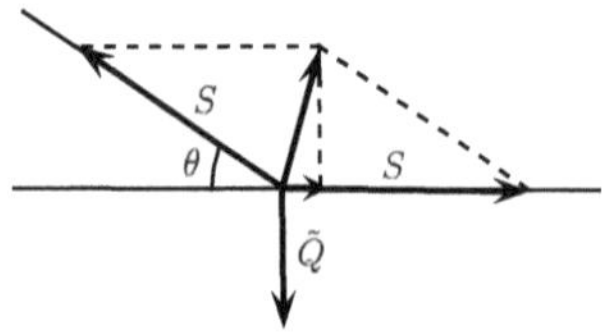

by its edge. In this way, the tape gradually peels off from the plane. The boundary between the attached and detached parts of the tape moves, and hence it is the free boundary in this problem.

Let us calculate the potential energy in the stationary case. The ground state is chosen so that the tape is completely attached to the surface. By choosing a suitable coordinate system and assuming that the tape is very thin, we suppose that the tape is a surface that is a graph of a scalar function, u. The tension force of the tape is assumed to be always the same (it is true exactly in the case of a soapy film). Under these assumptions, the energy coming from stretching the tape is described by $\int_{\{u>0\}} S(\sqrt{1+|\nabla u|^2}-1)dx$ (Fig. 1).

We also assume that the adhesion force has a potential, i.e., the energy of the peeling process. As depicted in the following picture, we can determine the angle at the free boundary by finding the equilibrium between the resulting tension force and the maximum adhesion force in the vertical direction. We call this force $\tilde{Q}$. The equilibrium condition is $S\sin\theta = \tilde{Q}$, then the force which we need to peel the tape off is $S(1-\cos\theta)$. It is proportional to the area on which the tape is peeled off.

Consequently, the energy can be expressed in the following way:

$$\begin{aligned} &\int_{\{u>0\}} S\big((\sqrt{1+|\nabla u|^2}-1)+(1-\sqrt{1-(\tilde{Q}/S)^2})\big)dx \\ &= S\int_{\Omega}\big(\sqrt{1+|\nabla u|^2}+(1-\sqrt{1-(\tilde{Q}/S)^2})\chi_{u>0}\big)dx - S|\Omega|, \end{aligned}$$

and the stationary state is described by the minimizer of the above functional [7].

1.2 Approximation by the Dirichlet Integral

Here we assume that the contact angle θ is small enough. In this case, we can approximate the energy by the Dirichlet integral:

$$I(u) := \frac{S}{2}\int_{\Omega}(|\nabla u|^2 + Q^2\chi_{u>0})dx \tag{1}$$

which is introduced by H.W. Alt, L.A. Caffarelli and A. Friedman [1, 2] and Omata and Yamaura [5, 6]. Here, $Q^2 = (\tilde{Q}/S)^2$ in this approximation.

In [1], it was shown that a minimizer of (1) exists in $\{u \in L^1_{\mathrm{loc}}(\Omega); \nabla u \in L^2(\Omega), u = u_0 \text{ on } \partial\Omega\}$ and Δu becomes a Radon measure whose support is on $\partial\{u>0\}$, and which is absolutely continuous with respect to the $n-1$-dimensional

Hausdorff measure. They also discuss the regularity of the free boundary, $\partial\{u > 0\}$. They show that the minimizer of this functional satisfies

$$\Delta u = 0 \quad \text{in } \{u > 0\}, \qquad |\nabla u| = Q \quad \text{on } \partial\{u > 0\}.$$

2 A Time-Dependent Problem

In this section, we consider a time-dependent problem. To this end, we introduce a Lagrangian and an action integral for this phenomenon. In the usual manner, the action can be expressed as

$$J(u) = \int_0^T \int_\Omega \Big((u_t)^2 \chi_{u>0} - |\nabla u|^2 - Q^2 \chi_{u>0} \Big) dx dt, \tag{2}$$

where T is a positive constant. The first term is the kinetic energy of the tape in the vertical direction. To obtain the momentum equation, we need to find a critical point of this action integral. Since it includes the characteristic function, its discontinuity prevents us from calculating the first variation in the usual manner. To overcome this, at first, we assume a regularity of a critical point u and its free boundary $\partial\{u > 0\}$. Moreover, the critical point u is assumed to be nonnegative. This can be rigorously justified using the maximum principle for the Laplacian. Indeed, if there is a set $\{u < 0\}$, we observe that $\Delta u = 0$ in $\{u < 0\}$ for almost every time t and then u must be equal to zero in the negative part, a contradiction. Therefore, we can consider a nonnegative critical point u of J in $C^0(\Omega \times (0, T)) \cap W^{1,2}(\Omega \times (0, T))$. For all $\zeta \in C_0^\infty(\Omega \times (0, T) \cap \{u > 0\})$. Then we get

$$\int_0^T \int_{\Omega \cap \{u>0\}} (\nabla u \cdot \nabla \zeta - u_t \zeta_t) \, dx dt = 0. \tag{3}$$

This means that a critical point, u, satisfies wave equation in $\{u > 0\}$ and $u = 0$ otherwise. On the other hand, to obtain the free boundary condition, we calculate the so-called inner variation. Let us assume that $u \in C^2(\Omega \times (0, T) \cap \{u > 0\})$ and that the free boundary, $\partial\{u > 0\}$ is a C^1 surface. For all $\eta \in C_0^\infty(\Omega \times (0, T); \mathbf{R}^n \times \mathbf{R})$ and for all small ϵ satisfying $0 < \epsilon < \operatorname{dist}(\operatorname{spt} \eta, \partial(\Omega \times (0, T)))$, we consider a map: $\tau_\epsilon : \Omega \times (0, T) \to \Omega \times (0, T)$ and we define $\tau_\epsilon(z) := z + \epsilon\eta(z)$. Here z is a variable in space and time, $\Omega \times \mathbf{R}$ (i.e., $z_j = x_j$ $(j = 1, \ldots, n)$, $z_{n+1} = t$). Here we use $u_\epsilon(z) = u \circ \tau_\epsilon^{-1}(z)$ and calculate the inner variation, $\frac{d}{d\epsilon} J(u_\epsilon)|_{\epsilon=0} = 0$. We shall formally calculate the right-hand side of the first variation in the above, at first,

$$J(u_\epsilon) = -\int_{\Omega \times (0,T) \cap \{u>0\}} \Big\{ \langle \nabla_h u (D\tau_\epsilon)^{-1}, \nabla_z u (D\tau_\epsilon)^{-1} \rangle + Q^2 \Big\} |\det D\tau_\epsilon| dz,$$

where $\nabla_z f = (f_{z_1}, \cdots, f_{z_n}, f_{z_{n+1}})$ and $\nabla_h f = (f_{z_1}, \cdots, f_{z_n}, -f_{z_{n+1}})$. Here, $\langle\cdot,\cdot\rangle$ is the inner product on $\mathbf{R}^{n+1}$ and $D\tau_\epsilon$ is the Jacobi matrix for the map τ_ϵ.

We have

$$J(u_\epsilon) - J(u) = \epsilon \int_{\Omega\times(0,T)\cap\{u>0\}} \{\langle\nabla_h u, \nabla_z u\rangle + Q^2\}\nabla_z \cdot \eta dz - \epsilon \int_{\Omega\times(0,T)\cap\{u>0\}} 2\nabla_h u D\eta \nabla_z u dz + o(\epsilon).$$

Using the equality

$$\begin{aligned} &\nabla_z \cdot \Big(\eta\langle\nabla_h u, \nabla_z u\rangle - 2\langle\eta, \nabla_z u\rangle\nabla_h u\Big) \\ &= \nabla_z \cdot \eta\langle\nabla_h u, \nabla_z u\rangle - 2\nabla_h u D\eta\nabla_z u - 2\langle\eta, \nabla_z u\rangle(\Delta u - u_{tt}) \end{aligned}$$

we have

$$\begin{aligned} \frac{d}{d\epsilon}J(u_\epsilon)|_{\epsilon=0} &= \int_{\Omega\times(0,T)\cap\{u>0\}} \nabla_z \cdot \Big(\eta\langle\nabla_h u, \nabla_z u\rangle - 2\langle\eta, \nabla_z u\rangle\nabla_h u + Q^2\eta\Big)dz \\ &= \int_{\Omega\times(0,T)\cap\partial\{u>0\}} \big((\langle\nabla_h u, \nabla_z u\rangle\eta - 2\langle\eta, \nabla_z u\rangle\nabla_h u)\cdot\nu \\ &\qquad + Q^2\langle\eta, \nu\rangle\big)d\mathcal{H}^n, \end{aligned}$$

where $\mathcal{H}^n$ is the n-dimensional Hausdorff measure. Here, ν is the outer unit normal vector to $\Omega\times(0,T)\cap\{u>0\}$.

On the other hand, since $\nabla_z u = -\nu|\nabla_z u|$, we get

$$\frac{d}{d\epsilon}J(u_\epsilon)|_{\epsilon=0} = -\int_{\Omega\times(0,T)\cap\partial\{u>0\}} (\langle\nabla_h u, \nabla_z u\rangle - Q^2)\langle\eta, \nu\rangle d\mathcal{H}^n.$$

This is equal to 0, thus we get the free boundary condition. Rewriting it in the strong form, we have

$$\begin{cases} u_{tt} - \Delta u = 0 & \text{in} \quad \Omega\times(0,T)\cap\{u>0\}, \\ |\nabla u|^2 - (u_t)^2 = Q^2 & \text{on} \quad \Omega\times(0,T)\cap\partial\{u>0\}. \end{cases} \tag{4}$$

For this problem, we have shown the existence of a strong solution in the one-dimensional case.

Theorem 1. *K. Kikuchi, S. Omata [4]. Let $\Omega = (0,\infty)$. Suppose that a boundary condition $u(0,t) = g(t)$ is given and assume that the initial data u_0 is positive in $(0,l)$. Moreover, at $x = 0$ and $x = l$, the initial data satisfies the appropriate compatibility condition and $g'(t) > 0$. Then the equation (4) has a unique time local classical solution satisfying $u \in C^2(Q_T \cap \{u>0\})$ and $\partial\{u>0\}$ is a C^1-curve in (x,t).*

3 Regularized Equation and Its Limit

3.1 General Treatment for the Equation

In this section, we consider a regularized equation. As in the previous section, if the energy is conserved, we get (4). However, in the case that the tape touches plane and attaches to it, the energy is not preserved. To establish the kinetics equation, we shall start from the momentum equation. But in our energy functional, it includes the characteristic function, $\chi_{u>0}$, depending on the unknown function u. To obtain the equations, we need to compute the first variation of the characteristic function. But it seems hopeless to justify this computation rigorously. To overcome this difficulty, in an elliptic case, a smoothing of the characteristic function is used and its formal limit leads to the free boundary conditions. We will follow this approach for this type of hyperbolic equations.

Here, we treat a more general form of (4). We introduce a new symbol $\mathcal{Q} \subset \mathbb{R}^n$ and $z = (z_1, \cdots, z_n) \in \mathcal{Q}$. Let us consider the following hyperbolic free boundary problem:

$$\begin{cases} \operatorname{div}(\mathscr{D}u) = 0 \text{ in } \mathcal{Q} \cap \{u > 0\} \\ \nabla u \cdot \mathscr{D}u = 2 \text{ on } \mathcal{Q} \cap \partial\{u > 0\}. \end{cases} , \tag{5}$$

where $\mathscr{D}$ is a linear first-order differential operator which satisfies $\nabla u_{x_k} \cdot \mathscr{D}u = \nabla u \cdot \mathscr{D}u_{x_k}$. Examples of this are $\mathscr{D}u = \nabla u$ or $\mathscr{D}u = (u_{z_1}, \cdots, u_{z_{n-1}}, -u_{z_n})$.

For (5), we define a mild equation in the following way:

$$\operatorname{div}(\mathscr{D}u^\varepsilon) = \beta_\varepsilon(u^\varepsilon) \quad \text{in } \mathcal{Q} \tag{6}$$

where β_ε is defined as a smooth function satisfying $\beta \geq 0$, $\beta = 0$ outside $[0, 1]$ and the following conditions (see [3]):

$$\begin{cases} 0 < \exists a < \exists b < 1 \text{ s.t. } \beta_\varepsilon \geq \dfrac{K}{\varepsilon} \quad \text{in } [a\varepsilon, b\varepsilon], \\ \beta_\varepsilon(u) = \dfrac{1}{\varepsilon}\beta\left(\dfrac{u}{\varepsilon}\right), \\ \displaystyle\int_0 \beta(s)\, ds = 1. \end{cases}$$

Let β_ε be a function defined above, and let B_ε be

$$B_\varepsilon(u) := \int_0^u \beta_\varepsilon(s)\, ds \left(= \int_0^{u/\varepsilon} \beta(s)\, ds \right).$$

Clearly $B_\varepsilon(u^\varepsilon) = 1$ $(u \geq \varepsilon)$. We get the following proposition.

Proposition 1. *Let β_ε and B_ε be the functions defined above. We assume the existence of u^ε, a solution to (6). Moreover, we assume that $u^\varepsilon \to v$,* $\mathrm{div}(\mathscr{D}v) = 0$ in $\mathcal{Q} \cap \{v > 0\}$ *in some appropriate topology. Then we have*

$$\nabla v \cdot \mathscr{D}v = 2 \quad \text{on } \mathcal{Q} \cap \partial\{v > 0\}.$$

We will show this by a formal calculation. Multiply both sides of equation (6) by $\zeta u_k \left(\equiv \zeta u_{x_k}\right) \left(\zeta \in C_0^\infty(\mathcal{Q})\right)$ and integrate over $\mathcal{Q}$. We have

$$\int_{\mathcal{Q}} \zeta\, u_k^\varepsilon \,\mathrm{div}(\mathscr{D}u^\varepsilon)\, dz = \int_{\mathcal{Q}} \zeta\, u_k^\varepsilon\, \beta_\varepsilon(u^\varepsilon)\, dz. \tag{7}$$

Here we note that $\left[B_\varepsilon(u)\right]_{x_k} = \beta_\varepsilon(u) u_k$. Using the integration by parts, the right-hand side of (7) can be calculated as

$$\begin{aligned}
(r.h.s.(7)) &= -\int_{\mathcal{Q}} \zeta_k B_\varepsilon(u^\varepsilon) dz \\
&\underset{\varepsilon\to 0}{\longrightarrow} -\int_{\mathcal{Q}\cap\{v>0\}} \zeta_k dz \quad (u^\varepsilon \to v,\ B_\varepsilon(u^\varepsilon) \to \chi_{v>0} \text{ are assumed}) \\
&= -\int_{\mathcal{Q}\cap\partial\{v>0\}} \zeta\, \nu_k dS.
\end{aligned}$$

Here, $\nu = (\nu_1 \cdots \nu_n)$ is the unit outer normal to $\{v > 0\} \subset \mathcal{Q}$ and ν_k is its k-th element. On the other hand, the left-hand side of (7) can be calculated as

$$\int_{\mathcal{Q}} \zeta u_k^\varepsilon \,\mathrm{div}(\mathscr{D}u^\varepsilon)\, dz = -\int_{\mathcal{Q}} \left(\nabla(\zeta u_k^\varepsilon) \cdot \mathscr{D}u^\varepsilon\right) dz.$$

Moreover, we can assume that outside of $\{u^\varepsilon > 0\}$ the function u^ε is zero. By the assumption on $\mathscr{D}$, and the equality

$$\begin{aligned}
\int_{\mathcal{Q}} \zeta\, \nabla u_k^\varepsilon \cdot \mathscr{D}u^\varepsilon\, dz &= \frac{1}{2}\int_{\mathcal{Q}} \zeta\, \frac{\partial}{\partial x_k}(\nabla u^\varepsilon \cdot \mathscr{D}u^\varepsilon)\, dz \\
&= -\frac{1}{2}\int_{\mathcal{Q}} \zeta_k\, \nabla u^\varepsilon \cdot \mathscr{D}u^\varepsilon\, dz
\end{aligned}$$

the left-hand side of (7) can be calculated in the following way:

$$
\begin{aligned}
&-\int_{\mathcal{Q}}[\mathscr{D}u^{\varepsilon}\cdot\nabla\zeta]u_k^{\varepsilon}-\int_{\mathcal{Q}}\zeta\nabla u_k^{\varepsilon}\cdot\mathscr{D}u^{\varepsilon}dz\\
&=-\int_{\mathcal{Q}}\Big(\big[\mathscr{D}u^{\varepsilon}\cdot\nabla\zeta\big]u_k^{\varepsilon}-\frac{1}{2}\zeta_k\nabla u^{\varepsilon}\cdot\mathscr{D}u^{\varepsilon}\Big)dz\\
&\underset{\varepsilon\to 0}{\longrightarrow}-\int_{\mathcal{Q}}\Big(\big[\mathscr{D}v\cdot\nabla\zeta\big]v_k-\frac{1}{2}\zeta_k\nabla v\cdot\mathscr{D}v\Big)dz\\
&=-\int_{\mathcal{Q}}\Big(\nabla(\zeta v_k)\cdot\mathscr{D}v\Big)\chi_{v>0}\,dz+\frac{1}{2}\int_{\mathcal{Q}}(\zeta_k\nabla v\cdot\mathscr{D}v)\,\chi_{v>0}\,dz\\
&=\int_{\mathcal{Q}\cap\{v>0\}}\zeta v_k\operatorname{div}(\mathscr{D}v)\,dz-\int_{\mathcal{Q}\cap\partial\{v>0\}}\zeta v_k\mathscr{D}v\cdot\nu\,dS+\frac{1}{2}\int_{\mathcal{Q}\cap\partial\{v>0\}}\nabla v\cdot\mathscr{D}v\zeta\nu_k\,dS\\
&=-\int_{\mathcal{Q}\cap\partial\{v>0\}}\zeta v_k\mathscr{D}v\cdot\nu\,dS+\frac{1}{2}\int_{\mathcal{Q}\cap\partial\{v>0\}}\nabla v\cdot\mathscr{D}v\zeta\nu_k\,dS.
\end{aligned}
$$

Since, the unit outer normal to $\{v>0\}$ can be written as $\nu=-\nabla v/|\nabla v|$, we have $v_k=-\nu_k|\nabla v|$ on $\mathcal{Q}\cap\partial\{u>0\}$.

The left-hand side of (7) converges to

$$
-\frac{1}{2}\int_{\mathcal{Q}\cap\partial\{v>0\}}\nabla v\cdot\mathscr{D}v\zeta\nu_k\,dS.
$$

Thus we arrive at

$$
\int_{\mathcal{Q}\cap\partial\{v>0\}}\zeta\nu_k\,dS=\frac{1}{2}\int_{\mathcal{Q}\cap\partial\{v>0\}}\nabla v\cdot\mathscr{D}v\zeta\nu_k\,dS.
$$

This immediately leads to

$$
\nabla v\cdot\mathscr{D}v=2 \quad\text{on } \mathcal{Q}\cap\partial\{v>0\}. \tag{8}
$$

3.2 Regularized Variational Problem

Here we apply the above treatment to our problem. Let Γ be a positive constant which controls the contact angle. We change the action integral (2) to the smoothed action

$$
J_{\epsilon}(u)=\int_0^T\int_{\Omega}\Big((u_t)^2\chi_{u>0}-|\nabla u|^2-\Gamma B_{\epsilon}(u)\Big)dxdt
$$

and we calculate the first variation, which written in the strong form reads

$$
\begin{cases}
\chi_{\overline{u>0}}u_{tt}=\Delta u-\Gamma\beta_{\varepsilon}(u)\ \text{ in } Q_T\\
u(x,0)=u_0(\geq 0)\\
u_t(x,0)=v_0\\
u(x,t)|_{\partial\Omega}=\psi(x,t)\text{ with }\psi(x,0)=u_0\text{ on }\partial\Omega.
\end{cases} \tag{9}
$$

Comparing this with formal calculation done in the previous subsection, we have

$$\int_{Q_T \cap \partial\{v>0\}} \Gamma \zeta \nu_k \, dS = \frac{1}{2} \int_{Q_T \cap \partial\{v>0\}} \left[|\nabla v|^2 - (v_t)^2 \right] \zeta \nu_k \, dS.$$

This leads to

$$|\nabla v|^2 - (v_t)^2 = 2\Gamma \quad \text{on } \partial\{v > 0\}. \tag{10}$$

We conclude that we can write $2\Gamma = Q^2$. For this reason, (9) is a nice approximation of the tape peeling problem.

We calculate the limit of β_ε when $\varepsilon \to 0$. Here we assume that the angle $|\nabla v|$ is nonzero on $\partial\{u > 0\}$. If u^ε converges to v in an appropriate sense, and moreover $\mathscr{H}^{n-1}(\partial\{u > 0\})$ is finite, we can deduce from the above calculation that

$$\beta_\varepsilon(u_\varepsilon) \to \frac{C}{|\nabla v|} d\mathscr{H}^{n-1} \llcorner \partial\{u > 0\},$$

for some positive constant C. Now, the right-hand side of (7) becomes

$$\int_{\mathcal{Q}} \zeta u_k^\varepsilon \beta_\varepsilon(u^\varepsilon) \, dz \underset{\varepsilon \to 0}{\longrightarrow} \int_{\mathcal{Q} \cap \partial\{v>0\}} \zeta \nu_k \frac{C}{|\nabla v|} d\mathscr{H}^{n-1} \llcorner \partial\{v > 0\}.$$

It is reasonable if $C = \frac{1}{2}$. We can say that

$$\frac{1}{2}\beta_\varepsilon(u_\varepsilon) \to \frac{1}{|\nabla v|} d\mathscr{H}^{n-1} \llcorner \partial\{v > 0\}.$$

4 Strong Formula

By the above consideration, we can write the (degenerate) hyperbolic type equation as

$$\chi_{\overline{u>0}} u_{tt} - \Delta u = -\frac{Q^2}{|Du|} \mathscr{H}^n \llcorner \partial\{u > 0\}, \tag{11}$$

where $Du = (\frac{\partial u}{\partial x_1}, \cdots, \frac{\partial u}{\partial x_n}, \frac{\partial u}{\partial t})$. We understand that when $u > 0$, u satisfies the wave equation and when $u < 0$, $\Delta u = 0$ a.e. $t \in (0, T)$. By the maximum principle for the Laplace operator, we can say on $u \leq 0$, u is identically zero, $u \equiv 0$. Thus, we see that when $u > 0$, u satisfies the wave equation. If we write equation (11) in a weak form, we have

$$-\int_0^T \int_\Omega \left(\chi_{\overline{u>0}} \zeta\right)_t u_t \, dx \, dt + \int_0^T \int_\Omega \nabla u \nabla \zeta \, dx \, dt = -\int_{\partial\{u>0\}} \frac{Q^2}{|Du|} \zeta \, d\mathscr{H}^n.$$

We split the integral on the left-hand side into four parts, $\Omega \times (0, T) \cap \{u > 0\}, \Omega \times (0, T) \cap \partial\{u > 0\}, \Omega \times (0, T) \cap \partial\{u = 0\}^{\circ}$ and $\Omega \times (0, T) \cap \{u = 0\}^{\circ}$. For each, we integrate by parts, and we conclude that except for $\Omega \times (0, T) \cap \partial\{u > 0\}$, all the integrals vanish.

Since $\chi_{\overline{u>0}} = 1$ on $\partial\{u > 0\}$, the left-hand side is

$$\int_{\Omega\times(0,T)\cap\partial\{u>0\}} \left(-\chi_{\overline{u>0}}\zeta \cdot u_t \cdot \frac{-u_t}{|Du|} + \zeta \nabla u \cdot \frac{-\nabla u}{|Du|} \right) d\mathscr{H}^n$$
$$= \int_{\Omega\times(0,T)\cap\partial\{u>0\}} \left(\zeta \frac{u_t^2}{|Du|} + \zeta \frac{-|\nabla u|^2}{|Du|} \right) d\mathscr{H}^n.$$

We formally recover $|\nabla u|^2 - u_t^2 = Q^2$.

References

1. H.W. Alt, L.A. Caffarelli, Existence and regularity for a minimum problem with free boundary. J. Reine Angew. Math. **325**, 105–144 (1981)
2. H.W. Alt, L.A. Caffarelli, A. Friedman, A free boundary problem for quasi-linear elliptic equations. Ann. Scuola Norm. Sup. Pisa Cl. Sci. **11**(4), 1–44 (1984)
3. H. Berestycki, L.A. Cafafrelli, L. Nirenberg, Uniform estimates for regularization of free boundary problems, *Analysis and Partial Differential Equation* (Marcel Dekker, New York, 1990)
4. K. Kikuchi, S. Omata, A free boundary problem for a one dimensional hyperbolic equation. Adv. Math. Sci. Appl. **9**(2), 775–786 (1999)
5. S. Omata, Y. Yamaura, A free boundary problem for quasilinear elliptic equations. Proc. Jpn. Acad. Ser. A Math. **21**, 281–286 (1990)
6. S. Omata, Y. Yamaura, A free boundary problem for quasilinear elliptic equations Part II: $C^{1,\alpha}$-regularity of free boundary. Funkcialaj Ekvacioj **42**(1), 9–70 (1999)
7. Y. Yamaura, The regularity of minimizers of a radially symmetric free boundary problem. Ann. Univ. Ferrara - Sez. VII - Sc. Mat. **XXXVIII**, 177–192 (1992)

Epi-Two-Dimensional Flow and Generalized Enstrophy

Zensho Yoshida and Philip J. Morrison

Abstract The conservation of the enstrophy (L^2 norm of the vorticity ω) plays an essential role in the physics and mathematics of two-dimensional (2D) Euler fluids. Generalizing to compressible ideal (inviscid and barotropic) fluids, the generalized enstrophy $\int_{\Sigma(t)} f(\omega/\rho)\rho \mathrm{d}^2x$ (f an arbitrary smooth function, ρ the density, and $\Sigma(t)$ an arbitrary 2D domain co-moving with the fluid) is a constant of motion, and plays the same role. On the other hand, for the three-dimensional (3D) ideal fluid, the helicity $\int_M \boldsymbol{V}\cdot\boldsymbol{\omega}\, \mathrm{d}^3x$ ($\boldsymbol{V}$ the flow velocity, $\boldsymbol{\omega} = \nabla \times \boldsymbol{V}$, and M the three-dimensional domain containing the fluid) is conserved. Evidently, the helicity degenerates in a 2D system, and the (generalized) enstrophy emerges as a compensating constant. This transition of the constants of motion is a reflection of an essential difference between 2D and 3D systems, because the conservation of the (generalized) enstrophy imposes stronger constraints, than the helicity, on the flow. In this paper, we make a deeper inquiry into the helicity-enstrophy interplay: the ideal fluid mechanics is cast into a Hamiltonian form in the phase space of Clebsch parameters, generalizing 2D to a wider category of epi-2D flows (2D embedded in 3D has zero-helicity, while the converse is not true – our epi-2D category encompasses a wider class of zero-helicity flows); how helicity degenerates and is substituted by a new constant is delineated; and how a further generalized enstrophy is introduced as a constant of motion applying to epi-2D flow is described.

Keywords Hamiltonian fluid mechanics · Clebsch parameterization · Lie dragging Enstrophy · Barotropic flow · Turbulence

---

Z. Yoshida (✉)
Department of Advanced Energy, University of Tokyo, Chiba 277-8561, Japan
e-mail: yoshida@ppl.k.u-tokyo.ac.jp

P.J. Morrison
Department of Physics and Institute for Fusion Studies, University of Texas at Austin, Austin, TX 78712-1060, USA
e-mail: morrison@physics.utexas.edu

Y. Maekawa and S. Jimbo (eds.), *Mathematics for Nonlinear Phenomena—Analysis and Computation*, Springer Proceedings in Mathematics & Statistics 215,
DOI 10.1007/978-3-319-66764-5_13

1 Introduction

The aim of this paper is to elucidate, from the perspective of Hamiltonian dynamics [11], how two-dimensional (2D) flow is different from general three-dimensional (3D) flow. Phenomenologically, 2D flow is often very different from 3D flow in that the former is less-turbulent and is more capable of generating and sustaining large-scale vortical structures—typhoons, jet streams, polar vortexes being spectacular examples of such structures created in atmospheric 2D flow. If we could delineate the root cause of such special behavior in 2D, we might be able to obtain a flow "intermediate" between 2D and 3D, where the "regularity" of 2D flow is maintained. As we will show, such is indeed possible.

We invoke the *helicity* as the key parameter for characterizing the transition from 3D to 2D (see Section 2). As is well known, the helicity is a constant of motion in an "ideal" flow (in this paper ideal means inviscid and barotropic). In 2D geometry, however, the helicity degenerates to zero; but as a compensation, the *enstrophy* (or its generalization, cf. Remark 1) becomes a nontrivial constant (see Sections 2.2 and 3.3). The conservation of the (generalized) enstrophy is a most essential property for distinguishing 2D from 3D. The enstrophy is a higher order functional in comparison with the helicity, and its conservation is deemed to be the reason for the aforementioned difference between 2D and 3D systems. Even when the constancy of the enstrophy or the helicity is broken by the inclusion of dissipation, the macroscopic structure of the fluid system is strongly influenced by these ideal constants of motion (cf. [4]).

Needless to say, zero-helicity flow is not necessarily 2D. In Section 4, we introduce our category of "epi-2D" flow that maintains the basic properties of zero-helicity and enstrophy conservation, while not necessarily being 2D. Having cast ideal fluid mechanics into a Hamiltonian form in the phase space of Clebsch parameters (e.g. [1, 5, 8, 10, 14, 18]), the category of epi-2D flow is, then, defined as a reduction of the phase space. One of the reduced parameter used is a *phantom* [15, 16], by which we define a generalized enstrophy. In Section 5, we introduce the notion of an epi-2D particle to elucidate our theory.

2 Preliminaries

2.1 Three-Dimensional Fluid Mechanics

We start by reviewing the basic equations of fluid mechanics and the associated conservation laws. Here we use the conventional notation of 3D vector analysis, with vector fields denoted by bold-face symbols.

Let M be a 3D domain containing an ideal fluid. For simplicity, we assume $M = T^3$, the 3-torus, and ignore the effect of boundaries. We denote by ρ the mass density, $\boldsymbol{V}$ the fluid velocity, and P the pressure. Thus, the governing equations are

$$\partial_t \rho = -\nabla \cdot (\boldsymbol{V}\rho), \tag{1}$$
$$\partial_t \boldsymbol{V} = -(\boldsymbol{V} \cdot \nabla)\boldsymbol{V} - \rho^{-1}\nabla P. \tag{2}$$

Assuming a barotropic relation $P = P(\rho)$, with $\rho^{-1}\nabla P = \nabla h$ for $h = h(\rho)$, the energy of the system is

$$H = \int_M \left[\frac{1}{2}|\boldsymbol{V}|^2 + \varepsilon(\rho)\right] \rho \, \mathrm{d}^3x, \tag{3}$$

where $\varepsilon(\rho)$ is the specific thermal energy, satisfying $\partial(\rho\varepsilon(\rho))/\partial\rho = h(\rho)$. Evidently, $\mathrm{d}H/\mathrm{d}t = 0$.

The vorticity $\boldsymbol{\omega} = \nabla \times \boldsymbol{V}$ obeys the vorticity equation, obtained by taking the curl of (2), i.e.,

$$\partial_t \boldsymbol{\omega} = \nabla \times (\boldsymbol{V} \times \boldsymbol{\omega}). \tag{4}$$

Evidently, the total mass $N = \int_M \rho \, \mathrm{d}^3x$ is a constant of motion, along with another conserved quantity, the helicity:

$$C = \int_M \boldsymbol{V} \cdot \boldsymbol{\omega} \, \mathrm{d}^3x. \tag{5}$$

Using (2) and (4), we easily verify $\mathrm{d}C/\mathrm{d}t = 0$.[1]

2.2 *Two-Dimensional Fluid Mechanics*

To compare 2D and 3D systems, it is convenient to immerse a 2D system into 3D space. For simplicity, we consider a flat torus T^2, on which we define Cartesian coordinates x and y. We add a "perpendicular" coordinate z and extend T^2 to T^3, with $\boldsymbol{e}_z = \nabla z$, which we call the perpendicular vector. Now we may define a 2D system by the reduction of the 3D system with $\boldsymbol{e}_z \cdot \ = 0$ and $\partial_z = 0$. Indeed, a 2D fluid model is formulated by such a reduction.[2] We interpret a 2D flow $\boldsymbol{v} = (v_x, v_y)^{\mathrm{T}}$ as a special 3D flow such that $\boldsymbol{V} = (v_x, v_y, 0)^{\mathrm{T}}$. The vorticity can be defined as $\boldsymbol{\omega} = \nabla \times \boldsymbol{V} = \omega \boldsymbol{e}_z$ with $\omega = \partial_x v_y - \partial_y v_x$. In which case the vorticity equation (4) reduces to a single-component equation:

$$\partial_t \omega = -\nabla \cdot (\boldsymbol{v}\omega). \tag{6}$$

[1]In this work, we do not argue for the existence of regular solutions of the model equations. The conservation laws discussed are, therefore, *a priori* relations satisfied by all regular solutions if they exist.

[2]We note that $\partial_z = 0$ does not mean that the system extends in the z-direction homogeneously; instead, we consider a thin system in which variation of physical quantities in the z-direction is much larger than in the x and y directions. Thus, ∂_z can be separated from ∂_x and ∂_y.

Because $\boldsymbol{V} \cdot \boldsymbol{\omega} = 0$, the helicity conservation is now trivial, $C \equiv 0$; however, interestingly, a new constant emerges that replaces the degenerated helicity.[3] By (6) and the mass conservation law, which now reads $\partial_t \rho = -\nabla \cdot (\boldsymbol{v}\rho)$, we obtain, the following equation for the *potential vorticity*, $\vartheta = \omega/\rho$:

$$\partial_t \vartheta = -\boldsymbol{v} \cdot \nabla \vartheta, \tag{7}$$

and we define the *generalized enstrophy* by

$$Q = \int_M f(\vartheta)\rho \, \mathrm{d}^2 x, \tag{8}$$

where f is an arbitrary C^1-class function. Using (7) and the mass conservation law, we can easily verify that $\mathrm{d}Q/\mathrm{d}t = 0$.

Remark 1. For an incompressible flow ($\nabla \cdot \boldsymbol{v} = 0$), we may assume $\rho =$ constant, and then, (8) has a special form of $\int_M \omega^2 \mathrm{d}^2 x$, which is the usual *enstrophy*.

We end this introductory section by drawing attention to the fact that all constants of motion, i.e., the total mass N, the helicity C, and the generalized enstrophy Q are defined by the spatial integrals over the 3D or 2D domain. This means that the integrand of a constant of motion defines an n-form (n the spatial dimension) in the language of differential geometry. In the following analysis, this fact guides our formulation of generalized enstrophy.

3 Topological Invariants in Fluid Motion

3.1 Hamiltonian Formalism of Ideal Fluid Motion

For the study of geometrical properties of fluid mechanics, we reformulate the governing equations in the framework of differential geometry. We first introduce a *phase space* X that hosts the underlying state vectors $\boldsymbol{\xi}$; the physical quantity $\boldsymbol{u} = (\rho, \boldsymbol{V})^{\mathrm{T}}$ ($\in \mathscr{V}$, the space of physical variables) is some function parameterized by $\boldsymbol{\xi}$. Let

$$\boldsymbol{\xi} = (\varrho, \varphi, p, q, r, s)^{\mathrm{T}} \in X, \tag{9}$$

where $\xi_1 = \varrho, \xi_3 = p, \xi_5 = r$ are n-forms and $\xi_2 = \varphi, \xi_4 = q, \xi_6 = s$ are 0-forms in the base space $M = T^3$. We assume ξ_j $(j = 1, \cdots, 6)$ are smooth (i.e., C^∞-class) functions. The dual space X^* is the Hodge-dual of X, i.e., the odd number components of $\boldsymbol{\eta} \in X^*$ are 0-forms and the even number components are n-forms.

[3]Fukumoto [3] points out that the helicity and the generalized enstrophy can be unified by the concept of *cross helicity*.

The pairing of X^* and X is [4]

$$\langle \boldsymbol{\eta}, \boldsymbol{\xi} \rangle = \sum_j \int_M \eta_j \wedge \xi_j, \quad \boldsymbol{\eta} \in X^*, \; \boldsymbol{\xi} \in X. \tag{10}$$

On the space $C^\infty(X)$ of *observables*, we define a canonical Poisson bracket

$$\{F, G\} = \langle \partial_{\boldsymbol{\xi}} F, J \partial_{\boldsymbol{\xi}} G \rangle, \tag{11}$$

where $F, G \in C^\infty(X)$, $\partial_{\boldsymbol{\xi}} F$ is the gradient of F defined by

$$F(\boldsymbol{\xi} + \varepsilon \boldsymbol{\zeta}) - F(\boldsymbol{\xi}) = \varepsilon \langle \partial_{\boldsymbol{\xi}} F, \boldsymbol{\zeta} \rangle + O(\varepsilon^2) \quad (\forall \boldsymbol{\zeta} \in X), \tag{12}$$

and $J : X^* \to X$ is the symplectic operator

$$J = J_c \oplus J_c \oplus J_c, \quad J_c = \begin{pmatrix} 0 & I \\ -I & 0 \end{pmatrix}. \tag{13}$$

We denote by $C^\infty_{\{\,,\,\}}(X)$ the Poisson algebra of observables on X. The adjoint representation of Hamiltonian dynamics is, for a given Hamiltonian H,

$$\frac{\mathrm{d}}{\mathrm{d}t} F = \{F, H\}, \tag{14}$$

which is equivalent to Hamilton's equation of motion

$$\frac{\mathrm{d}}{\mathrm{d}t} \boldsymbol{\xi} = J \partial_{\boldsymbol{\xi}} H. \tag{15}$$

We relate the physical quantity $\boldsymbol{u} \in \mathscr{V}$ and $\boldsymbol{\xi} \in X$ by $\rho \Leftrightarrow \varrho^*$ (i.e., $\varrho^* \mathrm{vol}^n = \varrho$ with the volume n-form vol^n; here $n = 3$),[5] and

$$\boldsymbol{V} \Leftrightarrow \wp = \mathrm{d}\varphi + \check{p}\mathrm{d}q + \check{r}\mathrm{d}s, \quad \left(\check{p} = p^*/\varrho^*, \; \check{r} = r^*/\varrho^*\right). \tag{16}$$

Writing a vector as (16) is called the *Clebsch parameterization.* The five Clebsch parameters $(\varphi, q, \check{p}, s, \check{r})$ are sufficient to represent every 3-vector (1-form in 3D space) [14]. Inserting (16) into the fluid energy (3), we obtain a Hamiltonian

---

[4]Here the phase space (function space) X may be viewed as a cotangent bundle of $X_q = \{(\xi_2, \xi_4, \xi_6)^{\mathrm{T}};\; \xi_j \in \Omega^0(M)\}$. For $F \in C^\infty(X)$, $\partial_{\boldsymbol{\xi}} F \in X^*$ (to be defined in (12)) may be regarded as a "1-form" on X. Hence, the duality of X^* and X corresponds to that of "co-vectors" and "vectors." At the same time, the components of the field $\boldsymbol{\xi} \in X$ are differential forms (0-forms and n-forms) on the "base space" M; the "Hodge-duality" of X^* and X is in the sense of the differential forms on M, while the duality (10) is in the sense of "co-vectors" and "vectors" on the function space X.

[5]Here we denote by α^* the Hodge-dual of a differential form α.

$$H(\boldsymbol{\xi}) = \int_M \left[\frac{1}{2} \left| \mathrm{d}\varphi + (p^*/\varrho^*)\mathrm{d}q + (r^*/\varrho^*)\mathrm{d}s \right|^2 + \varepsilon(\varrho^*) \right] \varrho. \tag{17}$$

With this H, the equation of motion (15) reads

$$\widetilde{\mathscr{L}}_V \varphi = h - V^2/2, \tag{18}$$

$$\widetilde{\mathscr{L}}_V q = 0, \quad \widetilde{\mathscr{L}}_V s = 0, \tag{19}$$

$$\widetilde{\mathscr{L}}_V \varrho = 0, \quad \widetilde{\mathscr{L}}_V p = 0, \quad \widetilde{\mathscr{L}}_V r = 0, \tag{20}$$

where we denote

$$\widetilde{\mathscr{L}}_V = \partial_t + \mathscr{L}_V, \tag{21}$$

with $\mathscr{L}_V$ being the conventional Lie derivative along the vector $\boldsymbol{V} \in TM$.[6,7]

The first equation of (20) is nothing but the mass conservation law (1). Evaluating $\partial_t \boldsymbol{V}$ by inserting (16) and using (18)–(20), we obtain (2). Hence, Hamilton's equation (15) with the Hamiltonian (17) describes the fluid motion obeying (1) and (2).[8]

3.2 Gauge Symmetry and Helicity

In (16), the Clebsch parameters are apparently a redundant representation of a 3-vector $\boldsymbol{V}$. In fact, the map $X \to \mathscr{V}$ is not an injection (although a surjection) [14]. For example, the transformation

$$\varphi \mapsto \varphi + \varepsilon \quad (\varepsilon \in \mathbb{R}) \tag{22}$$

does not change the physical quantity $\boldsymbol{u} \in \mathscr{V}$. Such a map is called a *gauge transformation*. We find that the map (22) is a Hamiltonian flow generated by the constant of motion $N = \int_M \varrho$, i.e., the map (22) may be written as $(I + \varepsilon J \partial_{\boldsymbol{\xi}} N)$. Or, the co-adjoint orbit $\mathrm{Ad}_N^*(\varepsilon)$ is a gauge transformation group of the Clebsch parameterization.

The helicity C, which now reads

$$C = \int_M \wp \wedge \mathrm{d}\wp, \tag{23}$$

---

[6]Here $\boldsymbol{V}$ is regarded as a vector $\in TM$ through the following identification. By (16), $\boldsymbol{V} \Leftrightarrow \wp \in T^*M$. In the Hamiltonian (17), $|\wp|^2 \Leftrightarrow \boldsymbol{V}^\dagger \cdot \boldsymbol{V}$ with the dual $\boldsymbol{V}^\dagger = \boldsymbol{V} \in TM$.

[7]When considering a relativistic fluid, we generate a diffeomorphism group $\mathrm{e}^{\tau U}$ (τ the proper time), acting on a space-time manifold $\tilde{M} = \mathbb{R} \times M$, by a space-time velocity $U \in T\tilde{M}$. Then, the space-time derivative $\widetilde{\mathscr{L}}_V$ is replaced by the natural Lie derivative $\mathscr{L}_U$. When $\mathscr{L}_U$ applies to a differential form α on $\tilde{M}$, the temporal and spatial components are mixed up (cf. [17]).

[8]See [6] for the underlying action principle that yields the canonical system of Hamilton's equation (18)–(20). See also [10] for the comparison of the noncanonical bracket in terms of the fluid variables and the canonical bracket in terms of the Clebsch (potential) variables.

yields a different gauge group $\mathrm{Ad}_C^*(\varepsilon)$ (see [13] for the explicit form the corresponding gauge transformation).

Remark 2. If we denote by $\{F, G\}_{\mathscr{V}}$ the Poisson bracket of (11) evaluated only for observables F and G in which $\boldsymbol{\xi}$ appears in terms of the Clebsch parameterized $\boldsymbol{u} \in \mathscr{V}$, we obtain

$$\{G, N\}_{\mathscr{V}} = 0, \quad \{G, C\}_{\mathscr{V}} = 0 \quad \forall G. \tag{24}$$

Hence, N and C are the Casimir elements of the reduced Poisson algebra $C^\infty_{\{\ ,\ \}_{\mathscr{V}}}(\mathscr{V})$. See e.g. [9] for general discussion on *reduction* of Poisson brackets. The existence of Casimir elements is characteristic of *noncanonical Poisson brackets* [11, 12]. We note that $C^\infty_{\{\ ,\ \}_{\mathscr{V}}}(\mathscr{V})$ has much more (in fact, infinitely many) topological constraints (constants of motions) that are not integrable, i.e., do not define Casimir elements (see [15]).

3.3 Two-Dimensional System and Generalized Enstrophy

In a 2D system ($M = T^2$), we can parameterize a general 2D velocity as

$$\boldsymbol{V} \Leftrightarrow \wp = \mathrm{d}\varphi + \check{p}\mathrm{d}q. \tag{25}$$

Now only three Clebsch parameters φ, $\check{p}$, and q suffice [14]. Hence, the phase space is

$$Z = \{\boldsymbol{\zeta} = (\varrho, \varphi, p, q)^{\mathrm{T}};\ \varphi, q \in \Omega^0(T^2),\ \varrho, p \in \Omega^2(T^2)\}. \tag{26}$$

All other formalisms are the same as the case of 3D systems. However, because the 3-form $\wp \wedge \mathrm{d}\wp$ cannot be defined in 2D space we do not have the helicity conservation law.

As mentioned in Section 2.2, a different constant of motion emerges in 2D, the generalized enstrophy, which is a spatial (2D) integral of a 2-form that involves the vorticity $\omega = \mathrm{d}\wp$. In preparation for the development of the next section, we reformulate the generalized enstrophy in the language of differential geometry (with a slight extension), and re-prove its conservation. Let

$$Q = \int_{\Sigma(t)} f(\omega^*/\varrho^*)\varrho, \tag{27}$$

where f is an arbitrary smooth function, and $\Sigma(t) \subset M$ is a co-moving "volume" (in fact, a 2D surface). Notice that the integral is evaluated on a subset $\Sigma(t)$ that is moved by the group action of $\mathrm{e}^{t\boldsymbol{v}}$.

By the following Lemma 1 and the mass conservation law $\tilde{\mathscr{L}}_{\boldsymbol{v}}\varrho = 0$, we find

$$\frac{\mathrm{d}}{\mathrm{d}t}Q = \int_{\Sigma(t)} \tilde{\mathscr{L}}_v[f(\omega^*/\varrho^*)\varrho]$$
$$= \int_{\Sigma(t)} \left[f'\varrho\tilde{\mathscr{L}}_v(\omega^*/\varrho^*) + f\tilde{\mathscr{L}}_v\varrho\right] = 0. \tag{28}$$

Lemma 1. *Let α and β be a pair of n-forms defined on a smooth manifold M of dimension n. Denoting $\alpha = \alpha^*\mathrm{vol}^n$ and $\beta = \beta^*\mathrm{vol}^n$, we define $\vartheta = \alpha^*/\beta^*$. If $\tilde{\mathscr{L}}_V\alpha = 0$ and $\tilde{\mathscr{L}}_V\beta = 0$ for a vector $V \in TM$, then $\tilde{\mathscr{L}}_V\vartheta = 0$.*

Proof. By the definition,

$$\tilde{\mathscr{L}}_V\alpha = (\tilde{\mathscr{L}}_V\alpha^*)\mathrm{vol}^n + \alpha^*(\tilde{\mathscr{L}}_V\mathrm{vol}^n) = (\tilde{\mathscr{L}}_V\alpha^*)\mathrm{vol}^n + \alpha^*(\mathrm{div}\, V)\mathrm{vol}^n.$$

When $\tilde{\mathscr{L}}_V\alpha = 0$, we may write $\tilde{\mathscr{L}}_V\alpha^* = -\alpha^*\mathrm{div}\, V$. The same formula applies to $\tilde{\mathscr{L}}_V\beta^*$. We thus have

$$\tilde{\mathscr{L}}_V\left(\frac{\alpha^*}{\beta^*}\right) = \frac{\tilde{\mathscr{L}}_V\alpha^*}{\beta^*} - \frac{\alpha^*\tilde{\mathscr{L}}_V\beta^*}{\beta^{*2}} = \frac{-\alpha^*\mathrm{div}\, V}{\beta^*} + \frac{\alpha^*\beta^*\mathrm{div}\, V}{\beta^{*2}} = 0.$$

□

We want to generalize Q to a class of 3D systems by considering a 3-form integral of the form

$$Q = \int_{V(t)} f(\vartheta)\varrho, \tag{29}$$

for some scalar ϑ that reflects ω.

4 Epi-Two-Dimensional Flow

4.1 Reduction of the Phase Space

A thought, drawn from the foregoing observation, is that the degeneration of one constant of motion (i.e., the helicity) must be compensated by a new constant of motion (i.e., the generalized enstrophy). Although we have seen that the degeneracy of the helicity is usual for 2D systems, it may occur in a more general situation. Then, it is conceivable that the compensation should also occur simultaneously. If so, a generalized enstrophy may exist as a topological constraint in a wider class of ideal flows, which we call *epi-2D flows*.

Definition 1 (epi-2D flow). Let Y be a phase space of smooth Clebsch parameters such that

$$Y = \{\eta = (\varrho, \varphi, p, q)^{\mathrm{T}};\ \varphi, q \in \Omega^0(T^3),\ \varrho, p \in \Omega^3(T^3)\}. \tag{30}$$

The corresponding physical fields $\rho \Leftrightarrow \varrho^*$ and

$$V \Leftrightarrow \wp = \mathrm{d}\varphi + \left(\frac{p^*}{\varrho^*}\right)\mathrm{d}q \tag{31}$$

are called epi-two-dimensional (epi-2D) flows.

Notice the difference between Y of (30) and Z of (26); in particular, the epi-2D flows are defined on the 3D domain T^3. The reduced phase space Y is a closed subset of X. We denote by $\langle\ ,\ \rangle_Y$ the reduced pairing of Y^* and Y (cf. (10)). By restricting observables in Y, we define a canonical Poisson algebra $C^\infty_{\{\ ,\ \}_Y}(Y)$, where the Poisson bracket is

$$\{F, G\}_Y = \langle \partial_\eta F, J_Y \partial_\eta G\rangle_Y, \quad J_Y = J_c \oplus J_c. \tag{32}$$

Epi-2D flow may have a finite vorticity $\mathrm{d}\wp = \mathrm{d}\check{p} \wedge \mathrm{d}q$, where $\check{p} = p^*/\varrho^*$. However, we observe

$$\wp \wedge \mathrm{d}\wp = \mathrm{d}\varphi \wedge \mathrm{d}\check{p} \wedge \mathrm{d}q = \mathrm{d}(\varphi \wedge \mathrm{d}\check{p} \wedge \mathrm{d}q),$$

i.e., the helicity density $\wp \wedge \mathrm{d}\wp$ is an exact 3-form. Hence, we have

Proposition 1. *Epi-2D flow has zero-helicity, i.e.,* $C = \int_M \wp \wedge \mathrm{d}\wp = 0$.

A *vortex line* is a curve determined by

$$\frac{\mathrm{d}}{\mathrm{d}\tau}\boldsymbol{x} = \boldsymbol{\omega}(\boldsymbol{x}), \tag{33}$$

where $\boldsymbol{\omega}$ is the vorticity. For epi-2D flow, $\boldsymbol{\omega} = \nabla\check{p} \times \nabla q$ ($\Leftrightarrow \mathrm{d}\wp = \mathrm{d}\check{p} \wedge \mathrm{d}q$). Evidently, the level-sets of $\check{p}$ and q are the "integral surfaces" of vortex lines:

$$\begin{aligned}
\frac{\mathrm{d}}{\mathrm{d}\tau}\check{p}(\boldsymbol{x}(\tau)) &= \nabla\check{p} \cdot \frac{\mathrm{d}}{\mathrm{d}\tau}\boldsymbol{x} = \nabla\check{p} \cdot \boldsymbol{\omega}(\boldsymbol{x}) = 0,\\
\frac{\mathrm{d}}{\mathrm{d}\tau}q(\boldsymbol{x}(\tau)) &= \nabla q \cdot \frac{\mathrm{d}}{\mathrm{d}\tau}\boldsymbol{x} = \nabla q \cdot \boldsymbol{\omega}(\boldsymbol{x}) = 0.
\end{aligned}$$

This well-known fact can be stated as

Proposition 2. *The vortex line equation of epi-2D flow is integrable; two Clebsch parameters* $\check{p}$ *and* q *define the integral surfaces. We call the surface spanned by* $\mathrm{d}\check{p}$ *and* $\mathrm{d}q$ *the* vortex surface.

The epi-2D flow generated by a reduced Hamiltonian

$$H(\eta) = \int_M \left[\frac{1}{2}\left|\mathrm{d}\varphi + (p^*/\varrho^*)\mathrm{d}q\right|^2 + \varepsilon(\varrho^*)\right]\varrho \tag{34}$$

satisfies the 3D fluid equations (1) and (2).

We may observe the epi-2D dynamics in the larger phase space X. Since the reduced Hamiltonian $H(\eta)$ does not include the variables r and s, the flow velocity $\boldsymbol{V}$ is independent of r and s. However, they obey the same equations (19) and (20), i.e.,

$$\widetilde{\mathscr{L}}_V r = 0, \quad \widetilde{\mathscr{L}}_V s = 0. \tag{35}$$

Such fields, co-moving with the epi-2D flow, are called *phantom fields* [15, 16]. Every functional such as $F(r, s)$ is a constant of motion: $\{F(r, s), H(\eta)\} = 0$.

Remark 3. The epi-2D class absorbs the conventional 2D flows as well as their moderate generalizations pertinent to some symmetry group. For example, we may consider "helical flows" in an axisymmetric domain.[9] Let ι be a real constant, and

$$T(\delta)\begin{pmatrix} x \\ y \\ z \end{pmatrix} = \left[\iota\begin{pmatrix} \cos\delta & \sin\delta & 0 \\ -\sin\delta & \cos\delta & 0 \\ 0 & 0 & 1 \end{pmatrix} + \begin{pmatrix} 0 & 0 & 0 \\ 0 & 0 & 0 \\ 0 & 0 & \delta \end{pmatrix}\right]\begin{pmatrix} x \\ y \\ z \end{pmatrix},$$

which we call the helical transformation. We denote the cylindrical coordinates $(r, \theta, z) = (\sqrt{x^2 + y^2}, \arctan(y/x), z)$. Evidently, r and $\psi = m\theta + kz$ ($m \in \mathbb{Z}$ and $k = m\iota$) are invariant under the group action of $T(\delta)$. When all field variables are functions of only r and ψ, we say the system has helical symmetry. The vector $\boldsymbol{h} = (\iota y, -\iota x, 1)^{\mathrm{T}}$ is tangent to the symmetry lines. Every helical flow vector $\boldsymbol{V}(r, \psi)$ that is perpendicular to $\boldsymbol{h}$ (i.e., $\boldsymbol{h} \cdot \boldsymbol{V} = 0$) may be cast into the form of $\mathrm{d}\varphi + \check{p}\mathrm{d}\psi$, i.e., it is epi-2D. Such helical flows enjoy the 2D properties, especially the vanishing of vortex stretching; see [2].

Remark 4. When a flow in a 3D domain has some continuous symmetry (i.e., invariant under some Lie group action), it is often called a 2.5D flow [2]. For example, a general helical flow (see Remark 3), not necessarily restricted by the orthogonality condition $\boldsymbol{h} \cdot \boldsymbol{V} = 0$, is 2.5D (when $\iota = 0$, it is the combination of the usual 2D flow $\boldsymbol{v}(x, y)$ and a vertical flow $V_z(x, y)\boldsymbol{e}_z$). Such 2.5D flows are not included in the epi-2D class (evidently, a general 2.5D flow has a finite helicity). However, there are some similarities between 2D and 2.5D flows; see [7] for discussion of helical flows in the Navier–Stokes system.

4.2 Generalized Enstrophy of Epi-2D Flow

In light of the above, it is not surprising that we have a family of conservation laws for epi-2D fluid motion:

---

[9] While we have started the discussion with the assumption of the boundary-less domain $M = T^3$ (in order to avoid complexities in formulating the Hamiltonian mechanics), the definition of the epi-2D flow is independent of the boundary condition, so we may consider an arbitrary domain.

Theorem 1. *Let $\eta(t)$ be an epi-2D flow generated by the reduced Hamiltonian $H(\eta)$ of (34), and $s(t)$ be a co-moving phantom. We define a generalized enstrophy (denoting $\omega = \mathrm{d}\check{p} \wedge \mathrm{d}q$)*

$$Q = \int_{\Omega(t)} f(\vartheta)\, \varrho, \quad \vartheta = \frac{(\omega \wedge \mathrm{d}s)^*}{\varrho^*} \tag{36}$$

with an arbitrary smooth function f and an arbitrary co-moving 3D volume element $\Omega(t) \subset M$. Then, $\mathrm{d}Q/\mathrm{d}t = 0$.

Proof. Since $\widetilde{\mathscr{L}}_V \varrho = 0$, what we have to prove is $\widetilde{\mathscr{L}}_V \vartheta = 0$. We have

$$\begin{aligned}\widetilde{\mathscr{L}}_V(\mathrm{d}\check{p} \wedge \mathrm{d}q \wedge \mathrm{d}s) &= (\widetilde{\mathscr{L}}_V \omega) \wedge \mathrm{d}s + \omega \wedge \widetilde{\mathscr{L}}_V \mathrm{d}s \\ &= \omega \wedge \mathrm{d}(\widetilde{\mathscr{L}}_V s) = 0.\end{aligned}$$

By Lemma 1, we obtain $\widetilde{\mathscr{L}}_V \vartheta = 0$. □

The generalized enstrophy (36) is a three-dimensional generalization of the two-dimensional one (27). About its application, we have the following remarks:

Remark 5. In a general 3D flow, Q is also a constant of motion. However, it does not fully characterize the vorticity, since ω must be inflated to $\mathrm{d}\wp = \mathrm{d}\check{p} \wedge \mathrm{d}q + \mathrm{d}\check{r} \wedge \mathrm{d}s$.

Remark 6. In the case of 2D flow, we may first immerse the system into 3D by adding a perpendicular coordinate z (see Section 2.2), and take the phantom $s = z$ (this s is stationary). Then, Theorem 1 reproduces the result of Section 3.3.

Remark 7. Suppose that $\omega \neq 0$. Then, we may choose the initial value of s so that

$$(\mathrm{d}\check{p} \wedge \mathrm{d}q \wedge \mathrm{d}s)^* = \frac{D(\check{p}, q, s)}{D(x, y, z)} \neq 0. \tag{37}$$

Hence, the generalized enstrophy can be made nontrivial. Analogous to the 2D generalized enstrophy, such an s is a coordinate co-moving with the fluid, which penetrates the vortex surface (see Proposition 2).

5 A Particle Picture

5.1 Epi-2D "Particles"

We can exploit *local* epi-2D regions in order to define particle-like behavior. In the general 3D parameterization $V \Leftrightarrow \wp = \mathrm{d}\varphi + \check{p}\mathrm{d}q + \check{r}\mathrm{d}s$, a region in which $\check{r} = 0$ may be called an *epi-2D domain*. Since $\check{r}$ co-moves with the fluid, every infinitesimal volume element (denoted by $\Omega_j(t)$ with j an index for each such volume element) included in an *epi-2D domain* may be viewed as a quasiparticle, which we call an *epi-2D particle*. The generalized enstrophy evaluated for $\Omega_j(t)$, which we denote by $Q_1(\Omega_j)$ is a constant of motion, characterizing the vorticity included there. We call $Q_1(\Omega_j)$ the *charge* of the epi-2D particle Ω_j.

A symmetric epi-2D particle can be defined by a domain in which $\check{p} = 0$, with the corresponding generalized enstrophy given by

$$Q_2(\Omega_j) = \int_{\Omega_j(t)} f(\vartheta_2)\,\varrho, \quad \vartheta_2 = \frac{(\omega_2 \wedge \mathrm{d}q)^*}{\varrho^*},$$

measuring the vorticity $\omega_2 = \mathrm{d}\check{r} \wedge \mathrm{d}s$.

As noted in Remark 5, both Q_1 and Q_2 can be evaluated in a general co-moving domain (particle) $\Omega(t) \subset M$, and they are ubiquitous constants. However, they do not represent the "enstrophy" of an actual vorticity when the vorticity exists in a mixed state $\mathrm{d}\check{p} \wedge \mathrm{d}q + \mathrm{d}\check{r} \wedge \mathrm{d}s$. Hence, we may interpret Q_1 and Q_2 as "potential" quantities, which become "observable" when one of $\omega_1 = \mathrm{d}\check{p} \wedge \mathrm{d}q$ or $\omega_2 = \mathrm{d}\check{r} \wedge \mathrm{d}s$ alone occupies a domain.

5.2 Discovering the Epi-2D Particle

In the preceding subsection, the notion of an epi-2D particle (or domain) was introduced using the Clebsch parameters which are the potential fields lying beneath the observables. Here we make an attempt to discover an epi-2D particle only from the physical variable $\boldsymbol{u}$.

We start by remembering the well-known relation:

Lemma 2 (Frobenius). *Let $\wp$ be a C^1-class 1-form on a smooth manifold M of dimension n (≥ 3). The following two conditions are equivalent:*

1. *$\wp$ has* zero-helicity density, *i.e.,*

$$\wp \wedge \mathrm{d}\wp = 0. \tag{38}$$

2. *$\wp$ is locally (i.e., in a neighborhood Ω of every point of M)* integrable, *i.e., there exist two scalars α and β by which $\wp$ can be written as*

$$\wp = \alpha \mathrm{d}\beta. \tag{39}$$

Then, the Pfaffian equation $\wp = 0$ foliates Ω by the level-sets of β.

We define a quotient space

$$V_s = \Omega^1(M)/\mathrm{d}\Omega^0(M), \tag{40}$$

which may be identified as the space of *solenoidal vector fields*. If we identify a 1-form $\wp \in V_s$ with a 3-vector field $\boldsymbol{V}$, the integral $\int_M \wp \wedge \mathrm{d}\wp$ evaluates the helicity $C = \int_M \boldsymbol{V} \cdot \nabla \times \boldsymbol{V}\, \mathrm{d}^3x$. Notice that the transformation $\boldsymbol{V} \mapsto \boldsymbol{V} + \nabla\phi\ (\forall\phi)$ does not change the helicity, while the integrand (helicity density) $\boldsymbol{V} \cdot \nabla \times \boldsymbol{V}$ is modified. But by defining the helicity density $\wp \wedge \mathrm{d}\wp$ on V_s, the gauge ($\mathrm{d}\phi$) dependence has been removed.

If $\wp \wedge \mathrm{d}\wp = 0$ in $\Omega \subset M$, we say that $\wp$ is "helicity free" in Ω. By Lemma 2, a helicity-free $\wp$ can be represented as $\wp = \check{p}\mathrm{d}q\ (\exists\check{p}, q)$ in some $\Omega' \subset \Omega$, which implies that $\wp' = \wp + \mathrm{d}\varphi\ (\exists\varphi)$ can be identified with the flow velocity $\boldsymbol{V}$ in Ω'. To put it in another way, we have

Proposition 3. *Given that the projection onto V_s of a flow velocity $\boldsymbol{V}$ is helicity free in $\Omega \subset M$, i.e., there exists $\nabla\phi$ by which we can make $\boldsymbol{V} - \nabla\phi \sim \wp \in V_s$ such that $\wp \wedge \mathrm{d}\wp = 0$ in Ω, then such a $\boldsymbol{V}$ is epi-2D in some $\Omega' \subset \Omega$.*

6 Conclusion

Diverse structures generated in fluids are not attributed to features of some nontrivial energy functional. In fact, the energy of a usual fluid is quite simple, it being the equivalent of a *norm* on the phase space of physical variables such as that given by (3). This is in marked contrast to the usual situation in condensed-matter physics where, for example, phase transitions or spinodal decompositions are modeled by bumpy energies. The key role of fluids is, then, played by "constraints" that forbid the dynamics from obeying simple orbits that might be determined by the energy alone. In the ideal (no-dissipation) limit, such constraints are manifested as conservation laws. Indeed, ideal fluid mechanics has infinitely many such constants of motion, and some of them are essential for controlling bifurcations of diverse structures or maintaining stability of some vertical motion. In the present work, we focused on two well-known constants of motion: the helicity of 3D flow and the (generalized) enstrophy of 2D flow, and we studied the basic mechanism of their creation.

In physics, a constant of motion is expected to be the product of some *symmetry*. However, the energy (Hamiltonian) of the fluid, represented in terms of the usual physical (Eulerian) variables, does not bear such symmetries to produce the helicity or enstrophy. Therefore, we are led to consider a set of underlying basic parameters beneath the physical quantities, and assume that some specific combinations of them appear as observables. Here, we invoked Clebsch parameters, and showed that the helicity is the product of gauge symmetry of the Clebsch parameterization.

We have observed that the phase space X of general 3D flows is hierarchically foliated into submanifolds, where the smallest subsystem hosts vorticity-free (irrotational) flows. The next hierarchy Y hosts the epi-2D flows, which is a subset of the zero-helicity leaf (Proposition 1). The subsystem Y is foliated by the generalized enstrophy (Theorem 1), which is a 3D generalization of the conventional one for 2D systems. Notice that the reduction from X (general 3D flow) to Y (epi-2D flow) is not a geometrical constraint (cf. the 2D system of Section 3.3). However, epi-2D systems have intrinsic vortex surfaces (Proposition 2), which parallels the *a priori* base space of the 2D system. The generalized enstrophy is the measure of the circulation on such vortex surfaces.

Acknowledgements ZY acknowledges discussions with Professor Y. Fukumoto and Y. Kimura. The work of ZY was supported by JSPS KAKENHI Grant Number 23224014 and 15K13532, while that of PJM was supported by USDOE contract DE-FG02-04ER-54742.

References

1. A. Clebsch, Über eine allgemeine Transformation der hydrodynamischen Gleichungen. Z. Reine Angew. Math. **54**, 293–312 (1857)
2. B. Ettinger, E.S. Titi, Global existence and uniqueness of weak solutions of three-dimensional Euler equations with helical symmetry in the absence of vorticity stretching. SIAM J. Math. Anal. **41**, 269–296 (2009)
3. Y. Fukumoto, A unified view of topological invariants of fluid flows. Topologica **1**, 003 (12 pp) (2008)
4. A. Hasegawa, Self-organization processes in continuous media. Adv. Phys. **34**, 1–42 (1985)
5. R. Jackiw, *Lectures on Fluid Dynamics - A Particle Theorist's View of Supersymmetric, Non-Abelian, Noncommutative Fluid Mechanics and d-Branes* (Springer, New York, 2002)
6. C.C. Lin, Hydrodynamics of Helium II, in *Proceedings of the International School of Physics "Enrico Fermi" XXI* (Academic Press, New York, 1963), pp. 93–146
7. A. Mahalov, E.S. Titi, S. Leibovich, Invariant helical subspaces for the Navier–Stokes equations. Arch. Rational Mech. Anal. **112**, 193–222 (1990)
8. J. Marsden, A. Weinstein, Coadjoint orbits vortices, and Clebsch variables for incompressible fluids. Phys. D **7**, 305–323 (1983)
9. J.E. Marsden, T. Ratiu, A. Weinstein, Semidirect products and reduction in mechanics. Trans. Am. Math. Soc. **281**, 147–177 (1984)
10. P.J. Morrison, Poisson brackets for fluid and plasmas. AIP Conf. Proc. **88**, 13–46 (1982)
11. P.J. Morrison, Hamiltonian description of the ideal fluid. Rev. Mod. Phys. **70**, 467–521 (1998)
12. P.J. Morrison, J.M. Greene, Noncanonical Hamiltonian density formulation of hydrodynamics and ideal magnetohydrodynamics. Phys. Rev. Lett. **45**, 790–794 (1980)
13. K. Tanehashi, Z. Yoshida, Gauge symmetries and Noether charges in Clebsch-parameterized magnetohydrodynamics. J. Phys. A: Math. Theor. **48**, 495501 (20 pp) (2015)
14. Z. Yoshida, Clebsch parameterization: basic properties and remarks on its applications. J. Math. Phys. **50**, 113101 (16 pp) (2009)
15. Z. Yoshida, P.J. Morrison, A hierarchy of noncanonical Hamiltonian systems: circulation laws in an extended phase space. Fluid Dyn. Res. **46**, 031412 (21 pp) (2014)
16. Z. Yoshida, P.J. Morrison, Hierarchical structure of noncanonical Hamiltonian systems. Phys. Scr. **91**, 024001 (7 pp) (2016)
17. Z. Yoshida, Y. Kawazura, T. Yokoyama, Relativistic helicity and link in Minkowski space-time. J. Math. Phys. **55**, 043101 (19 pp) (2014)
18. V.E. Zakharov, E.A. Kuznetsov, Variational principle and canonical variables in magnetohydrodynamics. Sov. Phys. Dokl. **15**, 913–914 (1971)

Analysis of a Living Fluid Continuum Model

Florian Zanger, Hartmut Löwen and Jürgen Saal

Abstract Generalized Navier–Stokes equations which were proposed recently to describe active turbulence in living fluids are analyzed rigorously. Results on well-posedness and stability in the $L^2(\mathbb{R}^n)$-setting are derived. Due to the presence of a Swift-Hohenberg term, global wellposedness in a strong setting for arbitrary initial data in $L^2_\sigma(\mathbb{R}^n)$ is available. Based on the existence of global strong solutions, results on linear and nonlinear (in-) stability for the disordered steady state and the manifold of ordered polar steady states are derived, depending on the involved parameters.

Keywords Generalized Navier–Stokes equations · Active turbulence
Global wellposedness · (in-) stability

1 Introduction

There is a need to study analytical properties of generalized Navier–Stokes equations which were recently proposed [7, 8, 33] for active soft matter (for recent reviews see [3, 21, 22, 25]) to describe the dynamics of "living fluids" such as dense bacterial suspensions at low Reynolds number [24]. On a continuum scale, a living fluid flows with an internal speed that is set by the internal self-propulsion velocity of the bacteria. Generalized incompressible Navier–Stokes equations were designed to describe the spontaneous formation of fluid vortices on the mesoscale by including higher order derivatives in the velocities entering into the stress tensor. Indeed, at high

---

Dedicated to Professor Yoshikazu Giga on the occasion of his 60th birthday

F. Zanger · J. Saal (✉)
Mathematisches Institut, Angewandte Analysis, Heinrich-Heine-Universität Düsseldorf, 40204 Düsseldorf, Germany
e-mail: saal@math.uniduesseldorf.de

H. Löwen
Institut für Theoretische Physik II - Soft Matter, Heinrich-Heine-Universität Düsseldorf, Universitätsstraße 1, 40225 Düsseldorf, Germany
e-mail: hlowen@thphy.uni-duesseldorf.de

Y. Maekawa and S. Jimbo (eds.), *Mathematics for Nonlinear Phenomena—Analysis and Computation*, Springer Proceedings in Mathematics & Statistics 215,
DOI 10.1007/978-3-319-66764-5_14

density of the bacterial suspension, experiments reveal the spontaneous formation of meso-scale vortices [33], which is confirmed by particle-resolved simulations of self-propelled particles [34] and is consistent with the predictions of the Navier–Stokes equations generalized to living fluids. Therefore these continuum equations constitute an important general framework for the flow of living or active fluids and provide a minimal continuum model for swirling. Though different to ordinary turbulence, which occurs at high Reynolds number, this phenomenon is often called "active turbulence" [37]. Active turbulence, which occurs at small Reynolds number, is characterized by scaling laws different to ordinary turbulence [4, 33].

Therefore a thorough mathematical study of this generalized Navier–Stokes system is highly desirable both from a physical and a mathematical point of view. This paper concerns the following minimal hydrodynamic model to describe the bacterial velocity in the case of highly concentrated bacterial suspensions with negligible density fluctuations considered on the domain $(0, \infty) \times \mathbb{R}^n$:

$$\begin{aligned} v_t + \lambda_0 v \cdot \nabla v &= f - \nabla p + \lambda_1 \nabla |v|^2 - (\alpha + \beta |v|^2) v + \Gamma_0 \Delta v - \Gamma_2 \Delta^2 v \\ \operatorname{div} v &= 0 \\ v(0) &= v_0 \end{aligned} \tag{1}$$

Here $v : (0, \infty) \times \mathbb{R}^n \to \mathbb{R}^n$ is the (vectorial) bacterial velocity field and $p : (0, \infty) \times \mathbb{R}^n \to \mathbb{R}$ the (scalar) pressure. The first equation is the conservation of momentum and the equation $\operatorname{div} v = 0$ results from conservation of mass and the assumption of constant density.

The generalized Navier–Stokes equations defined in (1) were originally proposed by Wensink et al. in [33] and then considered in Refs. [7, 8]. Clearly, for $\lambda_0 = 1$, $\lambda_1 = \beta = \Gamma_2 = 0$, and $\Gamma_0 > 0$, the model reduces to the incompressible Navier–Stokes equations in n spatial dimensions. Let us briefly discuss the physics behind the various terms entering in (1). The parameter λ_0 describes advection and nematic interactions and λ_1 is a prefactor in front of an active pressure contribution [7]. The two parameters λ_0 and λ_1 depend on the hydrodynamic nature of the swimmer, i.e., whether they are pushers or pullers [9], and on the dimension $n \in \{2, 3\}$. The term involving the parameters α and β pushes the system towards rest with velocity $v = 0$ if $\alpha > 0$ and towards a characteristic nonvanishing velocity of $\sqrt{-\alpha/\beta}$ if $\alpha < 0$ and $\beta > 0$ as in the Toner–Tu model [31] corresponding to a quartic Landau-type velocity potential. If the parameter Γ_0 is positive, it determines the suspension's viscosity similar to the pure Navier–Stokes case. If it is negative, $\Gamma_2 > 0$ is required for stability reasons as in the traditional Swift–Hohenberg equation [30], for a review see [10]. By this fact, here we always assume $\Gamma_2 > 0$.

The main objective of this note is to provide an analytical approach to the generalized Navier–Stokes equations (1) in the $L^2(\mathbb{R}^n)$-setting. This will be performed in Section 3. There we will consider the general system (5), which includes the transformed systems about the steady states given in Section 2. These are the *disordered isotropic state* and the manifold of *globally ordered polar states*. The main results of Section 3 are as follows. Subsection 3.1 provides an approach to the linearized equations. The corresponding linear operator admits a bounded H^∞-calculus, see Lemma 1. Propositions 1 and 2 give precise information on linear (in-) stability of the steady states depending on the values of the involved parameters. In Subsection 3.2 we will prove global (strong) wellposedness for the generalized Navier–Stokes equations (1) in the L^2-setting, see Theorem 2. Based on the global solvability, Subsection 3.3 concerns nonlinear (in-) stability. Theorem 3 transfers the linear stability results for the disordered state to the nonlinear situation. Theorem 4 then proves a nonlinear instability result for the ordered polar state.

We note that the physically relevant steady states (see Section 2) were given earlier in [33]. There is also a linear instability analysis performed on a non-rigorous level. Here we provide a rigorous analysis and go beyond linear instability.

The fact that we can prove the existence of a global unique strong solution for arbitrary initial data in $L^2_\sigma(\mathbb{R}^n)$ for system (1) of course is due to the presence of the Swift–Hohenberg term $\Gamma_2\Delta^2 u$. It causes the nonlinear terms to appear less strong compared to the classical second order Navier–Stokes equations. We refer to [2, 5, 11, 29, 35] and the references cited therein for more information on the classical Navier–Stokes equations.

We also remark that the purpose of this note is not to present best possible results in every direction. The L^2-approach given here is merely a first step towards a thorough analysis of the active fluids continuum model (1) in a variety of further significant situations. Further developments and future projects, e.g., including fluid boundaries, are addressed in Section 4.

2 Steady States

We assume $\Gamma_2, \beta > 0$ and $\alpha \in \mathbb{R}$. Then the following physically relevant stationary solutions appear [33]:

$$(v, p) = (0, p_0) \tag{2}$$

with a pressure constant p_0 and, if $\alpha < 0$, additionally

$$(v, p) = (V, p_0), \tag{3}$$

where $V \in B_{\alpha,\beta} := \{x \in \mathbb{R}^n : |V| = \sqrt{-\alpha/\beta}\}$, i.e., V denotes a constant vector with arbitrary orientation and fixed swimming speed $|V| = \sqrt{-\alpha/\beta}$.

The steady state (2) corresponds to a *disordered isotropic state* and (3) to the manifold $B_{\alpha,\beta}$ of *globally ordered polar states.*

Note that mathematically there is a further manifold of stationary solutions given by

$$v(x) = v_0, \qquad p(x) = p_0 - (\alpha + \beta|v_0|^2)v_0 \cdot x, \qquad x \in \Omega,\ p_0 \in \mathbb{R}, \tag{4}$$

with $v_0 \in \mathbb{R}^3$ arbitrary. For $v_0 = 0$ or $|v_0| = \sqrt{-\alpha/\beta}$ these solutions correspond to the above steady states (2) and (3), respectively. For all other values of v_0 they are, however, physically not relevant since their pressure takes arbitrary large negative values. Thus, in the sequel, we will only consider (2) and (3).

3 Wellposedness and Stability

We perform an approach to the hydrodynamic model (1) in the $L^2(\mathbb{R}^n)$-setting. In order to include the steady states in our analysis we consider the following generalized system:

$$\begin{aligned} u_t + \lambda_0 [(u+V)\cdot\nabla]u + (M+\beta|u|^2)u - \Gamma_0\Delta u + \Gamma_2\Delta^2 u + \nabla q &= f + N(u), \\ \operatorname{div} u &= 0, \\ u(0) &= u_0. \end{aligned} \tag{5}$$

Here $q = p - \lambda_1|v|^2$, $M \in \mathbb{R}^{n\times n}$ is a symmetric matrix, and $N(u) = \sum_{j,k} a_{jk}u^j u^k$ with $(a_{jk})_{j,k=1}^n \subset \mathbb{R}^n$ is a quadratic nonlinear term. By setting

$$V = 0, \quad M = \alpha, \quad N(u) = 0 \tag{6}$$

we obtain (1) for $u = v$, i.e., the system corresponding to the steady state (2) and by setting

$$V \in B_{\alpha,\beta}, \quad M = 2\beta VV^t, \quad N(u) = -\beta|u|^2V - 2\beta(u\cdot V)u \tag{7}$$

we obtain the system for $u = v - V$ corresponding to (3). Note that for the appearing parameters we always assume that $\lambda_0, \lambda_1, \Gamma_0, \alpha \in \mathbb{R}$ and that $\Gamma_2, \beta > 0$. Furthermore, the dimension is always assumed to be $n \in \{2,3\}$.

For a domain $\Omega \subset \mathbb{R}^n$, a Banach space X, and $1 \le p \le \infty$ in the sequel $L^p(\Omega, X)$ denotes the standard Bochner–Lebesgue space with norm

$$\|u\|_{L^p(X)} = \left(\int_\Omega \|u(x)\|_X^p\,dx\right)^{1/p},$$

if $1 \le p < \infty$ and $\|u\|_{L^\infty(X)} = \operatorname{ess\,sup}_{x\in\Omega}\|u\|_X$ if $p = \infty$. In case that $\Omega = X = \mathbb{R}^n$ its subspace of solenoidal functions is denoted by

$$L^p_\sigma(\mathbb{R}^n) := \{v \in L^p(\mathbb{R}^n);\ \operatorname{div} v = 0\}.$$

Note that for $1 < p < \infty$ we have the Helmholtz decomposition

$$L^p(\mathbb{R}^n) = L^p_\sigma(\mathbb{R}^n) \oplus G_p(\mathbb{R}^n)$$

with $G_p(\mathbb{R}^n) = \{\nabla p;\ p \in \mathscr{D}'(\mathbb{R}^n),\ \nabla p \in L^p(\mathbb{R}^n)\}$. The associated Helmholtz projector onto $L^p_\sigma(\mathbb{R}^n)$ is represented as

$$P = \mathscr{F}^{-1}\left(I - \frac{\xi\xi^T}{|\xi|^2}\right)\mathscr{F},$$

where $\mathscr{F}$ denotes the Fourier transformation and I the identity matrix in $\mathbb{R}^n$.

The symbol $W^{k,p}(\Omega, X)$, $k \in \mathbb{N}_0$, $1 \le p \le \infty$, stands for the standard Sobolev space of k-times differentiable functions in $L^p(\mathbb{R}^n, X)$. Its norm is given as

$$\|f\|_{W^{k,p}(X)} := \Big(\sum_{|\alpha|\le k} \|\partial^\alpha f\|^p_{L^p(X)}\Big)^{1/p}$$

with the usual modification if $p = \infty$. The fractional order Sobolev resp. Besov spaces are defined by complex resp. real interpolation as

$$W^{t,p}(\Omega, X) := \big[W^{k,p}(\Omega, X), W^{k+1,p}(\Omega, X)\big]_{t/k},$$
$$B^t_p(\Omega, X) := \big(W^{k,p}(\Omega, X), W^{k+1,p}(\Omega, X)\big)_{t/k,p}$$

for $t \in (0, k)$ and $k \in \mathbb{N}_0$. For $p = 2$ and $s \ge 0$ we use the notation $H^s(\Omega, X) := W^{s,2}(\Omega, X)$ and we frequently write $L^p(\Omega)$, $W^{s,p}(\Omega)$, and $B^s_p(\Omega)$ if $X = \mathbb{R}^n$. Also note that $H^s(\Omega) = W^{s,2}(\Omega) = B^s_2(\Omega)$, but that $W^{s,p}(\Omega, X) \ne B^s_p(\Omega, X)$ in general. Finally, $\mathscr{L}(X, Y)$ denotes the space of all bounded and linear operators from the space X into the space Y, we write $\mathscr{L}(X)$ if $X = Y$, and $\sigma(A)$ denotes the spectrum of a linear operator $A : D(A) \subset X \to X$.

3.1 Linear Theory

In this subsection, we consider the linearized system

$$\begin{aligned} u_t + \lambda_0(V\cdot\nabla)u + Mu - \Gamma_0\Delta u + \Gamma_2\Delta^2 u + \nabla q = f \ &\text{ in } (0,\infty)\times\mathbb{R}^n,\\ \operatorname{div} u = 0 \ &\text{ in } (0,\infty)\times\mathbb{R}^n,\\ u(0) = u_0 \ &\text{ in } \mathbb{R}^n. \end{aligned} \tag{8}$$

Thanks to $\Gamma_2 > 0$ the operator

$$A_{SH}u := \Gamma_2\Delta^2 u, \quad u \in D(A_{SH}) := W^{4,p}(\mathbb{R}^n) \cap L^p_\sigma(\mathbb{R}^n),$$

admits a bounded H^∞-calculus on $L^p_\sigma(\mathbb{R}^n)$ with H^∞-angle $\phi^\infty_{A_{SH}} = 0$ for $p \in (1,\infty)$. This follows as an easy consequence of Mikhlin's multiplier theorem, for instance. See e.g., [6, 13, 19] for an introduction to the notion of a bounded H^∞-calculus. Since every other term appearing in (8), more precisely the operator

$$Bu := \lambda_0(V\cdot\nabla)u + PMu - \Gamma_0\Delta u, \tag{9}$$

is of lower order, by a standard perturbation argument we immediately deduce

Lemma 1. *Let $1 < p < \infty$. There is an $\omega > 0$ such that the operator $\omega + A_{LF}$, where*

$$A_{LF}u := (A_{SH} + B)u, \quad u \in D(A_{LF}) := W^{4,p}(\mathbb{R}^n) \cap L^p_\sigma(\mathbb{R}^n), \tag{10}$$

admits a bounded H^∞-calculus on $L^p_\sigma(\mathbb{R}^n)$ with H^∞-angle $\phi^\infty_{\omega+A_{LF}} < \pi/2$.

As a consequence, cf. [6], $-A_{LF}$ is the generator of an analytic C_0-semigroup on $L^p_\sigma(\mathbb{R}^n)$ and it has maximal regularity:

Corollary 1. *Let $1 < p < \infty$, $T \in (0, \infty)$. For $f \in L^p((0,T), L^p_\sigma(\mathbb{R}^n))$ and $u_0 \in B_p^{4-4/p}(\mathbb{R}^n) \cap L^p_\sigma(\mathbb{R}^n)$ there exists a unique solution (u, q) of (8) satisfying*

$$\begin{aligned}&\|u\|_{W^{1,p}((0,T),L^p)} + \|u\|_{L^p((0,T),W^{4,p})} + \|\nabla q\|_{L^p((0,T),L^p)}\\ &\le C\left(\|f\|_{L^p((0,T),L^p)} + \|u_0\|_{B_p^{4-4/p}}\right)\end{aligned}$$

with $C > 0$ independent of u, q, f, u_0.

To obtain precise information on the spectrum of the operator A_{LF} we apply Fourier transformation to (10) to the result that

$$\sigma_{A_{LF}}(\xi) = \mathscr{F} A_{LF} \mathscr{F}^{-1} = \Gamma_2|\xi|^4 + \Gamma_0|\xi|^2 + \sigma_P(\xi)M + i\lambda_0 V \cdot \xi, \quad \xi \in \mathbb{R}^n,$$

with $\sigma_P(\xi) = (1 - \xi\xi^t/|\xi|^2)$ the symbol of the Helmholtz projector P. We first consider the disordered state (2). We set $A_d := A_{LF}$ in this case. Then according to (6) the above expression takes the form

$$\sigma_{A_d}(\xi) = \Gamma_2|\xi|^4 + \Gamma_0|\xi|^2 + \alpha, \quad \xi \in \mathbb{R}^n.$$

Calculating the intersection points of the parabola in $s = |\xi|^2$ we obtain

$$s_\pm^2 = \frac{-\Gamma_0}{\Gamma_2}\left(\frac{1}{2} \pm \sqrt{\frac{1}{4} - \frac{\alpha\Gamma_2}{\Gamma_0^2}}\right).$$

Consequently, if $\Gamma_0 < 0$ there is an unstable band of modes for $s^2 \in (s_-^2, s_+^2)$ provided this interval is non-empty. In this case, the spectral bound $s(A_d) = \sup\{\operatorname{Re} z;\ z \in \sigma(A_d)\}$ of $-A_d$ is positive. Since it is well-known that for analytic C_0-semigroups spectral bound of the generator and growth bound of the generated semigroup coincide [23], we deduce that $\exp(-tA_d)$ is exponentially unstable precisely if $\Gamma_0 < 0$ and $4\alpha < \Gamma_0^2/\Gamma_2$, or if $\Gamma_0 \ge 0$ and $\alpha < 0$. This leads to the following result.

Proposition 1. *Let $\Gamma_2 > 0$, $\beta > 0$, and $1 < p < \infty$. If $\Gamma_0 < 0$, then the disordered state (2) is linearly stable if and only if $4\alpha \ge \Gamma_0^2/\Gamma_2$. If $\Gamma_0 \ge 0$, then the disordered state (2) is stable if and only if $\alpha \ge 0$. To be precise, the semigroup $(\exp(-tA_d))_{t\ge 0}$ on $L^p_\sigma(\mathbb{R}^n)$ corresponding to the disordered state (2) is*

(1) *exponentially stable if* $\Gamma_0 < 0$ *and* $4\alpha > \Gamma_0^2/\Gamma_2$, *or if* $\Gamma_0 \geq 0$ *and* $\alpha > 0$;
(2) *asymptotically stable if* $\Gamma_0 < 0$ *and* $4\alpha = \Gamma_0^2/\Gamma_2$, *or if* $\Gamma_0 \geq 0$ *and* $\alpha = 0$;
(3) *exponentially unstable if* $\Gamma_0 < 0$ *and* $4\alpha < \Gamma_0^2/\Gamma_2$, *or if* $\Gamma_0 \geq 0$ *and* $\alpha < 0$.

Proof. It remains to prove (2), the other assertions are obvious by the discussion above. On the other hand, in the situation of (2), we see that $(\exp(-tA_d))_{t\geq 0}$ is a bounded analytic C_0-semigroup on $L^p_\sigma(\mathbb{R}^n)$ such that A_d has dense range, which are known to be asymptotically stable, see [23].

Next, we consider the manifold $B_{\alpha,\beta}$ of ordered polar states (3). In this case, we set $A_o := A_{LF}$ and the symbol of this operator according to (7) reads as

$$\sigma_{A_o}(\xi) = \Gamma_2|\xi|^4 + \Gamma_0|\xi|^2 + 2\beta\sigma_P(\xi)VV^t + i\lambda_0 V\cdot\xi, \quad \xi \in \mathbb{R}^n,$$

with $V \in B_{\alpha,\beta}$. Note that the matrix $\sigma_P(\xi)VV^t$ is positive semidefinite and that zero is an eigenvalue by the fact that $V^t x = 0$ if $x \in \mathbb{R}^n \setminus \{0\}$ is perpendicular to V. Choosing $x, \xi \in \{V\}^\perp$ such that $|x| = 1$ and that $|\xi|$ is small enough, we can always achieve that

$$x^t\sigma_{A_o}(\xi)x = \Gamma_2|\xi|^4 + \Gamma_0|\xi|^2 < 0,$$

provided that $\Gamma_0 < 0$. Thus, here we obtain the following result.

Proposition 2. *Let* $1 < p < \infty$, $\Gamma_2 > 0$, $\beta > 0$, *and* $\alpha < 0$. *The ordered polar state (3) is linearly stable if and only if* $\Gamma_0 \geq 0$. *To be precise, the semigroup* $(\exp(-tA_o))_{t\geq 0}$ *corresponding to the ordered state (3) is*

(1) *exponentially unstable on* $L^p_\sigma(\mathbb{R}^n)$ *if* $\Gamma_0 < 0$;
(2) *asymptotically stable on* $L^2_\sigma(\mathbb{R}^n)$ *if* $\Gamma_0 \geq 0$.

Proof. Assertion (1) is clear. To see (2) first observe that due to the occurrence of the term $i\lambda_0 V\cdot\xi$, $(e^{-A_o t})_{t\geq 0}$ is not a bounded analytic semigroup. Hence we cannot argue as for A_d to deduce asymptotic stability. Instead, for (2) we restrict ourselves to the case $p = 2$ and proceed as in [15]: For $v_0 \in H^1(\mathbb{R}^n) \cap L^2_\sigma(\mathbb{R}^n)$, $v(t) := \exp(-tA_o)v_0$ solves

$$v_t(t) + \Gamma_2\Delta^2 v(t) - \Gamma_0\Delta v(t) + \lambda_0(V\cdot\nabla)v(t) + 2\beta PVV^t v(t) = 0.$$

Multiplication in $L^2(\mathbb{R}^n)$ with $v(t)$ and integration from $t = 0$ to T yields

$$\|v(T)\|^2_{L^2} + 2\int_0^T \left(\Gamma_2\|\Delta v(t)\|^2_{L^2} + \Gamma_0\|\nabla v(t)\|^2_{L^2} + 2\beta\|V\cdot v(t)\|^2_{L^2}\right)dt = \|v_0\|^2_{L^2}$$

for every $T > 0$. This has two consequences: First, the semigroup $\exp(-tA_o)$ is contractive since it is strongly continuous and since $H^1(\mathbb{R}^n) \cap L^2_\sigma(\mathbb{R}^n)$ is dense in $L^2_\sigma(\mathbb{R}^n)$, and second, we have

$$\int_0^\infty \|\Delta v(t)\|^2_{L^2} dt < \infty. \tag{11}$$

If $v_0 \in H^4(\mathbb{R}^n)$, using the contractiveness we obtain

$$\begin{aligned}\left|\frac{d}{dt}\|\Delta v(t)\|_{L^2}^2\right| &= |2\langle \Delta v(t), \Delta v_t(t)\rangle| \\ &= \left|-2\langle \Delta^2 \exp(-tA_o)v_0, A_o v(t)\rangle\right| \le C\|v_0\|_{H^4}^2,\end{aligned}$$

i.e., $\|\Delta v(\cdot)\|_{L^2}^2 \in BC^1(0,\infty)$. Together with (11) this implies

$$\lim_{t\to\infty} \|\exp(-tA_o)\Delta v_0\|_{L^2} = \lim_{t\to\infty} \|\Delta v(t)\|_{L^2} = 0. \tag{12}$$

In order to prove asymptotic stability we have to show that for every $u_0 \in L^2_\sigma(\mathbb{R}^n)$, $u(t) := \exp(-tA_o)u_0$ satisfies

$$\lim_{t\to\infty} \|u(t)\|_{L^2} = 0.$$

Considering that $\{\Delta w;\ w \in H^4(\mathbb{R}^n) \cap L^2_\sigma(\mathbb{R}^n)\}$ is dense in $L^2_\sigma(\mathbb{R}^n)$ we can always find $v_0 \in H^4(\mathbb{R}^n) \cap L^2_\sigma(\mathbb{R}^n)$ with $\|u_0 - \Delta v_0\|_{L^2}$ arbitrary small. Making once more use of the contractiveness of the semigroup we obtain

$$\begin{aligned}\|u(t)\|_{L^2} &\le \|\exp(-tA_o)u_0 - \exp(-tA_o)\Delta v_0\|_{L^2} + \|\exp(-tA_o)\Delta v_0\|_{L^2} \\ &\le \|u_0 - \Delta v_0\|_{L^2} + \|\exp(-tA_o)\Delta v_0\|_{L^2},\end{aligned}$$

and (12) yields the asymptotic stability.

3.2 Local and Global Strong Solvability

We first consider local-in-time wellposedness. For $T > 0$ we define relevant function spaces as

$$\begin{aligned}&\mathbb{E}_T := W^{1,p}((0,T), L^p_\sigma(\mathbb{R}^n)) \cap L^p((0,T), W^{4,p}(\mathbb{R}^n)), \\ &\mathbb{F}^1_T := L^p((0,T), L^p_\sigma(\mathbb{R}^n)), \quad \mathbb{F}^2 := B^{4-4/p}_p(\mathbb{R}^n), \\ &\mathbb{F}_T := \mathbb{F}^1_T \times \mathbb{F}^2,\end{aligned}$$

and the linear operator

$$L : \mathbb{E}_T \to \mathbb{F}_T, \quad Lu := (\partial_t u + A_{L_F} u, u(0)).$$

If we also set

$$H(u) := \beta P|u|^2 u + \lambda_0 P(u\cdot\nabla)u - PN(u) \tag{13}$$

and

$$F(u) := Lu + (H(u), 0), \tag{14}$$

then the full system (5) is rephrased as

$$F(u) = (f, u_0).$$

Lemma 2. *Let $p > (4+n)/4$. We have $H \in C^1(\mathbb{E}_T, \mathbb{F}_T)$ and its Fréchet derivative is represented as*

$$DH(v)u = P \sum_{|\alpha| \le 1} b_\alpha \partial^\alpha u + \lambda_0 P(u \cdot \nabla)v, \quad u, v \in \mathbb{E}, \tag{15}$$

with matrices $b_\alpha = b_\alpha(v) \in L^\infty((0,T) \times \mathbb{R}^n, \mathbb{R}^{n \times n})$.

Proof. First observe that by [1, Proposition 1.4.2] we have

$$\mathbb{E}_T \hookrightarrow L^\infty((0,T), B_p^{4-4/p}(\mathbb{R}^n)). \tag{16}$$

The assumption $p > (4+n)/4$ in combination with the Sobolev embedding yields

$$\mathbb{E}_T \hookrightarrow L^\infty((0,T) \times \mathbb{R}^n). \tag{17}$$

Utilizing this fact we obtain

$$\begin{aligned} \|(u \cdot \nabla)u\|_{\mathbb{F}_T^1} &\le C\|u\|_\infty \|\nabla u\|_{\mathbb{F}_T^1} \le C\|u\|_{\mathbb{E}_T}^2, \\ \||u|^2 u\|_{\mathbb{F}_T^1} &\le C\||u|^2\|_\infty \|u\|_{\mathbb{F}_T^1} \le C\|u\|_{\mathbb{E}_T}^3, \\ \|N(u)\|_{\mathbb{F}_T^1} &\le C\|u\|_\infty \|u\|_{\mathbb{F}_T^1} \le C\|u\|_{\mathbb{E}_T}^2, \end{aligned}$$

hence $H : \mathbb{E}_T \to \mathbb{F}_T$. By the fact that H consists of bi- and trilinear terms it is obvious that $H \in C^1(\mathbb{E}_T, \mathbb{F}_T)$ (even $H \in C^\infty(\mathbb{E}_T, \mathbb{F}_T)$). The Fréchet derivative reads as

$$\begin{aligned} DH(v)u = {} & \beta P|v|^2 u + 2\beta P(u \cdot v)v + \lambda_0 P(u \cdot \nabla)v \\ & + \lambda_0 P(v \cdot \nabla)u - 2P \sum_{i,k=1}^n a_{jk}(u^j v^k + u^k v^j). \end{aligned}$$

From this and (17) representation (15) easily follows.

Remark 1. Note that the lower bound $p > (4+n)/4$ is not optimal. But, since here we are mainly interested in an L^2-approach for dimension $n = 2, 3$, it is sufficient for our purposes.

Lemma 3. *Let $p > (4+n)/4$, $T \in (0, \infty)$, and $v \in \mathbb{E}_T$. Then we have*

$$L + (DH(v), 0) \in \mathscr{L}_{is}(\mathbb{E}_T, \mathbb{F}_T).$$

Proof. By employing representation (15) for $B(t) := DH(v(t))$ we will show that $B(\cdot)$ is a lower order perturbation of L. To this end, observe that $p > (4+n)/4$ yields

$$W^{3,p}(\mathbb{R}^n) \hookrightarrow L^\infty(\mathbb{R}^n)$$

and thanks to (16) also $v \in L^\infty((0,T), W^{1,p}(\mathbb{R}^n))$. This implies

$$\begin{aligned}\|(u\cdot\nabla)v(t)\|_{L^p(\mathbb{R}^n)} &\le \|\nabla v(t)\|_p \|u\|_\infty \\ &\le C\|v\|_{L^\infty((0,T),W^{1,p})} \frac{1}{\mu^{1/4}} \|(\mu + A_{LF})u\|_p\end{aligned}$$

for all $u \in D(A_{LF})$ and all $\mu > \mu_0$ with $\mu_0 > 0$ large enough. We estimate the second term in (15) as

$$\begin{aligned}\|\sum_{|\alpha|\le 1} b_\alpha \partial^\alpha u\|_{L^p(\mathbb{R}^n)} &\le C\|v\|_\infty \|\nabla u\|_p \\ &\le C\|v\|_\infty \frac{1}{\mu^{3/4}} \|(\mu + A_{LF})u\|_p\end{aligned}$$

for all $u \in D(A_{LF})$ and all $\mu > \mu_0$. This shows that

$$\|B(t)u\|_p \le \frac{C(v)}{\mu^{1/4}} \|(\mu + A_{LF})u\|_p \quad (u \in D(A_{LF}),\ \mu > \mu_0).$$

Thus, choosing μ large enough and due to Corollary 1, we can apply [26, Theorem 2.5] to the result that

$$L + (\mu + DH(v), 0) \in \mathscr{L}_{is}(\mathbb{E}_T, \mathbb{F}_T).$$

Since $L + DH(v)$ is linear, we can remove the shift $\mu > 0$ and the assertion follows.

Appealing to the local inverse theorem we can now prove the following result.

Theorem 1 (local wellposedness). *Let $\Gamma_2, \beta > 0$, $\Gamma_0, \alpha \in \mathbb{R}$, and $p > (4+n)/4$. For every $u_0 \in B_p^{4-4/p}(\mathbb{R}^n) \cap L_\sigma^p(\mathbb{R}^n)$ and $f \in L^p((0,T), L_\sigma^p(\mathbb{R}^n))$ there exists a $T > 0$ and a unique solution (u, q) of (5) such that*

$$\begin{aligned}&u \in W^{1,p}((0,T), L_\sigma^p(\mathbb{R}^n)) \cap L^p((0,T), W^{4,p}(\mathbb{R}^n)),\\ &\nabla q \in L^p((0,T), L^p(\mathbb{R}^n)).\end{aligned}$$

Proof. We fix $(f, u_0) \in \mathbb{F}_T$ and define a reference solution as

$$u^* := L^{-1}(f, u_0) \in \mathbb{E}_T.$$

For the Fréchet derivative of the nonlinear operator $F \in C^1(\mathbb{E}_T, \mathbb{F}_T)$ given in (14) we obtain in view of Lemma 3 that

$$DF(u^*) = L + DH(u^*) \in \mathscr{L}_{is}(\mathbb{E}_T, \mathbb{F}_T).$$

Hence the local inverse theorem yields neighborhoods $U \subset \mathbb{E}_T$ of u^* and $V \subset \mathbb{F}_T$ of $F(u^*)$ such that $F : U \to V$ is bijective.

Now, taking $T' > 0$ small enough, we find a solution. To see this, for $0 < T' < T$ we define $f_{T'} \in \mathbb{F}_T^1$ by

$$f_{T'}(t) := \begin{cases} f(t), & t \in (0, T') \\ f(t) + H(u^*)(t), & t \in [T', T). \end{cases}$$

The continuity of the integral implies

$$f_{T'} \xrightarrow{T' \to 0} f + H(u^*) \text{ in } \mathbb{F}_T^1.$$

Thus, since $F(u^*) = (f + H(u^*), u_0)$, there is $T' > 0$ with $(f_{T'}, u_0) \in V$. The unique function $u \in U$ with $F(u) = (f_{T'}, u_0)$ satisfies $F(u) = (f, u_0)$ on $(0, T')$. Thus, recovering the pressure via

$$\nabla q := -(I - P)\left[\lambda_0 [u \cdot \nabla] u + (M + \beta |u|^2)u - N(u)\right] \in L^p((0, T), L^p(\mathbb{R}^n))$$

we find that $(u, q)|_{(0,T')}$ is a strong solution to (5).

The just constructed local solution extends to a global one, at least if $p = 2$.

Theorem 2. (global wellposedness). *Let $\Gamma_2, \beta > 0$, $\Gamma_0, \alpha \in \mathbb{R}$, $T \in (0, \infty)$. For every $u_0 \in H^2(\mathbb{R}^n) \cap L^2_\sigma(\mathbb{R}^n)$ and $f \in L^2((0, T), L^2_\sigma(\mathbb{R}^n))$ there exists a unique solution (u, q) of (5) such that*

$$\begin{aligned} u &\in H^1((0, T), L^2_\sigma(\mathbb{R}^n)) \cap L^2((0, T), H^4(\mathbb{R}^n)), \\ \nabla q &\in L^2((0, T), L^2(\mathbb{R}^n)). \end{aligned}$$

Proof. We derive a priori bounds in the strong class which will give the result. To this end, we multiply (5) with u and integrate over $(0, t) \times \mathbb{R}^n$. This yields

$$\begin{aligned} \frac{1}{2}\|u(t)\|_2^2 + \Gamma_2 \int_0^t \|\Delta u\|_2^2 \, ds + \beta \int_0^t \|u\|_4^4 \, ds &= \frac{1}{2}\|u_0\|_2^2 + \int_0^t \int_{\mathbb{R}^n} f u \, dx ds \\ + \Gamma_0 \int_0^t \int_{\mathbb{R}^n} u \Delta u \, dx ds - \int_0^t \int_{\mathbb{R}^n} u M u \, dx ds &+ \int_0^t \int_{\mathbb{R}^n} u N(u) \, dx ds \end{aligned}$$

for $t \in (0, T)$. By applying Cauchy–Schwarz' and Young's inequality we can estimate as

$$\int_0^t \int_{\mathbb{R}^n} u\Delta u \, dxds \leq \frac{\Gamma_2}{2|\Gamma_0|} \|\Delta u\|^2_{L^2((0,t),L^2)} + C\int_0^t \|u\|_2^2 \, dt$$

and

$$\int_0^t \int_{\mathbb{R}^n} uN(u) \, dxds \leq \frac{\beta}{2} \|u\|^4_{L^4((0,t),L^4)} + C\int_0^t \|u\|_2^2 \, dt. \tag{18}$$

Plugging this into the above equality we arrive at

$$\begin{aligned} &\|u(t)\|_2^2 + \|u\|^2_{L^2((0,t),H^2)} + \|u\|^4_{L^4((0,t),L^4)} \\ &\leq C\left(\|u_0\|_2^2 + \|f\|^2_{L^2((0,t),L^2)}\right) + C\int_0^t \|u(s)\|_2^2 \, ds \quad (t \in (0,T)). \end{aligned}$$

Hence, Gronwall's lemma yields

$$\begin{aligned} &\|u\|^2_{L^\infty((0,T),L^2)} + \|u\|^2_{L^2((0,T),H^2)} + \|u\|^4_{L^4((0,T),L^4)} \\ &\leq C(1 + Te^{\omega T})\left(\|u_0\|_2^2 + \|f\|^2_{L^2((0,T),L^2)}\right) \end{aligned} \tag{19}$$

for some $C, \omega > 0$.

Next we multiply (5) with $-\Delta u$ and obtain by utilizing integration by parts

$$\begin{aligned} &\frac{1}{2}\|\nabla u(t)\|_2^2 + \Gamma_2 \int_0^t \|\nabla\Delta u\|_2^2 \, ds + \beta\int_0^t \int_{\mathbb{R}^n} (\nabla|u|^2 u)\nabla u \, dxds \\ &= \frac{1}{2}\|\nabla u_0\|_2^2 - \int_0^t \int_{\mathbb{R}^n} f\Delta u \, dxds - \Gamma_0\|\Delta u\|^2_{L^2((0,t),L^2)} \\ &\quad - \int_0^t \int_{\mathbb{R}^n} (\nabla u)M\nabla u \, dxds - \int_0^t \int_{\mathbb{R}^n} (\Delta u)N(u) \, dxds \\ &\quad - \lambda_0 \int_0^t \int_{\mathbb{R}^n} [(V + u)\cdot\nabla]u\Delta u \, dxds. \end{aligned} \tag{20}$$

Concerning the third term on the left-hand side we calculate

$$\begin{aligned} (\nabla|u|^2 u)\nabla u &= \sum_{j,k,\ell=1}^n (\partial_k u^j)\partial_k (u^\ell)^2 u^j \\ &= \sum_{j,k,\ell=1}^n (u^\ell)^2(\partial_k u^j)^2 + 2\sum_{k=1}^n (u\cdot\partial_k u)^2. \end{aligned}$$

This shows that this term is nonnegative, hence it drops out. For the fifth term on the right-hand side analogously to (18) we obtain

$$\int_0^t \int_{\mathbb{R}^n} (\Delta u)N(u) \, dxds \leq C\left(\|u\|^4_{L^4((0,t),L^4)} + \int_0^t \|\Delta u\|_2^2 \, ds\right),$$

whereas the last term on the right-hand side can be estimated utilizing integration by parts and div $(V+u)=0$ as

$$\begin{aligned}&\int_0^t\int_{\mathbb{R}^n}[(V+u)\cdot\nabla]u\,\Delta u\,dxds\\&\leq C\left(\|u\|^2_{L^2((0,t),L^2)}+\|u\|^4_{L^4((0,t),L^4)}\right)+\frac{\Gamma_2}{2|\lambda_0|}\int_0^t\|\nabla\Delta u\|_2^2\,dt.\end{aligned}$$

Plugging this into (20) yields in combination with (19) that

$$\begin{aligned}&\|u\|^2_{L^\infty((0,T),H^1)}+\|u\|^2_{L^2((0,T),H^3)}+\|u\|^4_{L^4((0,T),L^4)}\\&\quad\leq C(1+Te^{\omega T})\left(\|u_0\|^2_{H^1}+\|f\|^2_{L^2((0,T),L^2)}\right).\end{aligned}\tag{21}$$

In the third step, we multiply with $\Delta^2 u$ to obtain

$$\begin{aligned}&\frac{1}{2}\|\Delta u(t)\|_2^2+\Gamma_2\int_0^t\|\Delta^2u\|_2^2\,ds\\&=\frac{1}{2}\|\Delta u_0\|_2^2+\int_0^t\int_{\mathbb{R}^n}f\Delta^2u\,dxds-\Gamma_0\|\Delta\nabla u\|^2_{L^2((0,t),L^2)}\\&\quad-\int_0^t\int_{\mathbb{R}^n}(\Delta u)M\Delta u\,dxds+\int_0^t\int_{\mathbb{R}^n}(\Delta^2u)N(u)\,dxds\\&\quad-\lambda_0\int_0^t\int_{\mathbb{R}^n}[(V+u)\cdot\nabla]u\Delta^2u\,dxds-\beta\int_0^t\int_{\mathbb{R}^n}|u|^2u\Delta^2u\,dxds.\end{aligned}\tag{22}$$

It is enough to focus on the last two terms on the right-hand side. The remaining terms can be handled very similar as before. The first one of those two terms can be controlled by utilizing the estimate

$$\begin{aligned}\int_0^t\|(u\cdot\nabla)u\|_2^2\,ds&\leq C\int_0^t\|u\|_4^2\|\nabla u\|_4^2\,ds\\&\leq C\|u\|^2_{L^\infty(H^1)}\int_0^t\|u\|^2_{H^2}\,ds\\&\leq C(1+Te^{\omega T})^2\left(\|u_0\|^2_{H^1}+\|f\|^2_{L^2((0,T),L^2)}\right)^2,\end{aligned}$$

which is valid thanks to (21) and the Sobolev embedding $H^1(\mathbb{R}^n)\hookrightarrow L^4(\mathbb{R}^n)$.

For the last term, using complex interpolation [32] we obtain

$$L^\infty((0,T),H^1(\mathbb{R}^n))\cap L^2((0,T),H^3(\mathbb{R}^n))\hookrightarrow L^{2/s}((0,T),H^{2s+1}(\mathbb{R}^n))$$

for $s \in [0,1]$. Taking into account $n \leq 3$ the Sobolev embedding yields

$$H^{5/3} \hookrightarrow L^6(\mathbb{R}^n). \tag{23}$$

Hence, by setting $s = 1/3$ we obtain

$$L^\infty((0,T), H^1(\mathbb{R}^n)) \cap L^2((0,T), H^3(\mathbb{R}^n)) \hookrightarrow L^6((0,T), L^6(\mathbb{R}^n)).$$

On the other hand, from Cauchy–Schwarz' and Young's inequality, we see that

$$|\int_0^t \int_{\mathbb{R}^n} |u|^2 u \Delta^2 u \, dx ds| \leq C\|u\|^6_{L^6((0,t),L^6)} + \frac{\Gamma_2}{2\beta}\|\Delta^2 u\|^2_{L^2((0,t),L^2)}.$$

Here the second term on the right-hand side is absorbed by the left-hand side of (22) and the first term is again controlled by estimate (21). Summarizing, we arrive at

$$\begin{aligned}
&\|u\|^2_{L^\infty((0,T),H^2)} + \|u\|^2_{L^2((0,T),H^4)} \\
&\quad \leq C(1 + Te^{\omega T})^3 \left(\|u_0\|^2_{H^1} + \|f\|^2_{L^2((0,T),L^2)}\right)^3.
\end{aligned}$$

Using equations (5) it is straight forward to derive similar bounds for the quantities $\|u\|_{H^1((0,T),L^2)}$ and $\|\nabla q\|_{L^2((0,T),L^2)}$ as well. Thus the assertion is proved.

3.3 Nonlinear (In-) Stability

Most of the outcome on linear (in-) stability in the L^2-setting transfers to the corresponding nonlinear situation. For the transfer of the stability results we apply energy methods and for the transfer of the results on instability we employ Henry's instability theorem [14, Corollary 5.1.6], which we reformulate suitably for our purposes.

Proposition 3. *Let $-A$ be the generator of a holomorphic C_0-semigroup in a Banach space X and let $f: U \to X$, where U is an open neighborhood in $X^\gamma := D(A^\gamma)$ for some $\gamma \in (0,1)$, be locally Lipschitz. Let $x_0 \in D(A) \cap U$ be an equilibrium point of*

$$\dot{w}(t) + Aw(t) = f(w(t)), \tag{24}$$

i.e., $Ax_0 = f(x_0)$. Suppose

$$\begin{aligned}
f(x_0 + z) &= f(x_0) + Bz + g(z), \quad g(0) = 0, \\
\|g(z)\| &= O(\|z\|^s_{X^\gamma}), \quad \text{as } z \to 0 \text{ in } X^\gamma,
\end{aligned}$$

for some $s > 1$, $B \in \mathscr{L}(X^\gamma, X)$, and $\sigma(-A+B) \cap \{z \in \mathbb{C} : \operatorname{Re} z > 0\} \neq \emptyset$. Then x_0 is nonlinearly unstable in the following sense: there is a constant $\varepsilon_0 > 0$ such that *for any* $\delta > 0$ *there exists* $x \in X^\gamma$ *with* $\|x - x_0\|_\gamma < \delta$ *such that there is some finite time* $t_0 > 0$ *with*

$$\|w(t_0, x) - x_0\|_{X^\gamma} \geq \varepsilon_0,$$

where $w(\cdot, x)$ *denotes the solution of (24) with initial value* $w(0, x) = x$.

For instability also the following lemma will be helpful.

Lemma 4. *Let the nonlinearity* H *be given as in (13). Then for every* $\sigma \geq 5/4$ *we have* $H \in C^1(H^\sigma(\mathbb{R}^n), L^2_\sigma(\mathbb{R}^n))$ *and*

$$\|H(u)\|_2 \leq C\|u\|^2_{H^\sigma} \quad (\|u\|_{H^\sigma} \leq 1).$$

Proof. Employing Hölder's inequality we obtain

$$\|(u \cdot \nabla)u\|_2 \leq \|u\|_p \|\nabla u\|_q$$

for $1/p + 1/q = 1/2$. It is easily checked that the Sobolev embeddings $H^\gamma(\mathbb{R}^n) \hookrightarrow L^p(\mathbb{R}^n)$ and $H^{\gamma-1}(\mathbb{R}^n) \hookrightarrow L^q(\mathbb{R}^n)$ are sharp for $p = 12$, $q = 12/5$, and $\gamma = 5/4$. By the fact that

$$\||u|^2 u\|_2 = \|u\|^3_6 \leq C\|u\|^3_{H^1}$$

and since H consists of bi- and trilinear terms the assertion follows.

Remark 2. It is clear that due to better Sobolev embeddings the lower bound on σ can be improved if $n = 2$.

As before, first, we consider the disordered state (2).

Theorem 3. *Let* $\Gamma_2 > 0$ *and* $\beta > 0$. *Then the disordered state (2) is nonlinearly*

(1) (globally) exponentially stable in $L^2_\sigma(\mathbb{R}^n)$ *if* $\Gamma_0 < 0$ *and* $4\alpha > \Gamma_0^2/\Gamma_2$, *or if* $\Gamma_0 \geq 0$ *and* $\alpha > 0$;
(2) stable in $L^2_\sigma(\mathbb{R}^n)$ *if* $\Gamma_0 < 0$ *and* $4\alpha = \Gamma_0^2/\Gamma_2$, *or if* $\Gamma_0 \geq 0$ *and* $\alpha = 0$;
(3) unstable in $H^\gamma(\mathbb{R}^n) \cap L^2_\sigma(\mathbb{R}^n)$ *for* $\gamma \in [5/16, 1)$ *if* $\Gamma_0 < 0$ *and* $4\alpha < \Gamma_0^2/\Gamma_2$, *or if* $\Gamma_0 \geq 0$ *and* $\alpha < 0$.

Proof. Suppose u and q with regularity as in Theorem 2 solve the nonlinear system (5) corresponding to the disordered state (2), that is

$$u_t + \Gamma_2 \Delta^2 u - \Gamma_0 \Delta u + \lambda_0 (u \cdot \nabla)u + (\alpha + \beta|u|^2)u + \nabla q = 0.$$

Testing this equation with u we obtain

$$\frac{1}{2}\frac{d}{dt}\|u\|_2^2 + \Gamma_2\|\Delta u\|_2^2 + \Gamma_0\|\nabla u\|_2^2 + \alpha\|u\|_2^2 + \beta\|u\|_4^4 = 0.$$

If $\Gamma_0 \geq 0$ and $\alpha \geq 0$, all coefficients are nonnegative and we deduce

$$\frac{d}{dt}\|u\|_2^2 \leq -2\alpha\|u\|_2^2,$$

which yields

$$\|u(t)\|_2^2 \leq e^{-2\alpha t}\|u_0\|_2^2 \qquad (t \geq 0),$$

i.e., stability if $\alpha = 0$ and exponential stability if $\alpha > 0$.

If $\Gamma_0 < 0$, we use the Plancherel theorem, Hölder's inequality, and Young's inequality with ε to estimate the term

$$\begin{aligned}\|\nabla u\|_2^2 &= \int_{\mathbb{R}^n} |\xi|^2|\hat{u}(\xi)|^2 d\xi \leq \||\xi|^2\hat{u}(\xi)\|_2\|\hat{u}\|_2 \\ &= \|\Delta u\|_2\|u\|_2 \leq \frac{\varepsilon^2}{2}\|\Delta u\|_2^2 + \frac{1}{2\varepsilon^2}\|u\|_2^2.\end{aligned}$$

The $\varepsilon^2/2$-term can be absorbed by the Γ_2-term if we choose $\varepsilon^2 = 2\Gamma_2/|\Gamma_0|$. Dropping the β-term as well we are left with

$$\frac{1}{2}\frac{d}{dt}\|u\|_2^2 + \alpha\|u\|_2^2 \leq \frac{\Gamma_0^2}{4\Gamma_2}\|u\|_2^2.$$

As before, this implies stability if $4\alpha = \Gamma_0^2/\Gamma_2$ and exponential stability if $4\alpha > \Gamma_0^2/\Gamma_2$. Thus assertions (1) and (2) are proved.

To see instability we apply Proposition 3. In our situation we have $x_0 = 0$, $B = 0$, $x = u$, $A = A_d$, and $f(u) = H(u)$. From this, we also see that $H(0) = 0$, i.e., that $g = f = H$ in Henry's notation. Note that Proposition 1(3) under the assumption (3) above implies that $\sigma(-A_d) \cap \{z \in \mathbb{C} : \operatorname{Re} z > 0\} \neq \emptyset$. Next, the fact that $\omega + A_d$ admits a bounded H^∞-calculus for some $\omega > 0$ (Lemma 1) yields

$$D(A_d^\gamma) = \left[L_\sigma^2(\mathbb{R}^n), D(A_d)\right]_\gamma = H^{4\gamma}(\mathbb{R}^n) \quad (\gamma \in [0, 1]),$$

where $[\cdot, \cdot]_s$ denotes the complex interpolation space, cf. [32]. Lemma 4 then implies that the assertions of Proposition 3 are fulfilled for $\gamma \in [5/16, 1)$ and $s = 2$. Hence the disordered state is unstable.

Finally, we consider instability for the ordered polar state.

Theorem 4. *Let $\Gamma_2, \beta > 0$ and $\Gamma_0, \alpha < 0$. Then the ordered polar state (3) is nonlinearly unstable in $H^\gamma(\mathbb{R}^n) \cap L_\sigma^2(\mathbb{R}^n)$ for $\gamma \in [5/16, 1)$.*

Proof. Based on Lemma 4 and Proposition 2(1) the proof is analogous to the proof of Theorem 3(3).

Remark 3. We have seen in Theorem 3 that for the disordered steady state the results on linear (in-) stability in principle completely transfers to the nonlinear situation.

Note that for the time being it is not clear if for $\Gamma_0 \geq 0$ the linear stability for the ordered polar state given by Proposition 2(2) transfers to the nonlinear situation as well. Proceeding as for the disordered state, i.e., employing energy methods, it appears that for the ordered state "disturbing" terms on the right-hand side can not easily be absorbed by the "good" terms on the left-hand side by just applying Young, Sobolev, and Hölder. It seems that here a refined analysis is required which, however, is left as a future challenge.

4 Conclusions and Future Developments

In conclusion, we have shown that the model proposed by Wensink et al. in [33] gives rise to a mathematically wellposed system which reflects the asymptotic behavior observed in simulations and experiments. In detail we have proved:

(i) existence of a unique local-in-time solution for initial data in $B_p^{4-4/p}(\mathbb{R}^n) \cap L_\sigma^p(\mathbb{R}^n)$;
(ii) existence of a unique global strong solution for arbitrary initial data in $H^2(\mathbb{R}^n) \cap L_\sigma^2(\mathbb{R}^n)$;
(iii) results on stability and instability of the ordered and the disordered steady states in the L^2-setting depending on the values of the occurring physically relevant parameters.

Note that (ii) is in contrast to the (mathematical) situation of the classical Navier–Stokes system. The fact that we can prove existence of a unique global strong solution here, of course essentially relies on the presence of the fourth order term in (1) which provides sufficient regularity.

The intention of this note is to give an analytical approach in the L^2-setting which serves as a first step for further thorough examinations in several directions. Future work should address the following problems: first of all, it should also be mentioned that the generalized Navier–Stokes framework was augmented even further by including further higher derivatives in the velocities entering into the stress tensor, see the recent work by Slomka and Dunkel [28]. We anticipate that our analysis can also be performed in this more general case provided the highest order term has the correct stabilizing sign.

Second, another more complex problem is that of boundary conditions for the fluid velocity field. This is important to take into account walls and obstacles which confine the bacterial flow. In fact, recent experiments with bacterial turbulence were performed with mobile wedge-like obstacles or carriers [16–18] which were powered by activity or in static meso-structured environments [36] where there is an interesting competition between the geometric structure of the boundaries and the swirling. Recent observations have also addressed spheres as passive additives to steer bacterial turbulence. For all these interesting set-ups the mathematical analysis should include boundary conditions. The latter are typically stick or slip or involve a finite slipping length. Therefore the wellposedness and stability of the generalized Navier–Stokes equation in nontrivial boundaries should be addressed in future studies.

Third, some bacteria move on the surface of an emulsion droplet, which has motivated recent studies of active particles on a compact manifold, such as a sphere [12, 20, 27]. It would be interesting to generalize the hydrodynamic model (1) towards a nonplanar geometry and to prove wellposedness and stability for the problem on curved space.

References

1. H. Amann, *Linear and Quasilinear Parabolic Problems*. vol. I. Birkhäuser (1995)
2. H. Amann, On the strong solvability of the Navier–Stokes equations. J. Math. Fluid Mech. **2**(1), 16–98 (2000)
3. C. Bechinger, R. di Leonardo, H. Löwen, C. Reichhardt, G. Volpe Active Brownian motion in complex and crowded environments. (submitted to Reviews of Modern Physics in 2016)
4. V. Bratanov, F. Jenko, E. Frey, New class of turbulence in active fluids. PNAS **112**, 15048–15053 (2015)
5. M. Cannone, Harmonic analysis tools for solving the incompressible Navier-Stokes equations. in *H*andbook of Mathematical Fluid Dynamics. vol. III. (Elsevier, North Holland, 2004)
6. R. Denk, M. Hieber, J. Prüss, *R-boundedness, Fourier multipliers and problems of elliptic and parabolic type*. Memoirs of the American Mathematical Society. **166**, viii+114 (2003)
7. J. Dunkel, S. Heidenreich, M. Bär, R.E. Goldstein, Minimal continuum theories of structure formation in dense active fluids. N. J. Phys. **15**, 045016 (2013)
8. J. Dunkel, S. Heidenreich, K. Drescher, H.H. Wensink, M. Bär, R.E Goldstein, Fluid dynamics of bacterial turbulence. Phys. Rev. Lett. **110**, 228102 (2013)
9. J. Elgeti, R.G. Winkler, G. Gompper, Physics of microswimmers–single particle motion and collective behavior: a review. Rep. Prog. Phys **78**, 056601 (2015)
10. H. Emmerich, H. Löwen, R. Wittkowski, T. Gruhn, G.I. Tóth, G. Tegze, L. Gránásy, Phase-field crystal models for condensed matter dynamics on atomic length and diffusive time scales: an overview. Adv. Phys. **61**, 665–743 (2012)
11. G. Galdi, *An Introduction to the Mathematical Theory of the Navier–Stokes Equations*, 2nd edn. (Springer, Verlag, 2011)
12. R. Großmann, F. Peruani, M. Bär, A geometric approach to self-propelled motion in isotropic and anisotropic environments. Eur. Phys. J. Spec. Top. **224**, 1377–1394 (2015)
13. M. Haase, *The Functional Calculus for Sectorial Operators*, Operator theory: advances and applications. (Birkhäuser Verlag Basel, 2006)
14. D. Henry, *Geometric Theory of Semilinear Parabolic Equations* (Springer, Berlin, 1981)
15. M. Hess, M. Hieber, A. Mahalov, J. Saal, Nonlinear stability of Ekman layers. Bull. Lond. Math. Soc. **42**(4), 691–706 (2010)
16. A. Kaiser, A. Peshkow, A. Sokolov, B. ten Hagen, H. Löwen, I.S. Aranson, Transport powered by bacterial turbulence. Phys. Rev. Lett. **112**, 158101 (1–5) (2014)
17. A. Kaiser, A. Sokolov, I.S. Aranson, H. Löwen, Mechanisms of carrier transport induced by a microswimmer bath. IEEE Trans. Nanobiosci. **14**, 260–266 (2015)
18. A. Kaiser, A. Sokolov, I.S. Aranson, H. Löwen, Motion of two micro-wedges in a turbulent bacterial bath. Eur. Phys. J. Spec. Top. **224**, 1275–1286 (2015)
19. P. Kunstmann, L. Weis, Maximal L_p-regularity for parabolic equations, Fourier multiplier theorems, and H^∞-functional calculus, Lecture Notes in Mathematics, vol. 1855. (Springer, Berlin 2003)
20. W. Li, Collective motion of swarming agents evolving on a sphere manifold: A fundamental framework and characterization. Scientific reports **5**, 13603 (2015)

21. M.C. Marchetti, J.F. Joanny, S. Ramaswamy, T.B. Liverpool, J. Prost, M. Rao, A. Sima, Hydrodynamics of soft active matter. Rev. Mod. Phys. **85**, 1143–1189 (2013)
22. A.M. Menzel, Tuned, driven, and active soft matter. Physics reports **554**, 1–45 (2015)
23. R. Nagel, *One-parameter Semigroups of Positive Operators*, Lecture Notes in Mathematics, vol. 1184. (Springer, Berlin, 1986)
24. E.M. Purcell, Life at low Reynolds number. Am. J. Phys. **45**, 3–11 (1977)
25. P. Romanczuk, M. Bär, W. Ebeling, B. Lindner, L. Schimansky-Geier, Active brownian particles. from individual to collective stochastic dynamics. Eur. Phys. J. Special Topics **202**, 1–162 (2012)
26. J. Saal, Maximal regularity for the Stokes equations in non-cylindrical space-time domains. J. Math. Soc. Japan **58**(3), 617–641 (2006)
27. R. Sknepnek, S. Henkes, Active swarms on a sphere. Phys. Rev. **91**, 022306 (2015)
28. J. Slomka, J. Dunkel, Generalized Navier-Stokes equations for active suspensions. Eur. Phys. J. Spec. Top. **224**, 1349–1358 (2015)
29. H. Sohr, *The Navier-Stokes Equations - An Elementary Functional Analytic Approach*, Birkhäuser, 2001
30. J. Swift, P.C. Hohenberg, Hydrodynamic fluctuations at the convective instability. Phys. Rev. A **15**, 319–328 (1977)
31. J. Toner, Y. Tu, Long-Range order in a two-dimensional dynamical XY model: how birds fly together. Phys. Rev. Lett. **75**, 4326–4329 (1995)
32. H. Triebel, *Interpolation Theory, Function Spaces, Differential Operators* (North Holland, 1978)
33. H.H. Wensink, J. Dunkel, S. Heidenreich, K. Drescher, R.E. Goldstein, H. Löwen, J.M. Yeomans, Meso-scale turbulence in living fluids. Proc. Natl. Acad. Sci. USA **109**, 14308–14313 (2012)
34. H.H. Wensink, H. Löwen, Emergent states in dense systems of active rods: from swarming to turbulence. J. Phys. Condens. Matter **24**, 464130 (2012)
35. M. Wiegner, The Navier-Stokes equations - a neverending challenge? Jahresber. Dtsch. Math.-Ver. **101**(1), 1–25 (1999)
36. H. Wioland, F.G. Woodhouse, J. Dunkel, R.E. Goldstein, Ferromagnetic and antiferromagnetic order in bacterial vortex lattices. Nat. Phys. (2016)
37. J.M. Yeomans, Active matter: playful topology. Nat. Mater. **13**, 1004 (2014)

MIX
Papier aus verantwortungsvollen Quellen
Paper from responsible sources
FSC® C105338

If you have any concerns about our products,
you can contact us on
ProductSafety@springernature.com

In case Publisher is established outside the EU,
the EU authorized representative is:
Springer Nature Customer Service Center GmbH
Europaplatz 3, 69115 Heidelberg, Germany

Printed by Libri Plureos GmbH
in Hamburg, Germany